90 0619091 1

Shaofan Li · Wing Kam Liu

Meshfree Particle Methods

Shaofan Li · Wing Kam Liu

Meshfree Particle Methods

With 189 figures, 74 in colour

 Springer

Professor Shaofan Li

University of California
Department of Civil and Environmental Engineering
783 Davis Hall
Berkeley, CA 94720-1710
USA
li@ce.berkeley.edu

Wing Kam Liu

Walter P. Murphy Professor
Director of NSF Summer Institute
on Nano Mechanics and Materials
Northwestern University
Department of Mechanical Engineering
2145 Sheridan Road
Evanston, IL 60208-3111
USA
w-liu@northwestern.edu

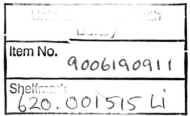
ISBN 3-540-22256-1 Springer Berlin Heidelberg New York

Library of Congress Control Number: 2004108217

Springer. Part of Springer Science+Business Media
springeronline.com

© Springer-Verlag Berlin Heidelberg 2004
Printed in Germany

Typesetting: data delivered by authors
Cover design: E. Kirchner, Springer-Verlag Heidelberg
Printed on acid free paper   62/3020/M - 5 4 3 2 1 0

Preface

The origin of meshfree methods could be traced back to a few decades, but it was not until after the early 1990s that substantial and significant advances were made in this field.

Mesh based numerical methods, e.g. finite element methods, have been the primary computational methodologies in engineering computations for more than half a century. One of the main limitations of finite element approximations is that they can only live on a prearranged topological environment — a mesh, which is an artificial constraint to ensure compatibility of finite element interpolation. A mesh itself is not, and is often in conflict with, the real physical compatibility condition that a continuum possesses. When conflicts between mesh and a physical compatibility occur, remeshing becomes inevitable, which is not only a time consuming process, but also leads to degradation of computational accuracy — a process of gradually losing numerical accuracy, producing significant pollution, and tainting computational results.

The resurgent interest in meshfree methods was to develop meshless interpolant schemes that can relieve the burden of remeshing and successive mesh generation, which have posed serious technical impediments in some finite element procedures such as adaptive refinement, simulations of a solid with progressive strong and weak discontinuities, and solid objects moving in a fluid field. However, the development of meshfree interpolants (e.g. Nayroles et al. (1992), Belytschko et al. (1994), and Liu et al. (1993ab,1995)) have lead to a comprehensive understanding of meshfree discretization, associated variational formulations, and the birth of a new class of meshfree methods and partition of unity methods. It turns out that a continuum model represented by a meshfree interpolation is fundamentally better than a continuum model approximated by mesh based interpolations in a number of aspects, such as smoothness, isotropy, nonlocal characters in interpolation (this is vitally important in nano-scale simulations), flexible connectivity, and refinement and enrichment procedures.

During the past decade, meshfree technology has achieved many successes. It has attracted much attention because of its applications to computational failure mechanics, multiscale computations, and nano-technology. There have been three special issues published in three major international archival journals in computational mechanics, which are devoted to meshfree particle methods (*Computer Methods in Applied Mechanics and Engineering*, [1996], Vol. 139; *Computational Mechanics*, [2000] vol. 25; and *International Journal for Numerical Methods and Engineering* [2000], Vol. 48). Since then, there have been several review articles surveying the field, e.g. Li and Liu [2002][274] and Babuška, Banerjee, and Osborn [2004].[25]

It appears to us that a comprehensive exposition on meshfree methods and their applications will not only provide an opportunity to re-examine previous contributions in retrospect, but also to re-organize them in a coherent fashion, in anticipating a new leap forward in advancing computational technology.

Most theories, computational formulations, and simulation results presented in this book are recent developments in meshfree methods, and they were either just published in the literature a few years ago or have not been published yet. Many of them are the outgrowth of the authors' personal contributions in this field. In order to meet the need of a broad audience, the presentation of the technical materials is heuristic and explanatory in style with a tradeoff between mathematical rigor and engineering practice. The objectives of this book are to be a pedagogical tool for novices, serving as a (graduate) textbook, to be a comprehensive source for researchers, providing the state-of-the-art technical documentation on meshfree particle methods. It is our hope that readers may find the presentation coherent, easy to digest, and insightful.

Meshfree particle methods are today's cutting-edge technology in computations. It has been demonstrated that meshfree methods are powerful computational apparatus to solve difficult problems such as crack growth and propagation; strain localization and dynamic shear band propagation; wave scattering and propagation; projectile penetration and target fragmentation, underwater explosion, compressible and incompressible flow pass obstacles, and high resolution shock capturing, particle motion in Stokes flow, etc. As a growing field, the meshfree method has shown promising potential to become a major numerical methodology in scientific and engineering computations. New theories in formulations and new algorithms in implementations are to be invented, and new applications are to be explored. The research field of meshfree particle methods is an open arena for both scientists and engineers to display their talent and creativity. It is expected that meshfree methods will provide a major technological thrust in advancing computational physics, computational material science, computational biophysics and biochemistry, computational nano-mechanics, and multi-scale modeling and simulation.

This book consists of 9 chapters: Chapter 1 is the introduction, which includes some definitions of the notations as well as terminologies that are used in the rest of the book; Chapter 2 is devoted to Smoothed Particle Hydrodynamics (SHP) method; Chapter 3 introduces some major mechfree Galerkin methods that are used in applications; Chapter 4 discusses the approximation theory of meshfree Galerkin method; Chapter 5 has compiled some main applications of meshfree Galerkin methods; Chapter 6 focus on the lates development on meshfree method — *Reproducing Kernel Element Method*, a hybrid meshfree/FE method, Chapter 7 discusses molecular dynamics (MD) and multiscale methods, which includes the recent development in quasi-continume method and bridging scale method; Chapter 8 presents the basic theory and applications of the immersed meshfree/finite element methods and their applications in bio-mechanics and bio-engineering, and the last Chapter, Chapter 9, outlines a few miscellaneous meshfree methods that have also been frequently used in applications.

In short, it is our hope that this book can provide a valuable technical resource for potential investigators to start their own research, and it may help initiate and promote further in-depth studies on those subjects that are discussed in this book.

The authors would like to acknowledge the contribution made by the following individuals in writing this book:

1. Harold Park and E. G. Karpov: Chapter 7;
2. Yaling Liu: Chapter 8;

Moreover, the authors would also like to express thanks to many of their friends, colleagues, and students who read the preliminary version of this book and who offered numerous suggestions, proof-readings, and corrections, especially Rita Tongpituk, Lucy Zhang, Xiaohu Liu, Daniel C. Simkins, Jr., Hunsheng Lu, and Dong Qian.

At last, we would like to express our thanks to National Science Foundation (NSF) for its sponsorship to authors' research. In particular, we would like to acknowledge the support from NSF Grant No. CMS-0239130 to University of California (SL), NSF-IGERT program to Northwestern University (WKL), and the NSF Summer Institute on Nano-Mechanics and Materials (WKL).

Shaofan Li and Wing Kam Liu
Berkeley, California and Evanston, Illinois
Summer, 2004

Contents

1. Introduction

1.1 Why Do We Need Meshfree Particle Methods ?

Almost half a century ago, Turner, Clough, Martin, and Topp (Turner et al.[433]) published their seminal work in finite element (FE) approximation in structural mechanics. Today, finite element method (FEM) based computational mechanics, as a far-reaching field ranging from basic science over applied research to applications, plays a prominent role in technology advancement. Model-based simulations are changing the face of scientific and engineering inquiry.

A salient feature of the finite element method is that it divides a continuum into discrete elements, which form a non-overlapping covering (subdivision) for the domain of interest. This subdivision is often called FE discretization. In FEM implementation, individual elements are connected together by a topological map, which is usually called a mesh. Finite element interpolation functions are then built upon the mesh, which ensures the compatibility of the FE interpolation.

In principle, any valid discretization technique has to provide a numerical compatibility for the corresponding interpolation field. In the past few decades, finite element subdivision procedure is the dominant discretization technique in numerical computations that provide compatible interpolation fields. Nevertheless, FEM subdivision is not the only possible means for numerical discretization. Moreover, the FEM subdivision procedure is not always advantageous in computations:

1. FEM numerical compatibility is only an approximation of the real physical compatibility of a continuum. This approximation is not accurate in many cases. For instance, in Lagrangian type of computations, one may experience mesh distortion, which can either terminate the computation or result in drastic deterioration in accuracy.
2. FEM computations often require a very fine mesh in problems with high gradients or a distinct local character, which can be computationally expensive. For this reason, adaptive FEM has become a necessity. The successive h-type FEM adaptive refinement may not be always approachable, if a structured mesh is required in each refinement. This is because many domains of practical applications have complex geometry, and structured

mesh refinement has its limit. In many cases, an infinite successive refinement is merely a platonic dream. In fact, adaptive remeshing has become a challenging and formidable task to undertake in simulations of impact/penetration, explosion/fragmentation, flow pass obstacles, and fluid-structure interactions. The difficulties involved are not only with remeshing, but also mapping state variables (thermodynamics variables) from the old mesh to the new mesh. This process often introduces numerical errors, and frequent remeshing is thus not desirable.

3. In computational failure mechanics, the main task is to simulate surface separation and material disintegration. In FEM computations, simulated material disintegration is in fact the disintegration of FEM subdivision, therefore the simulated disintegration pattern has been embedded in the FEM subdivision before the simulation even started, which may not capture the real material disintegration pattern. When different meshes are used in numerical simulations of the same physical problem, different material disintegration patterns may be obtained. This is the so-called mesh sensitivity in crack growth simulations. In some other applications, a mesh may be the source of inherent bias in numerical simulations, and its presence can become a nuisance in computations. A well-known example is the simulation of the strain localization problem, which is notorious for its mesh alignment pathology.[349, 350] An ultimate call for meshless discretization is from computational micro-mechanics and multi-scale computation, where length scale is so small that almost all the physical processes involved are nonlocal in character. The finite element based local approximation becomes irrelevant in most cases.

4. In most applications, FEM interpolation fields are primarily $C^0(\Omega)$ functions. Higher order interpolation fields are difficult to construct for domains with arbitrary geometry in multiple dimensions. This severely restricts the ability of finite element based computations being used in practical problems with high order derivatives, such as simulations of plates and shells and high gradient elasticity and plasticity.

Since 1970, much attention has been devoted in research to seek better alternatives in special applications. The early effort is to tailor the FEM subdivision to adapt topological and geometric changes undergone by the material. For instance, the so-called Arbitrary Lagrangian Eulerian (ALE) finite element formulations have been developed (e.g. Liu et al.[216, 287–289]) [1], whose objective is to move the mesh independently from material motion so that mesh distortion can be minimized. Unfortunately, in simulations of fluid flow or large strain continuum deformation, even with the ALE formulation, mesh distortion is still present, and it introduces severe errors in numerical computations. Furthermore, convective transport effects in ALE often lead

---

[1] For a comprehesive treatment on ALE, readers may consult Chapter 7 of the book by Belytschko, Liu and Moran.[52]

to spurious oscillation that needs to be stabilized by artificial diffusion or a Petrov-Galerkin stabilization.

To find a better approximation of continuum compatibility, new discretization methods and principles have been sought. Meshfree particle methods were born under such circumstances, and they are designed to improve the inadequacy of FEM discretizations. On a philosophical level, it is believed that it may be computationally efficient and physically accurate to discretize a continuum by only a set of nodal points, or particles, without additional mesh constraints. This is the *leitmotif* of contemporary meshfree Galerkin methods.

On the other hand, the booming of meshfree methods is intimately related to generalization of FEM. The contemporary computational paradigm is based on the notation of partition of unity (Bubuška and Melenk[23]), which is born as a principle of meshfree interpolation. The meshfree partition of unity approximation theory and adaptive refinement methodologies may become a successor to the popular status that FEM related methodologies now occupy. This is because meshfree methods are the natural extension of finite element methods, they provide a perfect habitat for a more general and more appealing computational paradigm — the partition of unity. The advantages of the meshfree particle methods can be summarized as follows:

- They can easily handle simulations of very large deformations, since the connectivity among nodes is generated as part of the computation and can change with time;[96,97,99,227,270,279]
- The data structure of meshfree discretization can be linked more easily with a CAD database than finite element discretization, since it is not necessary to generate mesh;[246,402,460–462]
- The method is designed to adapt the changes of the topological structure of a continuum, and it provides computational efficiency to handle damage or disintegration of continua such as crack growth, shear banding, underwater explosion, etc., which is proven to be instrumental in modeling of material failures;[40–43,272,273,278,386,387,389,461]
- It can easily provide a smooth higher order interpolation field in any dimension, which makes the implementation of both single primary variable Galerkin variational formulations and mixed Galerkin variational formulations easy;
- It has the ability to include a priori knowledge about the local behavior of the solution in the meshfree interpolation space;[23,45,136,166,389,445]
- It supports a very flexible adaptive refinement procedure, such that computational accuracy can be easily controlled, because in areas where more refinement is needed, particles can be added without any restraints (h-adaptivity).[217,248,269,306,443] In fact, meshfree h-type refinement may be considered as *infinitely approachable*, meaning that there is no practical limit on successive h-refinement, which is not true for structured finite element h-refinement.

- It allows researchers to use three-dimensional continuum models to simulate large deformation of plates and shells, such as thin shell structures and nanotubes;[250, 268, 355, 356, 378]
- The meshfree method is a better computational paradigm for multi-scale computations than mesh based discretization method, because it is a non-local interpolation and it can easily incorporate an enrichment of fine scale solutions of features into the coarse scale;[304, 305, 308, 309, 378]
- Meshfree discretization can provide accurate representation of geometric objects.[271]

In general, particle methods can be classified based on two different criteria: physical principles; and computational formulations. According to physical modeling, they may be categorized into two classes: those based on deterministic models; and those based on probabilistic models. On the other hand, according to computational modeling, they may be categorized into two different classes as well: those serving as approximations of the strong form of a partial differential equation (PDE); and those serving as approximations of the weak form of a PDE.

When meshfree particle methods are used to approximate the strong form of a PDE, the PDE is often discretized by a specific collocation technique. Examples are: smoothed particle hydrodynamics (SPH),[55, 176, 284, 329, 331] vortex method,[62, 111–113, 260, 261] meshfree finite difference method,[280, 281] and others. It is worth mentioning that some particle methods, such as SPH and vortex methods, were initially developed as probabilistic methods,[111, 284] and it turns out that both SPH and vortex method are now most frequently used as deterministic tools.

Nevertheless, many particle methods in this category are based on probabilistic principles, or used as probabilistic simulation tools. There are three major methods in this category: (1) molecular dynamics (both quantum molecular dynamics[197, 203, 243, 247, 249] and classical molecular dynamics[8, 9, 91, 116, 117, 390]); (2) direct simulation Monte Carlo (DSMC), or Monte Carlo method based molecular dynamics (such as quantum Monte Carlo methods[6, 26, 63–65, 152, 183, 362, 434]); and (3) the lattice gas automaton (LGA), or lattice gas cellular automaton[169, 199, 231–233] and its later derivative, the Lattice Boltzmann equation method (LBE).[107, 109, 381–383] It may be pointed out that Lattice Boltzmann equation method is not a meshfree method, and it requires a grid; this example shows that particle methods are not always meshfree.

The second class of meshfree particle methods are used with various Galerkin weak formulations, which are called meshfree Galerkin methods. Examples are: Diffuse element method (DEM),[80, 81, 85, 348] Element free Galerkin method (EFGM),[40, 44, 47, 48, 315] Reproducing kernel particle method (RKPM),[96, 103, 295–297, 299, 300, 302, 303] h-p Cloud method,[153, 154, 282, 357] Partition of unity method,[23, 24, 323] Meshless local Petrov-Galerkin method (MLPG),[16–19] Finite point method,[359–361] Finite sphere method,[136] Free mesh method,[173, 402, 460–462] Moving particle finite element method,[193] Natural element method,[416,417] Reproducing kernel element method,[275,311,314,404] and others.

Since the early 1990s, meshfree Galerkin methods have been successfully used in solving many difficult engineering problems. For instance, one of the early applications of meshfree Galerkin methods is to utilize their flexibility in interpolation to simulate crack growth—a critical issue in computational fracture mechanics. Belytschko and his co-workers have systematically applied the EFG method to simulate crack growth/propagation problems.[41–44,46,251,252] Special techniques, such as the visibility criterion, are developed in modeling a discontinuous field.[44,251] Subsequently, the partition of unity method is also exploited in crack growth simulations in both 2D and 3D.[133,151] Meshfree procedure offers considerable advantages over the traditional finite element methods in crack growth simulation, because remeshing is avoided.

Another successful example in using meshfree methods is simulations of strain localization problems. In a series of publications by Li and Liu,[268–270,272,273] and others (e.g.[228]) have shown that the meshless method has the ability to sustain large mesh distortion. By using a meshfree interpolant, one can effectively avoid numerical pathology due to the notorius mesh-alignment sensitivity in simulations of strain localization problems. This is because there is no mesh involved in meshfree simulations, whereas in finite element simulations the numerical shear band tends to grow along a finite element boundary instead of the real physical path.

The on-going research on meshfree methods is intimately related to the pressing issues of emerging technologies. Subjects such as meshfree-based quasi-continuum discretization, meshfree multi-scale methods, nodal integration technique, hybrid meshfree element methods (e.g. RKEM), and mixed meshfree-finite element refinement technique, are the key technical components that have far-reaching implications in nano-technology, computational biology, computational material science, and multi-scale modeling.

Today there is a strategic shift in applied research. Computational mechanics is facing a grand challenge — multi-scale computation. In computational fluid and solid mechanics, many of the most difficult and important problems are characterized by the presence of a wide range of length and time scales. This multi-scale nature might be due to the strong non-linearity of the dynamics, as in turbulent fluid flow, or to evolving inhomogeneities in the material such as crack growth. One of the most intriguing questions in multi-scale computation is: how to connect micromechanics at the atomic scale with material behaviors at the macro-scale ? Conceptually there are two multi-scale methods. Traditionally, the multi-scale computation is referred to as hierarchical simulations, in which information at one scale, obtained from simulation, analysis, or experiment, is incorporated in a consistent way with a simulation at a larger scale. However, when the micro-structure of the medium is evolving, such as in fracture or turbulence flow problems, the dynamics is controlled by interscale interaction that cannot be captured by conventional homogenization based hierarchical methods. The proper coupling and resolving of these scales requires concurrent multi-scale computations in a single simulation, which is the essence of contemporary multi-scale method and the future of computational mechanics.

The affinity of the meshfree particle method to multi-scale computation rests upon its two distinguished characteristics. First, most meshfree particle methods are either based on non-local interpolations, or physical laws of non-local interactions. Most physical phenomena at the micro-scale are also governed by physical laws of non-local interaction. The local numerical approximations, such as FEM, FDM, and FVM, are intrinsically associated with macro/phenomenological physical models, and they often break down at the micro-scale level. Second, another salient feature of meshfree particle methods is their multi-scale character and their interdisciplinary character. In fact, meshfree particle methods encompass a class of first-principle based methods: Molecular dynamics (both ab initio and classical), Monte Carlo method, Bridging scale methods, and Lattice Boltzmann equation method, whose applications are spanning a number of disciplines such as astrophysics, solid state physics, bio-chemistry, material science, continuum mechanics, and nano-science and technology. From this perspective, a meshfree multi-scale paradigm is a perfect embodiment of concurrent multi-scale modeling and simulation, which is the focus of current research in computational mechanics.

It is the authors' belief that a systematic exposition on meshfree particle methods will foster interdisciplinary interests on multi-scale computation, and bridge the efforts in different scientific disciplines and bring further advancement in computational biology, computational physics and chemistry, computational material science, computational mechanics and micromechanics, and multi-scale modeling in general.

1.2 Preliminary

Meshfree particle methods are numerical methods. They share many common threads with other numerical methods, e.g. finite element methods and finite difference methods. The readers of this book are assumed to have basic knowledge in finite element methods and finite difference methods. On the other hand, to be self-sufficient and to be consistent in notation, a few important topics in numerical analysis are reviewed in the following.

1.2.1 Notation

In this book, the letter d is a positive integer and is used for the spatial dimension. We denote $\Omega \subset \mathbb{R}^d$ to be a nonempty, open bounded set with a

Lipschitz continuous boundary. Einstein convention on index summation is followed throughout the book,

$$x_i y_i := x_1 y_1 + x_2 y_2 + x_3 y_3 \ . \tag{1.1}$$

A point in $\mathbb{R}^d$ is denoted by a vector in Cartisian coordinates $\mathbf{x} = x_i \mathbf{e}_i$. We use the Euclidean norm to measure the vector length:

$$\|\mathbf{x}\| = \left(\sum_{i=1}^{d} x_i x_i \right)^{1/2} . \tag{1.2}$$

The boldface letter, Roman or Greek, usually refers to either a vector or a tensor. However, for position vector $\mathbf{x} = x_i \mathbf{e}_i$, the following customary convention is implicitly adopted: for both scalar and vector functions

$$A(\mathbf{x}) := A(x_1, x_2, x_3) \tag{1.3}$$
$$\mathbf{A}(\mathbf{x}) := \mathbf{A}(x_1, x_2, x_3) \tag{1.4}$$

The operation designated by the tensor product $\otimes$ between two vectors $\mathbf{A} = A_i \mathbf{e}_i$ and $\mathbf{B} = B_i \mathbf{e}_i$ is defined as

$$\mathbf{A} \otimes \mathbf{B} = A_i B_j \mathbf{e}_i \otimes \mathbf{e}_j \tag{1.5}$$

The gradient operator is only used in Cartesian coordinates, and it is defined as

$$\nabla := \frac{\partial}{\partial x_i} \mathbf{e}_i. \tag{1.6}$$

The usual convention for time derivative is also adopted,

$$\frac{d}{dt} \{ \ \} = \{ \ \dot{} \ \} . \tag{1.7}$$

Let p be a non-negative integer. We use the notation $\mathcal{P}_p = \mathcal{P}_p(\Omega)$ for the space of the polynomials of degree less than or equal to p on Ω. The dimension of the polynomial space is

$$N_p := dim \mathcal{P}_p = \binom{p+d}{d} = \frac{(p+d)!}{p! d!} \tag{1.8}$$

The essence of meshfree interpolation is its nonlocal character. For $\bar{\mathbf{x}} \in \mathbb{R}^d$ and $\varrho > 0$, a spherical ball (closed) with radius ϱ is defined as the domain of influence of $\bar{\mathbf{x}}$,

$$B_\varrho(\bar{\mathbf{x}}) = \left\{ \mathbf{x} \ \Big| \ \|\mathbf{x} - \bar{\mathbf{x}}\| \leq \varrho, \ \mathbf{x} \in \mathbb{R}^d \right\} \tag{1.9}$$

Consider a bounded open set $\Omega \subset \mathbb{R}^d$. An effective domain of influence $\bar{\mathbf{x}}$ may be defined as

$$B_\varrho^e(\bar{\mathbf{x}}) = B_\varrho(\bar{\mathbf{x}}) \cap \Omega \tag{1.10}$$

Definition 1.2.1 (The C^k spaces). *Let Ω be a bounded domain in $\mathbb{R}^d$ with piecewise continuous boundary $\partial\Omega$. We let $C^0(\Omega)$ denote the space of all continuous functions on $\bar{\Omega}$ with the norm*

$$\|u\|_{C^0(\Omega)} = \max_{\boldsymbol{x}\in\bar{\Omega}}|u(x)| . \tag{1.11}$$

where $u(x) \in \mathbb{R}$. We denote

$$C^k(\Omega) := \left\{ v \in C^0(\Omega) \,\Big|\, \|v\|_{C^k(\Omega)} < \infty \right\} \tag{1.12}$$

where

$$\|v\|_{C^k(\Omega)} := \sum_{0\leq j\leq k} \sum_{m_{1j}+m_{2j}+\cdots m_{dj}=j} \left\| \frac{\partial^j u}{\partial x_1^{m_{1j}} \partial x_2^{m_{2j}} \cdots \partial x_d^{m_{dj}}} \right\|_{C^0(\Omega)} . \tag{1.13}$$

We also denote $C_0(\Omega)$ as the space of all continuous functions on Ω that also vanish at $\partial\Omega$.

1.2.2 Partition of Unity

A more academic and more general name for meshfree interpolants is the so-called "*partition of unity*", which is even speculated to replace the status of the finite element shape function. A rigorous but restricted definition of partition of unity is given below:

Definition 1.2.2 (Partition of unity). *Let $\Omega \subset \mathbb{R}^d (d = 1,2,3)$ be an open bounded domain. Let $\Omega_1, \Omega_2, \cdots\cdots, \Omega_{NP}$ be a family of open sets in $\mathbb{R}^d$, and*

1. *The family of a open set $\{\Omega_I\}_{I\in\Lambda}$ generates a covering for domain Ω,*

$$\Omega \subset \bigcup_{I\in\Lambda} \Omega_I \tag{1.14}$$

2. *There exists a family of functions, $\Phi_I \in C_0^s(\mathbb{R}^d)$, $s \geq 0$, and $\mathrm{supp}\{\Phi_I\} \subset \bar{\Omega}_I$*
3.

$$0 \leq \Phi_I(\mathbf{x}) \leq 1 \;\;, \quad \forall\, \mathbf{x} \in \Omega_I \tag{1.15}$$

4. *The summation*

$$\Phi_1(\mathbf{x}) + \Phi_2(\mathbf{x}) + \cdots\cdots + \Phi_{NP}(\mathbf{x}) = 1, \quad \forall\, \mathbf{x} \in \Omega \tag{1.16}$$

The family of generating function, or the interpolation basis, $\{\Phi_I\}_{I\in\Lambda}$ is called a partition of unity subordinate to the open cover $\{\Omega_I\}_{I\in\Lambda}$.

The last property (1.16) suggests the name of *partition of unity*. The other distinguished property of the partition of unity is that the set of open supports, $\{\Omega_I\}_{I \in \Lambda}$, can be overlapping, and they do not necessarily form a sub-division (mesh) of Ω, as long as they generate a covering for Ω (see Eq. (1.14)). Melenk and Babuska call such a covering patches, Duarte and Oden call them "clouds", and many people (e.g. De and Bathe) call them "spheres".

Often in practice, the second condition (1.15) may not be satisfied, and $\Phi_I(\mathbf{x})$ may be negative in some region. The interpolation basis is then called the *signed partition of unity*.

1.2.3 Window Function and Mollifier

To construct the meshfree partition of unity, special generating functions are chosen to meet the following requirements:

1. $\Phi \in C^k(\mathbb{R}^d)$ $k \geq 1$;
2. $supp\{\Phi\} = B_1$;
3. $\Phi(\mathbf{x}) > 0$ for $\|\mathbf{x}\| < 1$;
4. $\displaystyle\int_{B_1} \Phi(\mathbf{x})d\Omega = 1.$

Scaling $supp\{\Phi(\mathbf{x})\}$, one may define a new function

$$\Phi_\varrho(\mathbf{x}) := \frac{C_d}{\varrho^d}\Phi\left(\frac{\mathbf{x}}{\varrho}\right) \tag{1.17}$$

where the constant C_d is determined by the condition

$$\int_{B_\varrho} \Phi_\varrho(\mathbf{x})d\Omega = 1 \tag{1.18}$$

The value of C_d depends on the shape of its compact support and dimension d of the space. For instance, for 2D cubic spline function, $C_d = 1$ for 2D rectangular domain, $C_2 = \dfrac{15}{7\pi}$, if the compact support is a circular domain.

The function $\Phi_\varrho(\mathbf{x})$ is usually called a C^k-*mollifier* in mathematics literature, which has the following properties:

1. $\Phi_\varrho \in C^k(\mathbb{R}^n)$ $k \geq 1$;
2. $supp\{\Phi_\varrho\} = B_\varrho$;
3. $\Phi_\varrho(\mathbf{x}) > 0$ for $\|\mathbf{x}\| < \varrho$;
4. $\displaystyle\int_{B_\varrho} \Phi_\varrho(\mathbf{x})d\Omega = 1.$

1.2.4 Hilbert Space

Concepts in linear functional analysis are central to a qualitative understanding of convergence of any numerical method. For the most part, knowledge in elementary functional analysis including Hilbert space and Galerkin variational form of a boundary value problem (BVP) will suffice. In a few cases, we may venture to discuss the results in general Sobolev space (Adams 1975). The notion of function spaces in describing the topological structure of a set of functions in $\mathbb{R}^d$ is analogous to that of vector spaces in linear algebra. The abstract notation, $Au = f$, is used to describe a partial differential equation (PDE). It is noted that the differential operator $A : U \to V$ maps functions from one function space U to another function space V. All the discussions conducted in this book are restricted to real valued functions.

Let Ω be an open bounded domain in $\mathbb{R}^d$ with a piecewise continuous boundary Γ (Γ is a Lipschitz boundary). The function space $L^2(\Omega)$ is defined as

$$L^2(\Omega) := \left\{ f \mid \int_\Omega f^2 d\Omega < +\infty \right\} \tag{1.19}$$

the space is equiped with the L^2 norm

$$\|u\|_{L^2(\Omega)} := (u, u)_{L^2(\Omega)}^{1/2} \tag{1.20}$$

where the operator $(,)$ is a scalar product defined as

$$(u, v)_{L^2(\Omega)} := \int_\Omega uv d\Omega, \quad u, v \in L^2(\Omega) \tag{1.21}$$

Multi-index Notation. It is convenient to use multi-index notation to express partial derivatives in multiple dimensions.

Let $\mathbf{Z}^d$ denote the set of all ordered d-tuples of non-negative integers. A multi-index is an ordered collection (d-tuple) of d non-negative integers, $\alpha = (\alpha_1, \cdots, \alpha_d)$, and its length is defined as

$$|\alpha| = \sum_{i=1}^d \alpha_i \tag{1.22}$$

We write $\alpha! = \alpha_1! \alpha_2! \cdots \alpha_d!$ and $\varrho^\alpha = \varrho_1^{\alpha_1} \varrho_2^{\alpha_2} \cdots \varrho_d^{\alpha_d}$ and $\mathbf{x}^\alpha = x_1^{\alpha_1} x_2^{\alpha_2} \cdots x_d^{\alpha_d}$. For a differentiable function $u(\mathbf{x})$ and any α with $|\alpha| \le p$,

$$D^\alpha u(\mathbf{x}) = \frac{\partial^{|\alpha|} u(\mathbf{x})}{\partial x_1^{\alpha_1} \cdots \partial x_d^{\alpha_d}} \tag{1.23}$$

is the αth order partial derivative. As usual, $D^0 u(\mathbf{x}) = u(\mathbf{x})$.

Linear Functional. We consider a linear functional $l : V \to \mathbb{R}$ that maps real-valued functions from a Hilbert space V to real numbers in $\mathbb{R}$. The functional l is defined by its action on an arbitrary member $v \in V$ by

$$< l, v >= (u, v)_{L^2(\Omega)} \tag{1.24}$$

for a given u and arbitrary v. The notation $\langle \cdot, \cdot \rangle$ is used to denote the duality pair. The space of bounded linear functionals on V is denoted by the dual space V', and for $l \in V'$ the norm $\|l\|_{V'}$ is defined as

$$\|l\|_{V'} = \sup_{v \in V \; \{0\}} \frac{\langle l, v \rangle}{\|v\|_V} . \tag{1.25}$$

The functional l is said to be continuous, if there exists a positive constant M such that

$$|\langle l, v \rangle| \le M \|v\|_V . \tag{1.26}$$

Weak Derivatives. We denote $C_0^\infty(\Omega)$ as the space of infinitely differentiable functions which are non-zero only on a compact subset of Ω.

If $u \in L^2(\Omega)$ is a locally integrable function, we say that u possesses the (weak) derivative $v = D^\alpha u$ in $L^2(\Omega)$, if and only if $v \in L^2(\Omega)$ and

$$\langle v, w \rangle = (-1)^{|\alpha|} \langle u, D^\alpha w \rangle, \quad \forall \, w \in C_0^\infty(\Omega) . \tag{1.27}$$

Introduce Hilbert space $H^m(\Omega)$. The function space, $H^m(\Omega)$, is defined as the space consisting of those functions in $L^2(\Omega)$ that have their weak partial derivatives exist up to order m, i.e.

$$H^m(\Omega) = \{u \;\big|\; D^\alpha u \in L^2(\Omega), \; \forall \, |\alpha| \le m\} \tag{1.28}$$

$H^m(\Omega)$ is a Hilbert space, because its norm is induced by the following inner product

$$\|u\|_{H^m(\Omega)} := (u, u)_{H^m(\Omega)}^{1/2} \tag{1.29}$$

where the inner product

$$(u, v)_{H^m(\Omega)} = \sum_{|\alpha| \le m} (D^\alpha u, D^\alpha v)_{L^2(\Omega)} , \quad \forall \, u, v \in H^m(\Omega) . \tag{1.30}$$

We note that $H^0(\Omega) = L^2(\Omega)$ and the family of semi-norm on $H^m(\Omega)$ are defined as

$$|u|_{H^p(\Omega)} := \left\{ \sum_{|\alpha| = p} (D^\alpha u, D^\alpha u)_{L^2(\Omega)} \right\}^{1/2} \tag{1.31}$$

where $1 \le p \le m$. We denote function space $H_0^m(\Omega)$ as a subspace of $H^m(\Omega)$ in which all the functions are compactly supported in a subset of Ω. For example,

$$H_0^1(\Omega) = \left\{ u \mid u \in H^1(\Omega), u \mid_{\boldsymbol{x} \in \Gamma} = 0 \right\} \tag{1.32}$$

$$H_0^2(\Omega) = \left\{ u \mid u \in H^1(\Omega), u \mid_{\boldsymbol{x} \in \Gamma} = 0, \ \frac{\partial u}{\partial n} \mid_{\boldsymbol{x} \in \Gamma} = 0 \right\} \tag{1.33}$$

where n is the outward normal of Γ.

1.2.5 Variational Weak Formulation

Before discussing the variational formulation, a few definitions are in order. $a(u, v) : H \times H \to \mathbb{R}$ that maps a pair of functions from a Hilbert space to real number in $\mathbb{R}$. Since the operator $a(\cdot, \cdot)$ is linear in both slots, it is termed as bilinear form.

Definition 1.2.3 (Continuity). *Let V be a Hilbert space. A bilinear form $a(u, v) : V \times V \to \mathbb{R}$ is continuous if there exists a constant $K > 0$ such that*

$$|a(u, v)| \le K \|u\|_V \|v\|_V, \quad \forall \, u, v \in V \tag{1.34}$$

Definition 1.2.4 (Ellipticity). *Let V be a Hilbert space. A continuous bilinear form is said to be V-elliptic, or coercive, if there exists a constant $\alpha > 0$ such that*

$$a(v, v) \ge \alpha \|v\|^2, \quad \forall \, v \in V \tag{1.35}$$

Consider a Hilbert space H and let U and V be convex subspaces of H. Let $a(u, v) : U \times V \to \mathbb{R}$ be a continuous, symmetric, V-elliptic bilinear form and let $l : V \to \mathbb{R}$ be a continuous linear functional on V. We consider the following abstract variational formulation,

$$a(u, v) = \langle l, v \rangle, \quad \forall \, v \in V \tag{1.36}$$

where any function $u \in U$ is usually called a trial function, and functions $v \in V$ are called test functions.

From a physical viewpoint, the above variational statement is equivalent to the following minimization problem

$$\inf_{u \in U} J(u) = \inf_{u \in U} \frac{1}{2} a(u, u) - \langle l, u \rangle \tag{1.37}$$

To illustrate how to formulate a Galerkin variational weak form we consider the following Dirichlet boundary value problem,

$$-\nabla^2 u + u = f, \quad \forall \, \boldsymbol{x} \in \Omega \tag{1.38}$$

$$u = 0, \quad \forall \, \boldsymbol{x} \in \Gamma \tag{1.39}$$

Choose $V = H_0^1(\Omega)$. Multiplying Eq. (1.38) by an arbitrary function $v \in V$ and integrating them over Ω, one may obtain that

$$\int_\Omega \left(-\nabla^2 u + u\right)v\,d\Omega = \int_\Omega fv\,d\Omega \tag{1.40}$$

Integration by parts yields

$$\int_\Omega v\nabla^2 u\,d\Omega = -\int_\Omega \nabla u \cdot \nabla v\,d\Omega + \int_\Gamma v\frac{\partial u}{\partial n}\,d\Gamma \tag{1.41}$$

where n is the outward normal of Γ. Considering the boundary condition (1.39), one may derive

$$\int_\Omega \left(\nabla u \cdot \nabla v + uv\right)d\Omega = \int_\Omega fv\,d\Omega \tag{1.42}$$

which is the variational weak form of (1.38)–(1.39).

Comparing Eq. (1.42) with the abstract variational form (1.36), one can identify that

$$a(u, v) = \int_\Omega (\nabla u \cdot \nabla v + uv)d\Omega \tag{1.43}$$

$$\langle l, v \rangle = \int_\Omega fv\,d\Omega \tag{1.44}$$

In this case, $a(u, u) = \|u\|^2_{H^1(\Omega)}$. It is obvious that the bilinear form is continuous, symmetric, and coercive. Since the solution of Eq. (1.42) belongs $H^1(\Omega)$ while the solution of Eqs. (1.38) – (1.39) belong at least to $C_0^1(\Omega)$, the solution of (1.42) is referred to as *weak solution*, because it has a lesser requirement on the solution's differentiability.

1.2.6 Galerkin Methods

In the Galerkin variational statement (1.36), both function spaces, U and V are infinite dimensional subspaces of a Hilbert space H. In computational practice, an approximated solution is sought in a finite dimensional subspace of U, which is built on discretization of the continuum. Let $U^h \subset U$ and $V^h \subset V$. They are both spanned by a set of shape functions, i.e.

$$U^\varrho = span\{\Psi_I\}_{I \in \Lambda} \tag{1.45}$$

$$V^\varrho = span\{\Phi_I\}_{I \in \Lambda} \tag{1.46}$$

where parameter ϱ refers to a measure that is related to spacing of the discretization. In the context of finite element methods, we usually label $\varrho = h$, and the Galerkin variational statement becomes:

Find $u^h \in U^h$ such that

$$a(u^h, v^h) = <l, v^h>, \quad \forall \, v^h \in V^h \tag{1.47}$$

If the trial and test functions spaces are identical, $U^h \equiv V^h$, the Galerkin method is called Ritz-Galerkin method, or Bubnov-Galerkin method. On the other hand, if the trial and test function spaces are different, $U^h \neq V^h$, the associated Galerkin method is called Petrov-Galerkin method, in which the trial function and test function belong to different function spaces.

Consider a Bubnov-Galerkin formulation, $U^h = V^h$. Both the trial function and the test function can be expressed by the same interpolant

$$u^h(\boldsymbol{x}) = \sum_{I \in \Lambda} \Phi_I(\boldsymbol{x}) u_I \tag{1.48}$$

$$v^h(\boldsymbol{x}) = \sum_{I \in \Lambda} \Phi_I(\boldsymbol{x}) v_I \tag{1.49}$$

Substituting (1.48) and (1.49) into (1.47) yields a system of algebraic equations

$$\sum_{I \in \Lambda} \sum_{J \in \Lambda} a(\Phi_I, \Phi_J) u_I v_J = \sum_{J \in \Lambda} <l, \Phi_J> v_J, \quad \forall \, v_J \in V^h \tag{1.50}$$

which can be recast into the form

$$\sum_{J \in \Lambda} v_J \left(\sum_{I \in \Lambda} K_{IJ} u_I - f_J \right) = 0 \tag{1.51}$$

where $K_{IJ} = a(\Phi_I, \Phi_J)$ and $f_J = <l, \Phi_J>$. Since $v_J \in V^h$ is arbitrary, the following finite dimension linear algebraic equations can be obtained,

$$[K_{IJ}][u_J] = [f_I] \tag{1.52}$$

where $[K_{IJ}]$ is referred to as the stiffness matrix because of its early application in structural mechanics.[433]

1.2.7 Time-stepping Algorithms

The time integration algorithm is a critical part of meshfree particle methods. It is a challenge today to integrate a multiple time scale system with accuracy, which occurs in many particle method based simulations. To be self-sufficient, the conventional time-stepping algorithms are listed as follows. Special issues on time integration relating to specific particle methods will be treated separately in related chapters.

To outline time-stepping algorithms, we consider the following semi-discrete equation of motion, which may be formed by any spatial discretization process,

$$\mathbf{M}\ddot{\mathbf{d}} + \mathbf{C}\dot{\mathbf{d}} + \mathbf{K}\mathbf{d} = \mathbf{F} \tag{1.53}$$

where $\mathbf{M}$ is the mass matrix, $\mathbf{C}$ is the viscous damping matrix, $\mathbf{K}$ is the stiffness matrix, $\mathbf{F}$ is the external force vector, and $\ddot{\mathbf{d}}, \dot{\mathbf{d}}$ and $\mathbf{d}$ are acceleration, velocity, and displacement vectors. The initial-value problem for (1.53) consists of finding a displacement, $\mathbf{d} = \mathbf{d}(t)$ satisfying (1.53) and the given initial data,

$$\mathbf{d}(0) = \mathbf{d}_0, \quad \text{and} \quad \dot{\mathbf{d}} = \mathbf{v}_0 \tag{1.54}$$

Let

$$\mathbf{y} = \left\{ \begin{matrix} \dot{\mathbf{d}} \\ \mathbf{d} \end{matrix} \right\} \tag{1.55}$$

and

$$\mathbf{f}(\mathbf{y}, t) = \left\{ \begin{matrix} \dot{\mathbf{d}} \\ \mathbf{M}^{-1}(\mathbf{F}(t) - \mathbf{C}\dot{\mathbf{d}} - \mathbf{K}\mathbf{d}) \end{matrix} \right\} \tag{1.56}$$

One may convert Eq. (1.53) into a first order form,

$$\dot{\mathbf{y}} = \mathbf{f}(\mathbf{y}, t) \tag{1.57}$$

Newmark Integration Algorithm. A very useful and popular method to solve (1.53) and (1.54) is the Newmark method. The Newmark algorithm can be written as follows

$$\mathbf{M}\mathbf{a}_{n+1} + \mathbf{C}\mathbf{v}_{n+1} + \mathbf{K}\mathbf{d}_{n+1} = \mathbf{F}_{n+1} \tag{1.58}$$

$$\mathbf{d}_{n+1} = \mathbf{d}_n + \Delta t \mathbf{v}_n + \frac{\Delta t^2}{2}\{(1 - 2\beta)\mathbf{a}_n + 2\beta\mathbf{a}_{n+1}\} \tag{1.59}$$

$$\mathbf{v}_{n+1} = \mathbf{v}_n + \Delta t\{(1 - \gamma)\mathbf{a}_n + \gamma\mathbf{a}_{n+1}\} \tag{1.60}$$

By choosing $\beta = \frac{1}{4}$ and $\gamma = \frac{1}{2}$, the Newmark method yields the trapezoidal rule, which is based on central difference scheme.

$$\mathbf{d}_{n+1} = \mathbf{d}_n + \Delta t \mathbf{v}_{n+1/2} \tag{1.61}$$

$$\mathbf{v}_{n+1/2} = \frac{\mathbf{v}_{n+1} + \mathbf{v}_n}{2} \tag{1.62}$$

$$\mathbf{v}_{n+1} = \mathbf{v}_n + \Delta t \mathbf{a}_{n+1/2} \tag{1.63}$$

$$\mathbf{a}_{n+1/2} = \frac{\mathbf{v}_{n+1} + \mathbf{v}_n}{2} \tag{1.64}$$

Apply trapezoidal rule to the first order system (1.57). The same result follows

$$\mathbf{z}_n := \mathbf{f}(\mathbf{y}_n, t_n) \tag{1.65}$$

$$\mathbf{y}_{n+1} = \mathbf{y}_n + \frac{\Delta t}{2}(\mathbf{z}_n + \mathbf{z}_{n+1}) \tag{1.66}$$

Let $\beta = 0$ and $\gamma = \frac{1}{4}$. The Newmark method yields the classical central difference scheme, i.e. central difference on regular grid,

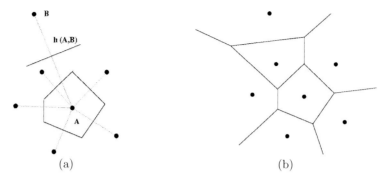

Fig. 1.1. Voronoi diagram of a set of seven nodes: (a) Voronoi cell for node A, and (b) Voronoi diagram V(N).

$$\mathbf{v}_n = \frac{(\mathbf{d}_{n+1} - \mathbf{d}_{n-1})}{2\Delta t} \tag{1.67}$$

$$\mathbf{a}_n = \frac{(\mathbf{d}_{n+1} - 2\mathbf{d}_n + \mathbf{d}_{n+1})}{\Delta t^2} \tag{1.68}$$

Time Integration for Hyperbolic Equations. Consider the first order wave equation (Euler equation) of the form,

$$\frac{\partial u}{\partial t} + a\frac{\partial u}{\partial x} = 0, \quad a > 0 \tag{1.69}$$

Euler's forward time and forward space approximation

$$\frac{u_i^{n+1} - u_i^n}{\Delta t} = -a\frac{u_{i+1}^n - u_i^n}{\Delta x} \tag{1.70}$$

where n is the time step, and i is a typical spatial nodal point.

Midpoint Leapfrog Method The so-called leapfrog method uses three time levels of information, n-1, n, and n+1.

$$\frac{u_i^{n+1} - u_i^{n-1}}{2\Delta t} = -a\frac{u_{i+1}^n - u_{i-1}^n}{2\Delta x} \tag{1.71}$$

1.2.8 Voronoi Diagram and Delaunay Tessellation

The Voronoi diagram is a fundamental structure in computational geometry, because the Voronoi diagram and its dual Delaunay tessellation are one of the most fundamental and useful topological (geometric) structures that connect a set of randomly distributed particles. Today, some of the best finite element mesh generators, or mesh generation algorithms are based on the Voronoi diagram and Delaunay tessellation. Apparently, some of the mesh-free discretizations are also based on local properties of the Voronoi diagram

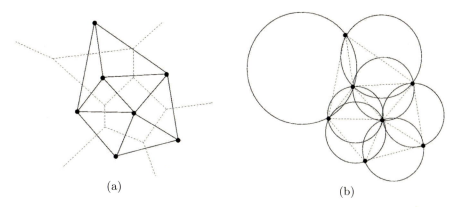

(a)

(b)

Fig. 1.2. Voronoi diagram of a set of seven nodes: (a) Delaunay triangulation DT(N), and (b) Natural neighbor circum-circles.

and Delaunay tessellation. For simplicity, we only outline the main properties of the Voronoi diagram and Delaunay tessellation in a two-dimensional Euclidean space $\mathbb{R}^2$, though the theory is, however, applicable in a general d-dimensional framework.

Consider a set of distinct nodes $\Lambda = \{1, 2, \cdots, I, \cdots, NP\}$ in $\mathbb{R}^2$. The Voronoi diagram (or 1st-order Voronoi diagram) of the point set Λ is a subdivision of the plane into regions (tiles) $\mathbb{R}^2 \subset \bigcup_{I \in \Lambda} \{T_I\}$, where each T_I (tile) is either closed and convex (in interior), or unbounded (at boundary). The tile T_I is defined as a region that contains a node n_I, such that any point in T_I is closer to I than to any other nodal point $J \in \Lambda$ $(J \neq I)$,

$$T_I = \left\{ \mathbf{x} \in \mathbb{R}^2 : d(\mathbf{x}, \mathbf{x}_I) \leq d(\mathbf{x}, \mathbf{x}_J), \ \forall I, J \in \lambda \} \right\} \tag{1.72}$$

where $d(\mathbf{x}, \mathbf{x}_I) := \|\mathbf{x} - \mathbf{x}_I\|$ is the Euclidean norm.

The Voronoi cell for node A and the Voronoi diagram for a set of seven nodes are shown in Fig. 1.2. In Fig. 1.2, it is shown that each Voronoi cell T_I is the intersection of finitely many open half-spaces, each being delimited by the perpendicular bisector of the line joining nodes I and J $(J \neq I)$. Denote the convex hull, $CH(\Lambda)$, as the smallest convex set containing the NP nodes. For all nodes I that are inside the convex hull, Voronoi cells are closed and convex, while the cells associated with nodes on the boundary of the convex hull are unbounded.

The Delaunay triangulation, which is the straight-line dual of the Voronoi diagram, is constructed by connecting the nodes whose Voronoi cells have common boundaries. The duality between the Voronoi diagram and Delaunay triangulation refers to the fact that there is a Delaunay edge between two nodes in the plane if only their Voronoi cells share a common edge. Among all triangulations, the Delaunay triangulation maximizes the minimum angle in each individual triangle, which is a desired property in finite element triangulation. Another important property of Delaunay triangles is the *empty circumcircle criterion* (Lawson [1977]), which states that if $DT(I, J, K)$ is any Delaunay triangle of the nodal set Λ, then the circumcircle of DT contains only three nodes (I,J,K) in Λ that are the vortices of the Delaunay triangle. In natural neighbor interpolation, these circles are known as natural neighbor circumcircles. Fig. 1.2 (b) shows natural neighbor circumcircles and the associated Delaunay triangulation.

A detailed description of the properties and applications of the Voronoi diagrams can be found in Boots [1966], and Okabe, Boots, and Sugihara [1992].

A discussion on randomized algorithms to compute Voronoi diagrams and Delaunay tessellations can be found in Mulmuley (1994). For applying Voronoi diagram to finite element mesh generation, readers may consult Fortune (1995).

2. Smoothed Particle Hydrodynamics (SPH)

In 1977, Lucy[284] and Gingold and Monaghan[176] simultaneously formulated the so-called Smoothed Particle Hydrodynamics, or SPH. Early contributions have been reviewed in several articles (e.g.[55,331,338]).

Initially, SPH was formulated to solve astrophysics problems, such as the formation and evolution of proto-stars or galaxies. In contrast to the concept of discretization methods which discretize a continuum into a finite set of nodal points, SPH consolidates a set of discrete particles into a quasi-continuum. Since the collective movement of astrophysical particles at a large scale is similar to the movement of a liquid, or gas flow, it is modeled by SPH as a quasi-fluid governed by the equations of classical Newtonian hydrodynamics.

In astrophysical and quantum mechanics applications, the real physical system is always discrete. In order to avoid singularities, a local continuous field is generated to represent the collective behavior of the discrete system. This is accomplished by introducing a localized kernel function, which serves as a smoothing interpolation field. If one wishes to interpret the physical meaning of the kernel function as the probability of a particle's position, one then has a probabilistic method. Otherwise, SPH is a smoothing technique-based particle method. Today, SPH is being extensively used in simulations of supernova,[215] collapse as well as formation of galaxies,[58–60,337] coalescence of black holes with neutron stars,[258,259] single and multiple detonations in white dwarfs,[175] and even "*Modeling the Universe*".[336]

Because of its distinct advantages of being a particle method, SPH has been widely adopted as one of the efficient computational techniques to solve applied mechanics problems. Instead of viewing SPH as a consolidation, computational mechanicians use it in the reverse order: SPH is considered as a discretization method that uses a set of particles to approximate a continuum. Furthermore, the term "hydrodynamics" should really be interpreted as "mechanics" in general, if the collective motion of the particles are modeled as a continuum dynamics rather than classical hydrodynamics. In fact, to make such distinction, some authors, e.g. Kum, Hoover, and Posch,[253,376] called it *Smoothed Particle Applied Mechanics*. In this book, the viewpoint of computational mechanicians is taken, and SPH is presented as a meshless discretization method.

2.1 SPH Interpolation

2.1.1 Delta Function

As a meshfree method, SPH interpolation is built on a set of disordered points in a continuum without a grid or mesh. Its interpolation is based on the following simple concept

$$A(\mathbf{x}) = \int_{\mathbb{R}^d} \delta(\mathbf{x}' - \mathbf{x}) A(\mathbf{x}') d\Omega_{\mathbf{x}'} , \qquad \forall\, \mathbf{x} \in \Omega \ (\text{or } \mathbf{x} \in \mathbb{R}^d) \qquad (2.1)$$

where $\delta(\mathbf{x})$ is the Dirac delta function (see Fig. 2.1), which is the limit of the following function, $\delta(x) = \lim_{\epsilon \to 0} \delta_\epsilon(x)$,

$$\delta_\epsilon(x) = \lim_{\epsilon \to 0} \begin{cases} 0; & x < -\epsilon/2 \\ 1/\epsilon; & -\epsilon/2 < x < \epsilon/2 \\ 0; & x > \epsilon/2 \end{cases} \qquad (2.2)$$

The Dirac delta function has some useful properties

$$(1) \quad \int_{-\infty}^{\infty} \delta(x)dx = \int_{-\epsilon/2}^{\epsilon/2} \frac{1}{\epsilon}dx = 1 \qquad (2.3)$$

$$(2) \quad \int_{-\infty}^{\infty} \delta(\zeta - x) f(\zeta)d\zeta = f(x) \qquad (2.4)$$

The discrete counterpart of (2.3) is the property of partition of unity, and the discrete counterpart of (2.4) is the Kronecker delta property.

However, the Dirac delta function can not be used in either interpolation or a collocation process, because it is a generalized function, or a "pathological function", and hence it lacks some "normal" properties of a well behaved function, such as continuity and differentiability. To retain the desirable properties of the Dirac delta function and to remedy its pathology, the main technical ingredient of SPH is to choose a smooth kernel, $\Phi(\mathbf{x}, h)$ (h is the smoothing length) to mimic the valuable part of the Dirac delta function and to fix the pathology, such that it can be used in an interpolation or a collocation process.

2.1.2 SPH Averaging Operator

Define the SPH averaging operator as

$$< A(\mathbf{x}) > = \int_{\mathbb{R}^d} \Phi_h(\mathbf{x} - \mathbf{x}') A(\mathbf{x}') d\Omega_{\mathbf{x}'}$$

$$\approx \sum_{I \in \Lambda} \Phi_h(\mathbf{x} - \mathbf{x}_I) A(\mathbf{x}_I) \Delta V_I \qquad (2.5)$$

where $< \cdot >$ denotes the averaging operator, and h is referred to as **the smoothing length**, which dictates the content of $A(\mathbf{x})$ that is confined in

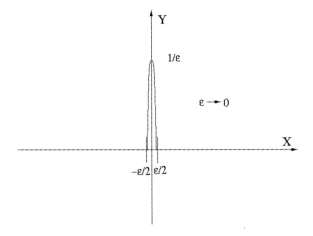

Fig. 2.1. The Dirac delta function

$< A(\mathbf{x}) >$ under a projection determined by a spatial filter $\varPhi$ (high frequency filter). If $\varPhi$ is compactly supported, h is referred to as the radius of the support, or the support size.

It may be noted that first $< A(\mathbf{x}) >$ is a function of the spatial variable $\mathbf{x}$, and in fact $< A(\mathbf{x}) >$ may be viewed as a non-local representation of $A(\boldsymbol{x})$. Second, in much of the SPH literature, $< A(\mathbf{x}) >$ is taken as $A(\mathbf{x})$ without further explanation. In this book, the approximation, $A(\mathbf{x}) \approx < A(\mathbf{x}) >$, is also used from time to time. Third, SPH kernel function $\varPhi_h(\boldsymbol{x}) = \varPhi(\boldsymbol{x}, h)$ has the following properties:

(i) $$\int_{\mathbb{R}^3} \varPhi_h(\boldsymbol{x}) d\Omega_{\boldsymbol{x}} = 1 \qquad (2.6)$$

(ii) $$\varPhi_h(\boldsymbol{x}) \rightarrow \delta(\boldsymbol{x}), \quad h \rightarrow 0 \qquad (2.7)$$

(iii) $$\varPhi_h(\boldsymbol{x}) \in C_0^k(\mathbb{R}^d), \quad k \geq 1 \qquad (2.8)$$

The first two properties (2.6) and (2.7) are the reminiscence of the properties of the Dirac delta function (2.3) and (2.4). The last property comes from the requirement that the smoothing kernel be differentiable more than once, which is the first word "smoothed" referred to in the coined term *smoothed particle hydrodynamics*. It may be noted that most SPH window functions, or kernel functions, [1] are compactly supported, except that the Gaussian function is not compactly supported, but its exponential decay property is virtually "equivalent" to the third requirement in a physical sense.

---

[1] In this book, a window function is referred to as either a primary kernel or the core of a kernel function, and the term kernel function is reserved for *the kernel function*.

2.1.3 Kernel Functions

Several examples of kernel function that are commonly used in computations are listed in the following.

1. The Gaussian function:

$$\Phi_h(\mathbf{u}) = \frac{1}{(\pi h^2)^{n/2}} \exp[-\mathbf{u}^2/h^2] \tag{2.9}$$

where n is the dimension of the space, and $\mathbf{u}^2 = u_i u_i$.

2. The cubic spline function:

$$\Phi_h(r) = \frac{C}{h^n} \begin{cases} 1 - 3/2q^2 + 3/4q^3 ; & 0 \le q \le 1 \\[2mm] \dfrac{1}{4}(2-q)^3 ; & 1 \le q \le 2 \\[2mm] 0 ; & \text{otherwise} \end{cases} \tag{2.10}$$

where $q = r/h$ and $r = (x_i x_i)^{1/2}$, and C is the normalization factor,

$$C = \begin{cases} 2/3; & 1d \\[2mm] 10/7\pi; & 2d \text{ (circular)} \\[2mm] 1/\pi; & 3d \text{ (spherical)} \end{cases} \tag{2.11}$$

3. The quartic spline function

$$\Phi_h(r) = \frac{C}{h^n} \begin{cases} 1 - 6q^2 + 8q^3 - 3q^4, & 0 < q < 1 \\[2mm] 0, & q \ge 1 \end{cases} \tag{2.12}$$

where the constant C is determined by the normalization condition (2.6).

In most cases, the kernel function is chosen as symmetric, i.e.

$$\Phi_h(\boldsymbol{x}_{IJ}) = \frac{1}{h^d} f\left(\frac{|\boldsymbol{x}_{IJ}|}{h}\right) =: \Phi_{IJ} = \Phi_{JI} \tag{2.13}$$

where $\boldsymbol{x}_{IJ} := \boldsymbol{x}_I - \boldsymbol{x}_J$, $x_{IJ} := |\boldsymbol{x}_{IJ}| = \sqrt{(x_{Ii} - x_{Ji})(x_{Ii} - x_{Ji})}$, and d is the spatial dimension.

If the kernel function has the form (2.13), by chain rule,

$$\nabla_I \Phi_{IJ} := \nabla \Phi_h(|\mathbf{x} - \mathbf{x}_J|)\Big|_{\boldsymbol{x}=\boldsymbol{x}_I} = \frac{\partial \Phi_h(|\boldsymbol{x} - \boldsymbol{x}_J|)}{\partial |\boldsymbol{x} - \boldsymbol{x}_J|} \frac{\partial |\boldsymbol{x} - \boldsymbol{x}_J|}{\partial x_I} \mathbf{e}_I \Big|_{\boldsymbol{x}=\boldsymbol{x}_I}$$

$$= \frac{\boldsymbol{x}_{IJ}}{|\boldsymbol{x}_{IJ}|} \frac{\partial \Phi_{IJ}}{\partial x_{IJ}} \tag{2.14}$$

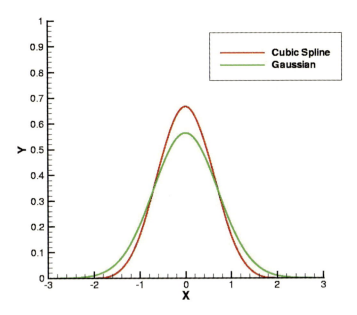

Fig. 2.2. Kernel function examples

where the gradient operator is defined as $\nabla_I = \dfrac{\partial}{\partial x_\alpha} \mathbf{e}_\alpha \Big|_{x_I}$.

By virtue of the fact that $x_{IJ} = |\boldsymbol{x}_I - \boldsymbol{x}_J| = x_{JI}$ and $\Phi_{IJ} = \Phi_{JI}$, a useful identity can be derived

$$\nabla_I \Phi_{IJ} = \frac{\boldsymbol{x}_{IJ}}{|\boldsymbol{x}_{IJ}|} \frac{\partial \Phi_{IJ}}{\partial x_{IJ}} = -\frac{\boldsymbol{x}_{JI}}{|\boldsymbol{x}_{JI}|} \frac{\partial \Phi_{JI}}{\partial x_{JI}} = -\nabla_J \Phi_{JI} \tag{2.15}$$

which can be used to verify many important properties of SPH formulations, such as conservation of linear momentum and energy, etc.

Recall Eq. (2.5),

$$< A(\mathbf{x}) > \approx \sum_{I \in \Lambda} \Phi_h(\mathbf{x} - \mathbf{x}_I) A(\mathbf{x}_I) \Delta V_I \tag{2.16}$$

and assume that the continuum has a distributed mass density. We may divide the continuum into NP small volume elements with associated masses, $m_1, m_2, \cdots, \cdots, m_{NP}$. Then each volume element with mass can be assigned to a particular particle,

$$\Delta V_I = \frac{m_I}{\rho_I} \qquad \Rightarrow \qquad m_I = \rho_I \Delta V_I \tag{2.17}$$

Remark 2.1.1.

1. In Lucy's original paper,[284] it was defined that

$$< A(\mathbf{x}) > := \int_{\mathbb{R}^d} \Phi(\mathbf{x} - \mathbf{x}', h) P(\mathbf{x}') A(\mathbf{x}') d\Omega_{\mathbf{x}'} \tag{2.18}$$

where $P(\mathbf{x}')$ is defined as the probability that at point $\mathbf{x}'$ to find $A(\mathbf{x}')$. According to Lucy, the formulation is derived from the Monte Carlo method.

2. Consider the SPH approximation

$$< A(\mathbf{x}) > = \sum_{I \in \Lambda} \Phi(\mathbf{x} - \mathbf{x}_I, h) A(\mathbf{x}_I) \frac{m_I}{\rho_I} \tag{2.19}$$

if one chooses $A(\mathbf{x}) = \rho(\mathbf{x})$, then

$$< \rho(\mathbf{x}) > = \sum_{I \in \Lambda} \Phi(\mathbf{x} - \mathbf{x}_I) \rho(\mathbf{x}_I) \frac{m_I}{\rho_I}$$

$$= \sum_{I \in \Lambda} \Phi(\mathbf{x} - \mathbf{x}_I) m_I \tag{2.20}$$

2.1.4 Choice of Smoothing Length h

The smoothing length, h, defines a region that confines the major part of a kernel function; it may be related to the radius of the effective support of the kernel function. In principle, the number of particles inside the effective support controls the accuracy of the interpolation. To judiciously choose smoothing length based on the prior accuracy requirement is an important issue in computer implementation. For a given accuracy, the number of particles inside the effective support usually is fixed. If the density of the particle distribution varies in space, the smoothing length should vary accordingly in order to maintain the same accuracy. Moreover, in dynamic simulations, most particles are moving and particle density varies with time. For given accuracy and computation efficiency, the smoothing length should vary with time as well. Thus, it is crucial to set a criterion to determine the smoothing length such that it can be adjusted during computations to maintain the designed numerical accuracy as well as computational efficiency. One way to do this is to assume that the kernel function is a Gaussian, and each particle has approximately the same amount of mass, and assume that the following equation holds based on certain probability assumption,

$$\sum_{I \in \Lambda} \sum_{J \in \Lambda} \exp\{-\frac{(\boldsymbol{x}_J - \boldsymbol{x}_I)^2}{h^2}\} = \gamma = const. \tag{2.21}$$

Choose $\Delta m = m_1 = m_2 \cdots = m_{NP}$. By virtue of SPH interpolation,

$$< \bar{\rho} > = \sum_{I \in \Lambda} \sum_{J \in \Lambda} \frac{1}{(\pi^{d/2} h^d)} \exp\{-\frac{(\boldsymbol{x}_J - \boldsymbol{x}_I)^2}{h^2}\}(\Delta m)^2$$

$$= \frac{\gamma (\Delta m)^2}{\pi^{d/2} h^d} \tag{2.22}$$

which leads to

$$h \propto 1/< \bar{\rho} >^{1/d} \tag{2.23}$$

The intuitive reasoning is: if every particle carries the same amount of mass and the same smoothing length, the smoothing length should be proportional to the reciprocal of the inverse d-th (dimension number) root of average density, which is because the d-th power of smoothing length is proportional to the volume that each particle carries, and it implies that the smoothing length is proportional to the particle separation length.

If it is further assumed that

$$\sum_{J \in \Lambda} \exp\{-(\boldsymbol{x}_I - \boldsymbol{x}_J)^2/h_J^2\} = \gamma_1 = const. \tag{2.24}$$

by virtue of

$$< \rho >_I = \sum_{J \in \Lambda} \frac{m_J}{\pi^{d/2} h_J^d} \exp\{-\frac{(\boldsymbol{x}_I - \boldsymbol{x}_J)^2}{h_J^2}\} \tag{2.25}$$

and $m_1 = m_2 = \cdots = m_{NP}$, $h_1 = h_2 = \cdots = h_{NP}$, one may expect

$$h_I \propto 1/< \rho >_I^{1/d} . \tag{2.26}$$

This can be deduced from the fact that $m_I =< \rho >_I \Delta V_I$ and $\Delta V_I \propto h_I^d$ as $h_I \to 0$.

2.2 Approximation Theory of SPH

As a numerical method, SPH is constructed based on a set of coherent approximation principles. Before proceeding to derive SPH dynamic equations, it would be expedient to lay out some approximation rules. By using them, SPH can collocate a set of continuous hydrodynamics conservation laws into discrete form. These rules of approximation are referred to as the **Golden Rules** of SPH by Monaghan,[339] and these rules are the main guidelines to construct discrete SPH equations.

2.2.1 SPH Approximation Rules

As a discrete representation (algebraic equations) of hydrodynamics conservation laws (partial differential equations), discrete SPH equations may obscure the original physical interpretation of the continuous form. To ensure that SPH equations have a clear physical footing, Monaghan recommended: *it is always best to assume that the kernel function is a Gaussian*, which is considered as **the first Golden Rule** of SPH. In other words, to analyze the physical coherence of an SPH model, Gaussian is always the best choice over other functions as SPH kernel, even though the Gaussian function is not compactly supported, but its fast decay behavior compensates that shortcoming to certain extent.

The second approximation rule of SPH states that the ensemble (average) of the product of two functions can be approximated by the product of the individual ensembles:

$$< A \cdot B > \approx < A > \cdot < B >$$
(2.27)

which may be viewed as an assumption on statistical average closure.

This is a very crude approximation. Let $A = u_i$ and $B = u_j$. This approximation takes

$$< u_i u_j > \approx < u_i > < u_j >,$$

which is generally not acceptable in multi-scale computations, since the difference between the two sides of (2.27), ($< u_i u_j > - < u_i > < u_j >$), contains the mechanical information in fine scale. For instance in large eddy simulation (LES) of turbulence flow, the stress in fine scale, i.e. the so-called *subgrid scale Reynold stress*, is calculated based on this difference, or residual,

$$\tau_{ij}^s = \rho \Big(< u_i u_j > - < u_i > < u_j > \Big)$$
(2.28)

for it captures the effect of the momentum flux in large scale due to the unresolved scale. From this standpoint, the classical SPH approximation may not be suitable in turbulence simulation, because of its limitation due to this approximation. This may also be the reason, the authors speculate, why the method is called Smoothed Particle Hydrodynamics (SPH), rather than Smoothed Particle Fluid Dynamics.

The third approximation rule of SPH is: for any scalar field A,

$$< \nabla A > = \nabla < A >$$
(2.29)

The third approximation rule may be an exact relation in an infinite domain. Utilizing integration by parts, one can show that in an unbounded space,

$$
\begin{aligned}
< \nabla A > &= \int_{\mathbb{R}^3} \Phi_h(\boldsymbol{x} - \boldsymbol{x}') \nabla_{\boldsymbol{x}'} A(\boldsymbol{x}') d\Omega_{\boldsymbol{x}'} \\
&= \int_{\mathbb{R}^3} \Big[\nabla_{\boldsymbol{x}'} \Big(\Phi_h(\boldsymbol{x} - \boldsymbol{x}') A(\boldsymbol{x}') \Big) - \nabla_{\boldsymbol{x}'} \Phi_h(\boldsymbol{x} - \boldsymbol{x}') A(\boldsymbol{x}') \Big] d\Omega_{\boldsymbol{x}'} \\
&= \int_{S_\infty} \Phi_h(\boldsymbol{x} - \boldsymbol{x}') A(\boldsymbol{x}') \mathbf{n} dS + \int_{\mathbb{R}^3} \nabla_{\boldsymbol{x}} \Phi_h(\boldsymbol{x} - \boldsymbol{x}') A(\boldsymbol{x}') d\Omega_{\boldsymbol{x}'} \\
&= \nabla < A >
\end{aligned}
$$
(2.30)

where S_∞ is a spherical ball with infinite radius, and on S_∞, $\Phi \to 0$.

A very useful property of the third approximation rule is:

$$< \nabla A >= \nabla < A > -A < \nabla \Phi > . \tag{2.31}$$

Since the kernel function is localized, the second term of (2.31) on the right is always zero: $< \nabla \Phi >= 0$. This fact can also be understood as

$$
\begin{aligned}
< \nabla \Phi > &\approx \nabla < \Phi > \\
&\approx \nabla \sum_{J \in \Lambda} \Phi_h(\boldsymbol{x} - \boldsymbol{x}_J) \Delta V_J \\
&= \nabla(1) = 0
\end{aligned}
\tag{2.32}
$$

since $\{\Phi_I(\mathbf{x})\}$ is a partition of unity.

2.2.2 SPH Approximations of Derivatives (Gradients)

Since SPH is used to approximate the strong form of a PDE, the key technical ingredients of constructing a discrete SPH dynamics is how to approximate the derivatives. Finite difference method does this by utilizing a grid or a stencil, whereas SPH, as a meshfree particle method, approximate the derivatives based on the above listed approximation rules.

SPH Gradient Formula I. Without any special treatment, one directly approximates the derivative of a continuous function by using SPH interpolation,

$$\nabla A(\boldsymbol{x}) = \sum_{I \in \Lambda} \nabla \Phi_h(\boldsymbol{x} - \boldsymbol{x}_I) A_I \Delta V_I \tag{2.33}$$

Such straightforward approximation is usually not accurate, and often destroys the conservation property of the associated continuous system. Nonetheless, when the approximation is combined with an additional term that contains a null expression $< \nabla \Phi >= 0$, it may produce better results. For instance, from the identity (2.31), one can obtain the approximated gradient formula of a scalar field

$$\boxed{\text{SPH gradient I (scalar)}: \quad \nabla A(\boldsymbol{x}_I) = \sum_{J \in \Lambda} (A_J - A_I) \nabla_I \Phi_{IJ} \Delta V_J} \tag{2.34}$$

Note that the second term $\sum_{J \in \Lambda} A_I \nabla_I \Phi_{IJ} \Delta V_J = A_I < \Phi >= 0$. The same approximation is valid for a vector field, $\mathbf{A}$,

$$\boxed{\text{SPH gradient I (vector)}: \quad \nabla \cdot \mathbf{A}(\boldsymbol{x}_I) = \sum_{J \in \Lambda} (\mathbf{A}_J - \mathbf{A}_I) \cdot \nabla_I \Phi_{IJ} \Delta V_J}$$

$$\tag{2.35}$$

SPH Gradient Formula II. In SPH formulation, density is an very important quantity, because its intimate relation with a particle's mass and volume. To obtain higher order accuracy of gradient formulas, it is suggested to first write the gradient of a scalar field as

$$\nabla A = \big(\nabla(\rho A) - A\nabla\rho\big)/\rho \tag{2.36}$$

and similarly the gradient of a vector field may be first written as

$$\nabla \cdot \mathbf{A} = \big(\nabla \cdot (\rho\mathbf{A}) - \mathbf{A} \cdot \nabla\rho\big)/\rho . \tag{2.37}$$

The gradient of SPH interpolants may then be approximated as follows

SPH gradient II (scalar)
$$\begin{cases} \rho_I <\nabla A>_I = \sum_{J\in\Lambda}(A_J - A_I)\nabla_I \Phi_{IJ}m_J \\[2mm] \qquad\quad = -\sum_{J\in\Lambda}A_{IJ}\nabla_I \Phi_{IJ}m_J \end{cases}$$

$$\tag{2.38}$$

where $<\nabla\rho>_I = \sum_{J\in\Lambda}\nabla_I \Phi_{IJ}m_J$ is used (Recall Eq. (2.20)).

Following similar procedures, one may be able to extend such an approximation to gradients of vector fields as well. The algorithmic expression for a dot product of a gradient operator and a vector field is,

SPH gradient II (dot product) $\rho_I <\nabla \cdot \mathbf{A}>_I = -\sum_{J\in\Lambda}\mathbf{A}_{IJ} \cdot \nabla_I \Phi_{IJ}m_J,$

$$\tag{2.39}$$

and the tensor product of gradient operator and a vector field is,

SPH gradient II (tensor product) $\rho_I <\nabla \otimes \mathbf{A}>_I = -\sum_{J\in\Lambda}\nabla_I \Phi_{IJ} \otimes \mathbf{A}_{IJ}m_J,$

$$\tag{2.40}$$

and the cross product between the gradient operator and a vector field is

SPH gradient II (cross product). $\rho_I <\nabla \times \mathbf{A}>_I = \sum_{J\in\Lambda}\mathbf{A}_{IJ} \times \nabla_I \Phi_{IJ}m_J .$

$$\tag{2.41}$$

Note that in (2.41), there is no minus sign at the right handside.

SPH Gradient Formula IIIa. Although (2.38) — (2.41) are acceptable approximations, a common deficiency shared with these approximations is: they are not symmetric with respect to index I and J, which may affect the conservation properties of the discrete SPH system. To obtain a symmetric gradient approximation for a scalar field, one needs another set of SPH approximation rules to approximate differentiation. Considering the identity

$$\frac{\nabla A}{\rho} = \frac{A}{\rho^\sigma}\nabla\left(\frac{1}{\rho^{1-\sigma}}\right) + \frac{1}{\rho^{2-\sigma}}\nabla\left(\frac{A}{\rho^{\sigma-1}}\right) \tag{2.42}$$

where σ is an integer, one can establish an SPH approximation,

$$\left(\frac{\nabla A}{\rho}\right)_I = \sum_{J\in\Lambda}\left(\frac{A_J}{\rho_I^{2-\sigma}\rho_J^\sigma} + \frac{A_I}{\rho_I^\sigma\rho_J^{2-\sigma}}\right)\nabla_I\Phi_{IJ}m_J \tag{2.43}$$

Consider a scalar field, $A(\mathbf{X})$, and take $\sigma = 1$. It yields

$$\boxed{\text{SPH gradient IIIa (scalar)}\quad (\nabla A)_I = \sum_{J\in\Lambda}(A_I + A_J)\nabla_I\Phi_{IJ}\Delta V_J \ ;}$$
$$\tag{2.44}$$

For the dot product between the gradient operator and a vector field $\mathbf{A}$, the III derivative approximation yields,

$$\boxed{\text{SPH gradient IIIa (vector)}\quad (\nabla\cdot\mathbf{A})_I = \sum_{J\in\Lambda}(\mathbf{A}_I + \mathbf{A}_J)\cdot\nabla_I\Phi_{IJ}\Delta V_J \ ;}$$

$$\tag{2.45}$$

The formulas for the cases of tensor product and cross product are left to the reader as an exercise.

SPH Gradient Formula IIIb. Letting $\sigma = 2$ in Eq. (2.42), one can construct another set of SPH gradient formulas.

For a scalar field, it yields

$$\boxed{\text{SPH gradient IIIb (scalar)}\quad \left(\frac{\nabla A}{\rho}\right)_I = \sum_{J\in\Lambda}\left(\frac{A_J}{\rho_J^2} + \frac{A_I}{\rho_I^2}\right)\nabla_I\Phi_{IJ}m_J \ ;}$$

$$\tag{2.46}$$

For the dot product between the gradient operator and a vector field $\mathbf{A}$, the approximated gradient is

$$\boxed{\text{SPH gradient IIIb (vector)}\quad \left(\frac{\nabla\cdot\mathbf{A}}{\rho}\right)_I = \sum_{J\in\Lambda}\left(\frac{\mathbf{A}_J}{\rho_J^2} + \frac{\mathbf{A}_I}{\rho_I^2}\right)\cdot\nabla_I\Phi_{IJ}m_J \ .}$$

$$\tag{2.47}$$

Again, the formulas for the cases of tensor product and cross product are left to the reader as an exercise.

2.3 Discrete Smooth Particle Hydrodynamics

The SPH is not only an interpolation scheme, but also provides a set of approximation rules to establish discrete dynamics by collocating the corresponding continuum dynamics. To illustrate how to use these approximation rules to construct discrete SPH dynamics, SPH equations in continuum mechanics are derived in detail.

2.3.1 Conservation Laws in Continuum Mechanics

Before constructing SPH equations, main conservation laws in continuum mechanics are outlined as follows. It is assumed that there is no body force, no mass and heat sources, no chemical potentials, no diffusion process, and no heat conduction involved.

- Continuity equation:

$$\frac{d\rho}{dt} + \rho \nabla \cdot \mathbf{v} = 0 \tag{2.48}$$

where ρ is the density, $\mathbf{v}$ is the velocity field $\mathbf{v} := d\mathbf{x}/dt$;
- Linear momentum equation:

$$\frac{d\mathbf{v}}{dt} = -\frac{1}{\rho} \nabla \cdot \boldsymbol{\sigma} \tag{2.49}$$

where $\boldsymbol{\sigma} := \sigma_{ij} \mathbf{e}_i \otimes \mathbf{e}_j$, and $\sigma_{ij} = P\delta_{ij} - S_{ij}$. Note that $\boldsymbol{\sigma}$ is actually the Cauchy stress with a minus sign, which has been a convention in SPH literature.
- Energy equation

$$\frac{dE}{dt} = -\frac{1}{\rho} \boldsymbol{\sigma} : \nabla \mathbf{v} \tag{2.50}$$

where E is specific internal energy. The energy equation (2.50) is obtained by subtracting the mechanical energy balance law,

$$\frac{\partial}{\partial t} \left(\rho \frac{1}{2} v^2 \right) + \nabla \cdot \left(\rho \mathbf{v} \frac{1}{2} v^2 \right) = -\mathbf{v} \cdot \nabla \cdot \boldsymbol{\sigma}, \tag{2.51}$$

from the general energy equation,

$$\frac{\partial}{\partial t} \left(\rho (E + \frac{1}{2} v^2) \right) + \nabla \cdot \left(\rho \mathbf{v} (E + \frac{1}{2} v^2) \right) = -\nabla \cdot (\mathbf{v} \cdot \boldsymbol{\sigma}) . \tag{2.52}$$

Based on different approximation rules, there are different ways to form SPH equations. Up till today, there has not been a unified approach that has gained public consensus yet, and as a matter of fact, these different versions of SPH equations have all been used in practical computations. In this book, a balanced approach is adopted to present these different algorithms, and hopefully it may encourage and bring new attempts to unify them.

2.3.2 SPH Continuity Equation

There are two SPH continuity equations used in computations, which are derived by different approximation rules.

First, following Eq. (2.48),

$$< \frac{d\rho}{dt} > = - < \rho \nabla \cdot \mathbf{v} >$$
$$\approx - <\rho><\nabla \cdot \mathbf{v}> \quad \Leftarrow \text{ Approximation rule 2}$$
$$\approx - <\rho> \nabla \cdot <\mathbf{v}> \quad \Leftarrow \text{ Approximation rule 3}$$
$$\approx - <\rho> \nabla \cdot <\mathbf{v}> +\rho\mathbf{v} \cdot <\nabla\Phi> \quad \Leftarrow \text{Approximation rule 4}$$

$$(2.53)$$

Consider

$$\left(\nabla \cdot <\mathbf{v}>\right)_I = \sum_{J=1}^{N} \Delta V_J \mathbf{v}_J \cdot \nabla_I \Phi_{IJ} = \sum_{J=1}^{N} \frac{m_J}{\rho_J}\mathbf{v}_J \cdot \nabla_I \Phi_{IJ} \qquad (2.54)$$

and

$$< \nabla\Phi >_I = \sum_{J=1}^{N} \frac{m_J}{\rho_J} \nabla_I \Phi_{IJ} \ . \qquad (2.55)$$

Substituting (2.54) and (2.55) into (2.53) yields

$$\frac{d\rho_I}{dt} = \rho_I \sum_{J=1}^{N} \frac{m_J}{\rho_J}(\mathbf{v}_I - \mathbf{v}_J) \cdot \nabla_I \Phi_{IJ} = \rho_I \sum_{J=1}^{N} \frac{m_J}{\rho_J}\mathbf{v}_{IJ} \cdot \nabla_I \Phi_{IJ} \qquad (2.56)$$

The technical ingredient of this derivation is the SPH gradient approximation rule I for a vector field. An alternative SPH continuity equation can be also derived based on SPH gradient approximation rule II for dot product

$$< \rho\nabla \cdot \mathbf{v} >_I \approx \sum_{J=1}^{N}(\mathbf{v}_J - \mathbf{v}_I) \cdot \nabla_I \Phi_{IJ} m_J \qquad (2.57)$$

Therefore

$$\frac{d\rho_I}{dt} = \rho_I \sum_{J=1}^{N} \frac{m_J}{\rho_I}(\mathbf{v}_I - \mathbf{v}_J) \cdot \nabla_I \Phi_{IJ} = \rho_I \sum_{J=1}^{N} \frac{m_J}{\rho_I}\mathbf{v}_{IJ} \cdot \nabla_I \Phi_{IJ} \qquad (2.58)$$

One can easily find the difference between (2.56) and (2.58).

2.3.3 SPH Momentum Equation

For balance of linear momentum, there are three different SPH equation of motion used in practice. The derivations are as follows:

1. Based on the SPH gradient approximation rule IIIa ($\sigma = 1$), one has

$$< \rho \frac{d\mathbf{v}}{dt} >_I = - < \nabla \cdot \boldsymbol{\sigma} >_I$$

$$\Rightarrow \quad \rho_I \frac{d\mathbf{v}_I}{dt} \approx - \sum_{J=1}^{N} (\boldsymbol{\sigma}_I + \boldsymbol{\sigma}_J) \cdot \nabla_I \Phi_{IJ} \Delta V_J$$

which leads to

$$\boxed{m_I \frac{d\mathbf{v}_I}{dt} = - \sum_{J=1}^{N} \Delta V_I \Delta V_J (\boldsymbol{\sigma}_J + \boldsymbol{\sigma}_I) \cdot \nabla_I \Phi_{IJ}}$$

$$(2.59)$$

2. Based on the SPH gradient approximation rule IIIb ($\sigma = 2$), one has

$$\left\langle \frac{d\mathbf{v}}{dt} \right\rangle_I = - \left\langle \frac{1}{\rho} \nabla \cdot \boldsymbol{\sigma} \right\rangle_I$$

$$\Rightarrow \quad \frac{d\mathbf{v}_I}{dt} \approx - \sum_{J=1}^{N} \left(\frac{\boldsymbol{\sigma}_I}{\rho_I^2} + \frac{\boldsymbol{\sigma}_J}{\rho_J^2} \right) \cdot \nabla_I \Phi_{IJ} m_J$$

which leads to

$$\boxed{m_I \frac{d\mathbf{v}_I}{dt} = - \sum_{J=1}^{N} m_I m_J \left(\frac{\boldsymbol{\sigma}_I}{\rho_I^2} + \frac{\boldsymbol{\sigma}_J}{\rho_J^2} \right) \cdot \nabla_I \Phi_{IJ}}$$

$$(2.60)$$

3. Based on SPH gradient approximation rule I for vector and tensorial field, one has

$$\left\langle \rho \frac{d\mathbf{v}}{dt} \right\rangle_I = - < \nabla \cdot \boldsymbol{\sigma} >_I \ = -\left(\nabla \cdot < \boldsymbol{\sigma} >_I - < \boldsymbol{\sigma} >_I < \nabla \Phi >_I \right)$$

$$\Rightarrow \quad \rho_I \frac{d\mathbf{v}_I}{dt} = \sum_{J=1}^{N} \Delta V_J (\boldsymbol{\sigma}_I - \boldsymbol{\sigma}_J) \nabla_I \Phi_{IJ} \qquad (2.61)$$

which leads to

$$\boxed{m_I \frac{d\mathbf{v}_I}{dt} = - \sum_{J=1}^{N} \Delta V_I \Delta V_J (\boldsymbol{\sigma}_J - \boldsymbol{\sigma}_I) \cdot \nabla_I \Phi_{IJ}}$$

$$(2.62)$$

The first two SPH momentum equations are symmetric with indices I and J, whereas the third SPH momentum equation is not, which leads to serious consequences for conservation properties that shall be discussed in later sections.

2.3.4 SPH Energy Equation

There are at least three SPH energy equations used in practice.

1. Following Eq. (2.50) and by using approximation rule 2, one will have

$$< \frac{dE}{dt} > = -\left\langle \frac{1}{\rho} \boldsymbol{\sigma} : \nabla \mathbf{v} \right\rangle \approx -\left\langle \frac{1}{\rho^2} \boldsymbol{\sigma} : (\rho \nabla \mathbf{v}) \right\rangle$$

$$\approx -\frac{1}{< \rho >^2} < \boldsymbol{\sigma} >:< \rho \nabla \mathbf{v} > \quad \Leftarrow \quad \text{Approximation rule 2}$$

$$\tag{2.63}$$

considering the SPH gradient approximation rule (Ib), one has

$$\boxed{\frac{dE_I}{dt} = \left(\frac{\boldsymbol{\sigma}_I}{\rho_I^2}\right) : \sum_{J=1}^{N} m_J \mathbf{v}_{IJ} \otimes \nabla_I \Phi_{IJ}}$$

$$\tag{2.64}$$

2. Considering

$$< \frac{dE}{dt} > = -\nabla \cdot \left\langle \frac{\boldsymbol{\sigma} \cdot \mathbf{v}}{\rho} \right\rangle + < \mathbf{v} > \cdot \left(\nabla \cdot \left\langle \frac{\boldsymbol{\sigma}}{\rho} \right\rangle \right) \tag{2.65}$$

and taking the derivative directly, one has

$$\boxed{\frac{dE_I}{dt} = \sum_{J=1}^{N} m_J \left(\frac{\boldsymbol{\sigma}_J}{\rho_J^2}\right) : \left(\mathbf{v}_{IJ} \otimes \nabla_I \Phi_{IJ} \right)}$$

$$\tag{2.66}$$

3. Averaging (2.64) and (2.66), one has

$$\boxed{\frac{dE_I}{dt} = \frac{1}{2} \sum_{J=1}^{N} m_J \left(\frac{\boldsymbol{\sigma}_J}{\rho_J^2} + \frac{\boldsymbol{\sigma}_I}{\rho_I^2}\right) : \left(\mathbf{v}_{IJ} \otimes \nabla_I \Phi_{IJ} \right)}$$

$$\tag{2.67}$$

Among the three SPH energy equations, we shall show that expression (2.67) conserves energy of the discrete system exactly.

2.3.5 SPH Artificial Viscosity

When the SPH method is used to simulate shock or convection dominated flows, spurious oscillations occur in both the velocity fields and the pressure field, which may ruin a simulation completely. The origins of such numerical instabilities are either due to discontinuity in velocity fields or algorithmic pathology in approximation of parabolic or hyperbolic systems. Various analyses have been devoted to this topic in the context of finite difference methods and finite element methods, and many measures have been taken to prevent such numerical instabilities.

In order to control numerical instability, in finite difference methods, a common technique used is to add artificial viscous terms in both the discrete momentum equation and the discrete energy equation to dampen or to "smear out" undesirable oscillations. As an approximation of the strong form of a PDE, SPH is similar to finite difference methods. Therefore, adding artificial viscous pressure into SPH equations is recommended to treat shock induced numerical instabilities.[330]

Two types of artificial viscosity are commonly used in computational fluid dynamics to stabilize numerical computations. They are: (a) the Von Neumann-Richtmyer viscous pressure which is proportional to the quadratic form of velocity gradient,

$$
q = \begin{cases} \alpha \rho h^2 (\nabla \cdot v)^2, & \nabla \cdot v < 0, \\ 0, & \nabla \cdot v > 0, \end{cases}
\tag{2.68}
$$

where α is a constant, and (b) a bulk viscous pressure which is proportional to the linear form of velocity gradient,

$$
q = \begin{cases} -\alpha \rho h c \nabla \cdot v, & \nabla \cdot v < 0, \\ = 0, & \nabla \cdot v > 0, \end{cases}
\tag{2.69}
$$

The SPH approximations of Eq. (2.69) are based on how to approximate the velocity gradient. In practice, the following two formulas are used,

$$
p_I = m_I \sum_{J \in \Lambda} \frac{\mathbf{v}_J \cdot \nabla_I W_{IJ}}{\rho_J}
\tag{2.70}
$$

or

$$
p_I = \frac{m_I}{\rho_I} \sum_{J \in \Lambda} \mathbf{v}_{IJ} \cdot \nabla_I W_{IJ}
\tag{2.71}
$$

where the latter is derived by using the relation (the third approximation rule),

$$
\nabla \cdot \mathbf{v} = \frac{1}{\rho} \left[\nabla \cdot (\rho \mathbf{v}) - \mathbf{v} \cdot \nabla \rho \right]
$$

Nevertheless, the standard artificial viscosity does not work very well in SPH shock simulation. The reason for this, as Monaghan and Gingold pointed out, is because in shock simulations, in order to maintain reasonable accuracy, the resolved length scale should be much greater than the particle spacing, and irregular motion on this latter scale is then only weakly affected by the artificial viscosity. In other words, in SPH shock simulations, the shock is simulated by inducing irregular particle motions on length scale corresponding to particle spacing, whereas the length scale of conventional artificial viscosity is controlled by the smoothing length of the kernel function, which is usually smaller than particle separation. Therefore, the conventional artificial viscosity is very inefficient in damping out irregular particle motions.

To suit the SPH character, Monaghan and Gingold proposed the following artificial viscosity for SPH equations. Consider the equation of motion with an added viscous term

$$\rho \dot{\mathbf{v}} = \nabla \cdot (\boldsymbol{\sigma}_{real} + \boldsymbol{\sigma}_{viscous}) \tag{2.72}$$

The corresponding SPH equation with artificial viscous stress is formed in[330] as

$$\frac{d\mathbf{v}_I}{dt} = -\sum_{J=1}^{N} m_J \left(\frac{\boldsymbol{\sigma}_J}{\rho_J^2} + \frac{\boldsymbol{\sigma}_I}{\rho_I^2} + \Pi_{IJ} \mathbf{e}_i \otimes \mathbf{e}_j \right) \cdot \nabla_I \Phi_{IJ} \tag{2.73}$$

The effect of artificial viscous stress is reflected in the term Π_{IJ}, which is given as

$$\Pi_{IJ} = \begin{cases} \dfrac{-\alpha \bar{c}_{IJ} \mu_{IJ} + \beta \mu_{IJ}^2}{\bar{\rho}_{IJ}}, & \mathbf{v}_{IJ} \cdot \boldsymbol{x}_{IJ} < 0 \\[2ex] 0, & \mathbf{v}_{IJ} \cdot \boldsymbol{x}_{IJ} > 0 \end{cases} \tag{2.74}$$

and

$$\mu_{IJ} = \frac{h \mathbf{v}_{IJ} \cdot \boldsymbol{x}_{IJ}}{\boldsymbol{x}_{IJ}^2 + \eta^2} \tag{2.75}$$

In Eq. (2.74), the linear part of μ_{IJ} corresponds to bulk viscosity, and the quadratic part of μ_{IJ} corresponds to von Neumann-Richtmyer viscosity.

Remark 2.3.1. First, the reason that the above artificial viscosity is better than the conventional artificial viscosity is because a particle spacing term, $\boldsymbol{x}_{IJ}$, is incorporated into the viscous stress expression, which reflects the length scale of the particle distribution, or particle spacing.

Second $\mathbf{v}_{IJ} \cdot \boldsymbol{x}_{IJ} > 0$ is equivalent to $\nabla \cdot \mathbf{v} > 0$ in SPH formulation. This can be shown as follows. Assume the kernel function is Gaussian (the first approximation rule)

$$\rho_I (\nabla \cdot \mathbf{v})_I = \sum_{J=1}^{N} m_J (\mathbf{v}_J - \mathbf{v}_I) \nabla_I \Phi_{IJ}$$

$$= \sum_{J=1}^{N} m_J (\mathbf{v}_J - \mathbf{v}_I) \cdot \left(-2 \frac{(\boldsymbol{x}_I - \boldsymbol{x}_J)}{h^2} \right) \Phi_{IJ}$$

$$= \sum_{J=1}^{N} \left(\frac{2 m_J}{h^2} \right) (\mathbf{v}_I - \mathbf{v}_J) \cdot (\boldsymbol{x}_I - \boldsymbol{x}_J) \Phi_{IJ} \tag{2.76}$$

Thus

$$(\nabla \cdot \mathbf{v})_I > 0 \Rightarrow \mathbf{v}_{IJ} \cdot \boldsymbol{x}_{IJ} > 0 \tag{2.77}$$

and vice versa.

2.3.6 Time Integration of SPH Conservation Laws

The SPH discretization is a spatial discretization. To integrate SPH dynamics equations in time, one has to use proper numerical integration schemes.

For simplicity, the majority of SPH codes use explicit time integration scheme to integrate discrete SPH equations.

A popular choice is the standard leap-frog algorithm, which is outlined as follows

$$\rho_I^{n+1} = \rho_I^n (1 - D_I^n \Delta t^n) \tag{2.78}$$

$$\mathbf{v}_I^{n+1/2} = \mathbf{v}_I^{n-1/2} + \frac{1}{2}\left(\Delta t^n + \Delta t^{n-1}\right)\mathbf{F}_I^n \tag{2.79}$$

$$E_I^{n+1} = E_I^n + \Delta t^n G_I^n \tag{2.80}$$

$$\mathbf{S}_I^{n+1} = \mathbf{S}_I^n + \Delta t^n \mathbf{H}_I^n \tag{2.81}$$

$$\boldsymbol{x}_I^{n+1} = \boldsymbol{x}_I^n + \mathbf{v}_I^{n+1/2} \Delta t^n \tag{2.82}$$

where

$$D_I^n = \sum_{J=1}^{N} \frac{m_J}{\rho_J}(\mathbf{v}_I(t^n) - \mathbf{v}_J(t^n)) \cdot \nabla_I \Phi_{IJ} \tag{2.83}$$

$$\mathbf{F}_I^n = -\sum_{J=1}^{N} m_J \left(\frac{\boldsymbol{\sigma}_I^n}{\rho_I^2(t^n)} + \frac{\boldsymbol{\sigma}_J^n}{\rho_J^2(t^n)} + \Pi_{IJ}^n \mathbf{I}\right) \cdot \nabla_I \Phi_{IJ} \tag{2.84}$$

$$G_I^n = -\sum_{J=1}^{N} m_J \mathbf{v}_{JI}^n \cdot \left(\frac{\boldsymbol{\sigma}_I^n}{\rho_I^2(t^n)} + \frac{1}{2}\Pi_{IJ}^n \mathbf{I}\right) \cdot \nabla_I \Phi_{IJ} \tag{2.85}$$

$$\mathbf{H}_I^n = \overset{\triangledown}{\mathbf{S}}_I(t^n) + \mathbf{S}_I(t^n) \cdot \mathbf{W}_I(t^n) + \mathbf{S}_I(t^n) \cdot \mathbf{W}_I^T(t^n) \tag{2.86}$$

and $\mathbf{v}_I^n$ is evaluated as

$$\mathbf{v}_I^n = \frac{1}{2}(\mathbf{v}_I^{n+1/2} + \mathbf{v}_I^{n-1/2}) \tag{2.87}$$

The stability condition of leap-frog scheme is controlled by Courant-Friedrichs-Lewy (CFL) condition. A recommended time step size is

$$\Delta t \leq C_{CFL}\frac{h}{c+s} \tag{2.88}$$

where c is sound speed, s is the particle speed (or maximum particle speed among all particles), and $C_{CFL} : 0.0 \sim 1.0$.

2.3.7 SPH Constitutive Update

In continuum media, constitutive relations may be time dependent, or rate-dependent. To update material constitutive behavior, one has to integrate the constitutive equation to keep track of stress-strain relations.

Consider the stress update of a Newtonian fluid. Let

$$\sigma_{ij} = P\delta_{ij} - S_{ij} \tag{2.89}$$

The hydrostatic presure update can be derived based on the equation of state. For instance, for ideal gas,

$$P = (\gamma - 1)\rho E \tag{2.90}$$

The SPH pressure update can then be obtained as

$$P_I = (\gamma - 1)\rho_I E_I \tag{2.91}$$

For nearly incompressible fluid,

$$P = P_0 \left\{ \left(\frac{\rho}{\rho_0} \right)^\gamma - 1 \right\} \Rightarrow P_I = P_0 \left\{ \left(\frac{\rho_I}{\rho_{I0}} \right)^\gamma - 1 \right\} \tag{2.92}$$

For a Newtonian fluid, deviatoric stress update is also needed. Since SPH is a Lagrangian method, based on the principle of material objectivity, or principle of material frame indifference, a constitutive relation should be independent of rigid body rotation (motion). Therefore, the objective rate is required for constitutive update in finite deformation. In practice, the Jaumann rate is the most popular choice used in computations,

$$\overset{\triangledown}{S}_{ij} = \dot{S}_{ij} - S_{ik} W_{kj} - S_{kj} W_{ik} \tag{2.93}$$

where

$$W_{ij} = \frac{1}{2} \left(\frac{\partial v_i}{\partial x_j} - \frac{\partial v_j}{\partial x_i} \right) \tag{2.94}$$

For viscous fluids,

$$\overset{\triangledown}{S}_{ij} = \mu(D_{ij} - \frac{1}{3}\delta_{ij}\dot{\epsilon}_{kk}) \tag{2.95}$$

where

$$D_{ij} = \frac{1}{2} \left(\frac{\partial v_i}{\partial x_j} + \frac{\partial v_j}{\partial x_i} \right) \tag{2.96}$$

By using the SPH gradient approximation rule I, one can derive that

$$\overset{\triangledown}{S}_{Iij} = \frac{\mu}{2} \sum_{J=1}^{N} \frac{m_J}{\rho_J} \left[(v_{Ji} - v_{Ii}) \partial_j \Phi_{IJ} + (v_{Jj} - v_{Ii}) \partial_i \Phi_{IJ} - \frac{1}{3} D_I \delta_{ij} \right] \quad (2.97)$$

where

$$D_I = \frac{\dot{\rho}_I}{\rho_I} = \sum_{J=1}^{N} \frac{m_J}{\rho_J} (\mathbf{v}_I - \mathbf{v}_J) \cdot \nabla_I \Phi_{IJ} \qquad (2.98)$$

2.4 Invariant Properties of SPH Equations

An important question is whether or not discrete SPH equations can still preserve physical quantities of the corresponding continuum system, such as energy, momentum, etc. This is a critical issue related to the quality and accuracy of numerical solutions.

Both discontinuous Galerkin finite element formulations and finite volume methods possess local as well as global conserving properties. Recently, Hughes et al.[213] pointed out that continuous Galerkin finite element formulation may preserve both local and global conserving properties as well.

However, because of the non-local character in meshfree interpolation scheme, conserving properties are not automatic for meshfree interpolants. Nevertheless, by careful design, some global conservation properties are preserved by SPH formulation in a discrete sense.

2.4.1 Galilean Invariance

Mathematically, conservative properties of a PDE are intimately related with the invariance properties of a PDE. This fact has been well documented in physics, mathematics, and mechanics literature.

Physically, the governing equations of Newtonian mechanics is a Galilean invariant. This fact is particularly significant in astrophysics problems.

Above all, computationally, a good interpolation scheme should be at least able to represent rigid body motion correctly, which means that the interpolation field can at least represent a constant. If an interpolant forms a partition of unity, this is a built-in property.

Unfortunately, the early SPH interpolant is not a partition of unity, which means that it can not properly represent rigid body motions, which could lead to serious setbacks in computations.

How can we do better under this situation? The thought process is as follows: rigid body motion consists of rigid body translation and rigid body rotation. Even if complete rigid body mode may not be represented properly by an interpolation field, but rigid body translation (Galilean invariance) can be represented or preserved in discrete computation, it is a merit.

We show first that the SPH continuity equation and energy equation derived previously are Galilean invariant. Under the coordinate translation

$$x' = x + at \quad \Rightarrow \quad v' = v + a \tag{2.99}$$

the SPH equations (2.56) and (2.64) become

$$\frac{d\rho_I}{dt} = \rho_I \sum_{J=1}^{N} \frac{m_J}{\rho_J} (v'_I - v'_J) \cdot \nabla_I \Phi_{IJ} \tag{2.100}$$

$$\frac{dE_I}{dt} = \frac{\sigma_{Iij}}{\rho_I^2} \sum_{J=1}^{N} m_J (v'_I - v'_J) \cdot \nabla_I \Phi_{IJ} \tag{2.101}$$

Obviously after transformation, the velocity difference remain the same,

$$(v'_I - v'_J) = (v_I - v_J)$$

Therefore the SPH equations remain unchanged.

2.4.2 Conservation of Mass

In the following, we show that (2.58) preserves mass conservation. Since

$$\rho_I = \sum_{J=1}^{N} m_J \Phi_{IJ},$$

it leads to

$$\frac{d\rho_I}{dt} = \sum_{J=1}^{N} m_J v_{IJ} \cdot \nabla_I \Phi_{IJ} = \rho_I \sum_{J=1}^{N} \frac{m_J}{\rho_I} v_{IJ} \cdot \nabla_I \Phi_{IJ} \tag{2.102}$$

Substituting (2.102) into (2.58) yields an identity:

$$\frac{D\rho_I}{Dt} + \rho_I (\nabla \cdot v)_I$$
$$= \sum_{J=1}^{N} m_J v_{IJ} \nabla_I \Phi_{IJ} - \sum_{J=1}^{N} m_J v_{IJ} \nabla_I \Phi_{IJ} = 0 . \tag{2.103}$$

This indicates that the discrete continuity equation (2.58) is satisfied at the local level (at each individual particle).

2.4.3 Conservation of Linear Momentum

Suppose that a continuum is discretized into a discete system of N-particles. The linear momentum of the discrete system may be denoted by G

$$\mathbf{G} = \sum_{I=1}^{N} m_I \mathbf{v}_I \tag{2.104}$$

Taking the time derivative of G and substituting

$$\frac{d\mathbf{v}_I}{dt} = -\sum_{J=1}^{N} \left(\frac{\boldsymbol{\sigma}_I}{\rho_I^2} + \frac{\boldsymbol{\sigma}_J}{\rho_J^2} \right) m_J \cdot \nabla_I \Phi_{IJ}$$

into Eq. (2.104), one has

$$\dot{\mathbf{G}} = \sum_{I=1}^{N} m_I \frac{d\mathbf{v}_I}{dt}$$

$$= -\sum_{I=1}^{N}\sum_{J=1}^{N} m_I m_J \left(\frac{\boldsymbol{\sigma}_I}{\rho_I^2} + \frac{\boldsymbol{\sigma}_J}{\rho_J^2} \right) \cdot \nabla_I \Phi_{IJ}$$

$$= -\frac{1}{2}\sum_{I=1}^{N}\sum_{J=1}^{N} \left\{ m_I m_J \left(\frac{\boldsymbol{\sigma}_I}{\rho_I^2} + \frac{\boldsymbol{\sigma}_J}{\rho_J^2} \right) \cdot \nabla_I \Phi_{IJ} \right.$$

$$\left. + m_J m_I \left(\frac{\boldsymbol{\sigma}_J}{\rho_J^2} + \frac{\boldsymbol{\sigma}_I}{\rho_I^2} \right) \cdot \nabla_J \Phi_{JI} \right\} \tag{2.105}$$

Since $\nabla_I \Phi_{IJ} = -\nabla_J \Phi_{JI}$, it is straightforward to show that

$$\dot{\mathbf{G}} = -\frac{1}{2}\sum_{I=1}^{N}\sum_{J=1}^{N} m_I m_J \left(\frac{\boldsymbol{\sigma}_I}{\rho_I^2} + \frac{\boldsymbol{\sigma}_J}{\rho_J^2} \right) \cdot \{ \nabla_I \Phi_{IJ} - \nabla_I \Phi_{IJ} \} = 0 \tag{2.106}$$

2.4.4 Conservation of Angular Momentum

Following similar arguments, one can show the conservation of angular momentum in a discrete SPH system. The angular momentum of a discrete SPH system with respect to a fixed point (the center of coordinate) may be written as

$$\mathbf{H} = \sum_{I=1}^{N} \mathbf{x}_I \times m_I \mathbf{v}_I \tag{2.107}$$

By taking time differentiation of (2.107) and considering (2.60),

$$\dot{\mathbf{H}} = \sum_{I=1}^{N} \mathbf{x}_I \times m_I \dot{\mathbf{v}}_I$$

$$= -\frac{1}{2} \sum_{I=1}^{N} \sum_{J=1}^{N} m_I m_J \left\{ \left(\frac{\boldsymbol{\sigma}_I}{\rho_I^2} + \frac{\boldsymbol{\sigma}_J}{\rho_J^2} \right) \cdot \{ \mathbf{x}_I \times \nabla_I \Phi_{IJ} \} \right.$$

$$\left. + m_J m_I \left(\frac{\boldsymbol{\sigma}_J}{\rho_J^2} + \frac{\boldsymbol{\sigma}_I}{\rho_I^2} \right) \cdot \{ \mathbf{x}_J \times \nabla_J \Phi_{JI} \} \right\}$$

$$= -\frac{1}{2} \sum_{I=1}^{N} \sum_{J=1}^{N} m_I m_J \left(\frac{\boldsymbol{\sigma}_I}{\rho_I^2} + \frac{\boldsymbol{\sigma}_J}{\rho_J^2} \right) \cdot \{ \mathbf{x}_{IJ} \times \nabla_I \Phi_{IJ} \} \qquad (2.108)$$

Recall

$$\nabla_I \Phi_{IJ} = \frac{\boldsymbol{x}_{IJ}}{|\boldsymbol{x}_{IJ}|} \frac{\partial \Phi_{IJ}}{\partial x_{IJ}}$$

and $\boldsymbol{x}_{IJ} \times \boldsymbol{x}_{IJ} = 0$, it leads to

$$\mathbf{x}_{IJ} \times \nabla_I \Phi_{IJ} = 0 \Rightarrow \dot{\mathbf{H}} = 0 \ . \qquad (2.109)$$

2.4.5 Conservation of Mechanical Energy

To show SPH formulation preserving global energy, one can sum (2.67) with index I and consider the fact $\nabla_I \Phi_{IJ} = -\nabla_J \Phi_{JI}$,

$$\sum_{I=1}^{N} m_I \frac{dE_I}{dt} = \frac{1}{2} \sum_{I=1}^{N} \sum_{J=1}^{N} m_I m_J \left(\frac{\boldsymbol{\sigma}_I}{\rho_I^2} + \frac{\boldsymbol{\sigma}_J}{\rho_J^2} \right) : \left((\mathbf{v}_I - \mathbf{v}_J) \otimes \nabla_I \Phi_{IJ} \right)$$

$$= \frac{1}{2} \sum_{I=1}^{N} \sum_{J=1}^{N} m_I m_J \left(\frac{\boldsymbol{\sigma}_I}{\rho_I^2} + \frac{\boldsymbol{\sigma}_J}{\rho_J^2} \right) : \left(\mathbf{v}_I \otimes \nabla_I \Phi_{IJ} \right)$$

$$- \frac{1}{2} \sum_{J=1}^{N} \sum_{I=1}^{N} m_J m_I \left(\frac{\boldsymbol{\sigma}_J}{\rho_J^2} + \frac{\boldsymbol{\sigma}_I}{\rho_I^2} \right) : \left(\mathbf{v}_I \otimes \nabla_J \Phi_{JI} \right)$$

$$= \sum_{I=1}^{N} \sum_{J=1}^{N} m_I m_J \left(\frac{\boldsymbol{\sigma}_I}{\rho_I^2} + \frac{\boldsymbol{\sigma}_J}{\rho_J^2} \right) : \left(\mathbf{v}_I \otimes \nabla_I \Phi_{IJ} \right)$$

$$= \sum_{I=1}^{N} m_I \mathbf{v}_I \cdot \left\langle \frac{\nabla \cdot \boldsymbol{\sigma}}{\rho} \right\rangle_I \approx \sum_{I=1}^{N} \mathbf{v}_I \cdot (\nabla \cdot \boldsymbol{\sigma}_I) \Delta V_I \qquad (2.110)$$

This indicates that the change of the internal energy equals the work done by mechanical force, which is the case when other sources of energy are absent.

2.4.6 Variational SPH Formulation

It is shown above that classical SPH enjoys Galilean invariance and other invariant properties if proper approximations on the derivative are chosen.

The consequences of these invariant properties lead to discrete conservation laws in the corresponding SPH formulations. The issue was reexamined by Bonet et al.[73] from a different angle. They set forth a discrete variational SPH formulation, or variational construction procedure. Following the variational construction procedure, one can derive SPH equations, which automatically yield discrete conservation laws.

Let J be the determinant of the Jacobian — the deformation gradient. By definition, it is the ratio between the initial and current volume element,

$$J = \frac{V}{V_0} = \frac{\rho_0}{\rho} \tag{2.111}$$

Consider that the deformation process is spherical and adiabatic and let the internal potential energy density in the referential configuration be $U(J)$, it is plausible to assume that the pressure may be evaluated as

$$p = \frac{dU}{dJ} \tag{2.112}$$

For instance, for elastic solids one can choose $U = 1/2\kappa(J-1)^2$ whereas for ideal gases $U = \kappa\rho_0^\gamma J^{1-\gamma}$ where κ and γ are material parameters.

Suppose the continuum is discretized by N particles distributed over the domain, the total internal energy of the mechanical system can then be expressed as

$$\Pi(\boldsymbol{x}) = \sum_I V_I^0 U(J_I) \tag{2.113}$$

At each particle

$$J_I = \frac{V_I}{V_I^0} = \frac{\rho_I^0}{\rho_I} \tag{2.114}$$

where ρ_I^0 and ρ_I are pointwise density in initial configuration and in current configuration.

Similarly, the pressure value at the position of any particle can be obtained from $p_I = \dfrac{\partial U_I}{\partial J}$.

Define the variation of an arbitrary functional via Gateaux derivative, $\Psi[\mathbf{v}]$,

$$D\Psi[\delta\mathbf{v}] = \delta\Psi = \frac{d}{d\epsilon}\bigg|_{\epsilon=0} \Psi(\mathbf{v} + \epsilon\delta\mathbf{v}) \tag{2.115}$$

Thus, the stationary condition of potential energy gives

$$\delta\Pi = D\Pi[\delta\mathbf{v}] = \sum_{I=1}^N V_I^0 DU_I[\delta\mathbf{v}] = \sum_{I=1}^N V_I^0 p_I\left(-\frac{\rho_I^0}{\rho_I^2}\right) D\rho_I[\delta\mathbf{v}]$$

$$= -\sum_{I=1}^N m_I\left(\frac{p_I}{\rho_I^2}\right) D\rho_I[\delta\mathbf{v}] \tag{2.116}$$

The directional derivative of the density can be found by different approaches. Considering first $\rho(\mathbf{x}) = \sum_{I=1}^{N} m_I \Phi_I(\mathbf{x})$

$$D\rho_I[\delta\mathbf{v}] = \sum_{J=1}^{N} m_J \nabla_I \Phi_{IJ} \cdot (\delta\mathbf{v}_I - \delta\mathbf{v}_J) \tag{2.117}$$

Substituting (2.117) into (2.116) yields the first variation of the potential energy

$$
\begin{aligned}
D\Pi[\delta\mathbf{v}] &= \sum_{I,J=1}^{N} m_I m_J \left(\frac{p_I}{\rho_I^2}\right) \nabla_I \Phi_{IJ} \cdot (\delta\mathbf{v}_J - \delta\mathbf{v}_I) \\
&= \sum_{I=1}^{N} \left\{ \sum_{J=1}^{N} m_I m_J \left(\frac{p_I}{\rho_I^2} + \frac{p_J}{\rho_J^2}\right) \nabla_I \Phi_{IJ} \right\} \delta\mathbf{v}_I
\end{aligned}
\tag{2.118}
$$

where m_I is the mass associated with particle I.

Then considering the continuity equation,

$$D\rho[\delta\mathbf{v}] = -\rho\, div\,\delta\mathbf{v} \tag{2.119}$$

one may derive

$$D\rho_I[\delta\mathbf{v}] = \rho_I \sum_{J=1}^{N} V_J (\delta\mathbf{v}_I - \delta\mathbf{v}_J) \cdot \nabla_I \Phi_{IJ} \tag{2.120}$$

Note the subtle difference between (2.117) and (2.120).

Substituting (2.120) into (2.116) yields another expression for the first variation of the potential energy,

$$
\begin{aligned}
D\Pi[\delta\mathbf{v}] &= \sum_{I,J=1}^{N} V_I V_J p_I \nabla_I \Phi_{IJ} \cdot (\delta\mathbf{v}_J - \delta\mathbf{v}_I) \\
&= \sum_{I=1}^{N} \left\{ \sum_{J=1}^{N} V_I V_J (p_I + p_J) \nabla_I \Phi_{IJ} \right\} \cdot \delta\mathbf{v}_I
\end{aligned}
\tag{2.121}
$$

On the other hand, by definition,

$$D\Pi[\delta\mathbf{v}] = \sum_I \frac{\partial \Pi}{\partial \mathbf{x}_I} \delta\mathbf{v}_I = \sum_I \mathbf{T}_I \cdot \delta\mathbf{v}_I \tag{2.122}$$

where $\mathbf{T}$ is the internal force (summation of stress).

Combining (2.122) with (2.118), one can identify that

$$\mathbf{T}_I = \sum_I m_I m_J \left(\frac{p_I}{\rho_I^2} + \frac{p_J}{\rho_J^2}\right) \nabla_I \Phi_{IJ} \tag{2.123}$$

Combining (2.122) with (2.121), one may identify that

$$\mathbf{T}_I = \sum_{J=1}^{N} V_I V_J (p_I + p_J) \nabla_I \Phi_{IJ} \qquad (2.124)$$

Based on these expressions, one may establish the following discrete SPH equations of motion,

$$m_I \frac{d\mathbf{v}_I}{dt} = -\sum_{J=1}^{N} m_I m_J \left(\frac{p_I}{\rho_I^2} + \frac{p_J}{\rho_J^2} \right) \nabla_I \Phi_{IJ} \qquad (2.125)$$

or alternatively,

$$m_I \frac{d\mathbf{v}_I}{dt} = \sum_{J=1}^{N} V_I V_J \left(p_I + p_J \right) \nabla_I \Phi_{IJ} \qquad (2.126)$$

2.5 Corrective SPH and Other Improvements on SPH

SPH methodology has many advantages in computations, e.g. simple in concept, easy to program, suitable for large deformation computation, meshfree in interpolation, etc. On the other hand, SPH has several main technical drawbacks,

- difficulty in enforcing essential boundary conditions;[344, 387, 425]
- tensile instability;[48, 155, 156, 343, 420]
- lack of interpolation consistency, or completeness;[145, 146, 297]

In the past twenty years, various new SPH techniques have been developed, aimed at improving its performance and eliminating pathologies in numerical computations (see:[56, 177, 332–334, 338, 340–342, 372]).

2.5.1 Enforcing the Essential Boundary Condition

In principle, non-local interpolants have difficulty incorporating prescribed boundary data on an essential boundary into the discrete governing equations. Most meshfree interpolants, e.g. the SPH interpolant, are intrinsically non-local. Therefore, SPH interpolant is unable to accommodate boundary interpolation without special treatment. Extra care has to be taken in order to enforce the essential boundary conditions. In the following, a so-called "ghost particle" approach is described, which has been often used in practical computations.

Suppose particle I is a boundary particle. All the other particles within its support, $\mathcal{N}(I)$, can be divided into three subsets:

1. $\mathcal{I}(I)$: all the interior points that are the neighbors of I;
2. $\mathcal{B}(I)$: all the boundary points that are the neighbors of I;

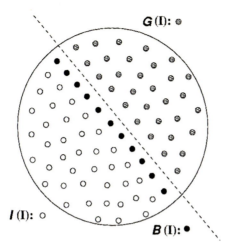

Fig. 2.3. "The Ghost particle" approach for boundary treatment

3. $\mathcal{G}(I)$: all the exterior points that are the neighbors of I, i.e. all the ghost particles.

Therefore $\mathcal{N}(I) = \mathcal{I}(I) \bigcup \mathcal{B}(I) \bigcup \mathcal{G}(I)$. Figure (2.3) illustrate such an arrangement.

Now we show how to use the "ghost particle approach" to calculate density function for particles on the essential boundary. By including exterior particles, the kernel function discretization remains as a partition of unity at position $\boldsymbol{x} = \boldsymbol{x}_I$,

$$1 = \sum_{J \in \mathcal{I}(I)} \Delta V_J \Phi_{IJ} + \sum_{J \in \mathcal{B}(I)} \Delta V_J \Phi_{IJ} + \sum_{J \in \mathcal{G}(I)} \Delta V_J \Phi_{IJ} \tag{2.127}$$

On the other hand, the kernel approximation for density at $\boldsymbol{x} = \boldsymbol{x}_I$ is

$$\rho_I = \sum_{J \in \mathcal{I}(I)} m_J \Phi_{IJ} + \sum_{J \in \mathcal{B}(I)} m_J \Phi_{IJ} + \sum_{J \in \mathcal{G}(I)} m_J \Phi_{IJ} \tag{2.128}$$

Assume that all the boundary particles and ghost particles, $J \in \mathcal{B}(I) \bigcup \mathcal{G}(I)$, have the same density, $\rho_J = \rho_I$. Multiplying Eq. (2.127) with ρ_I and subtracting it from Eq. (2.128), one can derive the following expression for ρ_I, which only depends on interior points,

$$\rho_I = \frac{\displaystyle\sum_{J \in \mathcal{I}(I)} m_J \Phi_{IJ}}{\displaystyle\sum_{J \in \mathcal{I}(I)} \Delta V_J \Phi_{IJ}} \tag{2.129}$$

Following a similar procedure, one may derive SPH boundary constraint equations for a general scalar field, f. Again assume $f_J = f_I$, $\forall J \in \mathcal{B}(I)$, and $\Delta V_J = \Delta V_I$, $f_J = f_{bc}$ $\forall J \in \mathcal{G}(I)$, here f_{bc} is the prescribed boundary value at $\boldsymbol{x} = \boldsymbol{x}_I$. Then from the kernel approximation,

$$f_I = \sum_{J \in \mathcal{I}(I)} f_J \Delta V_J \Phi_{IJ} + \sum_{J \in \mathcal{B}(I)} f_J \Delta V_J \Phi_{IJ} + \sum_{J \in \mathcal{G}(I)} f_{bc} \Delta V_I \Phi_{IJ} \qquad (2.130)$$

Multiplying Eq. (2.127) with f_{bc} and subtracting it from Eq. (2.130), one obtains the following boundary correction formula

$$f_I = f_{bc} + \frac{\displaystyle\sum_{J \in \mathcal{I}(I)} (f_J - f_{bc}) \Delta V_J \Phi_{IJ}}{\left(1 - \displaystyle\sum_{J \in \mathcal{B}(I)} \Delta V_J \Phi_{IJ}\right)} \qquad (2.131)$$

One of the advantages of the above formula is that the sampling formula only depends on interior particles.

2.5.2 Tensile Instability

Despite its growing popularity, the SPH method has a major deficiency: the numerical algorithm of SPH suffers from the so-called tensile instability. This refers to the numerical pathology that in a region with tensile stress state, a small perturbation on particles' positions will cause exponential growth in particles' velocity, and eventually results in particle clumping and oscillatory motions. Systematic study on its origin and prevention of tensile instability has been conducted by several authors, e.g.[48, 155, 156, 343, 420]

A brief discussion on the origin of tensile instability is presented in here based on the intuitive reasoning given by Swegle, et al.[420]

Consider the SPH momentum equation with only hydrostatic pressure and artificial viscosity

$$\frac{d\mathbf{v}_I}{dt} = -\sum_{J=1}^{N} m_J \sigma_{IJ} \nabla_I \Phi_{IJ} \qquad (2.132)$$

where

$$\sigma_{IJ} = \frac{p_I}{\rho_I^2} + \frac{p_J}{\rho_J^2} + \Pi_{IJ} \qquad (2.133)$$

To illustrate the concept, we examine an 1D model: a three points interaction problem. Assume that the three adjacent particles have the same amount mass, i.e. $m_{I-1} = m_I = m_{I+1}$. Considering the fact that $\dfrac{d}{dx_I}\Phi_{II} = 0$. The acceleration of particle I may be written as

$$\ddot{x}_I = -m \left\{ \sigma_{I(I-1)} \frac{d\Phi_{(I-1)I}}{dx_I} + \sigma_{I(I+1)} \frac{d\Phi_{I(I+1)}}{dx_I} \right\}$$

$$= m \left\{ \sigma_{I(I+1)} \Phi'_{I(I+1)} - \sigma_{(I-1)I} \Phi'_{(I-1)I} \right\} \tag{2.134}$$

making use of the fact that in 1D

$$\frac{d\Phi(|\boldsymbol{x} - \boldsymbol{x}_J|)}{dx_I} := \Phi'(|\boldsymbol{x}_I - \boldsymbol{x}_J|)\Big|_{\boldsymbol{x}=\boldsymbol{x}_I} = \begin{cases} \Phi'(\mathbf{x}_{IJ}) \ ; \ x_I > x_J \\ -\Phi'(\mathbf{x}_{IJ}) \ ; \ x_I < x_J \end{cases} \tag{2.135}$$

where $\mathbf{x}_{IJ} = x_I - x_J$.

Therefore, qualitatively

$$\ddot{x} \propto \Delta(\sigma\Phi') = \Delta\sigma(\Phi') + \sigma\Delta(\Phi') \tag{2.136}$$

Note that σ here is the negative Cauchy stress (see definition (2.133)).

Let $\tau = -\sigma$. (2.136) can be written as

$$\ddot{x} \propto \Delta(-\tau\Phi') = \Delta\tau(-\Phi') + \tau\Delta(-\Phi') \tag{2.137}$$

On the other hand,

$$\ddot{x} = \frac{1}{\rho_0} \frac{\partial\tau}{\partial x} \approx \frac{\Delta\dot{x}}{\Delta t} = \frac{\Delta\tau}{\rho_0\Delta x} = \frac{\Delta\tau}{\Delta m} \tag{2.138}$$

Therefore,

$$\left(\frac{1}{\Delta m} + \Phi' \right) \Delta\tau = -\tau\Delta\Phi' \tag{2.139}$$

At each particle $\nabla_I \Phi_{II} = \Phi'(0) \sim 0$. Hence $(1/\Delta m + \Phi') > 0$ and

$$\Delta\tau \propto -\tau\Delta\Phi' \tag{2.140}$$

From Fig. (2.4), one may find in the region $x > x_I$, $-\Phi'(x - x_I) > 0$, or $\Phi'(x - x_I) < 0$. Graphically, this fact is depicted in Fig. (2.5).

Then (2.140) provides a stability criterion that can be elaborated as follows:

1. if $\tau < 0$, (compression):
   a) When the slope of $\Phi' > 0$, $\Delta\tau > 0$, then τ increases as x increases (particles separate): i.e. the magnitude of compressive stress decreasese as the coordinate x increases, whereas the magnitude of compressive stress, increases as x decreases (particles come together). This means the system is stable.
   b) When the slope of $\Phi' < 0$, $\Delta\tau < 0$, the magnitude of compressive stress increases as x increases (particles separate); and the magnitude of compression stress decreases as x decreases (particles come together). This means the system is unstable.

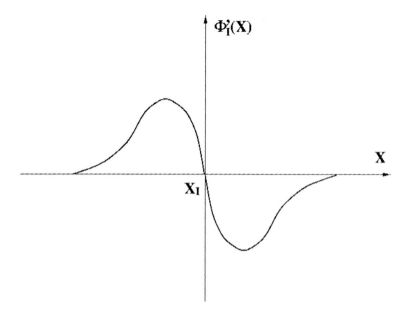

Fig. 2.4. Profile of the derivative of kernel function.

2. if $\tau > 0$ (tension);
   a) When the slope of $\Phi' < 0$ or $-\Phi' > 0$, $\Delta\tau > 0$; the tensile stress increases as x increases (particles separate); whereas the tensile stress decreases as x decreases (particles come together). This means the system is stable.
   b) When the slope of $\Phi' > 0$ or $-\Phi' < 0$, $\Delta\tau < 0$; the tensile stress decreases as x increases (particles separate), whereas the tensile stress increases as x decreases (particles come together). This means the system is unstable.

Consider a particle distribution in which the distance between two particles is one unit of the smoothing length away, which is a common situation. In such particle distribution, the particle $I + 1$ is positioned at $x_I + h$, and the point, x, where $\Phi'_I(x) = 0$ is on the left side of particle $I + 1$, or between particle I and $I + 1$ (see Fig. 2.5).

Suppose the system is under tension. Observing in Fig. (2.5) and based on the above arguments derived from (2.140), one may find that the system is unstable, because in this case the effective stress decreases, while the strain increases. This is the notorious "tensile instability" that often occurs in SPH computations.

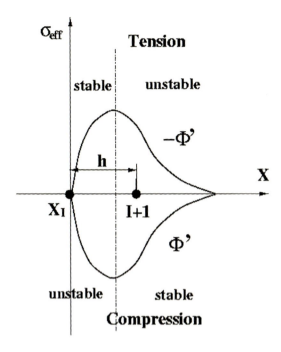

Fig. 2.5. Schematic illustration of tensile instability.

2.5.3 SPH Interpolation Error

The convergence of an SPH numerical solution to the analytical solution depends on several factors: interpolation error (smoothing error and truncation error), and collocation consistency error, etc.

To estimate the smoothing error of an SPH kernel approximation, we examine the following 1-d kernel approximation,

$$f(x)- <f>(x) = f(x) - \int_{-\infty}^{\infty} \Phi(x-y,h)f(y)dy \tag{2.141}$$

By Taylor expansion,

$$<f>(x) = \int_{-\infty}^{\infty} \Phi(u,h)\left\{ f(x) - uf'(x) + \frac{u^2}{2!}f''(x) - \cdots \right\} du \tag{2.142}$$

where $u = x - y$. Since

$$\int_{-\infty}^{\infty} \Phi(u,h)du = 1$$

$$\int_{-\infty}^{\infty} u\Phi(u,h)du = 0 \tag{2.143}$$

Therefore the smoothing error can be estimated

$$\epsilon_s := f(x) - \ <f> \ (x) = const. \ h^2 + \mathcal{O}(h^3) \tag{2.144}$$

Here we assume that $u \propto h$. By choosing proper kernel function, one may be able to push the smoothing error to even higher orders. For instance, for Super-Gaussian,

$$\int_{-\infty}^{\infty} u^k \Phi(u, h) du = 0, \quad k = 1, 2, 3 \tag{2.145}$$

where

$$\Phi(u, h) = \frac{1}{h\sqrt{\pi}} \exp\left(\frac{3}{2} - \frac{u^2}{h^2}\right) \tag{2.146}$$

The smoothing error in this case is now up to the order of $\mathcal{O}(h^4)$.

The SPH truncation error is defined as,

$$\epsilon_t := \int \Phi(x - x', h) f(x') dx' - \sum_{J=1} \Delta V_J f(x_J) \Phi(x - x_J, h), \tag{2.147}$$

It depends on the particle distribution, which is controlled by the accuracy of the nodal integration defined in (2.147). Several estimates on the accuracy of the nodal integration are given as follows:

1. Particles are orderly, and evenly distributed. In this case, the summation in (2.147) could be numerical integration based on Trapezoidal rule; thus, the truncation error is at least $\mathcal{O}(h^2)$ or higher.
2. Particles are completely randomly distributed, and disordered. Thus, the estimate of the numerical integration in (2.147) should be based on Monte Carlo estimation, which yields the truncation error as $N^{-1/2}$. Since $N \propto h^{-1}$, the Monte Carlo estimate of truncation error is $\sqrt{h}$;
3. Particle distribution is disordered, but it is evenly distributed, the so-called *disordered equi-distribution*. In this case, Niederreiter[353] showed that the truncation error is on the order of $N^{-1}(\log N)^{n-1}$, or $h|\log h|^{n-1}$.
4. A related, but improved result is given by Wozniakowski,[454] which was produced by a challenge with *an award of sixty-four dollars*. The improved result states that the average truncation error for numerical integral with disordered equi-distribution is on the order of $h|\log h|^{(n-1)/2}$.

where n is space dimension.

2.5.4 Correction Function (RKPM)

The early SPH interpolant does not form a partition of unity. This implies that SPH interpolant can not represent rigid body motion correctly, even though it is Galilean invariant (rigid body translation only). This problem was first noticed by Liu et al.[295–297] Then a key notion, *correction function*, is set forth to enforce consistency condition in interpolation, which has become the central theme of a class of SPH methods that are labeled as corrective SPH. The idea of corrective SPH is to construct a new corrective kernel to correct the original kernel function, such that the consistency and completeness of SPH interpolant can be satisfied. This correction procedure is to find a *corrective function* first and then multiply it with the original kernel function to form the new corrected kernel function. The resulting new interpolant is named the reproducing kernel particle interpolant.[295–297]

To understand corrective SPH, it is necessary to examine the theoretical foundation of SPH approximation. In SPH approximation, there are two types of interpolation errors: smoothing error and truncation error. For a continuous function, $f(x) \in C^0(\mathbb{R})$,

$$\epsilon_S := f(x) - <f>(x) \tag{2.148}$$

$$\epsilon_T := <f>(x) - \sum_I f_I \Phi_I(x) \Delta x_I \tag{2.149}$$

The total interpolation error is the sum of the smoothing error and truncation error,

$$\epsilon_I = \epsilon_S + \epsilon_T \tag{2.150}$$

Smoothing error is controlled by selection of SPH kernel function. Truncation error is controlled by the accuracy of the nodal integration. If particles are randomly distributed, or disordered, the truncation error will dictate the convergence process. Consider a one dimensional example. As shown previously, a valid SPH smoothing kernel function must satisfy the following conditions:

$$\int_R \Phi(x, h) dx = 1 \tag{2.151}$$

$$\int_R x^k \Phi(x, h) dx = 0, \quad k = 1, 2, \cdots \tag{2.152}$$

which can be viewed as the zero-th order moment and the higher order moment conditions.

However, these conditions only hold in continuous form. Due to truncation error, they may not be valid at all after discretization, i.e. the following moment conditions may not hold,

$$\sum_{I=1}^{NP} \Phi(x - x_I, h) \Delta x_I = 1 \tag{2.153}$$

$$\sum_{I=1}^{NP} (x - x_I)^k \Phi(x - x_I, h) \Delta x_I = 0, \quad k = 1, 2, \cdots \tag{2.154}$$

where NP is the total number of the particles. Note that condition (2.153) is the condition of partition of unity.

The conditions (2.153)–(2.154) are sometimes referred to as consistency conditions (smoothing is consistent with interpolation), or the completeness condition, which is referred to the completeness of interpolation function space. One way to satisfy these conditions is to find a suitable kernel function that does the job. Obviously, there is no analytical function that can satisfy all the discrete moment conditions for random particle distribution. A suitable kernel function has to take into account the information of particle distribution, which suggests that it would be better if the construction of suitable kernel function is an optimization process.

Since the original SPH kernel function can not satisfy the discrete moment conditions, a modified kernel function is introduced to enforce the discrete consistency conditions. Let

$$\mathcal{K}_h(x - x_I; x) = C_h(x - x_I; x)\Phi(x - x_I, h) \tag{2.155}$$

where $C_h(x - x_I; x)$ is the correction function, which can be expressed as

$$C_h(x - x_I; x) = b_0(x, h) + b_1(x, h)\left(\frac{x - x_I}{h}\right) + b_2(x, h)\left(\frac{x - x_I}{h}\right)^2$$
$$+ \cdots \cdots$$
$$= \mathbf{P}^T\left(\frac{x - x_I}{h}\right)\mathbf{b}(x, h) \tag{2.156}$$

where $\mathbf{P}^T(x) = \{1, x, x^2, \cdots, x^n\}$ is a polynomial basis and the components of the vector function, $\mathbf{b}^T = \{b_0(x, h), b_1(x, h), \cdots b_n(x, h)\}$, are unknown functions. By determining these unknown functions, one can correct the original kernel function to satisfy consistency conditions.

In order to find a set of criteria to determine the unknown functions $b_i(x, h)$, we consider an example involving a sufficiently smooth function $f(x)$. By Taylor expansion,

$$f_I = f(x_I) = f(x) + f'(x)\left(\frac{x_I - x}{h}\right)h + \frac{f''(x)}{2!}\left(\frac{x_I - x}{h}\right)^2 h^2 + \cdots \cdots \tag{2.157}$$

the modified kernel approximation can be written as,

$$f^h(x) = \sum_{I=1}^{NP} \mathcal{K}_h(x - x_I; x) f_I \Delta x_I$$

$$= \left(\sum_{I=1}^{NP} \mathcal{K}_h(x - x_I, x) \Delta x_I \right) f(x) \, h^0$$

$$- \left(\sum_{I=1}^{NP} (\frac{x - x_I}{h}) \mathcal{K}_h(x - x_I, x) \Delta x_I \right) f'(x) h + \cdots \cdots$$

$$+ \left(\sum_{I=1}^{NP} (-1)^n (\frac{x - x_I}{h})^n \mathcal{K}_h(x - x_I, x) \Delta x_I \right) \frac{f^n(x)}{n!} h^n$$

$$+ \mathcal{O}(h^{n+1}) \tag{2.158}$$

To obtain a n+1-th order truncation error, the moments of the modified kernel function must satisfy the following conditions:

$$\left. \begin{array}{l} \displaystyle\sum_{I=1}^{NP} \mathcal{K}_h(x - x_I, x) \Delta x_I = 1; \\[2mm] \displaystyle\sum_{I=1}^{NP} \left(\frac{x - x_I}{h}\right) \mathcal{K}_h(x - x_I, x) \Delta x_I = 0; \\[2mm] \vdots \\[2mm] \displaystyle\sum_{I=1}^{NP} \left(\frac{x - x_I}{h}\right)^n \mathcal{K}_h(x - x_I, x) \Delta x_I = 0; \end{array} \right\} \Rightarrow \left\{ \begin{array}{l} M_0(x) = 1 \\[4mm] M_1(x) = 0 \\[4mm] \vdots \\[4mm] M_n(x) = 0 \end{array} \right. \tag{2.159}$$

where the moments of modified kernel functions are defined as,

$$M_i(x) := \sum_{I=1}^{NP} \left(\frac{x - x_I}{h}\right)^i \mathcal{K}_h(x - x_I, x) \Delta x_I \tag{2.160}$$

Substituting $\mathcal{K}_h(x - x_I, x) = \mathbf{P}^T \left(\frac{x - x_I}{h}\right) \boldsymbol{b}(x, h) \Phi_h(x - x_I)$ into the moment conditions in Eq. (2.159), we can determine the n+1 unknown coefficient functions, $b_i(x, h)$, by solving the following **moment equations**:

$$\begin{pmatrix} m_0(x) & m_1(x) & \cdots & m_n(x) \\ m_1(x) & m_2(x) & \cdots & m_{n+1}(x) \\ & & & \\ \vdots & \vdots & \vdots & \vdots \\ & & & \\ m_n(x) & m_{n+1}(x) & \cdots & m_{2n}(x) \end{pmatrix} \begin{pmatrix} b_0(x, h) \\ b_1(x, h) \\ \\ \vdots \\ \\ b_n(x, h) \end{pmatrix} = \begin{pmatrix} 1 \\ 0 \\ \\ \vdots \\ \\ 0 \end{pmatrix} \tag{2.161}$$

or in vector form $\mathbf{M}_h(x) \boldsymbol{b}(x, \varrho) = \mathbf{P}(0)$, where

$$\mathbf{M}_h(x) := \begin{pmatrix} m_0(x) & m_1(x) & \cdots & m_n(x) \\ m_1(x) & m_2(x) & \cdots & m_{n+1}(x) \\ \vdots & \vdots & \vdots & \vdots \\ m_n(x) & m_{n+1}(x) & \cdots & m_{2n}(x) \end{pmatrix}, \tag{2.162}$$

$$m_k(x) := \sum_{I=1}^{NP} (\frac{x - x_I}{h})^k \varPhi_h(\frac{x - x_I}{h}) \Delta x_I \tag{2.163}$$

and $\mathbf{P}^T(0) = (1, 0, \cdots, 0)$.

Solving Eq. (2.161) yields the unknown vector $\boldsymbol{b}(x, h)$ and consider that the moment matrix is symmetrical,

$$\boldsymbol{b}\Big(x, h\Big) = \mathbf{M}_h^{-1}(x)\mathbf{P}(0) \Rightarrow \boldsymbol{b}^T\Big(x, h\Big) = \mathbf{P}^T(0)\mathbf{M}_h^{-1}(x) \tag{2.164}$$

Consequently, the correction function is obtained

$$\boxed{C_h(x - y) = \mathbf{P}^T(0)\mathbf{M}_h^{-1}(x)\mathbf{P}\Big(\frac{x - y}{h}\Big)} \tag{2.165}$$

and subsequently the RKPM kernel (shape) function

$$\boxed{\mathcal{K}_h(x - y) = \mathbf{P}^T(0)\mathbf{M}_h^{-1}(x)\mathbf{P}\Big(\frac{x - y}{h}\Big)\varPhi_h(x - y)} \tag{2.166}$$

It is worth mentioning that after introducing the correction function, the modified kernel function may not be a positive function anymore,

$$\mathcal{K}_h(x - x_I; x) \not\geq 0 . \tag{2.167}$$

Within the compact support, $\mathcal{K}_h(x - x_I; x)$ may become negative in some region of the domain.

There are other approaches to restore completeness of the SPH approximation. Their emphasis is not only consistency in interpolation, but also cost effectiveness in computation. Using RKPM, or moving-least-squares interpolant[145, 146] to construct modified kernels, one has to know all the neighboring particles that are adjacent to a spatial point where the kernel function is in evaluation. This will require additional CPU to search, update connectivity array, and to calculate a modified kernel function pointwise. It should be noted that the calculation of a modified kernel function requires pointwise matrix inversions at each time step, since particles are moving and the connectivity map is changing as well. Thus, using moving least square interpolant as a kernel function may not be cost-effective, nevertheless a moving least square interpolant based SPH is formulated by Dilts ([145]).

2.5.5 Moving Least Square Hydrodynamics (MLSPH)

One simple way to improve the accuracy of the SPH kernel function is to adopt a MLS interpolant as the SPH kernel function ([145]). By using a MLS interpolant, Dilts used a collocation/kernel Galerkin approximation to derive a so-called moving least square hydrodynamics (MLSPH). The critical steps of this particular kernel collocation is briefly outlined below. Assume

$$\textbf{(1)} \quad \Delta V_I \approx \int_\Omega \Phi(\boldsymbol{x} - \boldsymbol{x}_I, \boldsymbol{x}) d\Omega \tag{2.168}$$

$$\textbf{(2)} \quad \int_\Omega f(\boldsymbol{x})\Phi(\boldsymbol{x} - \boldsymbol{x}_I) d\Omega = \Delta V_I f(\boldsymbol{\zeta}_I) \approx \Delta V_I f(\boldsymbol{x}_I) \tag{2.169}$$

$$\textbf{(3)} \quad < \nabla f(\boldsymbol{x}) > = \sum_{I \in \Lambda} f_I \nabla \Phi(\boldsymbol{x} - \boldsymbol{x}_I) \tag{2.170}$$

Consider the conservation laws of hydrodynamics in Lagrangian form

$$\begin{aligned} \dot{\rho}/\rho &= -\nabla \cdot \mathbf{v} \\ \rho \dot{\mathbf{v}} &= \nabla \cdot \boldsymbol{\sigma} \\ \rho(\dot{e} + \mathbf{v} \cdot \mathbf{v}) &= \nabla \cdot (\boldsymbol{\sigma} \cdot \mathbf{v}) \end{aligned} \tag{2.171}$$

A kernel weighted residual form can be written as

$$\begin{aligned} &\int_\Omega \Phi(\boldsymbol{x} - \boldsymbol{x}_I) \left\{ \begin{pmatrix} \dot{\rho}/\rho \\ \rho \dot{\mathbf{v}} \\ \rho(\dot{e} + \mathbf{v} \cdot \mathbf{v}) \end{pmatrix} - \begin{pmatrix} < \dot{\rho}/\rho > \\ < \rho \dot{\mathbf{v}} > \\ < \rho(\dot{e} + \mathbf{v} \cdot \mathbf{v}) > \end{pmatrix} \right\} d\Omega \\ &= \int_\Omega \Phi(\boldsymbol{x} - \boldsymbol{x}_I) \left\{ \begin{pmatrix} \dot{\rho}/\rho \\ \rho \dot{\mathbf{v}} \\ \rho(\dot{e} + \mathbf{v} \cdot \mathbf{v}) \end{pmatrix} - \begin{pmatrix} - < \nabla \cdot \mathbf{v} > \\ < \nabla \cdot \boldsymbol{\sigma} > \\ < \nabla \cdot (\boldsymbol{\sigma} \cdot \mathbf{v}) > \end{pmatrix} \right\} d\Omega = 0 \end{aligned} \tag{2.172}$$

which leads to

$$\begin{aligned} &\int_\Omega \Phi(\boldsymbol{x} - \boldsymbol{x}_I) \begin{pmatrix} \dot{\rho}/\rho \\ \rho \dot{\mathbf{v}} \\ \rho(\dot{e} + \mathbf{v} \cdot \dot{\mathbf{v}}) \end{pmatrix} d\Omega \\ &= \int_\Omega \Phi(\boldsymbol{x} - \boldsymbol{x}_I) \sum_{J \in \Lambda} \begin{pmatrix} -\nabla \cdot \mathbf{v} \\ \nabla \cdot \boldsymbol{\sigma} \\ \nabla \cdot (\boldsymbol{\sigma} \cdot \mathbf{v}) \end{pmatrix} \cdot \nabla \Phi(\boldsymbol{x} - \boldsymbol{x}_J) d\Omega \end{aligned} \tag{2.173}$$

To discretize continuous integral forms, a collocation procedure is adopted. Specifically, the second approximation rule is used. By substituting the acceleration term in the energy equation with the gradient of Cauchy stress and canceling the factor ΔV_I in both sides of the equations, one will obtain the so-called MLSPH equations

$$\begin{pmatrix} \dot{\rho}_I/\rho_I \\ \rho_I \dot{\mathbf{v}}_I \\ \rho_I \dot{e}_I \end{pmatrix} = \sum_{J \in \Lambda} \begin{pmatrix} -\mathbf{v}_I \\ \boldsymbol{\sigma}_J \\ \boldsymbol{\sigma}_J \cdot (\mathbf{v}_J - \mathbf{v}_I) \end{pmatrix} \cdot \nabla_I \Phi_{IJ} \tag{2.174}$$

here $\{\varPhi(\boldsymbol{x} - \boldsymbol{x}_I)\}_{I \in \varLambda}$ are the MLS interpolants.

Several correction schemes have been proposed throughout the years, which are listed as follows,

1. Monaghan's symmetrization on derivative approximation;[334, 338]
2. Johnson-Beissel correction;[225]
3. Randles-Libersky correction;[387]
4. Krongauz-Belytschko correction;[47]
5. Chen-Beraun correction;[92, 94, 95]
6. Bonet-Kulasegaram integration correction;[74]
7. Aluru's collocation RKPM;[10]

Since the linear reproducing condition in the interpolation is equivalent to the constant reproducing condition in the derivative of the interpolant, some of the algorithms directly correct derivatives instead of the interpolant. The Chen-Beraun correction corrects even higher order derivatives, but it may require more computational effort in multi-dimensions.

2.5.6 Johnson-Beissel Correction

An early effort to correct the SPH kernel function was made by Johnson and Beissel, who proposed the well-known Johnson-Beissel correction in their work of simulating penetration and fragmentation of solids. In solid mechanics computations, requirements for the accuracy on strain rate are severe. To obtain an accurate velocity gradient, Johnson and Beissel proposed the following corrected, or normalized formula

$$D_x(\boldsymbol{x}_I) = v_{x,x}(\boldsymbol{x}_J) = -\sum_{I \in \varLambda} \beta_x \frac{\partial \varPhi_{JI}}{\partial x} \Delta V_I (v_{xI} - v_{xJ}) \qquad (2.175)$$

$$D_y(\boldsymbol{x}_I) = v_{y,y}(\boldsymbol{x}_J) = -\sum_{I \in \varLambda} \beta_y \frac{\partial \varPhi_{JI}}{\partial y} \Delta V_I (v_{yI} - v_{yJ}) \qquad (2.176)$$

where β_x and β_y are correction factors, or normalization factors. For each particle, the normalization factors are adjusted such that the kernel approximation of velocity gradient can preserve at least constant velocity gradient exactly. For constant velocity gradients,

$$
\left.\begin{aligned}
(x_I - x_J)D_x &= v_{xI} - v_{xJ} \\
(y_I - y_J)D_y &= v_{yI} - v_{yJ}
\end{aligned}\right\}
\Rightarrow
\begin{cases}
D_x = -\sum_{I \in \varLambda} \beta_x \dfrac{\partial \varPhi_{IJ}}{\partial x} \Delta V_I D_x (x_I - x_J) \\[2ex]
D_y = -\sum_{I \in \varLambda} \beta_y \dfrac{\partial \varPhi_{IJ}}{\partial y} \Delta V_I D_y (y_I - y_J)
\end{cases}
$$

$$(2.177)$$

Solving β_x and β_y, we have

$$\beta_x = -\frac{1}{\sum_{I \in \Lambda} \frac{\partial \Phi_{JI}}{\partial x} \Delta V_I (x_J - x_I)} \tag{2.178}$$

$$\beta_y = -\frac{1}{\sum_{I \in \Lambda} \frac{\partial \Phi_{JI}}{\partial y} \Delta V_I (y_J - y_I)} \tag{2.179}$$

However, the above correction is obtained by assuming a special velocity distribution in space. In general, it still does not reproduce general linear velocity distribution in space.

2.5.7 Randles-Libersky Correction

Randles and Libersky used the following corrected form in the linear momentum calculations,

$$(\nabla \cdot \boldsymbol{\sigma})_J = -\left(\sum_{I \in \Lambda} (\boldsymbol{\sigma}_I - \boldsymbol{\sigma}_J) \otimes \nabla_J \Phi_{JI} \Delta V_I \right) : \mathbf{B} \tag{2.180}$$

where the correction tensor $\mathbf{B}$ is expressed as

$$\mathbf{B} = \left(-\sum_{I \in \Lambda} (\boldsymbol{x}_I - \boldsymbol{x}_J) \otimes \nabla_J \Phi_{JI} \Delta V_I \right)^{-1} \tag{2.181}$$

An alternative form of the above correction is to use the Shepard interpolant $\{\Phi^S(\boldsymbol{x} - \boldsymbol{x}_I)\}_{I \in \Lambda}$. Since the Shepard interpolant reproduces constants, the term $\boldsymbol{x}_I - \boldsymbol{x}_J$ in $\mathbf{B}$ can be replaced by $\boldsymbol{x}_I$. Thus the corrected formula reads as

$$(\nabla \cdot \boldsymbol{\sigma})_J = -\left(\sum_{I \in \Lambda} (\boldsymbol{\sigma}_I - \boldsymbol{\sigma}_J) \otimes \nabla_J \Phi^S_{JI} \Delta V_I \right) : \mathbf{B}^S \tag{2.182}$$

$$\mathbf{B}^S = \left(-\sum_{I \in \Lambda} \boldsymbol{x}_I \otimes \nabla_J \Phi^S_{JI} \Delta V_I \right)^{-1} \tag{2.183}$$

2.5.8 Krongauz-Belytschko Correction

To construct a better SPH kernel function, Krongauz & Belytschko[47] proposed the following corrected interpolation formula. In 2D, it reads

$$u^h(\boldsymbol{x}) = \bar{\Phi}_I(\boldsymbol{x}) u_I \tag{2.184}$$
$$u^h_{,x}(\boldsymbol{x}) = G_{Ix}(\boldsymbol{x}) u_I \tag{2.185}$$
$$u^h_{,y}(\boldsymbol{x}) = G_{Iy}(\boldsymbol{x}) u_I \tag{2.186}$$

where

$$\bar{\Phi}_I(\boldsymbol{x}) = \alpha_{11}(\boldsymbol{x})\Phi_{I,x}^S(\boldsymbol{x}) + \alpha_{12}\Phi_{I,y}^S(\boldsymbol{x}) + \alpha_{13}(\boldsymbol{x})\Phi_I^S(\boldsymbol{x}) \tag{2.187}$$

$$G_{Ix}(\boldsymbol{x}) = \alpha_{21}(\boldsymbol{x})\Phi_{I,x}^S(\boldsymbol{x}) + \alpha_{22}(\boldsymbol{x})\Phi_{I,y}^S(\boldsymbol{x}) + \alpha_{23}(\boldsymbol{x})\Phi_I^S(\boldsymbol{x}) \tag{2.188}$$

$$G_{Iy}(\boldsymbol{x}) = \alpha_{31}(\boldsymbol{x})\Phi_{I,x}^S(\boldsymbol{x}) + \alpha_{32}(\boldsymbol{x})\Phi_{I,y}^S(\boldsymbol{x}) + \alpha_{33}(\boldsymbol{x})\Phi_I^S(\boldsymbol{x}) \tag{2.189}$$

The unknown coefficients, or the coefficient matrix, $\boldsymbol{\alpha}(\boldsymbol{x})$, are obtained by enforcing the conditions that the kernel function, $\bar{\Phi}_I$, and G_{Ix}, G_{Iy}, reproduce linear functions. These conditions lead to

$$\mathbf{A}\boldsymbol{\alpha} = \mathbf{I} \tag{2.190}$$

where

$$\mathbf{A} = \sum_{I \in \Lambda} \begin{pmatrix} \Phi_{I,x}^S(\boldsymbol{x}) & \Phi_{I,y}^S(\boldsymbol{x}) & \Phi_I^S(\boldsymbol{x}) \\ \Phi_{I,x}^S(\boldsymbol{x})x_I & \Phi_{I,y}^S(\boldsymbol{x})x_I & \Phi_I^S(\boldsymbol{x})x_I \\ \Phi_{I,x}^S(\boldsymbol{x})y_I & \Phi_{I,y}^S(\boldsymbol{x})y_I & \Phi_I^S(\boldsymbol{x})y_I \end{pmatrix} \tag{2.191}$$

and

$$\mathbf{I} = \begin{pmatrix} 1 & 1 & 1 \\ 1 & 1 & 1 \\ 1 & 1 & 1 \end{pmatrix} \tag{2.192}$$

This procedure is equivalent to use the meshfree wavelet shape function to approximate the derivative of RKPM shape function.[265, 266]

2.5.9 Chen-Beraun Correction

Chen & Beraun[94, 95] proposed another set of correction formulas for both the SPH kernel function and the derivatives of the SPH kernel function. The novelty of the proposal is to generalize the idea of the Shepard interpolant to evaluate derivatives of the kernel function.

They argued that any smooth function can be expanded in the following Taylor expansion,

$$\int_\Omega f(x)\Phi_I(x)dx = f(x_I)\int_\Omega \Phi_I(x)dx + f_{,x}(x_I)\int_\Omega (x - x_I)\Phi_I(x)dx$$
$$+ \frac{f_{,xx}(x_I)}{2!}\int_\Omega (x - x_I)^2\Phi_I(x)dx + \cdots \tag{2.193}$$

To evaluate $f(x_I)$, one may neglect all the derivative terms in 2.193, and obtain

$$f(x_I) \approx \frac{\int_\Omega f(x)\Phi_I(x)dx}{\int_\Omega \Phi_I(x)dx} \quad \Rightarrow \quad f_I \approx \frac{\sum_{J \in \Lambda} \Delta x_J f_J \Phi_{IJ}}{\sum_{J \in \Lambda} \Delta x_J \Phi_{IJ}} \tag{2.194}$$

which is basically a Shepard interpolant.

To evaluate the first order derivative, one may neglect all the derivatives that are higher than the second order derivatives, and obtain

$$f_{,x}(x_I) \approx \frac{\int_\Omega [f(x) - f(x_I)]\Phi_{I,x}(x)dx}{\int_\Omega (x - x_I)\Phi_{I,x}(x)dx} \tag{2.195}$$

which leads the corrected approximation for the derivative,

$$f_{I,x} \approx \frac{\sum_{J\in\Lambda} \Delta x_J (f_J - f_I)\Phi_{IJ,x}}{\sum_{J\in\Lambda} \Delta x_J (x_J - x_I)\Phi_{IJ,x}} \tag{2.196}$$

in which $f_I = f(x_I)$, $f_{I,x} = \dfrac{df}{dx}(x_I)$, $\Phi_{Ij} = \Phi(x_I - x_J; h)$, and $\Phi_{IJ,x} = \dfrac{d\Phi}{dx}(x - x_J)\Big|_{x=x_I}$.

Neglecting higher order derivatives, one may obtain a corrective formula for the second order derivative,

$$f_{,xx}(x_I) \approx \frac{\int_\Omega [f(x) - f(x_I)]\Phi_{I,xx}(x)dx - f_{I,x}\int_\Omega (x - x_I)\Phi_{I,xx}(x)dx}{\dfrac{1}{2}\int_\Omega (x - x_I)^2\Phi_{I,xx}(x)dx} \tag{2.197}$$

where $\Phi_{I,xx} = \partial^2\Phi_I(x)/\partial x^2$. The discrete version of approximation (2.197) becomes

$$f_{I,xx} \approx \frac{\sum_{J\in\Lambda} \Delta x_J (f_J - f_I)\Phi_{IJ,xx} - f_{I,x}\sum_{J\in\Lambda} \Delta x_J (x_J - x_I)K_{IJ,xx}}{\dfrac{1}{2}\sum_{J\in\Lambda} \Delta x_J (x_J - x_I)^2\Phi_{IJ,xx}} \tag{2.198}$$

where $f_{I,xx} = f_{,xx}(x_I)$, and $\Phi_{IJ,xx} = \partial^2\Phi(x - x_J)/\partial x^2\Big|_{x=x_I}$. The technical advantage of this correction is that it does not involve matrix inversion.

This approach follows the philosophy of Taylor expansion of an RKPM interpolation.[300, 303]

2.6 Remarks

Even though SPH method has achieved much success in engineering computations, the method has not been viewed as an accurate mathematical computation or numerical apparatus, which stems from the fact that it lacks a rigorous convergence theory as well as a successive refinement procedure. As a matter of fact, how to use SPH to solve a general PDE is still an open problem. This is because SPH is intrinsically a mechanics-related numerical method based on Lagrangian formulation, a generalization of SPH to solve general PDE seems to be irrelevent in many cases.

In order to develop general meshfree methods to solve various PDEs in engineering applications, people turned their attention to develop meshfree Galerkin methods in the begining of 1990s. In comparison with SPH method, meshfree Galerkin methods not only enjoy the flexibility to interpolate scattered data, but also have solid mathematical fundation in approximation theory.

Exercises

1.1 Show 1-d Super-Gaussian becomes a delta function as $h \to 0$, i.e.

$$\lim_{h \to 0} \Phi(u, h) \Rightarrow \delta(u) \tag{2.199}$$

where

$$\Phi(u, h) = \frac{1}{h\sqrt{\pi}} \exp\{-u^2/h^2\} \left(\frac{3}{2} - \frac{u^2}{h^2} \right) \tag{2.200}$$

1.2 Assuming the smoothing length h depending on time, i.e. $h(t)$. Show the 1-d continuity equation

$$\frac{\partial \rho}{\partial t} + \frac{\partial}{\partial x}\left(\rho v \right) = 0 \tag{2.201}$$

has the following non-local form

$$\frac{\partial \rho_K}{\partial t} - \dot{h}\frac{\partial \rho_K}{\partial h} + \frac{\partial}{\partial x}\left(\rho v \right)_K = 0 . \tag{2.202}$$

Show the discrete approximation

$$\rho_K(x) = \sum_{I=1}^{N} \Phi(x - x_I, h) m_I \tag{2.203}$$

satisfying the non-local continuity equation (2.202).

1.3 Assuming the smoothing length h is a function of time, i.e. $h = h(t)$. Show 1-d momentum equation

$$\frac{\partial}{\partial t}\left(\rho v\right) + \frac{\partial}{\partial x}\left(\rho v^2\right) = -\frac{\partial P}{\partial x}$$

(2.204)

has the non-local form

$$\frac{\partial}{\partial t}\left\langle \rho v \right\rangle + \frac{\partial}{\partial x}\left\langle \rho v^2 \right\rangle - \dot{h}\frac{\partial}{\partial h}\left\langle \rho v \right\rangle = -\left\langle \frac{\partial P}{\partial x} \right\rangle$$

(2.205)

Let

$$\left\langle \rho v \right\rangle \approx \sum_{I\in\Lambda} m_I v_I \Phi_h(x - x_I)$$

(2.206)

and

$$\left\langle \frac{\partial P}{\partial x} \right\rangle \approx \sum_{I\in\Lambda} m_I \frac{1}{\rho_I}\left(\frac{\partial P}{\partial x}\right)_I \Phi_h(x - x_I)$$

(2.207)

Verify that Eq. (2.205) is satisfied only if

$$\dot{v}_I = -\frac{1}{\rho_I}\left(\frac{\partial P}{\partial x}\right)_I$$

(2.208)

1.4 Utilizing the identity

$$\frac{\nabla P}{\rho} = \frac{2\sqrt{P}}{\rho}\nabla(\sqrt{P})$$

(2.209)

to show

$$\left(\frac{\nabla P}{\rho}\right)_I = 2\sum_{J\in\Lambda} m_J \frac{\sqrt{P_I}}{\rho_I}\frac{\sqrt{P_J}}{\rho_J}\nabla_I \Phi_h(\boldsymbol{x}_I - \boldsymbol{x}_J)$$

(2.210)

1.5 Show Case 2c and Case 2d of the fourth approximation rule

$$\rho_I < \nabla \mathbf{A} >_I = -\sum_{J\in\Lambda} \nabla_I \Phi_{IJ} \otimes \mathbf{A}_{IJ} m_J$$

(2.211)

$$\rho_I < \nabla \times \mathbf{A} >_I = \sum_{J\in\Lambda} \mathbf{A}_{IJ} \times \nabla_I \Phi_{IJ} m_J$$

(2.212)

3. Meshfree Galerkin Methods

In 1981, Lancaster and Salkauskas[255] published their seminal work in construction of smooth interpolation functions on a set of scattered or disordered data, which is called the Moving Least Square Interpolant (MLS). Until 1992, the methodology had remained a curve/surface fitting algorithm.

In 1992 Nayroles, Touzot, and Villon[348] rediscovered the moving least square interpolant. They suggested using it as a "diffused" (nonlocal) shape function in numerical computations, and named it the Diffuse Element Method (DEM). About the same time, in order to enforce the consistency conditions for SPH interpolant, Liu et al. ([295, 296]) introduced the notion of correction function to modify the primitive SPH kernel functions to enforce the consistency conditions on SPH interpolant. The development of the corrective function is motivated by the moment conditions employed in wavelet theories (e.g.[129]) and their ameniable properties in multi-scale analysis. In a 1994 landmark paper, Belytschko et al.[40] proposed the first Galerkin weak formulation to accomodate MLS interpolant in the simulation of crack growth in a linear elastic solid. Since then, a class of meshfree Galerkin methods have been invented in numerical computations. Many of these meshfree methods employ meshfree interpolation schemes that are essentially the moving least square (MLS) interpolant.

Moving least square method constructs a meshfree interpolant in an optimization (weighted least square) procedure, which guarantees an optimal interpolation error in the sense of least square. To emphasize this feature, the so-called Moving Least Square Reproducing Kernel Interpolant (MLSRK) is introduced here first (Liu, Li and Belytschko [1997]), which is a contemporary version of the classical MLS.

3.1 Moving Least Square Reproducing Kernel Interpolant

In the previous chapter, RKPM has been discussed as a remedy for SPH interpolant. The moving least square reproducing kernel interpolant is discussed here again from a different perspective.

There are two main technical ingredients in MLS methods. As suggested by its name, it has a *least square part*, which is a way of optimization, and it has a *moving* part, which is the way of local to global extension (globalization). The moving part was not all clearly stated until it was later elaborated by Liu, Li, and Belytschko[303] and again Li & Liu.[266]

Let $u(\boldsymbol{x})$ be a sufficiently smooth function [1] that is defined on a simply connected open set $\Omega \in \mathbb{R}^n$. Consider a fixed point $\bar{\boldsymbol{x}} \in \Omega$. One can define a local function,

$$
u^l(\boldsymbol{x}, \bar{\mathbf{x}}) := \begin{cases} u(\boldsymbol{x}) , & \forall \ \boldsymbol{x} \ \in \mathbf{B}_\varrho^e(\bar{\mathbf{x}}) \\ 0 , & \text{otherwise} \end{cases} \tag{3.1}
$$

where the set $\mathbf{B}_\varrho^e(\bar{\mathbf{x}})$ is the effective spherical ball, and ϱ is called the dilation parameter, which is similar to the smoothing length of SPH. In the one dimensional case, it is defined as the product of dilation coefficient and average particle spacing, $\varrho = a\Delta_x$ where $a = 1 \sim 3$ and Δ_x denotes the particle spacing.

In principle, one should be able to approximate $u(\boldsymbol{x})$ by a polynomial series locally according to the well-known Stone-Weierstrass theorem (Rudin [1976]). That is there exists a local polynomial operator, such that

$$
u^l(\boldsymbol{x}, \bar{\mathbf{x}}) \cong L_{\bar{\mathbf{x}}} u(\boldsymbol{x}) := \sum_{i=1}^{\ell} P_i(\frac{\boldsymbol{x} - \bar{\mathbf{x}}}{\varrho}) a_i(\bar{\mathbf{x}}, \rho)
$$

$$
= \mathbf{P}^T(\frac{\boldsymbol{x} - \bar{\mathbf{x}}}{\varrho}) \mathbf{a}(\bar{\mathbf{x}}, \varrho) \tag{3.2}
$$

The operator $L_{\bar{\mathbf{x}}}$ is a mapping

$$
L_{\bar{\mathbf{x}}} : C^0(\mathbf{B}_\varrho(\bar{\mathbf{x}})) \to C^m(\mathbf{B}_\varrho(\bar{\mathbf{x}})) \tag{3.3}
$$

and $\mathbf{a}(\bar{\mathbf{x}}, \varrho) := \{a_1(\bar{\mathbf{x}}, \varrho), a_2(\bar{\mathbf{x}}, \varrho), \cdots, a_\ell(\bar{\mathbf{x}}, \varrho)\}^T$ is an unknown vector function, and $\mathbf{P}(\boldsymbol{x}) := \{P_1(\boldsymbol{x}), P_2(\boldsymbol{x}), \cdots, P_\ell(\boldsymbol{x})\}^T$, $P_i(\boldsymbol{x}) \in C^m(\Omega)$ is an independent polynomial basis with $P_1(\boldsymbol{x}) = 1$.

The examples of $\mathbf{P}(\mathbf{x})$ in 1D, 2D, and 3D cases are

$$
\mathbf{P}(x) = (1, x, x^2, \cdots, x^{\ell-1})^T, \tag{3.4}
$$
$$
\mathbf{P}(\boldsymbol{x}) = (1, x_1, x_2, x_1^2, x_1 x_2, x_2^2)^T \tag{3.5}
$$
$$
\mathbf{P}(\boldsymbol{x}) = (1, x_1, x_2, x_3, x_1^2, x_2^2, x_3^2, x_1 x_2, x_2 x_3, x_3 x_1)^T \tag{3.6}
$$

Let $\Lambda := \{I \mid I = 1, \cdots, NP\}$. Define a discrete inner product,

$$
< f, g >_{\bar{\mathbf{x}}} := \sum_{I \in \Lambda} f(\bar{\mathbf{x}} - \boldsymbol{x}_I) g(\bar{\mathbf{x}} - \boldsymbol{x}_I) \Phi_\varrho(\bar{\mathbf{x}} - \boldsymbol{x}_I) w_I \tag{3.7}
$$

which may be viewed as a nodal integration (with integration weight w_I) of the following inner product in $L_2(\Omega)$,

---

[1] By this, we mean that $u(\boldsymbol{x}) \in C^0(\Omega)$ at least.

$$(f,g)_{\bar{\boldsymbol{x}}} := \int_\Omega f(\bar{\mathbf{x}} - \boldsymbol{x}) g(\bar{\mathbf{x}} - \boldsymbol{x}) \Phi_\varrho(\bar{\mathbf{x}} - \boldsymbol{x}) d\Omega \tag{3.8}$$

To quantify the interpolation error, a weighted interpolation residual form is constructed

$$J(\mathbf{a}(\bar{\mathbf{x}})) := \sum_{I \in \Lambda} \Phi_\varrho(\bar{\mathbf{x}} - \boldsymbol{x}_I) \left[\mathbf{P}^T \left(\frac{\bar{\mathbf{x}} - \boldsymbol{x}_I}{\varrho} \right) \mathbf{a}(\bar{\mathbf{x}}) - u(\boldsymbol{x}_I) \right]^2 w_I \tag{3.9}$$

where $r(\boldsymbol{x}_I, \bar{\mathbf{x}}) := \mathbf{P}^T \left((\bar{\mathbf{x}} - \boldsymbol{x}_I)/\varrho \right) \mathbf{a}(\bar{\mathbf{x}}) - u(\boldsymbol{x}_I)$ is the discrete local interpolation residual. The continuum counterpart of (3.9) is

$$J(\mathbf{a}(\bar{\mathbf{x}})) = \int_\Omega r^2(\boldsymbol{x}, \bar{\mathbf{x}}) \Phi_\varrho(\bar{\mathbf{x}} - \boldsymbol{x}) d\Omega \tag{3.10}$$

By minimizing the quadratic functional (3.9), one may obtain the optimal value for coefficient vector $\mathbf{a}(\bar{\mathbf{x}})$, which satisfies

$$\sum_{I \in \Lambda} \mathbf{P} \left(\frac{\bar{\mathbf{x}} - \boldsymbol{x}_I}{\varrho} \right) \left(u(\boldsymbol{x}_I) - \mathbf{P}^T \left(\frac{\bar{\mathbf{x}} - \boldsymbol{x}_I}{\varrho} \right) \mathbf{a}(\bar{\mathbf{x}}) \right) \Phi_\varrho(\bar{\mathbf{x}} - \boldsymbol{x}_I) w_I = \mathbf{0} \tag{3.11}$$

or equivalently

$$\mathbf{a}(\bar{\mathbf{x}}) = \mathbf{M}^{-1}(\bar{\mathbf{x}}) \sum_{I \in \Lambda} \mathbf{P} \left(\frac{\bar{\mathbf{x}} - \boldsymbol{x}_I}{\varrho} \right) u(\mathbf{x}_I) \Phi_\varrho(\bar{\mathbf{x}} - \boldsymbol{x}_I) w_I \tag{3.12}$$

where $\mathbf{M}(\bar{\mathbf{x}})$ is called the moment matrix that is defined as

$$\mathbf{M}(\boldsymbol{x}) = \sum_{I \in \Lambda} \mathbf{P} \left(\frac{\boldsymbol{x} - \boldsymbol{x}_I}{\varrho} \right) \mathbf{P}^T \left(\frac{\boldsymbol{x} - \boldsymbol{x}_I}{\varrho} \right) \Phi_\varrho(\boldsymbol{x} - \boldsymbol{x}_I) w_I \tag{3.13}$$

Its continuous counterpart may be derived by minimizing the functional (3.10),

$$\mathbf{a}(\bar{\mathbf{x}}) = \mathbf{M}^{-1}(\bar{\mathbf{x}}) \int_\Omega \mathbf{P} \left(\frac{\bar{\mathbf{x}} - \boldsymbol{y}}{\varrho} \right) u(\boldsymbol{y}) \Phi_\varrho(\bar{\mathbf{x}} - \boldsymbol{y}) d\Omega_y \tag{3.14}$$

where

$$\mathbf{M}(\boldsymbol{x}) = \int_\Omega \mathbf{P} \left(\frac{\boldsymbol{x} - \boldsymbol{y}}{\varrho} \right) \mathbf{P}^T \left(\frac{\boldsymbol{x} - \boldsymbol{y}}{\varrho} \right) \Phi_\varrho(\boldsymbol{x} - \boldsymbol{y}) d\Omega_y \tag{3.15}$$

Substituting (3.12) into (3.2), one can rewrite the local approximation formula as

$$\begin{aligned} u^l(\boldsymbol{x}, \bar{\mathbf{x}}) &\approx \mathbf{P}^T \left(\frac{\bar{\mathbf{x}} - \boldsymbol{x}}{\varrho} \right) \mathbf{a}(\bar{\mathbf{x}}) w_I \\ &= \mathbf{P}^T \left(\frac{\bar{\mathbf{x}} - \boldsymbol{x}}{\varrho} \right) \mathbf{M}^{-1}(\bar{\mathbf{x}}) \sum_{I \in \Lambda} \mathbf{P} \left(\frac{\bar{\mathbf{x}} - \boldsymbol{x}_I}{\varrho} \right) u(\boldsymbol{x}_I) \Phi_\varrho(\bar{\mathbf{x}} - \boldsymbol{x}_I) w_I \end{aligned}$$

$$\tag{3.16}$$

So far, the manipulation is the standard weighted least-square procedure. The expression (3.16) is optimal in a local region $\mathbf{B}_\varrho^e(\bar{\mathbf{x}}) \subset \Omega$. In order to extend (3.16) to the whole region, a *"moving"* process is ensured: the moving process is a global approximation that is defined as

$$G : C^0(\Omega) \to C^m(\Omega) \quad \Rightarrow \quad Gu(\mathbf{x}) := \lim_{\bar{\mathbf{x}} \to \mathbf{x}} L_{\bar{\mathbf{x}}} u^l(\mathbf{x}, \bar{\mathbf{x}}), \quad \forall \mathbf{x} \in \Omega \quad (3.17)$$

The moving least square reproducing kernel approximation is then established

$$u(\mathbf{x}) \approx Gu(\mathbf{x}) = \mathbf{P}^T(0)\mathbf{M}^{-1}(\mathbf{x}) \sum_{I \in \Lambda} \mathbf{P}\left(\frac{\mathbf{x} - \mathbf{x}_I}{\varrho}\right) u(\mathbf{x}_I)\Phi_\varrho(\mathbf{x} - \mathbf{x}_I)w_I \quad (3.18)$$

Define the so-called correction function

$$C_\varrho(\mathbf{x} - \mathbf{x}_I, \mathbf{x}) = \mathbf{P}^T(0)\mathbf{M}^{-1}(\mathbf{x})\mathbf{P}\left(\frac{\mathbf{x} - \mathbf{x}_I}{\varrho}\right)$$

$$= \mathbf{b}^T(\mathbf{x}, \varrho)\mathbf{P}\left(\frac{\mathbf{x} - \mathbf{x}_I}{\varrho}\right) = \mathbf{P}^T\left(\frac{\mathbf{x} - \mathbf{x}_I}{\varrho}\right)\mathbf{b}(\mathbf{x}, \varrho) \quad (3.19)$$

where the vector $\mathbf{b}(x)$ is an unknown function that is to be adjusted to suit the local particle distribution such that

$$\mathbf{b}(\mathbf{x}, \varrho) = \mathbf{M}^{-1}(\mathbf{x})\mathbf{P}(0) \quad (3.20)$$

The MLSRK kernel function is

$$\mathcal{K}_\varrho(\mathbf{x} - \mathbf{x}_I, \mathbf{x}) = C_\varrho(\mathbf{x} - \mathbf{x}_I, \mathbf{x})\Phi_\varrho(\mathbf{x} - \mathbf{x}_I)w_I \quad (3.21)$$

and

$$u_\varrho^h(\mathbf{x}) = \sum_{I \in \Lambda} \mathcal{K}_\varrho(\mathbf{x} - \mathbf{x}_I, \mathbf{x})u(\mathbf{x}_I)w_I \quad (3.22)$$

Again, one may argue that the MLSRK interpolation formula is an approximation (nodal integration) of continuous reproducing kernel formula [297],

$$u_\varrho(\mathbf{x}) = \int_\Omega \mathcal{K}_\varrho(\mathbf{x} - \mathbf{y}, \mathbf{x})u(\mathbf{y})d\Omega \quad (3.23)$$

First note that the second argument x that appears in the expressions of both correction function and modified kernel function is a parametric variable. This is to note that both correction function and the modified kernel function are dependent on the local particle distribution, and the measure of local particle distribution varies from point to point. Second, both (3.22) and (3.23) are summations on a finite domain, which is in contrast with the infinite domain representation of SPH.

Since Φ is symmetric, the moment matrix constructed in (3.15) is real and symmetric. If $x \in \Omega$, and there are enough particles inside the domain of influence of x (this condition shall be defined precisely in Chaper 4), one may show that $\mathbf{M}_\varrho(x)$ is positive definite, thus $\mathbf{M}_\varrho(x)$ is invertible.

3.1.1 Polynomial Reproducing Property

An important property of MLSRK interpolant is that it can reproduce all polynomial functions that are the components of the polynomial basis $\mathbf{P}(\boldsymbol{x})$. Because of the polynomial reproducing property, the name *reproducing kernel particle method* was coined (Liu et al. [1995]). In what follows, this property is proved in the one dimensional case.

Lemma 3.1.1. *Suppose* $\mathbf{P}(x) = \{1, x, \cdots, x^{\ell-1}\}^T$. *Define RKPM shape function*

$$\{N_I(x, \varrho)\} := \{\mathcal{K}_\varrho(x - x_I, x)\Delta x_I\} . \tag{3.24}$$

For $u(x) = 1, x, x^2, \cdots, x^{\ell-1}$, *MLSRK interpolants reproduce* $u(x)$ *exactly, i.e.*

$$\sum_{I=1}^{NP} N_I(x, \varrho)x_I^k = \sum_{I=1}^{NP} \left(\mathcal{K}_\varrho(x - x_I, x)\Delta x_I\right)x_I^k = x^k, \quad \forall\, k = 0, 1, \cdots, n \tag{3.25}$$

Proof:
We want to show for $0 \leq k \leq n$,

$$\sum_{I=1}^{NP} N_I(x, \rho)x_I^k = x^k, \quad \forall\, 0 \leq k \leq n \tag{3.26}$$

$$\sum_{I=1}^{NP} N_I(x, \varrho)x_I^k = \mathbf{P}^T(0)\mathbf{M}_\varrho^{-1}(x) \sum_{I=1}^{NP} \left(\mathbf{P}^T\left(\frac{x - x_I}{\varrho}\right)\Phi_\varrho(x - x_I)\Delta x_I\right)x_I^k \tag{3.27}$$

Let

$$z_I = \frac{x - x_I}{\varrho} \quad \Rightarrow \quad x_I^k = \left(x - \varrho z_I\right)^k \tag{3.28}$$

Therefore,

$$\sum_{I=1}^{NP} N_I(x, \varrho) x_I^k$$

$$= \sum_{I=1}^{NP} N_I(x, \varrho) \left(\sum_{j \leq k} \binom{k}{j} (-1)^{k-j} \varrho^{k-j} z_I^{k-j} x^j \right)$$

$$= \sum_{j \leq k} \binom{k}{j} (-1)^{k-j} \varrho^{k-j} x^j \mathbf{P}^T(0) \mathbf{M}_\varrho^{-1}(x) \sum_{I=1}^{NP} \left(\mathbf{P}^T(z_I) \varPhi(z_I) \Delta x_I \right) z_I^{k-j}$$

$$= \sum_{j \leq k} \binom{k}{j} (-1)^{k-j} \varrho^{k-j} x^j \mathbf{P}^T(0) \mathbf{M}_\varrho^{-1}(x) \sum_{I=1}^{NP} \begin{pmatrix} z_I^{k-j} \\ z_I^{k-j+1} \\ \vdots \\ \vdots \\ z_I^{k-j+m} \end{pmatrix} \varPhi(z_I) \Delta x_I$$

$$\tag{3.29}$$

in which

$$\mathbf{P}^T(0) \mathbf{M}_\rho^{-1}(x) \sum_{I=1}^{NP} \begin{pmatrix} z_I^{k-j} \\ z_I^{k-j+1} \\ \vdots \\ \vdots \\ z_I^{k-j+m} \end{pmatrix} \varPhi(z_I) \Delta x_I$$

$$= \mathbf{P}^T(0) \mathbf{M}_\varrho^{-1}(x) \begin{pmatrix} m_{k-j} \\ m_{k-j+1} \\ \vdots \\ \vdots \\ m_{k-j+m} \end{pmatrix} = \delta_{kj} \tag{3.30}$$

In the last step, the Laplace's expansion theorem in linear algebra (i.e.[327]) is used. Since $\mathbf{P}^T(0) \mathbf{M}_\varrho^{-1}(x)$ is the first row of $\mathbf{M}_\varrho^{-1}(x)$, the column vector next to it is the $k - j + 1$ column of $\mathbf{M}_\rho(x)$, the product of two does not equal to zero, unless $k = j$. It then follows

$$\sum_{I=1}^{N} N_I(x, \varrho) x_I^k = \sum_{j \leq k} \binom{k}{j} (-1)^{k-j} \varrho^{k-j} x^j \delta_{kj} = x^k \tag{3.31}$$

♣

Remark 3.1.1. After the RKPM kernel function has been determined, one can write down the RKPM interpolation, or sampling formula

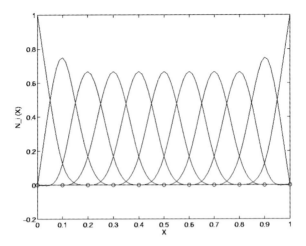

Fig. 3.1. 1D RKPM shape function.

$$f^R(x) = \sum_{I=1}^{NP} f_I N_I(x, \rho) \tag{3.32}$$

Nevertheless, in most of the mathematical literature, the term "interpolation" is exclusively reserved for the sampling representation that satisfies the condition,

$$f^R(x_I) = f_I, \quad \text{and} \quad N_I(x_J) = \delta_{IJ} \tag{3.33}$$

This is not the case for RKPM shape function, since

$$\mathcal{K}_\varrho(x_J - x_I)\Delta x_I \neq \delta_{IJ} \tag{3.34}$$

On the other hand, however, if $u(x) = 1, x, x^2, \cdots, x^{\ell-1}$, we will have $u(x_I) = u_I$.

3.1.2 The Shepard Interpolant

A special case of non-interpolating MLS interpolant is $\ell = 1$. In this case,

$$\mathbf{P}(x) = 1$$

$$\mathbf{M}(x) = \sum_{I \in \Lambda}[1] \cdot \Phi(x - x_I) \cdot [1] = \sum_{I \in \Lambda} \Phi(x - x_I) \tag{3.35}$$

$$\mathbf{M}^{-1}(x) = \frac{1}{\displaystyle\sum_{I \in \Lambda} \Phi(x - x_I)} \tag{3.36}$$

$$\text{and} \quad \mathcal{K}(x - x_I, x) = \frac{\Phi(x - x_I)}{\displaystyle\sum_{I \in \Lambda} \Phi(x - x_I)} \tag{3.37}$$

It is clear that

$$
\begin{cases}
(i) & 0 < \mathcal{K}(\boldsymbol{x} - \boldsymbol{x}_I, \boldsymbol{x}) < 1 \\[2mm]
(ii) & \displaystyle\sum_{I \in \Lambda} \mathcal{K}(\boldsymbol{x} - \boldsymbol{x}_I, \boldsymbol{x}) = 1 \\[3mm]
(iii) & \min\{f_1, f_2, \cdots, f_{NP}\} \le Gf(\boldsymbol{x}) \le \max\{f_1, f_2, \cdots, f_{NP}\}
\end{cases}
\tag{3.38}
$$

which means that the interpolation function at any point of the domain is a weighted average of given values at a set of discrete points. This particular interpolant was first studied by Shepard [1968], and it is named as *Shepard interpolant*. In short notation, we often denote

$$
Sf(\boldsymbol{x}) = Gf(\boldsymbol{x}) \Big|_{\ell=1}
\tag{3.39}
$$

Shepard interpolation of a continuous function, f, can be interpreted as the projection of f onto the basis $p_1(\boldsymbol{x})$, i.e.

$$
Sf(\boldsymbol{x}) = \frac{\displaystyle\sum_{I \in \Lambda} \phi(\boldsymbol{x} - \boldsymbol{x}_I) f_I}{\displaystyle\sum_{I \in \Lambda} \phi(\boldsymbol{x} - \boldsymbol{x}_I)} = (f, p_1)_{\boldsymbol{x}} \frac{p_1(\boldsymbol{x})}{(p_1, p_1)_{\boldsymbol{x}}}
\tag{3.40}
$$

3.1.3 Interpolating Moving Least Square Interpolant

As a non-local interpolation scheme, the moving least square interpolant, $\{N_I^\varrho(\boldsymbol{x})\}$, does not possess Kronecker delta property

$$
N_I^\varrho(\boldsymbol{x}_I) \ne 1
\tag{3.41}
$$
$$
N_I^\varrho(\boldsymbol{x}_J) \ne 0
\tag{3.42}
$$

In other words MLS interpolation scheme does not "interpolate", more precisely,

$$
Gf(\boldsymbol{x}_I) \ne f_I
\tag{3.43}
$$

In many situations, the interpolating property is desirable in order to enforce essential boundary conditions. A remedy for restoring the interpolation property, as suggested by Shepard as well as Lancaster and Salkauskas, is to use a singular weight function. For instance, if one chooses singular weight function

$$
\Phi_I(\boldsymbol{x}) = \frac{1}{|\boldsymbol{x} - \boldsymbol{x}_I|^\alpha} \,,
\tag{3.44}
$$

where α is a positive even integer. Then Shepard interpolant based on the singular weight function is

$$\mathcal{K}^S(\boldsymbol{x} - \boldsymbol{x}_I) = \frac{1}{|\boldsymbol{x} - \boldsymbol{x}_I|^\alpha \cdot \displaystyle\sum_{J \in \Lambda} \left\{ \frac{1}{(\boldsymbol{x} - \boldsymbol{x}_J)^\alpha} \right\}}$$

$$= \frac{1}{\displaystyle\sum_{J \in \Lambda \backslash \{I\}} \left\{ \dfrac{(\boldsymbol{x} - \boldsymbol{x}_I)^\alpha}{(\boldsymbol{x} - \boldsymbol{x}_J)^\alpha} \right\} + 1} \tag{3.45}$$

Therefore, if $\boldsymbol{x} = \boldsymbol{x}_I$,

$$\sum_{J \in \Lambda \backslash \{I\}} \left\{ \frac{(\boldsymbol{x}_I - \boldsymbol{x}_I)^\alpha}{(\boldsymbol{x}_I - \boldsymbol{x}_J)^\alpha} \right\} = 0 \Rightarrow \lim_{\boldsymbol{x} \to \boldsymbol{x}_I} \mathcal{K}^S(\boldsymbol{x} - \boldsymbol{x}_I) = 1 \tag{3.46}$$

And if $\boldsymbol{x} = \boldsymbol{x}_L$, $L \neq I$,

$$\sum_{J \in \Lambda \backslash \{I\}} \left\{ \frac{(\boldsymbol{x}_L - \boldsymbol{x}_I)^\alpha}{(\boldsymbol{x}_L - \boldsymbol{x}_J)^\alpha} \right\} \to \infty \Rightarrow \lim_{\boldsymbol{x} \to \boldsymbol{x}_L} \mathcal{K}^S(\boldsymbol{x} - \boldsymbol{x}_I) \to 0 \tag{3.47}$$

because $L \in \Lambda \backslash I$ and the denominator $(\mathbf{x}_L - \mathbf{x}_J) \to 0$.

It can be shown that

$$\begin{cases} (i) & \mathcal{K}^S(\boldsymbol{x}_J - \boldsymbol{x}_I, \boldsymbol{x}_J) = \delta_{IJ} \\\\ (ii) & 0 \leq \mathcal{K}^S(\boldsymbol{x} - \boldsymbol{x}_I, \boldsymbol{x}) \leq 1, \\\\ (iii) & \displaystyle\sum_{I \in \Lambda} \mathcal{K}^S(\boldsymbol{x} - \boldsymbol{x}_I, \boldsymbol{x}) = 1; \quad \forall\, \boldsymbol{x} \in \Omega \\\\ (iv) & \mathcal{K}^S(\boldsymbol{x} - \boldsymbol{x}_I, \boldsymbol{x}) \to \dfrac{1}{NP} \text{ as } |\boldsymbol{x}| \to \infty \end{cases} \tag{3.48}$$

The first three properties are obvious. The fourth property holds, because when $\boldsymbol{x} \to \infty$,

$$\frac{|\boldsymbol{x} - \boldsymbol{x}_I|^\alpha}{|\boldsymbol{x} - \boldsymbol{x}_J|^\alpha} \to 1 \;, \quad I, J \in \Lambda, \tag{3.49}$$

and

$$\sum_{I \in \Lambda} \frac{|\boldsymbol{x} - \boldsymbol{x}_I|^\alpha}{|\boldsymbol{x} - \boldsymbol{x}_J|^\alpha} = NP. \tag{3.50}$$

The above construction is simple, but there is a drawback in using Shepard interpolant to construct *interpolating* MLS interpolant. This is because the derivative of such interpolant is always zero at nodal point, i.e.

$$\nabla \mathcal{K}^S(\boldsymbol{x} - \boldsymbol{x}_I) \Big|_{\boldsymbol{x} = \boldsymbol{x}_J} = 0, \quad I, J \in \Lambda \tag{3.51}$$

This is because the nodal points are either at a local maximum point or at a local minimum point of a shape function. Therefore, around neighborhood of the particles, the shape functions are always "flat",

$$\nabla S f(\boldsymbol{x}) \Big|_{\boldsymbol{x}=\boldsymbol{x}_J} = \sum_{I \in \Lambda} f_I \nabla \mathcal{K}^S(\boldsymbol{x} - \boldsymbol{x}_I) \Big|_{\boldsymbol{x}=\boldsymbol{x}_J} = 0 \qquad (3.52)$$

This is a major limitation for such interpolants to be used in interpolation.

Eq. (3.51) can be shown by inspecting (3.38). The properties (i) and (ii) imply that $\mathcal{K}^S(\boldsymbol{x}_J - \boldsymbol{x}_I, \boldsymbol{x}_J)$ is either minimum or maximum, thus the continuity of $\mathcal{K}^S$ leads $\nabla \mathcal{K}^S_{(}\boldsymbol{x} - \boldsymbol{x}_I, \boldsymbol{x}) = 0$. In 1D case, it can be shown explicitly

$$\frac{d}{dx}\mathcal{K}^S(x - x_I, x) = \frac{\alpha}{\left(\sum_{L \in \Lambda} (\frac{x - x_I}{x - x_L})^\alpha\right)^2} \cdot \sum_{L \in \Lambda} \frac{(x - x_I)^{\alpha-1}}{(x - x_L)^{\alpha+1}}(x_I - x_L)$$

$$= \alpha(\mathcal{K}^S(x - x_I, x))^2 \cdot \sum_{L \in \Lambda} \frac{(x - x_I)^{\alpha-1}}{(x - x_L)^{\alpha+1}}(x_I - x_L)$$

$$(3.53)$$

If $x \to x_J$, one will have

$$\frac{d\mathcal{K}^S(x - x_I, x)}{dx} \Big|_{x=x_J} = \lim_{x \to x_J} \alpha \delta_{IJ} \sum_{L \in \Lambda} \frac{(x - x_I)^{\alpha-1}}{(x - x_L)^{\alpha+1}}(x_I - x_L) \to 0 \quad (3.54)$$

To construct a better interpolant, Lancaster and Salkauskas recommended the following approach. For a given linear independent basis $\mathbf{P}(\boldsymbol{x}) = \{p_0(\boldsymbol{x}), p_1(\boldsymbol{x}), \cdots, p_{n+1}(\boldsymbol{x})\}$ with $p_1(\boldsymbol{x}) = 1$, let

$$\beta_1(\boldsymbol{x}, \bar{\mathbf{x}}) = \frac{p_1(\boldsymbol{x})}{\|p_1\|_{\bar{\mathbf{x}}}} = \frac{1}{\left[\sum_{I \in \Lambda} \phi(\bar{\mathbf{x}} - \boldsymbol{x}_I)\right]^{1/2}} \qquad (3.55)$$

Subtracting the rest of $p_i(\boldsymbol{x}), i = 2, 3, \cdots, n+1$ from its $\bar{\mathbf{x}}$-projection on $\beta_1(\boldsymbol{x}, \bar{\mathbf{x}})$, we define

$$\gamma_i(\boldsymbol{x}, \bar{\mathbf{x}}) = p_i(\boldsymbol{x}) - (p_i(\boldsymbol{x}), \beta_1(\boldsymbol{x}, \bar{\mathbf{x}}))_{\bar{\mathbf{x}}} \beta_1(\boldsymbol{x}, \bar{\mathbf{x}})$$

$$= p_i(\boldsymbol{x}) - \sum_{I \in \Lambda} p_i(\boldsymbol{x}_I) \mathcal{K}^S(\bar{\mathbf{x}} - \boldsymbol{x}_I) \qquad (3.56)$$

such that

$$(\gamma_i(\boldsymbol{x}), \beta_1(\boldsymbol{x}, \bar{\mathbf{x}}))_{\bar{\mathbf{x}}} = (p_i(\boldsymbol{x}), \beta_1(\boldsymbol{x}, \bar{\mathbf{x}}))_{\bar{\mathbf{x}}}$$

$$-(p_i(\boldsymbol{x}), \beta_1(\boldsymbol{x}, \bar{\mathbf{x}}))_{\bar{\mathbf{x}}}(\beta_1(\boldsymbol{x}, \bar{\mathbf{x}}), \beta_1(\boldsymbol{x}, \bar{\mathbf{x}}))_{\bar{\mathbf{x}}}$$

$$= (p_i(\boldsymbol{x}), \beta_1(\boldsymbol{x}, \bar{\mathbf{x}}))_{\bar{\mathbf{x}}} - (p_i(\boldsymbol{x}), \beta_1(\boldsymbol{x}, \bar{\mathbf{x}}))_{\bar{\mathbf{x}}} = 0 \qquad (3.57)$$

Now consider the local least square procedure by using the new basis $\{\beta_1, \gamma_2, \gamma_3, \cdots, \gamma_\ell\}$. For any continuous function, $u(\boldsymbol{x})$, the local approximation can then be expressed as

$$L_{\bar{\mathbf{x}}} u(\boldsymbol{x}) = \beta_1(\bar{\mathbf{x}}, \boldsymbol{x}) \alpha_1(\bar{\mathbf{x}}) + \sum_{i=2}^{\ell} \gamma_i(\boldsymbol{x}, \bar{\mathbf{x}}) \alpha_i(\bar{\mathbf{x}}) \qquad (3.58)$$

where $\mathcal{A}(\bar{\mathbf{x}}) = \{\alpha_1(\bar{\mathbf{x}}), \alpha_2(\bar{\mathbf{x}}), \cdots, \alpha_\ell(\bar{\mathbf{x}})\}$ is the unknown vector. $\mathcal{A}(\bar{\mathbf{x}})$ can be determinated by the normal equations derived from the least square procedure,

$$(\beta_1, \beta_1)_{\bar{\mathbf{x}}} \alpha_1(\bar{\mathbf{x}}) + \sum_{i=2}^{\ell} (\gamma_i, \beta_1)_{\bar{\mathbf{x}}} \alpha_i(\bar{\mathbf{x}}) = (u, \beta_1)_{\bar{\mathbf{x}}}$$

$$(\beta_1, \gamma_j)_{\bar{\mathbf{x}}} \alpha_1(\bar{\mathbf{x}}) + \sum_{i=2}^{\ell} (\gamma_i, \gamma_j)_{\bar{\mathbf{x}}} \alpha_i(\bar{\mathbf{x}}) = (u, \gamma_j)_{\bar{\mathbf{x}}}$$

$$j = 2, 3, \cdots, \ell \qquad (3.59)$$

The orthogonality condition $(\beta_1, \gamma_j)_{\bar{\mathbf{x}}} = (\gamma_j, \beta_1)_{\bar{\mathbf{x}}} = 0$ yields

$$\alpha_1(\bar{\mathbf{x}}) = (f, \beta_1)_{\bar{\mathbf{x}}} \qquad (3.60)$$

$$\sum_{i=1}^{\ell} (\gamma_i, \gamma_j)_{\bar{\mathbf{x}}} \alpha_i(\bar{\mathbf{x}}) = (f, \gamma_j)_{\bar{\mathbf{x}}}, \quad j = 2, 3, \cdots, \ell \qquad (3.61)$$

Let

$$\Sigma(\bar{\mathbf{x}}) := \begin{pmatrix} (\gamma_2, \gamma_2)_{\bar{\mathbf{x}}}, & (\gamma_2, \gamma_3)_{\bar{\mathbf{x}}}, & \cdots, & (\gamma_2, \gamma_\ell)_{\bar{\mathbf{x}}} \\ (\gamma_3, \gamma_2)_{\bar{\mathbf{x}}}, & (\gamma_3, \gamma_3)_{\bar{\mathbf{x}}}, & \cdots, & (\gamma_3, \gamma_\ell)_{\bar{\mathbf{x}}} \\ \vdots & & \ddots & \vdots \\ (\gamma_\ell, \gamma_2)_{\bar{\mathbf{x}}}, & (\gamma_{n+1}, \gamma_3)_{\bar{\mathbf{x}}}, & \cdots, & (\gamma_\ell, \gamma_\ell)_{\bar{\mathbf{x}}} \end{pmatrix} \qquad (3.62)$$

and

$$\Gamma(\boldsymbol{x}, \bar{\mathbf{x}}) = \{\gamma_2(\boldsymbol{x}, \bar{\mathbf{x}}), \gamma_3(\boldsymbol{x}, \bar{\mathbf{x}}), \cdots, \gamma_\ell(\boldsymbol{x}, \bar{\mathbf{x}})\} \qquad (3.63)$$

The unknown vector $\mathcal{A}(\bar{\mathbf{x}})$ is determined as

$$\mathcal{A}(\bar{\mathbf{x}}) = \Sigma^{-1}(\bar{\mathbf{x}}) \sum_{I \in \Lambda} \Gamma^T(\boldsymbol{x}_I, \bar{\mathbf{x}}) \phi(\bar{\mathbf{x}} - \boldsymbol{x}_I) u_I \qquad (3.64)$$

and subsequently,

$$L_{\bar{\mathbf{x}}} u(\boldsymbol{x}) = S f(\bar{\mathbf{x}}) + \Gamma(\boldsymbol{x}, \bar{\mathbf{x}}) \Sigma^{-1}(\bar{\mathbf{x}}) \sum_{I \in \Lambda} \Gamma^T(\boldsymbol{x}_I, \bar{\mathbf{x}}) \phi(\bar{\mathbf{x}} - \boldsymbol{x}_I) u_I \qquad (3.65)$$

The associated global approximation is

$$G u(\boldsymbol{x}) = S f(\boldsymbol{x}) + \Gamma(\boldsymbol{x}, \boldsymbol{x}) \Sigma^{-1}(\boldsymbol{x}) \sum_{I \in \Lambda} \Gamma^T(\boldsymbol{x}_I, \boldsymbol{x}) \phi(\boldsymbol{x} - \boldsymbol{x}_I) u_I \qquad (3.66)$$

With this modified formulation, the flat spot around each particle due to the presence of singular window functions should vanish.

3.1.4 Orthogonal Basis for the Local Approximation

Increasing the order of the basis vector $\mathbf{P}(\boldsymbol{x})$ will significantly improve the interpolation error—an analogy to higher order finite element. By doing so, the inversion of moment matrix can be potentially time consuming; therefore, orthogonal basis would be a good choice in implementation. By construction of orthogonal basis for local approximation, we mean that for given linear independent basis vector, $\mathbf{P}^T(\boldsymbol{x}) = \{p_1(\boldsymbol{x}), p_2(\boldsymbol{x}), \cdots, p_\ell(\boldsymbol{x})\}$ with $p_1(\boldsymbol{x}) = 1$ and fixed point $\bar{\mathbf{x}} \in \Omega$, we seek a new orthogonal basis vector, $\mathbf{Q}(\boldsymbol{x}, \bar{\mathbf{x}}) = \{q_1(\boldsymbol{x}, \bar{\mathbf{x}}), q_2(\boldsymbol{x}, \bar{\mathbf{x}}), \cdots, q_\ell(\boldsymbol{x}, \bar{\mathbf{x}})\}$, such that the following condition holds, i.e. for the fixed $\bar{\mathbf{x}} \in \Omega$,

$$\sum_{I \in \Lambda} \phi(\bar{\mathbf{x}} - \boldsymbol{x}_I) q_j(\boldsymbol{x}_I, \bar{\mathbf{x}}) q_k(\boldsymbol{x}_I, \bar{\mathbf{x}}) = 0, \quad \Rightarrow \quad \left(q_j, q_k \right)_{\bar{\mathbf{x}}} = 0, \quad j \neq k \quad (3.67)$$

To do so, the standard Schmidt's orthogonalization procedure is followed.

$$q_1(\boldsymbol{x}, \bar{\mathbf{x}}) = p_1(\boldsymbol{x})$$

$$q_2(\boldsymbol{x}, \bar{\mathbf{x}}) = p_2(\boldsymbol{x}) - \frac{(p_2, q_1)_{\bar{\mathbf{x}}}}{(q_1, q_1)_{\bar{\mathbf{x}}}} q_1(\boldsymbol{x}, \bar{\mathbf{x}})$$

$$q_3(\boldsymbol{x}, \bar{\mathbf{x}}) = p_3(\boldsymbol{x}) - \frac{(p_3, q_1)_{\bar{\mathbf{x}}}}{(q_1, q_1)_{\bar{\mathbf{x}}}} q_1(\boldsymbol{x}, \bar{\mathbf{x}}) - \frac{(p_3, q_2)_{\bar{\mathbf{x}}}}{(q_2, q_2)_{\bar{\mathbf{x}}}} q_2(\boldsymbol{x}, \bar{\mathbf{x}})$$

$$\cdots \cdots \qquad (3.68)$$

In general,

$$q_k(\boldsymbol{x}, \bar{\mathbf{x}}) = p_k(\boldsymbol{x}) - \sum_{j=1}^{k-1} \frac{(p_k, q_j)_{\bar{\mathbf{x}}}}{(q_j, q_j)_{\bar{\mathbf{x}}}} q_j(\boldsymbol{x}, \bar{\mathbf{x}}), \quad k = 1, 2, \cdots, \ell \qquad (3.69)$$

Because of orthogonalization, the local moment matrix is diagonalized, its inversion is trivial

$$\mathcal{M}^{-1}(\bar{\mathbf{x}}) = \begin{pmatrix} (q_1, q_1)_{\bar{\mathbf{x}}}^{-1} & & & \\ & (q_2, q_2)_{\bar{\mathbf{x}}}^{-1} & & \\ & & \ddots & \\ & & & (q_\ell, q_\ell)_{\bar{\mathbf{x}}}^{-1} \end{pmatrix} \qquad (3.70)$$

Thus, the components of the local correction vector,

$$\mathbf{a}(\bar{\mathbf{x}}) = \mathcal{M}^{-1}(\bar{\mathbf{x}}) \sum_{I \in \Lambda} \mathbf{Q}^T(\boldsymbol{x}_I) \phi(\bar{\mathbf{x}} - \boldsymbol{x}_I) u_I, \qquad (3.71)$$

will have the form

$$a_j(\bar{\mathbf{x}}) = \frac{\sum_{I \in \Lambda} \phi(\bar{\mathbf{x}} - \boldsymbol{x}_I) q_j(\boldsymbol{x}_I, \bar{\mathbf{x}}) u_I}{(q_j, q_j)_{\bar{\mathbf{x}}}} \qquad (3.72)$$

After moving to the global approximation, one will have

$$Gu(\boldsymbol{x}) = \mathbf{Q}(\boldsymbol{x})\mathbf{a}(\boldsymbol{x}) = \sum_{I \in \Lambda} \left\{ \sum_{j=1}^{\ell} \left(\frac{\phi(\boldsymbol{x} - \boldsymbol{x}_I)q_j(\boldsymbol{x}, \boldsymbol{x})q_j(\boldsymbol{x}_I, \boldsymbol{x})}{(q_j, q_j)_{\boldsymbol{x}}} \right) u_I \right\}$$

$$= \sum_{I \in \Lambda} K_O(\boldsymbol{x} - \boldsymbol{x}_I, \boldsymbol{x})u_I \qquad (3.73)$$

where the orthogonal kernel function is the product of original window function and a sequence of correction functions:

$$\boxed{K_O(\boldsymbol{x} - \boldsymbol{x}_I, \boldsymbol{x}) = \phi(\boldsymbol{x} - \boldsymbol{x}_I) \sum_{j=1}^{n+1} C_j(\boldsymbol{x}_I, \boldsymbol{x})}$$
$$(3.74)$$

The correct function sequence is expressed in terms of the orthogonal basis as follows

$$\boxed{C_j(\boldsymbol{x}_I, \boldsymbol{x}) = \left(\frac{q_j(\boldsymbol{x}_I, \boldsymbol{x})}{\|q_j\|_{\boldsymbol{x}}^2} \right) q_j(\boldsymbol{x}, \boldsymbol{x}), \quad j = 1, 2, \cdots, n+1}$$
$$(3.75)$$

3.1.5 Examples of RKPM Kernel Function

1D Example. Consider a line segment with $x \in [0, 1]$. A cubic spline function is used as the window function. The dilation parameter ρ is chosen as $\rho = a\Delta x$ where a is a non-dimensional dilation parameter. The optimal value for a is in the range: $1.0 \le a \le 1.2$. Using linear base vector $\mathbf{P}(x) = (1, x)$, is moment matrix is then obtained as

$$\mathbf{M}_\varrho(x) = \begin{pmatrix} m_0 & m_1 \\ m_1 & m_2 \end{pmatrix} \qquad (3.76)$$

where

$$m_0(x) = \sum_{I=1}^{NP} \mathcal{K}_\varrho(x - x_I)\Delta x_I \qquad (3.77)$$

$$m_1(x) = \sum_{I=1}^{NP} \left(\frac{x - x_I}{\varrho} \right) \mathcal{K}_\varrho(x - x_I)\Delta x_I \qquad (3.78)$$

$$m_2(x) = \sum_{I=1}^{NP} \left(\frac{x - x_I}{\varrho} \right)^2 \mathcal{K}_\varrho(x - x_I)\Delta x_I \qquad (3.79)$$

Solving the moment equation

$$\begin{pmatrix} m_0 & m_1 \\ m_1 & m_2 \end{pmatrix} \begin{pmatrix} b_0 \\ b_1 \end{pmatrix} = \begin{pmatrix} 1 \\ 0 \end{pmatrix} \Rightarrow \begin{pmatrix} b_0 \\ b_1 \end{pmatrix} = \begin{pmatrix} \dfrac{m_2(x)}{D(x)} \\ -\dfrac{m_1(x)}{D(x)} \end{pmatrix} \qquad (3.80)$$

where $D(x) = det\{\mathbf{M}_\varrho(x)\} = m_0(x)m_2(x) - m_1(x)m_1(x)$.

Consequently, the correction function $C_\rho(x - y, x)$,

$$C_\rho(x - y, x) = b_0(x) + b_1(x)\left(\frac{x - y}{\varrho}\right)$$

$$= \frac{m_2(x)}{D(x)} - \frac{m_1(x)}{D(x)}\left(\frac{x - y}{\varrho}\right) \qquad (3.81)$$

2D Example. In this 2D example, the window function is chosen as a 2D cubic spline box, i.e. the Cartesian product of two 1D Cubic spline function, $\Phi(\boldsymbol{x}) = \phi(x_1)\phi(x_2)$, where $\boldsymbol{x} = x_1\mathbf{e}_1 + x_2\mathbf{e}_2$. A bilinear polynomial basis, $\mathbf{P}(\boldsymbol{x}) = (1, x_1, x_2, x_1x_2)$, is used in the construction of moment matrix,

$$\mathbf{M}_\varrho(x) = \begin{pmatrix} m_{00} & m_{10} & m_{01} & m_{11} \\ m_{10} & m_{20} & m_{11} & m_{21} \\ m_{01} & m_{11} & m_{02} & m_{12} \\ m_{11} & m_{21} & m_{12} & m_{22} \end{pmatrix} \qquad (3.82)$$

where

$$m_{ij}(\boldsymbol{x}) = \sum_{I \in \Lambda} \left(\frac{x_1 - x_{1I}}{\varrho_{x_1}}\right)^i \left(\frac{x_2 - x_{2I}}{\varrho_{x_2}}\right)^j \Phi_\varrho\left(\boldsymbol{x} - \boldsymbol{x}_I\right) \Delta V_I \qquad (3.83)$$

where $\varrho_{x_1} = a_{x_1}\Delta x_1$ and $\varrho_{x_2} = a_{x_2}\Delta x_2$.

It should be noted that the numerical integration of moments is carried out by nodal integration. Therefore, how to choose integration weight ΔV_I is a critical issue. In RKPM procedure, a generalized Trapezoidal rule is used in practice. Fig. 3.4 shows how the weight Δx_I is computed for 1D non-uniform particle distribution.

$$\begin{cases} \Delta x_1 = \dfrac{1}{2}\Delta x_1 \ ; \\[2mm] \Delta x_I = \dfrac{1}{2}(\Delta x_{I-1} + \Delta x_I) \qquad I = 2, \cdots, N - 1 \ ; \\[2mm] \Delta x_N = \dfrac{1}{2}\Delta x_{N-1} \ . \end{cases} \qquad (3.84)$$

Similar to the 1D case, Fig. 3.5 shows how the weight ΔV_I is computed for non-uniform particle distribution.

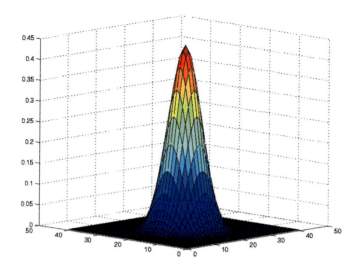

Fig. 3.2. 2D RKPM shape function.

$$
\begin{cases}
\Delta V_1 = \dfrac{1}{4}\Delta\Omega_1 \; ; \\[2mm]
\Delta V_2 = \dfrac{1}{4}(\Delta\Omega_1 + \Delta\Omega_2) \; ; \\[2mm]
\Delta V_4 = \dfrac{1}{4}(\Delta\Omega_1 + \Delta\Omega_3) \; ; \\[2mm]
\Delta V_5 = \dfrac{1}{4}(\Delta\Omega_1 + \Delta\Omega_2 + \Delta\Omega_3 + \Delta\Omega_4) \; .
\end{cases}
\tag{3.85}
$$

For 2D triangular background cell, ΔV_I is calculated based on the following formulas

$$
\begin{cases}
\Delta V_1 = \dfrac{1}{3}(\Delta\Omega_1 + \Delta\Omega_2) \; ; \\[2mm]
\Delta V_2 = \dfrac{1}{3}(\Delta\Omega_2 + \Delta\Omega_3) \; ; \\[2mm]
\Delta V_4 = \dfrac{1}{3}(\Delta\Omega_1 + \Delta\Omega_5) \; ; \\[2mm]
\Delta V_5 = \dfrac{1}{3}(\Delta\Omega_1 + \Delta\Omega_2 + \Delta\Omega_3 + \Delta\Omega_4 \\[1mm]
\qquad\quad + \Delta\Omega_5 + \Delta\Omega_6 + \Delta\Omega_7 + \Delta\Omega_8)
\end{cases}
\tag{3.86}
$$

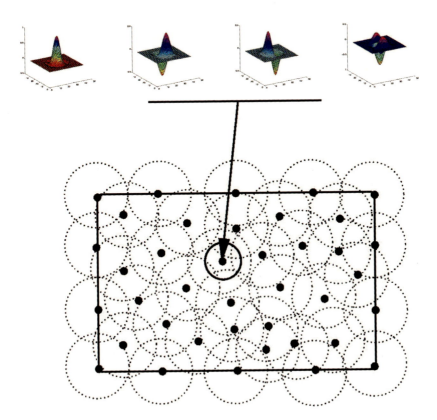

Fig. 3.3. Two-dimensional meshfree shape functions and their derivatives.

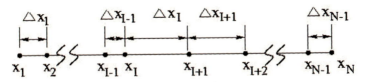

Fig. 3.4. Generalized trapezoidal rule for calculating Δx_I.

Based on the choice of ΔV_I, the following condition always holds

$$\sum_{I=1}^{NP} \Delta V_I = meas(\Omega) \tag{3.87}$$

One shall see that many other meshfree interpolants use different rules to choose integration weights, such that they do not satisfy (3.87).

3D Example. In this example a tri-linear polynomial basis is generally adopted to generate 3D RKPM shape function:

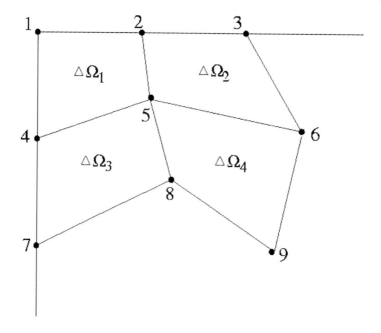

Fig. 3.5. Calculating 2D ΔV_I using the quadrilateral background cell.

$$\mathbf{P}^T(\mathbf{x}) = \{1, x_1, x_2, x_3, x_1x_2, x_2x_3, x_3x_1, x_1x_2x_3\} \ , \tag{3.88}$$

where $\mathbf{x} := (x_1, x_2, x_3)$. Recall that the shape function $N_I(\boldsymbol{x})$ can be explicitly written as:

$$N_I(\mathbf{x}, \rho) = \mathbf{P}^T\Big(\frac{\boldsymbol{x} - \boldsymbol{x}_I}{\varrho}\Big)\mathbf{b}\Big(\frac{\mathbf{x}}{\varrho}\Big)\Phi_\varrho(\mathbf{x} - \mathbf{x}_I)\Delta V_I \ . \tag{3.89}$$

The vector $\mathbf{b}(\boldsymbol{x}/\rho)$ is determined by solving the following algebraic equation:

$$\mathbf{M}_\varrho(\mathbf{x})\mathbf{b}\Big(\frac{\boldsymbol{x}}{\varrho}\Big) = \mathbf{P}(0) \ , \tag{3.90}$$

where $\mathbf{P}^t(0) = \{1, 0, \cdots, \cdots, 0, 0, 0\}$. In detail, one can write Eq. (9.32) as:

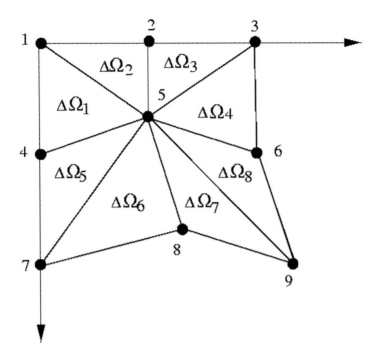

Fig. 3.6. Calculating 2D ΔV_I using the triangular background cell.

$$
\begin{pmatrix}
m^h_{000} & m^h_{100} & m^h_{010} & m^h_{001} & m^h_{110} & m^h_{011} & m^h_{101} & m^h_{111} \\[4pt]
m^h_{100} & m^h_{200} & m^h_{110} & m^h_{101} & m^h_{210} & m^h_{111} & m^h_{201} & m^h_{211} \\[4pt]
m^h_{010} & m^h_{110} & m^h_{020} & m^h_{011} & m^h_{120} & m^h_{021} & m^h_{111} & m^h_{121} \\[4pt]
m^h_{001} & m^h_{011} & m^h_{011} & m^h_{002} & m^h_{111} & m^h_{012} & m^h_{102} & m^h_{112} \\[4pt]
m^h_{110} & m^h_{210} & m^h_{120} & m^h_{111} & m^h_{220} & m^h_{121} & m^h_{211} & m^h_{221} \\[4pt]
m^h_{011} & m^h_{111} & m^h_{021} & m^h_{012} & m^h_{121} & m^h_{022} & m^h_{112} & m^h_{122} \\[4pt]
m^h_{101} & m^h_{201} & m^h_{111} & m^h_{102} & m^h_{211} & m^h_{112} & m^h_{202} & m^h_{212} \\[4pt]
m^h_{111} & m^h_{211} & m^h_{121} & m^h_{112} & m^h_{221} & m^h_{122} & m^h_{212} & m^h_{222}
\end{pmatrix}
\begin{pmatrix}
b_1(x/\varrho) \\[4pt]
b_2(x/\varrho) \\[4pt]
b_3(x/\varrho) \\[4pt]
b_4(x/\varrho) \\[4pt]
b_5(x/\varrho) \\[4pt]
b_6(x/\varrho) \\[4pt]
b_7(x/\varrho) \\[4pt]
b_8(x/\varrho)
\end{pmatrix}
=
\begin{pmatrix}
1 \\[4pt]
0 \\[4pt]
0 \\[4pt]
0 \\[4pt]
0 \\[4pt]
0 \\[4pt]
0 \\[4pt]
0
\end{pmatrix},
$$

$$(3.91)$$

where

$$m_{ijk}^{h}(\boldsymbol{x}) = \sum_{I \in \Lambda} \left(\frac{x_1 - x_{1I}}{\varrho_{x_1}}\right)^{i} \left(\frac{x_2 - x_{2I}}{\varrho_{x_2}}\right)^{j} \left(\frac{x_3 - x_{3I}}{\varrho_{x_3}}\right)^{k} \Phi_{\varrho}(\boldsymbol{x} - \boldsymbol{x}_I) \Delta V_I$$

$$\text{with} \quad i, j, k = 0, 1, 2 . \tag{3.92}$$

To visualize the spatial distribution of 3D RKPM shape function, a single shape function and its first three derivatives are plotted in Fig. 3.7. Even though the support size of the shape function is a rectangular box, one may observe that the domain of non-zero value of the shape function tends to be a sphere, and the domain of non-zero value of the derivatives of the shape function is formed by two connected spherical regions. This means that the spatial distribution of the RKPM shape function is almost "isotropic", which is a desired property in some numerical simulations, such as shear band simulation. In Fig. 3.7(a), the first octant is taken out from the quasi-sphere region, and one can see that the shape function reaches its maximum value at the corresponding particle, i.e. the central particle. In each of Figs. 3.7(b)–(d), one quadrant is taken out to see the orientation and the distribution of the shape function derivatives.

As may be seen later on, most meshfree interpolants (RKPM, EFG, h-p Clouds, finite Clouds, etc.) resemble the so-called Moving-least square interpolant (Lancaster and Salkauskas[255]). However, a major difference between RKPM interpolant and MLS interpolant is that RKPM interpolant uses the shifted basis ([297, 303]). In the shifted basis approach, a global meshfree shape function is proportional to,

$$N_I(\boldsymbol{x}) \sim \mathbf{P}^T(0) \mathbf{M}^{-1}(\boldsymbol{x}) \mathbf{P}(\boldsymbol{x} - \boldsymbol{x_I}) \Phi(\boldsymbol{x} - \boldsymbol{x_I}) \tag{3.93}$$

whereas for the non-shifted base, a meshfree shape function is actually proportional to

$$N_I(\boldsymbol{x}) \sim \mathbf{P}^T(\boldsymbol{x}) \mathbf{M}^{-1}(\boldsymbol{x}) \mathbf{P}(\boldsymbol{x_I}) \Phi(\boldsymbol{x} - \boldsymbol{x_I}) \tag{3.94}$$

Note that the calculations of the moment matrix is also different in these two approaches.

In a paper by Jin et al.,[221] the differences between shifted basis and non-shifted basis are studied. They found that taking into account the effect of correction function both shifted basis and non-shifted basis are mathematically equivalent. However, in discretized computations, the two bases show very different behaviors. When the particle number increases in a support size, the condition number of the non-shifted basis deteriorates drastically.

The table below is taken from the paper from Jin et al.,[221] in which a comparison in numerical computations between non-shifted basis and shifted basis is made in calculating the consistency condition of a meshfree interpolant.

One may find from Table (3.1) that as the number of particles inside a support increases the numerical consistency condition deteriorates drastically. Therefore, the shifted basis approach will yield a much better condition

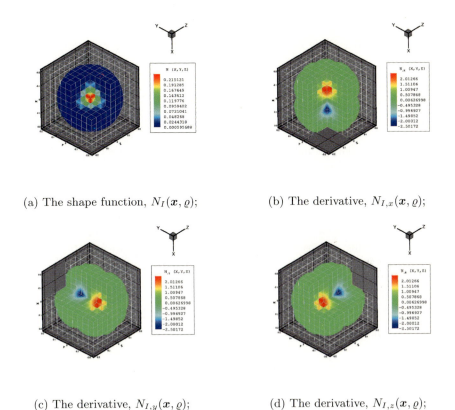

(a) The shape function, $N_I(\boldsymbol{x}, \varrho)$;

(b) The derivative, $N_{I,x}(\boldsymbol{x}, \varrho)$;

(c) The derivative, $N_{I,y}(\boldsymbol{x}, \varrho)$;

(d) The derivative, $N_{I,z}(\boldsymbol{x}, \varrho)$;

Fig. 3.7. 3D RKPM shape function and its first derivatives.

number of nodes	$\sum N_{I,x} x_I = 1.0$		$\sum N_{I,xx} x_I^2 = 2.0$		$\sum N_{I,xxx} x_I^2 = 6.0$	
	non-shifted	shifted	non-shifted	shifted	non-shifted	shifted
41	1.000	1.000	2.000	2.000	6.00035	6.000
81	1.000	1.000	1.99999	2.000	6.00473	6.000
161	0.999974	1.000	1.9994	2.000	8.32153	6.000
321	1.00185	1.000	2.01481	2.000	258.451	6.000
641	1.00185	1.000	2.00257	2.000	492.394	6.000
1281	1.00185	1.000	2.06534	2.000	492.394	5.999
2561	1.00185	1.000	0.387417	2.000	492.394	6.00161
5121	1.00185	1.000	0.6685.12	2.000	-25771	6.00752

Table 3.1. Comparison of numerical consistency condition calculated via shifted basis and via un-shifted basis (From Jin et al. [2001]).

number than that of un-shifted basis approach. Based on our computational experience ([297, 303]), the shifted RKPM shape function provides a better and convergent numerical solution than that of the un-shifted MLS interpolant.

3.1.6 Conservation Properties of RKPM Interpolant

In contrast to SPH, RKPM interpolant is used in a Galerkin weak formulation to solve partial differential equations. Since RKPM can be viewed as either a filter, or smoothed sampling procedure, i.e.

$$Ru : C^0(\Omega) \rightarrow C^n(\Omega) \tag{3.95}$$

Thus, the Galerkin formulation invoking RKPM interpolant can be deemed as smoothed Galerkin method (by which we mean that the requirement on the order of the continuity of interpolant is even stronger than that of continuous Galerkin). One of the setbacks of using such smoothed interpolant is that the formulation may lose its local conservative properties. Besides its smoothness, the non-local character of RKPM interpolant is another factor that destroys the local conservative properties of the Galerkin formulation. However, globally, it may still be able to preserve conservation properties of the PDE system that it solves. To demonstrate this point, we consider a solid mechanics problem,

$$\nabla \cdot \boldsymbol{\sigma} + \mathbf{b} = \rho \ddot{\mathbf{u}} \tag{3.96}$$

with only traction boundary condition

$$\boldsymbol{\sigma} \cdot \mathbf{n} = \bar{\mathbf{T}} , \quad \forall \, \boldsymbol{x} \in \Gamma_t \tag{3.97}$$

Denote the Gauss quadrature integration of a field $\mathcal{A}$ over a volume Ω and over a surface Γ_t as

$$\oint_\Omega \mathcal{A} d\Omega = \sum_{ik=1}^{mgk} \mathcal{A}(\boldsymbol{x}_{ik})\omega_{ik}, \quad \boldsymbol{x}_{ik} \in \Omega \tag{3.98}$$

$$\oiint_{\Gamma_t} \mathcal{A} dS = \sum_{ib=1}^{bgk} \mathcal{A}(\boldsymbol{x}_{ib})\omega_{ib}, \quad \boldsymbol{x}_{ib} \in \Gamma_t \tag{3.99}$$

where ω_{ik} and ω_{ib} are the Gauss quadrature weight for volume integration and surface integration respectively, and ik and ib are the index of Guass quadrature points in the domain Ω and on the traction boundary Γ_t respectively, and mgk, bgk are the total number of Gauss quadrature points employed within those domains accordingly.

Let

$$\mathbf{U}^\varrho = \{\mathbf{u} \,\big|\, \mathbf{u} \in H^1(\Omega) \,\&\, \mathbf{u} \in span\{N_I(\boldsymbol{x}, \varrho)\}, \; I \in \Lambda\} \tag{3.100}$$

The Galerkin variational formulation of the problem (3.96) and (3.97) can be stated as

Find $\mathbf{u}^\varrho$ *such that*

$$\oint_{\Omega} \left(\delta \mathbf{u}^{\varrho}\right)^T \cdot \rho \ddot{\mathbf{u}}^{\varrho} d\Omega + \oint_{\Omega} \left(\nabla_s \delta \mathbf{u}^{\varrho}\right)^T : \sigma^{\varrho} d\Omega - \oint_{\Omega} \left(\delta \mathbf{u}^{\varrho}\right)^T \cdot \mathbf{b} d\Omega$$
$$- \oint_{\Gamma_t} \left(\delta \mathbf{u}^{\varrho}\right)^T \cdot \bar{\mathbf{T}} dS = 0 , \qquad\qquad \forall\ \delta \mathbf{u}^{\varrho} \in \mathbf{U}^{\varrho} \qquad (3.101)$$

This can be further translated into a set of algebraic equations

$$\oint_{\Omega} \rho \ddot{u}_i^{\varrho}(\mathbf{x}) N_I(\mathbf{x}, \varrho) d\Omega = - \oint_{\Omega} \sigma_{ij}^{\varrho}(\mathbf{x}) N_{I,j}(\mathbf{x}, \varrho) d\Omega$$
$$+ \oint_{\Omega} b_i(\mathbf{x}) N_I(\mathbf{x}, \varrho) d\Omega + \oint_{\Gamma_t} \bar{T}(\mathbf{x}) N_I(\mathbf{x}, \varrho) dS ; \quad I \in \Lambda \qquad (3.102)$$

As part of the conservation properties, the exact solution of problem (3.96) – (3.97) preserves both linear momentum as well as angular momentum,

$$\frac{d}{dt} \int_{\Omega} \rho \mathbf{v} d\Omega = \int_{\Omega} \rho \frac{d\mathbf{v}}{dt} d\Omega = \int_{\Omega} \mathbf{b} d\Omega + \int_{\Gamma_t} \bar{\mathbf{T}} dS \qquad (3.103)$$

$$\frac{d}{dt} \int_{\Omega} \mathbf{x} \times \rho \mathbf{v} d\Omega = \int_{\Omega} \mathbf{x} \times \rho \frac{d\mathbf{v}}{dt} d\Omega = \int_{\Omega} \mathbf{x} \times \mathbf{b} d\Omega + \int_{\Gamma_t} \mathbf{x} \times \bar{\mathbf{T}} dS \quad (3.104)$$

We claim that the numerical solution of (3.101), or (3.102) preserves discrete linear momentum, if and only if $\sum_{I \in \Lambda} N_I(\mathbf{x}, \varrho) = 1$, i.e.

$$\oint_{\Omega} \rho \frac{d\mathbf{v}^{\varrho}}{dt} d\Omega = \oint_{\Omega} \mathbf{b} d\Omega + \oint_{\Gamma_t} \bar{\mathbf{T}} dS \qquad (3.105)$$

and that the numerical solution of (3.101), or (3.102) preserves discrete angular momentum, if and only if $\sum_{I \in \Lambda} \mathbf{x}_I^k N_I(\mathbf{x}, \varrho) = \mathbf{x}^k$, $|k| = 0, 1$, i.e.

$$\oint_{\Omega} \mathbf{x} \times \rho \frac{d\mathbf{v}^{\varrho}}{dt} d\Omega = \oint_{\Omega} \mathbf{x} \times \mathbf{b} d\Omega + \oint_{\Gamma_t} \mathbf{x} \times \bar{\mathbf{T}} dS \qquad (3.106)$$

We first show (3.105).

$$\oint_{\Omega} \rho \ddot{v}_i^{\varrho} d\Omega = \oint_{\Omega} \rho \ddot{v}_i^{\varrho} \cdot (1) \cdot d\Omega = \oint_{\Omega} \rho \ddot{v}_i^{\varrho} \sum_{I \in \Lambda} N_I(\mathbf{x}) d\Omega$$
$$= \sum_{I \in \Lambda} \oint_{\Omega} \rho \ddot{v}_i^{\varrho} N_I(\mathbf{x}) d\Omega = - \sum_{I \in \Lambda} \oint_{\Omega} \sigma_{ij}^{\varrho}(\mathbf{x}) N_{I,j}(\mathbf{x}) d\Omega$$
$$+ \sum_{I \in \Lambda} \oint_{\Omega} b_i(\mathbf{x}) N_I(\mathbf{x}) d\Omega + \sum_{I \in \Lambda} \oint_{\Gamma_t} \bar{T}(\mathbf{x}) N_I(\mathbf{x}) dS \qquad (3.107)$$

By virtue of $\sum_{I \in \Lambda} N_I(\mathbf{x}, \varrho) = 1$, one has that $\sum_{I \in \Lambda} N_{I,j}(\mathbf{x}, \varrho) = 0$. It then leads to

$$\sum_{I \in \Lambda} \oint_{\Omega} \sigma_{ij}^{\varrho}(\boldsymbol{x}) N_{I,j}(\boldsymbol{x}) d\Omega = \oint_{\Omega} \sigma_{ij}^{\varrho}(\boldsymbol{x}) \sum_{I \in \Lambda} N_{I,j}(\boldsymbol{x}) d\Omega = 0 \qquad (3.108)$$

Therefore

$$\oint_{\Omega} \rho \dot{v}_i^{\varrho} d\Omega = \oint_{\Omega} b_i(\boldsymbol{x}) \sum_{I \in \Lambda} N_I(\boldsymbol{x}) d\Omega + \oint_{\Gamma_t} \bar{\mathbf{T}}(\boldsymbol{x}) \sum_{I \in \Lambda} N_I(\boldsymbol{x}) dS$$

$$= \oint_{\Omega} b_i(\boldsymbol{x}) d\Omega + \oint_{\Gamma_t} \bar{\mathbf{T}}(\boldsymbol{x}) dS \qquad (3.109)$$

Now we show (3.106). From the reproducing condition

$$\sum_{I \in \Lambda} N_I(\boldsymbol{x}, \varrho) x_{jI} = x_j, \qquad (3.110)$$

we have

$$\oint_{\Omega} \epsilon_{ijk} x_j \rho \dot{v}_k d\Omega = \oint_{\Omega} \epsilon_{ijk} \Big(\sum_{I \in \Lambda} N_I(\boldsymbol{x}, \varrho) x_{jI} \Big) \rho \dot{v}_k d\Omega$$

$$= \sum_{I \in \Lambda} \epsilon_{ijk} x_{jI} \left\{ \oint_{\Omega} \rho \dot{v}_k N_I(\boldsymbol{x}, \varrho) d\Omega \right\}$$

$$= \sum_{I \in \Lambda} \epsilon_{ijk} x_{jI} \left\{ - \oint_{\Omega} \sigma_{k\ell}^{\varrho}(\boldsymbol{x}) N_{I,\ell}(\boldsymbol{x}, \varrho) d\Omega \right.$$

$$\left. + \oint_{\Omega} b_k(\boldsymbol{x}) N_I(\boldsymbol{x}, \varrho) d\Omega + \oint_{\Gamma_t} \bar{T}_k(\boldsymbol{x}) N_I(\boldsymbol{x}, \varrho) dS \right\}$$

$$= - \oint_{\Omega} \epsilon_{ijk} \sigma_{k\ell}^{\varrho}(\boldsymbol{x}) \Big(\sum_{I \in \Lambda} x_{jI} N_{I,\ell}(\boldsymbol{x}, \varrho) \Big) d\Omega$$

$$+ \oint_{\Omega} b_k(\boldsymbol{x}) \Big(\sum_{I \in \Lambda} x_{jI} N_I(\boldsymbol{x}, \varrho) \Big) d\Omega$$

$$+ \oint_{\Gamma_t} \bar{T}_k(\boldsymbol{x}) \Big(\sum_{I \in \Lambda} x_{jI} N_I(\boldsymbol{x}, \varrho) \Big) dS \qquad (3.111)$$

By virtue of

$$\sum_{I \in \Lambda} N_{I,\ell} x_{jI} = \delta_{j\ell} \qquad (3.112)$$

$$\oint_{\Omega} \epsilon_{ijk} \sigma_{k\ell}^{\varrho} \Big(N_{I,\ell} x_{jI} \Big) d\Omega = \oint_{\Omega} \epsilon_{ijk} \sigma_{k\ell}^{\varrho}(\boldsymbol{x}) \delta_{j\ell} d\Omega$$

$$= \oint_{\Omega} \epsilon_{ijk} \sigma_{kj}^{\varrho} d\Omega = 0 \qquad (3.113)$$

Considering (3.110), we finally have

$$\oint \epsilon_{ijk} x_j \rho \ddot{v}_k d\Omega = \oint \epsilon_{ijk} b_k(\boldsymbol{x}) \Big(\sum_{I \in \Lambda} x_{jI} N_I(\boldsymbol{x}, \varrho)\Big) d\Omega$$

$$+ \oint_{\Gamma_t} \epsilon_{ijk} \bar{T}_k(\boldsymbol{x}) \Big(\sum_{I \in \Lambda} x_{jI} N_I(\boldsymbol{x}, \varrho)\Big) dS$$

$$= \oint \epsilon_{ijk} b_k(\boldsymbol{x}) x_j d\Omega + \oint_{\Gamma_t} \epsilon_{ijk} \bar{T}_k(\boldsymbol{x}) x_j dS, \qquad (3.114)$$

which is the desired result.

3.1.7 One-dimensional Model Problem

Consider the following two-point boundary value problem which describes the wave propagation in a bar of length L. The governing equation is given as

$$\frac{d^2 F}{dx^2} + k^2 F = q(x) , \qquad (3.115)$$

where k is the wave number, and the boundary conditions are $F(0) = f_0$ and $F(L) = f_L$. Using integration by parts, the weak form can be written as

$$- \int_0^L \frac{d\delta F}{dx} \frac{dF}{dx} dx + \frac{dF}{dx} \delta F \Big|_0^L + \int_0^L k^2 \delta F \, F dx = \int_0^L \delta F \, q(x) dx , \quad (3.116)$$

Let

$$F \in \mathcal{U} = \{F \in H^1([0, L]) \mid F(0) = f_0, F(L) = f_L\} ; \qquad (3.117)$$

$$\delta F \in \mathcal{V} = \{\delta F \in H^1([0, L]) \mid \delta F(0) = \delta F(L) = 0\} . \qquad (3.118)$$

Eq. (3.116) can be simplified further

$$- \int_0^L \frac{d\delta F}{dx} \frac{dF}{dx} dx + \int_0^L k^2 \delta F \, F dx = \int_0^L \delta F \, F(x) dx , \qquad (3.119)$$

Since RKPM representation is non-interpolating in general, construction of a finite dimensional space, such that

$$F^\varrho \in \mathcal{U}^\varrho = \{F \in \ span\{N_I(x)\} \,\Big|\, F(0) = f_0, \ F(L) = f_L\} ; \qquad (3.120)$$

$$\delta F^\varrho \in \mathcal{V}^\varrho = \{\delta F \in \ span\{W_I(x)\} \mid \delta F(0) = \delta F(L) = 0\} , \qquad (3.121)$$

is not trivial.

We first assume that

$$F(x) = \sum_{I \in \Lambda} N_I(x) F_I \qquad (3.122)$$

where $\{N_I(x)\}$ are 1D RKPM shape functions. In general, $\{F_I\}_{I=1}^{N}$ are not completely independent because of non-interpolating characteristic on boundary [2]. To essential boundary conditions, we set

$$f_0 = N_1(L)F_1 + \sum_{I=2}^{N-1} N_I(0)F_I + N_N(0)F_N \; ; \tag{3.123}$$

$$f_L = N_1(L)F_1 + \sum_{I=2}^{N-1} N_I(L)F_I + N_N(L)F_N \; . \tag{3.124}$$

where $F_I \neq F(\boldsymbol{x}_I)$ in general.

Using the fact that $N_N(0) = N_1(L) = 0$, the nodal value F_1 and F_N are solved as

$$F_1 = \frac{f_0}{N_1(0)} - \sum_{I=2}^{N-1} \frac{N_I(0)}{N_1(0)} F_I \; ; \tag{3.125}$$

$$F_N = \frac{f_L}{N_N(L)} - \sum_{I=2}^{N-1} \frac{N_I(L)}{N_N(L)} F_I \; . \tag{3.126}$$

Substituting Eqs. (3.125) and (3.126) into Eq. (3.122) yields

$$
\begin{aligned}
F(x) &= N_1(x) \left[\frac{f_0}{N_1(0)} - \sum_{I=2}^{N-1} \frac{N_I(0)}{N_1(0)} F_I \right] + \sum_{I=2}^{N-1} N_I(x)F_I \\
&\quad + N_N(x) \left[\frac{f_L}{N_N(L)} - \sum_{I=2}^{N-1} \frac{N_I(L)}{N_N(L)} F_I \right] \\
&= \frac{f_0}{N_1(0)} N_1(x) + \sum_{I=2}^{N-1} \left[N_I(x) - \frac{N_1(x)N_I(0)}{N_1(0)} - \frac{N_N(x)N_I(L)}{N_N(L)} \right] F_I \\
&\quad + \frac{f_L}{N_N(L)} N_N(x) \\
&= tgZ \; N_1(x) + \sum_{I=2}^{N-1} W_I(x)F_I + tgL \; N_N(x) \; .
\end{aligned}
\tag{3.127}
$$

where $tgZ = \dfrac{f_0}{N_1(0)}$ and $tgL = \dfrac{f_L}{N_N(L)}$, and

$$W_I(x) = \left[N_I(x) - \frac{N_1(x)N_I(0)}{N_1(0)} - \frac{N_N(x)N_I(L)}{N_N(L)} \right] \tag{3.128}$$

Consequently,

---

[2] However, exceptions exist as shown in Fig. 3.1, the matter shall be further discussed in Chpater 4.

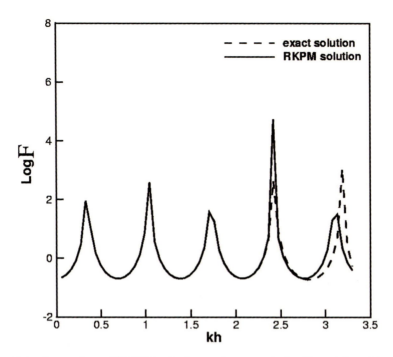

Fig. 3.8. Comparison of RKPM solution and exact solution for a 1D problem.

$$\delta F(x) = \sum_{I=2}^{N-1} W_I(x)\delta F_I \tag{3.129}$$

The final discrete equation can be obtained by substituting the weight function Eq. (3.129) and the trial function Eq. (3.122) into the weak form Eq. (3.119), which yields

$$K_{IJ}F_J = f_I , \tag{3.130}$$

where

$$K_{IJ} = \int_\Omega k^2 W_I(x)W_J(x)dx - \int_\Omega k^2 W_{I,x}(x)W_{J,x}(x)dx \tag{3.131}$$

$$f_I = \int_\Omega \bigg[W_{I,x}\big((tgZN_{1,x}(x) + tgLN_{N,x}(x)\big)$$

$$-k^2 W_I\big(tgZN_1(x) + tgLN_N(x)\big)\bigg] dx . \tag{3.132}$$

where the source term $q(x) = 0$ is assumed.

These integrals are evaluated by integrating over the domain of the problem using Gauss quadratures. The shape function and its derivatives are calculated at each quadrature point x_G, and then the contribution to the stiffness matrix $\mathbf{K}$ and the force vector $\mathbf{f}$ are assembled. After solving for the nodal coefficients F_I (note that this is not the same as the nodal value), the nodal value has to be recovered by interpolating the nodal coefficients.

The analytical solution of this one-dimensional problem is

$$F(x) = \frac{sin[k(L-x)]}{sin(kL)} . \qquad (3.133)$$

In the numerical study, the bar length L is taken to be 1.0, and 11 particles are used in the RKPM computation. Fig. 3.8 shows the comparison between the RKPM solution and the exact solution at $x = 0.5$. The RKPM solution yields excellent agreement for low wave number. However, as the wave number k increases, discrepancies start to build up from the resolution limit $k = \pi$. By adjusting the dilation parameter, the solution can be improved to certain extent. Detailed study on the performance of RKPM for this model problem can be found in Liu & Chen[299] and Li & Liu.[267]

3.1.8 Program Description

A computer program that solves this one-dimensional model problem is provided in Appendix A. The program is written in FORTRAN77 with certain degrees of generality. It serves as a platform for the readers to familiarize themselves with the whole implementation process of RKPM since it directly relates to the formulation presented in this chapter. To facilitate the understanding, a flow chart of the program is outlined as follows:

- Preprocessing that includes reading input file, calculating global control parameters, and initializing matrices;
- Setting up nodal coordinates, integration cells, nodal integration weights, and Jacobian;
- Loop over integration cells, compute shape function, its derivatives, and the corresponding stiffness term and force term, assemble them to the stiffness matrix and the force vector;
- Solve for the nodal coefficients;
- Recover the true nodal value and plot the results.

The program is structured following the flow chart, each individual subroutine contains self-explanatory comments to make it readable. After preprocessing, the nodal coordinates are set up. They are uniformly distributed to simplify the program which leads to a constant nodal dilation parameter. An alternative way of assigning nodal dilation parameters for the case of non-uniform nodal distribution is also included in the program (see Eq.(3.84) for details). Then, the nodal integration weights are calculated followed by setting up Gauss quadrature points and their weights. Before starting looping over the integration cells, the boundary shape functions $N_I(0)$ and $N_I(L)$ are computed and stored in vector **shpbZ(mnode)** and vector **shpbL(mnode)**. These vectors will be used later to modify the original shape function.

To assemble the discrete equations, the program loops over each integration point x_G, the shape functions $N_1(x_G)$ and $N_N(x_G)$ are evaluated first, then the modified shape functions $N_I(x_G)$, $N_J(x_G)$, and their derivatives are formed in subroutine **shrgd1a** in which the modification is done based on Eq. (3.128). The stiffness term k_{IJ} and the force term f_I are calculated according to Eq. (3.131) and Eq. (3.132). Once the stiffness matrix $\mathbf{K}$ and the force vector $\mathbf{f}$ are assembled, the standard equation solver (subroutines **dgeco** and **dgesl**) is used to solve for nodal coefficients F_I. In order to obtain the nodal value of $F(x)$, one has to interpolate the nodal coefficients. Note that the interpolation is done using the original shape functions since the nodal coefficients F_1 and F_N are recovered. Finally, the comparison is made between the exact solution and the RKPM numerical solution.

3.2 Meshfree Wavelet Interpolant

Moving least square reproducing kernel interpolant was generalized by Li and Liu.[266, 267] A class of spectral meshfree kernel functions are obtained. They can form a hierarchical partition of unity with increasing spectral contents. Such meshfree spectral basis is called as meshfree wavelets, which have been used in computational fluid dynamics, simulations of strain localization, numerical computation of wave propagations, etc.

3.2.1 Variation in a Theme: Generalized Moving Least Square Reproducing Kernel

As shown in ([303]), a local least square approximation of a continuous function, $u(x) \in C^0(\Omega)$, may be expressed as,

$$L_{\bar{x}} u(x) := \mathbf{P}\left(\frac{x - \bar{x}}{\varrho}\right) \mathbf{d}(\bar{x}) , \qquad \forall \, \bar{x} \in \Omega , \qquad (3.134)$$

where $\mathbf{P}$ is a polynomial basis with order p, and $\mathbf{d}$ is an unknown vector. Without loss of generality, the polynomial basis is assumed to have $\ell = N_p$ terms, namely,

$$\mathbf{P}(x) := (P_1(x), \cdots, P_i(x), \cdots, P_\ell(x)) , \qquad P_i(x) \in \pi_p(\Omega) , \qquad (3.135)$$

with $P_1 = 1$; $\quad P_i(0) = 0$, $\quad i \neq 1$, where $\pi_p(\Omega)$ denotes the collection of polynomials in $\Omega \subset \mathbb{R}^d$ of total degree $\leq p$. The unknown vector $\mathbf{d}(\bar{x})$ can be solved in the moving least square procedure, and subsequently Eq. (3.134) can be rewritten as (compare with Eq. (3.16)),

$$L_{\bar{x}} u(x) = \mathbf{P}\left(\frac{x - \bar{x}}{\varrho}\right) \mathbf{M}^{-1}(\bar{x}) \int_{\Omega_y} \mathbf{P}^t\left(\frac{y - \bar{x}}{\varrho}\right) u(y) \Phi_\varrho(y - \bar{x}) d\Omega_y \quad (3.136)$$

where the moment matrix $\mathbf{M}$ is

$$\mathbf{M}(x) := \int_{\Omega_y} \mathbf{P}^t\left(\frac{x - y}{\varrho}\right) \mathbf{P}\left(\frac{x - x}{\varrho}\right) \Phi_\varrho(x - y) d\Omega_y . \qquad (3.137)$$

Remark 3.2.1. In formula (3.136), the components of the polynomial vector $\mathbf{P}(x)$ can be any independent polynomial functions. In fact, the requirement of the polynomials is also not essential. The construction can be further generalized to include general linearly independent functions as basis, such as trigonometric functions, hyperbolic functions, and any other orthogonal or non-orthogonal basis functions.

Up to this point, all steps follow the moving least square reproducing procedure. To construct the new interpolant, instead of assigning the local approximation as what MLS does, i.e.

$$u^\ell(\boldsymbol{x}, \bar{\boldsymbol{x}}) := L_{\bar{\boldsymbol{x}}} u(\boldsymbol{x}) \ , \tag{3.138}$$

Let d denote the dimension of space. $\ell = N_p = \begin{pmatrix} p + d \\ d \end{pmatrix}$. Then

$$P_i(\boldsymbol{x}) = (\frac{\boldsymbol{x} - \bar{\boldsymbol{x}}}{\varrho})^\alpha \ , \qquad 0 \le |\alpha| \le p \ , \ 1 \le i \le \ell \tag{3.139}$$

with $P_1 = 1$ and $P_\ell(\boldsymbol{x}) = (\frac{\boldsymbol{x} - \bar{\boldsymbol{x}}}{\varrho})^\mu$, $|\mu| = p$.

Choose a different local approximation,

$$u^\ell(\boldsymbol{x}, \bar{\boldsymbol{x}}) := \sum_{|\alpha| \le p} \frac{C_\alpha(\bar{\boldsymbol{x}})}{\alpha!} D^\alpha \big(L_{\bar{\boldsymbol{x}}} u(\boldsymbol{x})\big) \varrho^\alpha$$

$$= L_{\bar{\boldsymbol{x}}} u(\boldsymbol{x}) + \sum_{1 \le |\alpha| \le p} \frac{C_\alpha(\bar{\boldsymbol{x}})}{\alpha!} D^\alpha \big(L_{\bar{\boldsymbol{x}}} u(\boldsymbol{x})\big) \varrho^\alpha \tag{3.140}$$

where $C_0 = 1$, and $C_\alpha(\bar{\boldsymbol{x}})$, $|\alpha| \le m$, are given functions.

Intuitively, the new local approximation (3.140) can be viewed as a truncated Taylor series by taking $C_\alpha(\boldsymbol{x}) = 1$, $\forall \alpha$. By substituting (3.136) into (3.140), the local approximation can explicitly be expressed as

$$u^\ell(\boldsymbol{x}, \bar{\boldsymbol{x}}) = \mathbf{P}(\frac{\boldsymbol{x} - \bar{\boldsymbol{x}}}{\varrho}) \mathbf{M}^{-1}(\bar{\boldsymbol{x}}) \int_{\Omega_y} \mathbf{P}^t(\frac{\boldsymbol{y} - \bar{\boldsymbol{x}}}{\varrho}) u(\boldsymbol{y}) \phi_\varrho(\boldsymbol{y} - \bar{\boldsymbol{x}}) d\Omega_y$$

$$+ \sum_{1 \le |\alpha| \le p} \frac{C_\alpha(\bar{\boldsymbol{x}}) \varrho^\alpha}{\alpha!} D^\alpha \Big(\mathbf{P}(\frac{\boldsymbol{x} - \bar{\boldsymbol{x}}}{\varrho}) \Big) \mathbf{M}^{-1}(\bar{\boldsymbol{x}})$$

$$\cdot \int_{\Omega_y} \mathbf{P}^T(\frac{\boldsymbol{y} - \bar{\boldsymbol{x}}}{\varrho}) u(\boldsymbol{y}) \phi_\varrho(\boldsymbol{y} - \bar{\boldsymbol{x}}) d\Omega_y \tag{3.141}$$

To globalize the approximation, we apply the moving procedure to (3.141),

$$Gu(\boldsymbol{x}) := \lim_{\bar{\boldsymbol{x}} \to \boldsymbol{x}} u^\ell(\boldsymbol{x}, \bar{\boldsymbol{x}}) \ , \tag{3.142}$$

which yields the following global approximation

$$u(\boldsymbol{x}) \approx Gu(\boldsymbol{x}) = \mathbf{P}^{(0)}(0) \mathbf{M}^{-1}(\boldsymbol{x}) \int_{\Omega_y} \mathbf{P}^t(\frac{\boldsymbol{y} - \boldsymbol{x}}{\varrho}) u(\boldsymbol{y}) \Phi_\varrho(\boldsymbol{x} - \boldsymbol{y}) d\Omega_y$$

$$+ C_1(\boldsymbol{x}) \mathbf{P}^{(1)}(0) \mathbf{M}^{-1}(\boldsymbol{x}) \int_{\Omega_y} \mathbf{P}^t(\frac{\boldsymbol{y} - \boldsymbol{x}}{\varrho}) u(\boldsymbol{y}) \Phi_\varrho(\boldsymbol{y} - \boldsymbol{x}) d\Omega_y$$

$$+ \quad \cdots\cdots\cdots$$

$$+ C_\mu(\boldsymbol{x}) \mathbf{P}^{(\mu)}(0) \mathbf{M}^{-1}(\boldsymbol{x}) \int_{\Omega_y} \mathbf{P}^t(\frac{\boldsymbol{y} - \boldsymbol{x}}{\varrho}) u(\boldsymbol{y}) \Phi_\varrho(\boldsymbol{y} - \boldsymbol{x}) d\Omega_y \ , \tag{3.143}$$

where $\mathbf{P}^{(\alpha)}(0) := \frac{1}{\alpha!} D^\alpha \mathbf{P}(\frac{\boldsymbol{x}}{\varrho}) \varrho^\alpha \big|_{\boldsymbol{x}=0}$, $0 \le |\alpha| \le p$. $\mathbf{P}^{(\alpha)}(0)$ has particular simple structure,

$$\mathbf{P}^{(0)}(0) = \big(\underbrace{1, 0, \cdots, 0, \cdots \cdots, 0}_{\ell}\big)$$

$$\mathbf{P}^{(1)}(0) = \big(0, \underbrace{1, 0, \cdots \cdots \cdots, 0}_{\ell-1}\big)$$

$$\cdots \cdots \cdots$$

$$\mathbf{P}^{(\alpha)}(0) = \big(0, \cdots, 0, \underbrace{1, 0, \cdots, 0}_{\ell-i}\big)$$

$$\cdots \cdots \cdots$$

$$\mathbf{P}^{(\mu)}(0) = \big(0, 0, \cdots, \cdots, 0, 1\big) \qquad (3.144)$$

The generalized reproducing kernel representation is then expressed as

$$\bar{\mathcal{R}}_\varrho^m u(\boldsymbol{x}) = Gu(\boldsymbol{x}) = \sum_{|\alpha| \leq p} C_\alpha(\boldsymbol{x}) \left\{ \int_\Omega \mathbf{P}(\frac{\boldsymbol{y}-\boldsymbol{x}}{\varrho}) u(\boldsymbol{y}) \Phi_\varrho(\boldsymbol{x}-\boldsymbol{y}) d\Omega_y \right\} \boldsymbol{b}^{(\alpha)}(\boldsymbol{x})$$

$$= \sum_{|\alpha| \leq p} C_\alpha(x) \int_\Omega \mathcal{K}_\varrho^{[\alpha]}(\boldsymbol{x}-\boldsymbol{y}, \boldsymbol{x}) u(\boldsymbol{y}) d\Omega_y \qquad (3.145)$$

where $\mathcal{K}_\varrho^{[\alpha]}$ is the α-th kernel,

$$\mathcal{K}_\varrho^{[\alpha]}(\boldsymbol{x}-\boldsymbol{y}, \boldsymbol{x}) := \mathbf{P}(\frac{\boldsymbol{y}-\boldsymbol{x}}{\varrho}) \boldsymbol{b}^{(\alpha)}(\boldsymbol{x}) \Phi_\varrho(\boldsymbol{x}-\boldsymbol{y}) \qquad \forall\, 0 \leq |\alpha| \leq m \qquad (3.146)$$

and $\boldsymbol{b}^{(\alpha)}(\boldsymbol{x})$ is determined by algebraic equations

$$\mathbf{M}(\boldsymbol{x}) \boldsymbol{b}^{(\alpha)}(\boldsymbol{x}) = \{\mathbf{P}^{(\alpha)}(0)\}^t \;. \qquad (3.147)$$

namely,

$$\boldsymbol{b}^{(\alpha)}(\boldsymbol{x}) = \frac{1}{\Delta} \left\{ (-1)^{1+i} A_{1i}(\boldsymbol{x}), (-1)^{2+i} A_{2i}(\boldsymbol{x}), \cdots, (-1)^{\ell+i} A_{\ell i}(\boldsymbol{x}) \right\}^t \quad (3.148)$$

where $\Delta = det\mathbf{M}$ and A_{ij} are the minors of the global moment matrix $\mathbf{M}(\boldsymbol{x})$. Note that since the moment matrix $\mathbf{M}(\boldsymbol{x})$ is symmetric

$$\mathbf{P}^{(\alpha)}(0)\mathbf{M}^{-1}(\boldsymbol{x})\mathbf{P}^t(\frac{\boldsymbol{y}-\boldsymbol{x}}{\varrho}) = \mathbf{P}(\frac{\boldsymbol{y}-\boldsymbol{x}}{\varrho})\mathbf{M}^{-1}(\boldsymbol{x})\{\mathbf{P}^{(\alpha)}(0)\}^t.$$

If $C_\alpha = 0$, $\forall\, |\alpha| \neq 0$, (3.145) recovers the regular RKPM representation,[303]

$$\mathcal{R}_\varrho^m u(\boldsymbol{x}) = \int_{\Omega_y} \mathcal{K}_\varrho^{[0]}(\boldsymbol{x}-\boldsymbol{y}, \boldsymbol{x}) u(\boldsymbol{y}) d\Omega_y \;. \qquad (3.149)$$

When $\alpha \neq 0$, a set of new kernel functions are derived, which are called meshfree wavelet functions, because they are actually pre-wavelet functions in continuous integral representation. To justify such a claim, we examine the moment conditions of meshfree kernel functions.

Equivalently, Eq. (3.147) can be interpreted as the following β-scale consistency conditions

$$\int_\Omega (\frac{\boldsymbol{x} - \boldsymbol{y}}{\varrho})^\beta \mathcal{K}_\varrho^{[\alpha]}(\boldsymbol{x} - \boldsymbol{y}, \boldsymbol{x}) d\Omega_y = \delta_{\alpha\beta} , \qquad 1 \leq |\beta| \leq m \qquad (3.150)$$

When $\Omega = \mathbb{R}^d$ and $\varrho = 1$, $\boldsymbol{b}^{(\alpha)}(\boldsymbol{x}) = const.$ and $\mathcal{K}_\varrho^{[\alpha]}(\cdot, \cdot) \equiv \mathcal{K}^{[\alpha]}(\cdot)$ [3] . Closely examining (3.150), one may find that

$$\int_{\mathbb{R}^n} \mathcal{K}^{[0]}(z) d\Omega_z = 1, \text{ and } \int_{\mathbb{R}^n} z^\beta \mathcal{K}^{[0]}(z) d\Omega_z = 0 , \quad 1 \leq |\beta| \leq m \quad (3.151)$$

Consequently, kernel $\mathcal{K}_\varrho^{[\alpha]}(\boldsymbol{x})$, $|\alpha| \neq 0$, satisfy $|\alpha| - 1$ order vanishing moment condition:

$$\int_{\mathbb{R}^n} z^\beta \mathcal{K}^{[\alpha]}(z) d\Omega_z = 0 , \quad 0 \leq |\beta| \leq |\alpha| - 1 \qquad (3.152)$$

and some other vanishing moment conditions as well,

$$\int_{\mathbb{R}^n} z^\beta \mathcal{K}^{[\alpha]}(z) d\Omega_z = 0 , \quad |\beta + 1| \leq |\alpha| \leq m. \qquad (3.153)$$

Consider only one-dimensional case. Assume that the original kernel function $\Phi \in H^{m+1}(\mathbb{R}) \cap C_c^m(\mathbb{R})$, definition (3.146) will guarantee that at least [4] $\mathcal{K}^{[\alpha]}(x) \in L^2(\mathbb{R}) \cap L^1(\mathbb{R})$, and

$$\int_{\mathbb{R}} |x|^\beta \left| \mathcal{K}^{[\alpha]}(x) \right| dx < +\infty , \qquad \beta > 0 \qquad (3.154)$$

which in turn, combining with (3.152), guarantees

$$C_{\mathcal{K}^{[\alpha]}} = (2\pi) \int_{\mathbb{R}_+} \left| \hat{\mathcal{K}}^{[\alpha]}(\zeta) \right|^2 \frac{d\zeta}{\zeta} = (2\pi) \int_{\mathbb{R}_-} \left| \hat{\mathcal{K}}^{[\alpha]}(\zeta) \right|^2 \frac{d\zeta}{|\zeta|} < +\infty . \qquad (3.155)$$

where $\hat{\mathcal{K}}^{[\alpha]}(\zeta)$ is the Fourier transform of $\mathcal{K}^{[\alpha]}(x)$,

$$\hat{\mathcal{K}}^{[\alpha]}(\zeta) := \frac{1}{\sqrt{2\pi}} \int_{-\infty}^\infty \exp(-i\zeta z) \mathcal{K}^{[\alpha]}(z) dz \qquad (3.156)$$

Eq. (3.155) is the admissible condition for the basis wavelet, or the mother wavelet (See Chui[114] pages 61-62; Daubechies[129] pages 7, 24-27; Meyer[324] page 16; Meyer and Ryan[325] pages 5, 27; and Kaiser[240] pages 61-72. [5]). The

---

[3] This is also true when $\boldsymbol{x}$ is in the interior domain of a finite domain Ω.

[4] The only exception in our examples is the Gaussian function; the associated kernel function, however, still satisfies this condition and Eq. (3.154).

[5] The definition was first introduced by Grossmann and Morlet in 1984.[184]

higher dimensional extension of the admissible condition (3.155) is discussed in[129] pages 33-34.

In the following, we show that this is true in one-dimensional case (the extension to higher dimensional cases can be readily followed.). Since $\mathcal{K}^{[\alpha]}(x) \in C_0^m(\mathbb{R})$ for $m \geq 1$, it follows immediately that $\mathcal{K}^{[\alpha]}(x) \in L^1(\mathbb{R}) \cap L^2(\mathbb{R})$, and consequently, $\hat{\mathcal{K}}^{[\alpha]}$ is continuous, and $\hat{\mathcal{K}}^{[\alpha]} \in L^2(\mathbb{R})$. Furthermore, since $\mathcal{K}^{[\alpha]}(x) \in C_0^m(\mathbb{R})$, it is obvious that

$$\int_{-\infty}^{\infty} |x| \left| \mathcal{K}^{[\alpha]}(x) \right| dx < +\infty \tag{3.157}$$

which implies that $\hat{\mathcal{K}}^{[\alpha]'}(\zeta) := \dfrac{d\hat{\mathcal{K}}^{[\alpha]}}{d\zeta}$ is also bounded. On the other hand, $\mathcal{K}^{[\alpha]}(x)$ is real,

$$\hat{\mathcal{K}}^{[\alpha]}(-\zeta) = \overline{\hat{\mathcal{K}}^{[\alpha]}}(\zeta) \quad \Rightarrow \quad \left| \hat{\mathcal{K}}^{[\alpha]}(-\zeta) \right|^2 \equiv \left| \hat{\mathcal{K}}^{[\alpha]}(\zeta) \right|^2 \tag{3.158}$$

Thus,

$$\int_{-\infty}^{\infty} \frac{\left| \hat{\mathcal{K}}^{[\alpha]}(\zeta) \right|^2}{|\zeta|} d\zeta = 2 \left\{ \int_0^a \frac{\left| \hat{\mathcal{K}}^{[\alpha]}(\zeta) \right|^2}{\zeta} d\zeta + \int_a^{+\infty} \frac{\left| \hat{\mathcal{K}}^{[\alpha]}(\zeta) \right|^2}{\zeta} d\zeta \right\} \tag{3.159}$$

where $a < 1$. Since $\hat{\mathcal{K}}^{[\alpha]}(\zeta) \in L^2(\mathbb{R})$ as shown early, $\exists C_1 > 0$ such that

$$\int_a^{+\infty} \frac{\left| \hat{\mathcal{K}}^{[\alpha]}(\zeta) \right|^2}{\zeta} d\zeta < C_1 < +\infty \tag{3.160}$$

The remaining concern is the term, $\displaystyle\int_0^a \frac{\left| \hat{\mathcal{K}}^{[\alpha]}(\zeta) \right|^2}{\zeta} d\zeta$. By Cauchy inequality,

$$\int_0^a \frac{\left| \hat{\mathcal{K}}^{[\alpha]}(\zeta) \right|^2}{\zeta} d\zeta \leq \sqrt{\int_0^a \left| \frac{\hat{\mathcal{K}}^{[\alpha]}(\zeta)}{\zeta} \right|^2 d\zeta} \sqrt{\int_0^a \left| \hat{\mathcal{K}}^{[\alpha]}(\alpha) \right|^2 d\zeta} \tag{3.161}$$

From the vanishing moment conditions (3.152), one has

$$\int_{-\infty}^{\infty} \mathcal{K}^{[\alpha]}(x) dx = 0 \quad \Rightarrow \quad \hat{\mathcal{K}}^{[\alpha]}(0) = 0 \tag{3.162}$$

By considering the facts that $\hat{\mathcal{K}}^{[\alpha]}(\zeta)$, and $\hat{\mathcal{K}}^{[\alpha]'}(\zeta)$ are continuous and bounded, there exists a constant, $C_2 > 0$, such that

$$\left| \frac{\hat{\mathcal{K}}^{[\alpha]}(\zeta)}{\zeta} \right| \leq \left| \frac{1}{\zeta} \left(\hat{\mathcal{K}}^{[\alpha]}(0) + \hat{\mathcal{K}}^{[\alpha]'}(\zeta^*)\zeta \right) \right| \leq \left| \hat{\mathcal{K}}^{[\alpha]'}(\zeta^*) \right| \leq C_2 , \quad 0 < \zeta^* < a$$

$$(3.163)$$

Inequality (3.161) is then under control, which leads to the desired result [6]

$$\int_{-\infty}^{0} \frac{\left| \hat{\mathcal{K}}^{[\alpha]}(\zeta) \right|^2}{|\zeta|} d\zeta = \int_{0}^{\infty} \frac{\left| \hat{\mathcal{K}}^{[\alpha]}(\zeta) \right|^2}{\zeta} d\zeta < +\infty \qquad (3.164)$$

Thereby, coincidentally and legitimately, the higher scale kernels, $\mathcal{K}^{[\alpha]}(x)$, $\alpha \neq 0$, are indeed a cluster of basic wavelet functions. It may be worth noting that there is a strong resemblance in the construction procedure between this class of wavelets and "*the coiflets*", a particular wavelet, constructed by Daubechies[129, 130] and Beylkin *et al.*[69]

3.2.2 Interpolation Formulas

To formulate a discrete interpolation scheme, a few definitions are in order. Let Λ be an index set of all particles. For a given bounded, simply connected region $\Omega \subset \mathbb{R}^n$, a particle distribution $\mathcal{D}$ within Ω is defined as

$$\mathcal{D} := \left\{ x_I \, \middle| \, x_I \in \Omega, \; I \in \Lambda \right\} \qquad (3.165)$$

For each $x_I \in \mathcal{D}$, there is an associated ball ω_I,

$$\omega_I := \left\{ x \in \mathbb{R}^n \, \middle| \, |x - x_I| \leq a_I \rho \right\} \qquad (3.166)$$

where $a_I \sim \mathcal{O}(1)$ and ρ is defined as the *dilation parameter*. As defined in,[303] for the admissible particle distribution, the dilation parameter is chosen so that it satisfies the following conditions
   1. for given constants N_{min}, N_{max}

$$N_{min} \leq card\{\Lambda_I\} \leq N_{max} \qquad (3.167)$$

where Λ_I is a subset of Λ, i.e.

$$\Lambda_I := \{J \, \middle| \, J \in \Lambda, \; \omega_J \cap \omega_I \neq \emptyset\} \qquad (3.168)$$

   2. the collection of all the balls,

$$\mathcal{F}_d := \left\{ \omega_I \, \middle| \, I \in \Lambda, \; x_I \in \mathcal{D}, \; \text{and} \; diam(\omega_I) \leq a_I \rho \right\} \qquad (3.169)$$

is a finite covering of domain Ω, i.e. $\bar{\Omega} \subset \bigcup_{I \in \Lambda} \omega_I$ we assume that there exists a constant C_d such that $\max_{I \in \Lambda}\{a_I\} \leq C_d$.

---

[6] As shown in,[129] condition $\int dx \psi(x) = 0$ and $\int dx(1 + |x|)^\beta |\psi(x)| < \infty$ for some $\beta > 0$ which guarantee $|\hat{\psi}(\zeta)| \leq C|\zeta|^\alpha$, with $\alpha = \min(\beta, 1)$ and then the admissible condition (3.155).

In what follows, we form the discrete interpolation formula by a straightforward discretization of the continuous integral representation, namely, Eqs. (3.145)-(3.146) and the moment equation (3.137), by Nyström quadrature method.[148]

For given window function, $\phi > 0$, around particle x_I, the polynomial basis takes the value $\mathbf{P}_{I\varrho}(\boldsymbol{x}) = \{P_{1I\varrho}, \cdots, P_{iI\varrho}, \cdots, P_{jI\varrho}, \cdots, P_{\ell I\varrho}\}$ with $P_{iI\varrho} = \left(\dfrac{\boldsymbol{x}-\boldsymbol{x}_I}{\varrho}\right)^{\alpha}$ and $P_{jI\varrho} = \left(\dfrac{\boldsymbol{x}-\boldsymbol{x}_I}{\varrho}\right)^{\beta}$, the discrete moment matrix (3.137) has the expression

$$\mathbf{M}^h(x) := \left\{M_{ij}^h(x)\right\}^{\ell\times\ell} = \left\{\sum_{I\in\Lambda}\left(\frac{\boldsymbol{x}-\boldsymbol{x}_I}{\rho}\right)^{\alpha+\beta}\Phi_\varrho(\boldsymbol{x}-\boldsymbol{x}_I)\Delta V_I\right\}^{\ell\times\ell} \quad (3.170)$$

Then, the α-th order discrete *correction function* is defined as

$$C_\varrho^{[\alpha]}(\boldsymbol{x}-\boldsymbol{x}_I,\boldsymbol{x}) := \mathbf{P}^{(\alpha)}(0)\{\mathbf{M}^h(\boldsymbol{x})\}^{-1}\mathbf{P}^t\left(\frac{\boldsymbol{x}-\boldsymbol{x}_I}{\varrho}\right) = \mathbf{P}\left(\frac{\boldsymbol{x}-\boldsymbol{x}_I}{\varrho}\right)\boldsymbol{b}^{(\alpha)}(x) \quad (3.171)$$

Accordingly, the discrete α-th scale kernel function is constructed as the modified window functions,

$$\mathcal{K}_\varrho^{[\alpha]}(\boldsymbol{x}-\boldsymbol{x}_I,\boldsymbol{x}) := C_\varrho^{[\alpha]}(\boldsymbol{x}-\boldsymbol{x}_I,\boldsymbol{x})\Phi_\varrho(\boldsymbol{x}-\boldsymbol{x}_I) . \quad (3.172)$$

Each kernel function generates a shape function sequence, i.e.

$$\{\Psi_I^{[\alpha]}(\boldsymbol{x})\}_{I\in\Lambda} := \left\{\alpha!\mathcal{K}_\varrho^{[\alpha]}(\boldsymbol{x}-\boldsymbol{x}_I,\boldsymbol{x})\Delta V_I\right\}_{I\in\Lambda} \quad (3.173)$$

The associated hierarchical interpolation is then set forth as

$$\mathcal{R}_{\varrho,h}^{m[\alpha]}u(\boldsymbol{x}) := \sum_{I=\Lambda}\Psi_I^{[\alpha]}(\boldsymbol{x})u(\boldsymbol{x}_I) = \alpha!\sum_{I=\Lambda}\mathcal{K}_\varrho^{[\alpha]}(\boldsymbol{x}-\boldsymbol{x}_I,\boldsymbol{x})u(\boldsymbol{x}_I)\Delta V_I \quad (3.174)$$

where $\{\Delta V_I\}_{I\in\Lambda}$ are the quadrature weights; they are so chosen such that

$$\Delta V_I \le a_I^n\rho^n , \quad \text{and} \quad \sum_{I\in\Lambda}\Delta V_I = meas(\Omega) \quad (3.175)$$

Eq. (3.175) is often referred to as the stability condition ([297, 303]). Note that in Eq. (3.171), the vector $\boldsymbol{b}^{(\alpha)}$ is determined by the discrete moment equations

$$\mathbf{M}^h(\boldsymbol{x})\boldsymbol{b}^{(\alpha)}(\boldsymbol{x}) = \{\mathbf{P}^{(\alpha)}(0)\} \quad (3.176)$$

One can readily verify that Eq. (3.176) is equivalent to the following discrete consistency condition

$$\sum_{I\in\Lambda}\left(\frac{\boldsymbol{x}-\boldsymbol{x}_I}{\varrho}\right)^\beta\mathcal{K}_\varrho^{[\alpha]}(\boldsymbol{x}-\boldsymbol{x}_I,\boldsymbol{x})\Delta V_I = \delta_{\alpha\beta} \quad (3.177)$$

3.2.3 Hierarchical Partition of Unity and Hierarchical Basis

From Eq. (3.177), one may find that the fundamental basis $\{\Psi_I^{[0]}(x)\}$ is a **signed partition of unity**, i.e.

$$\sum_{I\in\Lambda}\mathcal{K}_\varrho^{[0]}(x-x_I,x)\Delta V_I = \sum_{I\in\Lambda}\Psi_I^{[0]}(x) = 1 \qquad (3.178)$$

which is the original moving least square reproducing kernel basis; whereas the higher order bases, $\{\Psi_I^{[\alpha]}(x)\}$, $\alpha\neq 0$, are the **partition of nullity**, so to speak, because by construction,

$$\sum_{I\in\Lambda}\Psi_I^{[\alpha]}(x) = \sum_{I\in\Lambda}\alpha!\mathcal{K}_\varrho^{[\alpha]}(x-x_I,x)\Delta V_I = 0. \ \ 1\leq|\alpha|\leq m \qquad (3.179)$$

This is a very desirable property, because by inserting the higher order basis into the fundamental basis, one will still have a partition of unity, i.e.

$$\sum_{I\in\Lambda}\left(\mathcal{K}_\varrho^{[0]}(x-x-I,x)+1!\mathcal{K}_\varrho^{[1]}(x-x_I,x)+\cdots+\alpha!\mathcal{K}_\varrho^{[\alpha]}(x-x_I,x)\right)\Delta V_I$$

$$=\sum_{I\in\Lambda}\sum_{0\leq|\beta|\leq|\alpha|}\Psi_I^{[\beta]}(x) = 1 , \qquad\qquad |\alpha|\leq m \qquad (3.180)$$

In the rest of the paper, we denote the p-th order hierarchical partition of unity on the particle distribution $\mathcal{D}$ as $\mathcal{H}_p := \{\{\Psi_I^{[\alpha]}(x)\}_{I\in\Lambda} : \ 0\leq|\alpha|\leq p\}$. An example of such hierarchical partition of unity is displayed in Fig. 3.9.

Since discrete wavelet functions form a partition of nullity, they are not linearly independent because in a partition of nullity there are extra, or redundant shape functions. Thus, the hierarchical partition of unity is at most a frame in global sense. Nevertheless, by careful selection, one can still form a hierarchical basis.

Definition 3.2.1 (Hierarchical Basis).

Choose $\overset{\circ}{\Lambda}_{[\alpha]} \subset\subset \Lambda$, $\forall\ 1\leq|\alpha|\leq m$ and denote $n_{[\alpha]} := card\{\overset{\circ}{\Lambda}_{[\alpha]}\}$ such that $\forall\ f \in span\{\Psi_I^{[\alpha]}\}_{I\in\overset{\circ}{\Lambda}_{[\alpha]}}$, $\exists\ c_I$ and $c_I\neq 0$,

$$f(x) = \sum_{I\in\overset{\circ}{\Lambda}_{[\alpha]}}c_I\Psi_I^{[\alpha]}(x).$$

Define the global hierarchical basis

$$\{\Phi_J\}_{\Lambda_H} := \left\{\{\Psi_J^{[0]}\}_{J\in\Lambda}, \{\Psi_J^{[1]}\}_{J\in\overset{\circ}{\Lambda}_{[1]}},\cdots,\{\Psi_J^{[\alpha]}\}_{J\in\overset{\circ}{\Lambda}_{[\alpha]}}\right\}, \qquad (3.181)$$

where $\Lambda_H := \{J\ \big|\ J = 1,\cdots,np, np+1,\cdots,n_H\}$,$n_H := (np+n_{[1]}+\cdots+n_{[\alpha]})$ and

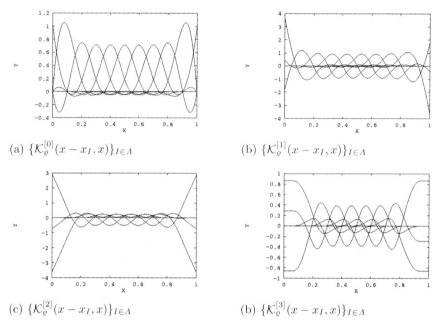

(a) $\{\mathcal{K}_{\varrho}^{[0]}(x - x_I, x)\}_{I \in \Lambda}$

(b) $\{\mathcal{K}_{\varrho}^{[1]}(x - x_I, x)\}_{I \in \Lambda}$

(c) $\{\mathcal{K}_{\varrho}^{[2]}(x - x_I, x)\}_{I \in \Lambda}$

(b) $\{\mathcal{K}_{\varrho}^{[3]}(x - x_I, x)\}_{I \in \Lambda}$

Fig. 3.9. An example of hierarchical partition of unity.

$$\mathbf{A}_H := \{\alpha_{IJ}\}^{n_H \times n_H} \quad , \quad \alpha_{IJ} = \int_{\Omega} \Phi_I \Phi_J d\Omega$$

If $\det\{\mathbf{A}_H\} > 0$, we say $\{\Phi_I\}_{I \in \Lambda_H}$ is a hierarchical basis for the finite dimensional space, $\mathcal{S}_H := span \left\{ \{\Psi_I^{[0]}(x)\}_{I \in \Lambda}, \{\Psi_I^{[1]}(x)\}_{I \in \overset{\circ}{\Lambda}_{[1]}}, \cdots, \{\Psi_I^{[\alpha]}(x)\}_{I \in \overset{\circ}{\Lambda}_{[\alpha]}} \right\}

♣

Remark 3.2.2. **1.** By properly choosing the size of the compact support of the window function, one can form a wavelet-like basis by taking some shape functions out of the partition of nullity, usually the ones that are on the boundary. In this way, in the interior region, the hierarchical basis remains as a partition of unity. **2.** In practice, by underintegration, it is possible that the stiffness matrix formed by hierarchical partition of unity is still invertible; in that case, however, spurious modes may occur. **3.** By taking out a certain number of extra shape functions from a partition of nullity, one may form an independent group of basis functions from the partition of nullity, but it does not automatically guaranty that (3.181) is an independent basis. In practice, exactly how many extra shape functions should be taken out is determined so far on a basis of trial and error.

Let $\mathbf{A}_H^{-1} := \{\beta_{IJ}\}^{n_H \times n_H}$ and

$$\sum_l \alpha_{IL}\beta_{LJ} = \begin{cases} 1, & I = J; \\ 0, & I \neq J. \end{cases}$$

One can then define the dual basis

$$\{\tilde{\Phi}_I(x)\}_{I\in\Lambda_H} \; : \quad \tilde{\Phi}_I(x) := \sum_{I\in\Lambda_H} \beta_{IJ}\Phi_J(x) \tag{3.182}$$

and subsequently the *reproducing kernel* of the hierarchical basis is:

$$K_H(y,x) := \sum_{I,J\in\Lambda_H} \beta_{IJ}\Phi_I(x)\Phi_J(y) \tag{3.183}$$

Thus, the generalized reproducing kernel formula becomes

$$\mathcal{R}_{\varrho,h}^{p[H]}f(x) :=< f(y), K_H(y,x) >_y= \sum_{I,J\in\Lambda_H} \beta_{IJ}\left(\int_\Omega f(y)\Phi_J(y)d\Omega_y\right)\Phi_I(x) \tag{3.184}$$

When $f \in span\{\Phi_I(x)\}_{\Lambda_H}$, one can readily verify that $\mathcal{R}_{\varrho,h}^{p[H]}f(x) = f(x)$.

In the following example, we illustrate how to construct a meshfree hierarchical partition of unity.

Example 3.2.1. In a 1-D segment $[-0.5, 0.5]$, let $m = 3$, $\ell = 3+1 = 4$. The hierarchical kernel functions are constructed in a *pointwise* fashion,

$$K_\varrho^{[\alpha]}(x_I - x, x) := \mathbf{P}(\frac{x_I - x}{\varrho})\mathbf{b}^{(\alpha)}(x)\phi_\varrho(x_I - x), \quad 0 \leq \alpha \leq 3 \tag{3.185}$$

The consistency conditions that the wavelet kernel packet satisfies are the following algebraic equation imposed on the vector $\mathbf{b}^{(\alpha)}(x)$,

$$\mathbf{M}^h(x)\mathbf{b}^{(\alpha)}(x) = \{\mathbf{P}^{(\alpha)}(0)\}^t , \quad \alpha = 0,1,2,3 \tag{3.186}$$

Or more explicitly,

$$\begin{pmatrix} m_0^h & m_1^h & m_2^h & m_3^h \\ m_1^h & m_2^h & m_3^h & m_4^h \\ m_2^h & m_3^h & m_4^h & m_5^h \\ m_3^h & m_4^h & m_5^h & m_6^h \end{pmatrix} \begin{pmatrix} b_1^{(\alpha)} \\ b_2^{(\alpha)} \\ b_3^{(\alpha)} \\ b_4^{(\alpha)} \end{pmatrix} = \begin{pmatrix} \delta_{\alpha 0} \\ \delta_{\alpha 1} \\ \delta_{\alpha 2} \\ \delta_{\alpha 3} \end{pmatrix} , \tag{3.187}$$

where the α-th discrete moment is defined as

$$m_\alpha^h(x) := \sum_{I=1}^{np}(\frac{x_I - x}{\varrho})^\alpha \phi_\varrho(x_I - x)\Delta x_I . \tag{3.188}$$

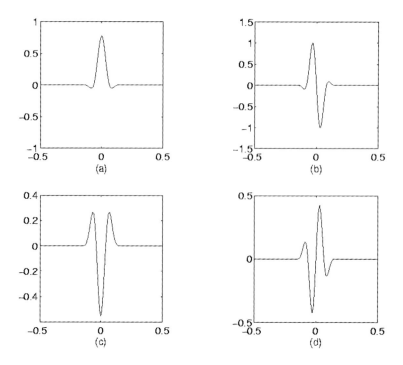

Fig. 3.10. The hierarchical kernels at point $x_I = 0$ for $\mathbf{P} = (1, x, x^2, x^3)$: (a) fundamental kernel; (b) The 1st order wavelet; (c) The 2nd order wavelet; (d) The 3rd order wavelet.

In Fig. 3.10, the constructed kernel function sequence is displayed at $x_I = 0$; and a uniform particle distribution (11 particles) is used in the computation. In computation, the parameter $\varrho = \Delta x$, and a fifth order spline is used as window function ($a_I = 3.3$). The second wavelet kernel in Fig. 3.10(c) also resembles an upside-down Mexican hat.

Example 3.2.2. In this example, the dimension of the space is $n = 2$, $|\alpha| = m = 2$, and $\ell = 6$, and $\mathcal{K}_\varrho^{[\alpha]}(x_I - x, x) := \mathbf{P}(\frac{x_I - x}{\varrho})\mathbf{b}^{(\alpha)}(x)\phi_\varrho(x_I - x)$, with $x_I = (x_{1I}, x_{2I})$, $x = (x_1, x_2)$, and $\alpha = (0, 0)$, $(1, 0)$, $(0, 1)$, $(2, 0)$, $(1, 1)$, $(0, 2)$. The vector $\mathbf{b}^{(\alpha)}(x)$ is determined by the global moment equation that is similar to (3.187), namely,

$$\begin{pmatrix} m_{00}^h & m_{10}^h & m_{01}^h & m_{20}^h & m_{11}^h & m_{02}^h \\ m_{10}^h & m_{20}^h & m_{11}^h & m_{30}^h & m_{21}^h & m_{12}^h \\ m_{01}^h & m_{11}^h & m_{02}^h & m_{21}^h & m_{12}^h & m_{03}^h \\ m_{20}^h & m_{30}^h & m_{21}^h & m_{40}^h & m_{21}^h & m_{22}^h \\ m_{11}^h & m_{21}^h & m_{12}^h & m_{31}^h & m_{22}^h & m_{13}^h \\ m_{02}^h & m_{12}^h & m_{03}^h & m_{22}^h & m_{13}^h & m_{04}^h \end{pmatrix} \begin{pmatrix} b_1^{(\alpha)} \\ b_2^{(\alpha)} \\ b_3^{(\alpha)} \\ b_4^{(\alpha)} \\ b_5^{(\alpha)} \\ b_6^{(\alpha)} \end{pmatrix} = \begin{pmatrix} \delta_{\alpha(0,0)} \\ \delta_{\alpha(1,0)} \\ \delta_{\alpha(0,1)} \\ \delta_{\alpha(2,0)} \\ \delta_{\alpha(1,1)} \\ \delta_{\alpha(0,2)} \end{pmatrix} \qquad (3.189)$$

Again, the above moment matrix is a full matrix for arbitrary particle distributions. By using a 2-D cubic spline as the window function, numerical computations have been carried out in a 2-D domain $[-1, 1] \times [-1, 1]$ on a uniform 21×21 particle distribution. In Fig. 3.11, the sequence of hierarchical kernel functions are displayed with respect to $x_I = (0, 0)$. In the computation, the dilation vector $\varrho = (\varrho_1, \varrho_2)$ is chosen as $\varrho = (\Delta x, \Delta y)$ and $\Delta x = \Delta y = h$. The window function is a direct product of two cubic spline functions.

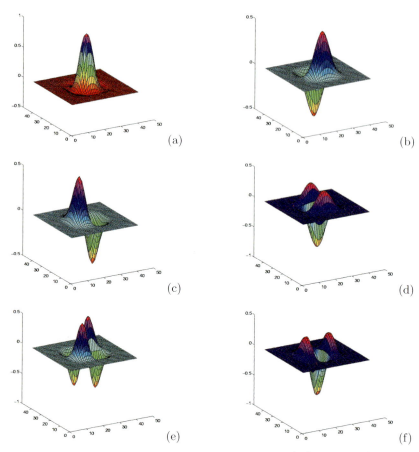

Fig. 3.11. A 2-D hierarchical kernel sequence; (a) $\Psi_I^{[0,0]}(x)$; (b) $\Psi_I^{[1,0]}(x)$; (c) $\Psi_I^{[0,1]}(x)$; (d) $\Psi_I^{[2,0]}(x)$; (e) $\Psi_I^{[1,1]}(x)$; (f) $\Psi_I^{[0,2]}(x)$.

3.3 MLS Interpolant and Diffuse Element Method

Directly calculating derivative of meshfree interpolant could be expensive, if particles are constantly changing their positions. Several approximations are proposed in the literature. The most well-known approximation is the so-called Diffuse Element Method.

3.3.1 Diffuse Element Method

Consider the original MLS interpolant,

$$N_I(\boldsymbol{x}) = \mathbf{P}^T(\boldsymbol{x})\mathcal{M}^{-1}(\boldsymbol{x})\mathbf{P}(\boldsymbol{x}_I)\phi(\boldsymbol{x} - \boldsymbol{x}_I) \tag{3.190}$$

where

$$\mathcal{M}(\boldsymbol{x}) := \sum_{I \in \Lambda} \mathbf{P}^T(\boldsymbol{x}_I)\phi(\bar{\mathbf{x}} - \boldsymbol{x}_I)\mathbf{P}(\boldsymbol{x}_I) \tag{3.191}$$

To evaluate the derivative of the MLS interpolant, Nayroles et al. made the following approximation in the name of the diffuse element method (DEM),

$$\nabla N_I(\boldsymbol{x}) \approx \left(\nabla \mathbf{P}(\boldsymbol{x})\right)\mathcal{M}^{-1}(\boldsymbol{x})\mathbf{P}^T(\boldsymbol{x}_I)\phi(\boldsymbol{x} - \boldsymbol{x}_I) \tag{3.192}$$

with the assumption that

$$\mathbf{P}(\boldsymbol{x})\nabla\mathcal{M}^{-1}(\boldsymbol{x})\mathbf{P}^T(\boldsymbol{x}_I)\phi(\boldsymbol{x}-\boldsymbol{x}_I) + \mathbf{P}(\boldsymbol{x})\mathcal{M}^{-1}(\boldsymbol{x})\nabla\phi(\boldsymbol{x}-\boldsymbol{x}_I) << 1 \tag{3.193}$$

which may not be true in general.

3.3.2 Evaluate the Derivative of MLS Interpolant

Let $\boldsymbol{b}(\boldsymbol{x}, \boldsymbol{x}_I) = \mathcal{M}^{-1}(\boldsymbol{x})\mathbf{P}(\boldsymbol{x}_I)$. Then $N_I(\boldsymbol{x}) = \mathbf{P}^T(\boldsymbol{x})\boldsymbol{b}(\boldsymbol{x}, \boldsymbol{x}_I)\phi(\boldsymbol{x}-\boldsymbol{x}_I)$ Thus, we have

$$\begin{aligned}
\nabla N_I(\boldsymbol{x}) = &\left(\nabla\mathbf{P}^T(\boldsymbol{x})\right)\mathcal{M}^{-1}(\boldsymbol{x})\mathbf{P}(\boldsymbol{x}_I)\phi(\boldsymbol{x} - \boldsymbol{x}_I) \\
&+ \mathbf{P}^T(\boldsymbol{x})\left(\nabla\boldsymbol{b}(\boldsymbol{x}, \boldsymbol{x}_I)\right)\phi(\boldsymbol{x} - \boldsymbol{x}_I) \\
&+ \mathbf{P}^T(\boldsymbol{x})\boldsymbol{b}(\boldsymbol{x}, \boldsymbol{x}_I)\nabla\phi(\boldsymbol{x} - \boldsymbol{x}_I)
\end{aligned} \tag{3.194}$$

Obviously, in (3.194), derivative of the first and the last term in the right-hand side are easy to be evaluated. The complicated part is how to evaluate $\nabla\boldsymbol{b}(\boldsymbol{x}, \boldsymbol{x}_I)$, or in general, $D_{\boldsymbol{x}}^{\alpha}\boldsymbol{b}(\boldsymbol{x}, \boldsymbol{x}_I)$, where $D_{\boldsymbol{x}}^{\alpha} = \partial_{x_1}^{\alpha_1}\partial_{x_2}^{\alpha_2}\cdots\partial_{x_n}^{\alpha_n}$. Since by definition,

$$\mathcal{M}(\boldsymbol{x})\boldsymbol{b}(\boldsymbol{x}, \boldsymbol{x}_I) = \begin{pmatrix} p_1(\boldsymbol{x}_I) \\ p_2(\boldsymbol{x}_I) \\ \vdots \\ p_\ell(\boldsymbol{x}_I) \end{pmatrix} \tag{3.195}$$

Therefore,

$$D_{\boldsymbol{x}}^{\alpha}\left(\mathbf{M}(\boldsymbol{x})\boldsymbol{b}(\boldsymbol{x},\boldsymbol{x}_I)\right) = 0 \quad \Rightarrow \sum_{\beta \leq \alpha} \binom{\alpha}{\beta} D_{\boldsymbol{x}}^{\beta}\mathbf{M}(\boldsymbol{x}) D_{\boldsymbol{x}}^{\alpha-\beta}\boldsymbol{b}(\boldsymbol{x}) = 0 \ , \quad (3.196)$$

or

$$\mathbf{M}D_{\boldsymbol{x}}^{\alpha}\boldsymbol{b}(\boldsymbol{x},\boldsymbol{x}_I) + \sum_{\left(\substack{\beta \leq \alpha \\ \beta \neq 0}\right)} (D_{\boldsymbol{x}}^{\beta}\mathbf{M}(\boldsymbol{x}))(D_{\boldsymbol{x}}^{\alpha-\beta}\boldsymbol{b}(\boldsymbol{x},\boldsymbol{x}_I)) = 0 \ . \quad (3.197)$$

where $1 \leq |\alpha| \leq k$. Thereby, one can solve $D_{\boldsymbol{x}}^{\alpha}\boldsymbol{b}(\boldsymbol{x},\boldsymbol{x}_I)$ recursively; for instance, in the 1-D case,

$$\begin{bmatrix} \mathbf{M} & 0 & 0 & \cdots & \cdots & 0 \\ \binom{2}{1}D_{\boldsymbol{x}}^{1}\mathbf{M} & \mathbf{M} & 0 & \cdots & \cdots & 0 \\ \binom{3}{2}D_{\boldsymbol{x}}^{2}\mathbf{M} & \binom{3}{1}D_{\boldsymbol{x}}^{1}\mathbf{M} & \mathbf{M} & \cdots & \cdots & 0 \\ \vdots & \vdots & & \ddots & & \vdots \\ \vdots & \vdots & & & \ddots & \vdots \\ \binom{k}{k-1}D_{\boldsymbol{x}}^{k-1}\mathbf{M} & \binom{k}{k-2}D_{\boldsymbol{x}}^{k-2}\mathbf{M} & \cdots & \cdots & & \mathbf{M} \end{bmatrix} \begin{pmatrix} D_{\boldsymbol{x}}^{1}\boldsymbol{b} \\ D_{\boldsymbol{x}}^{2}\boldsymbol{b} \\ D_{\boldsymbol{x}}^{3}\boldsymbol{b} \\ \vdots \\ \vdots \\ D_{\boldsymbol{x}}^{k}\boldsymbol{b} \end{pmatrix} = - \begin{pmatrix} (D_{\boldsymbol{x}}^{1}\mathbf{M})\boldsymbol{b} \\ (D_{\boldsymbol{x}}^{2}\mathbf{M})\boldsymbol{b} \\ (D_{\boldsymbol{x}}^{3}\mathbf{M})\boldsymbol{b} \\ \vdots \\ \vdots \\ (D_{\boldsymbol{x}}^{k}\mathbf{M})\boldsymbol{b} \end{pmatrix}$$

$$(3.198)$$

The coefficient matrix of the above equations is a lower triangular block matrix, which can be solved by forward substitution. Therefore, $D_{\boldsymbol{x}}^{\alpha}\boldsymbol{b}(\boldsymbol{x},\boldsymbol{x}_I)$ are uniquely determined.

3.4 Element-free Galerkin Method (EFGM)

The so-called Element-free Galerkin (EFG) Method is a Galerkin procedure using MLS interpolant to solve engineering problems, or partial differential equations in general (Belytschko et al. [1994],[1995],[1996], and Lu et al. [1996]).

A main difference between the meshfree Galerkin method and meshfree collocation methods, e.g. smoothed particle hydrodynamics, is that the former deals with meshfree/particle discretization of weak formulation of a PDE, and the latter deals with meshfree/particle discretization of strong form of a PDE.

In EFG, one uses MLS interpolant as both trial and test functions in a Galerkin procedure similiar to that of finite element methods. The difference is how to modify the Galerkin statement to accomodate MLS interpolant. One of the technical difficulties of implement meshfree Galerkin methods is how to impose essential boundary conditions for non-interpolating MLS shape function. Early on, Belytschko et al. used Lagrange multiplier method altering the variational statement to enforce essential boundary conditions. Later on, Zhu and Atluri used penalty method; Chen and others used transform method and boundary singular kernel method, Krongauz and Belytschko coupled meshfree and finite element at boundary, etc. A comprehesive account on how to treat essential boundary conditions is presented as follows.

3.4.1 Lagrangian Multiplier Method

Consider a solid occupying a domain Ω bounded by Γ:

$$\nabla \cdot \boldsymbol{\sigma} + \mathbf{f} = \mathbf{0}, \quad \boldsymbol{x} \in \Omega \tag{3.199}$$

where $\boldsymbol{\sigma}$ is the Cauchy stress tensor, $\mathbf{f}$ is the body force. The boundary conditions of the physical problem are given as follows

$$\boldsymbol{\sigma} \cdot \mathbf{n} = \bar{\mathbf{T}}, \quad \forall \boldsymbol{x} \in \Gamma_t \tag{3.200}$$

$$\mathbf{u} = \bar{\mathbf{u}}, \quad \forall \boldsymbol{x} \in \Gamma_u \tag{3.201}$$

in which $\mathbf{u}$ is the displacement, $\bar{\mathbf{u}}$ is the prescribed displacement on Γ_u, and $\bar{\mathbf{T}}$ is the prescribed traction on Γ_t and $\Gamma = \Gamma_u \bigcup \Gamma_t$. To accomodate the non-interpolating shape function, we introduce the reaction force, $\mathbf{R}$, on Γ_u as another unkown variable, which is complementary to the primary unknown, the displacement $\mathbf{u}$. A weak form of the original problem can be derived from the following weighted residual form

$$\int_\Omega \mathbf{v}^T \cdot \left(\nabla \cdot \boldsymbol{\sigma} + \mathbf{f} \right) d\Omega + \int_{\Gamma_u} \boldsymbol{\lambda}^T \cdot (\mathbf{u} - \bar{\mathbf{u}}) dS$$

$$+ \int_{\Gamma_u} \mathbf{v}^T \cdot (\mathbf{R} - \boldsymbol{\sigma} \cdot \mathbf{n}) dS = 0 \tag{3.202}$$

where $\mathbf{v}$ and $\boldsymbol{\lambda}$ are weighting function for displacement and reaction respectively. Note that both $\mathbf{u}, \mathbf{v} \in H^1(\Omega)$ and $\mathbf{v}(\boldsymbol{x}) \neq 0, \forall \, \boldsymbol{x} \in \Gamma_u$, and both $\mathbf{R}, \boldsymbol{\lambda} \in H^0(\Omega)$ and $\mathbf{R}(\boldsymbol{x}) \neq \boldsymbol{\sigma} \cdot \mathbf{n}$, $\forall \boldsymbol{x} \Gamma_u$. Integration by part yields

$$\int_\Omega (\nabla_s \mathbf{v}^T) : \boldsymbol{\sigma} d\Omega - \mathbf{v}^T : \mathbf{f} d\Omega - \int_{\Gamma_t} \mathbf{v}^T \cdot \bar{\mathbf{T}} dS$$

$$- \int_{\Gamma_u} \boldsymbol{\lambda}^T \cdot (\mathbf{u} - \bar{\mathbf{u}}) dS - \int_{\Gamma_u} \mathbf{v}^T \cdot \mathbf{R} d\Omega = 0,$$

$$\text{with} \quad \forall \, \mathbf{v} \in H^1(\Omega), \, \boldsymbol{\lambda} \in H^0(\Omega) \qquad (3.203)$$

where $H^0(\Omega), H^1(\Omega)$ are the Sobolev space of degree zero and one.

Let

$$\mathbf{u}^\varrho(\boldsymbol{x}) = \sum_{I \in \Lambda} N_I(\boldsymbol{x}) \mathbf{u}_I \qquad (3.204)$$

$$\mathbf{v}^\varrho(\boldsymbol{x}) = \sum_{I \in \Lambda} N_I(\boldsymbol{x}) \mathbf{v}_I \qquad (3.205)$$

Find the index subset of Λ, Λ_b, such that [7]

$$\Lambda_b = \{I \mid I \in \Lambda, N_I(\boldsymbol{x}) \neq 0, \boldsymbol{x} \in \Gamma_u\} \qquad (3.206)$$

And let

$$\mathbf{R}(\boldsymbol{x}) = \sum_{I \in \Lambda_b} \tilde{N}_I(\boldsymbol{x}) \mathbf{R}_I \qquad (3.207)$$

$$\boldsymbol{\lambda}(\boldsymbol{x}) = \sum_{I \in \Lambda_b} \tilde{N}_I(\boldsymbol{x}) \boldsymbol{\lambda}_I \qquad (3.208)$$

where $\tilde{N}_I(\boldsymbol{x})$ may be different from $N_I(\boldsymbol{x})$ in order to satisfy the stability condition. Substituting (3.204),(3.205),(3.207), and (3.208) into the weak form (3.203) yields the following algebraic equations,

$$\begin{pmatrix} \mathbf{K} & \mathbf{G} \\ \mathbf{G}^T & 0 \end{pmatrix} \begin{pmatrix} \mathbf{u} \\ \mathbf{R} \end{pmatrix} = \begin{pmatrix} \mathbf{h} \\ \mathbf{q} \end{pmatrix} \qquad (3.209)$$

For 2D plane stress linear elasticity problem, one may have explicit expressions on both stiffness matrix, as well as force terms

$$\mathbf{K}_{IJ} = \int_\Omega \mathbf{B}_I^T \mathbf{D} \mathbf{B}_J d\Omega \qquad (3.210)$$

$$\mathbf{G}_{IK} = - \int_{\Gamma_u} N_I \tilde{N}_K dS \qquad (3.211)$$

$$\mathbf{h}_I = \int_{\Gamma_t} N_I \bar{\mathbf{T}} dS + \int_\Omega N_I \mathbf{f} d\Omega \qquad (3.212)$$

$$\mathbf{q}_K = - \int_{\Gamma_u} N_K \bar{\mathbf{u}} dS \qquad (3.213)$$

---

[7] The original paper prefers $\Lambda_b = \{I \mid I \in \Lambda, \boldsymbol{x}_I \in \Gamma_u\}$

and

$$\mathbf{B}_I = \begin{pmatrix} N_{I,x} & 0 \\ 0 & N_{I,y} \\ N_{I,y} & N_{I,x} \end{pmatrix} \tag{3.214}$$

$$\mathbf{N}_K = \begin{pmatrix} N_k & 0 \\ 0 & N_k \end{pmatrix} \tag{3.215}$$

$$\mathbf{D} = \frac{E}{1 - \mu^2} \begin{pmatrix} 1 & \mu & 0 \\ \mu & 1 & 0 \\ 0 & 0 & (1 - \mu)/2 \end{pmatrix} \tag{3.216}$$

in which a comma in subscript designates a partial derivative with respect to the indicated spatial variable; E and μ are Young's modulus and Poisson's ratio.

The above Lagrangian multiplier formulation can be also derived from minimizing the following functional

$$\Pi_p = \frac{1}{2} \int_\Omega \boldsymbol{\epsilon}^T \mathbf{D} \boldsymbol{\epsilon} d\Omega - \int_\Omega \mathbf{u}^T \cdot \mathbf{f} d\Omega - \int_{\Gamma_t} \mathbf{u}^T \cdot \bar{\mathbf{T}} dS$$
$$+ \int_{\Gamma_u} \mathbf{R} \cdot (\mathbf{u} - \bar{\mathbf{u}}) dS \tag{3.217}$$

Take $\delta \Pi_p = 0$, and consider $\boldsymbol{\epsilon} = \nabla_s \mathbf{u}$ and $\boldsymbol{\sigma} = \mathbf{D} \boldsymbol{\epsilon}$. We will have

$$\int_\Omega (\nabla_s(\delta \mathbf{u}))^T : \boldsymbol{\sigma} d\Omega - (\delta \mathbf{u})^T : \boldsymbol{b} d\Omega - \int_{\Gamma_t} (\delta \mathbf{u})^T \cdot \bar{\mathbf{T}} dS$$
$$- \int_{\Gamma_u} (\delta \mathbf{R})^T \cdot (\mathbf{u} - \bar{\mathbf{u}}) dS - \int_{\Gamma_u} (\delta \mathbf{u})^T \cdot \mathbf{R} d\Omega = 0, \tag{3.218}$$

Integration by part yields

$$\int_\Omega (\delta \mathbf{u})^T \cdot \left(\nabla \cdot \boldsymbol{\sigma} + \mathbf{f} \right) d\Omega + \int_{\Gamma_u} (\delta \mathbf{R})^T \cdot (\mathbf{u} - \bar{\mathbf{u}}) dS$$
$$+ \int_{\Gamma_u} (\delta \mathbf{u})^T \cdot (\mathbf{R} - \boldsymbol{\sigma} \cdot \mathbf{n}) dS = 0 \tag{3.219}$$

Thus we identify that $\mathbf{R}$ is the reaction force on displacement prescribed boundary.

3.4.2 Penalty Method

The penalty method is a common alternative to impose essential boundary conditions. The version of penalty method presented as the following was initially suggested by Zhu and Atluri (1998) in an illustration of 2D linear elastostatics.

Suppose the prescribed essential boundary condition, and prescribed traction are as follows

$$\mathbf{u}(\boldsymbol{x}) = \bar{\mathbf{u}} \tag{3.220}$$

$$\boldsymbol{\sigma} \cdot \mathbf{n} = \bar{\mathbf{T}} \tag{3.221}$$

Let $\mathbf{u}^{\varrho}(\boldsymbol{x}) = \sum_{I \in \Lambda} N_I(\boldsymbol{x}, \varrho)\mathbf{u}_I$ and

$$\begin{aligned}
\Pi_{\varrho} = \frac{1}{2} \int_{\Omega} (\boldsymbol{\epsilon}^{\varrho})^T \cdot \mathbf{D} \cdot \boldsymbol{\epsilon}^{\varrho} d\Omega - \int_{\Omega} (\mathbf{u}^{\varrho})^T \cdot \mathbf{b} d\Omega \\
- \int_{\Gamma_t} (\mathbf{u}^{\varrho})^T \cdot \bar{\mathbf{T}} dS + \frac{\alpha}{2} \int_{\Gamma_u} (\mathbf{u}^{\varrho} - \bar{\mathbf{u}})^T \cdot (\mathbf{u}^{\varrho} - \bar{\mathbf{u}}) dS
\end{aligned} \tag{3.222}$$

Take $\delta\Pi_{\varrho} = 0$, we have

$$\int_{\Omega} (\nabla_s(\delta\mathbf{u}^{\varrho}))^T : \boldsymbol{\sigma} d\Omega - \int_{\Omega} (\delta\mathbf{u}^{\varrho})^T \cdot \mathbf{b} d\Omega - \int_{\Gamma_t} (\delta\mathbf{u}^{ha})^T \cdot \bar{\mathbf{T}} dS$$

$$+ \alpha \int_{\Gamma_u} (\delta\mathbf{u}^{\varrho})^T \cdot (\mathbf{u}^{\varrho} - \bar{\mathbf{u}}) dS = 0, \tag{3.223}$$

And consequently,

$$(\mathbf{K} + \alpha\mathbf{K}^u)\mathbf{U} = \mathbf{f} + \alpha\mathbf{f}^u \tag{3.224}$$

where the stiffness matrix $\mathbf{K}$ and the force vector $\mathbf{f}$ are the same as (3.210) and (3.212); however, the additional terms due to essential boundary conditions are

$$\mathbf{K}^u_{IJ} = \int_{\Gamma_u} N_I \mathbf{S} N_J dS \tag{3.225}$$

$$\mathbf{f}^u_I = \int_{\Gamma_u} N_I \mathbf{S} \bar{\mathbf{u}} dS \tag{3.226}$$

where

$$S_i = \begin{cases} 1 & \text{if } u_i \text{ is prescribed on } \Gamma_u, \\ 0 & \text{if } u_i \text{ is not prescribed on } \Gamma_u, \ i = 1, 2 \end{cases} \tag{3.227}$$

In computations, the penalty parameter is taken in the range $\alpha = 10^3 \sim 10^7$.

3.4.3 Nitsche's Method

In 1970, J. Nitsche, known for his contribution in finite element convergence theory (Nitsche's trick), proposed a variational approach to enforcing Dirichlet boundary condition for non-interpolating interpolant. In a certain sense, the method resembles Lagrange multiplier method, but it possesses better convergence property, ensures the existence and uniqueness of the solution, and provides guideline to select auxilary parameter or multiplier. It is the authors' opinion that in the future Nitsche's method may become the standard treatment to enforcing essential boundary condition for meshfree Galerkin methods.

To illustrate the method, we consider the following boundary value problem,

$$-\Delta u = f, \quad \forall \, \mathbf{x} \in \Omega$$
$$u = g, \quad \forall \, \mathbf{x} \in \partial\Omega \tag{3.228}$$

For interpolating interpolant, the Galerkin variational formulation is: let $\mathcal{V} \subset H^1(\Omega)$ be a finite dimensional interpolation space. Find $u^h \in \mathcal{V}$ such that

$$a(u^h, v^h) - (f, v^h) = 0, \quad \forall \, v^h \in \mathcal{V} \tag{3.229}$$

where

$$a(u, v) = \int_\Omega \nabla u \cdot \nabla v d\Omega$$
$$(f, v) = \int_\Omega f v d\Omega \tag{3.230}$$

For non-interpolating interpolant (interpolant does not have Kronecker delta property), Nitsche proposed the following Galerkin variational formulation: Find $u^h \in \mathcal{V}$ such that

$$a_N(u^h, v^h) - (f, v^h)_N = 0, \quad \forall \, v^h \in \mathcal{V} \tag{3.231}$$

where the subscript N under $a_N(,)$ and $(,)_N$ denoting Nitsche's bilinear form and Nitsche' linear form, which are defined as

$$a_N(u, v) := \int_\Omega \nabla u \cdot \nabla v d\Omega - \int_{\partial\Omega} u v_n dS - \int_{\partial\Omega} u v_n dS$$
$$+ \alpha_N \int_{\partial\Omega} u v dS \tag{3.232}$$

$$(f, v)_N = \int_\Omega f v d\Omega - \int_{\partial\Omega} g v_n dS + \alpha_N \int_{\partial\Omega} g v dS \tag{3.233}$$

where α_N is a constant; $u_n := \nabla u \cdot n$ and $v_n := \nabla v \cdot n$ and n is the outward normal of $\partial\Omega$.

The key technical ingredient here is how to choose the constant α_N to ensure the ellipticity of the bilinear form $a_N(,)$. To guaranty such property, α_N has to satisfy the following condition,

$$\frac{\alpha_N}{2} \geq \frac{\|v_n\|_{L_2(\partial\Omega)}}{\|\nabla v\|_{L_2(\Omega)}} \tag{3.234}$$

It depends on the order of the polynomial basis in $\mathbf{P}(\boldsymbol{x})$ and depends on the shape and the size of the compact support, and the regularity of the boundary. In general, a choice of large α_N will guaranty the positive definiteness of stiffness matrix, and ensure a numerical solution, but it may result in the bad condition number of the stiffness matrix or other adverse effects on certain linear solvers. Schweitzer[401] developed a procedure to determine α_N by solving a generalized eigenvalue problem. Since its variational consistency, guaranteed convergence, and stability, overall, Nitsche's method is a better candidate to enforce essential boundary condition for meshfree Galerkin method.

3.4.4 Transform Method

In this section, a consistent method is presented to enforce the essential boundary condition for RKPM, or MLS interpolant.

For simplicity, the following boundary value problem is considered,

$$\sigma_{ji,j} + b_i = 0, \tag{3.235}$$
$$\sigma_{ji} n_j = T_i^0 , \qquad \forall \, \boldsymbol{x} \in \Gamma_T \tag{3.236}$$
$$u_i = u_i^0 , \qquad \forall \, \boldsymbol{x} \in \Gamma_u \tag{3.237}$$

The corresponding weak formulation is

$$\int_\Omega \sigma_{ji} \delta u_{(j,i)} d\Omega - \int_\Omega b_i \delta u_i d\Omega$$
$$- \int_{\Gamma_T} T_i^0 \delta u_i d\Gamma - \int_{\Gamma_u} R_i \delta u_i d\Gamma = 0 . \tag{3.238}$$

Assume that the discrete trial, and test functions have the form

$$u_i^h(\mathbf{X},t) = \sum_{I=1}^{NP} N_I(\mathbf{X}) d_{iI}(t) . \tag{3.239}$$

$$\delta u_i^h(\mathbf{X},t) = \sum_{I=1}^{NP} N_I(\mathbf{X}) \delta d_{iI}(t) . \tag{3.240}$$

Substituting (3.239)-(3.240) into (3.238), a set of algebraic equations may be obtained, which provides the numerical solution of the problem.

Unlike FEM approximation, the RKPM or MLS interpolant has a shortcoming: that is its inability to represent essential boundary condition via boundary value interpolation, i.e.

$$u_i^h(\boldsymbol{x},t) \neq u_i^0(\boldsymbol{x},t) , \qquad \forall \, \boldsymbol{x} \in \Gamma_u \tag{3.241}$$

This is reflected in the weak form (3.238) as the extra term, $\int_{\Gamma_u} t_i \delta u_i d\Gamma$, which is a nuisance because the reaction force, R_i, is unknown on the essential boundary.

By modifying the meshfree interpolant, the essential boundary condition can be enforced in the interpolation scheme. To do so, one may distribute N_b number of particles along the boundary Γ^u to enforce the meshfree interpolant such that:

$$u^h(\boldsymbol{x}_I) = u^0(\boldsymbol{x}_I) =: g(\boldsymbol{x}_I) = g_I , \qquad I = 1, \cdots, N_b . \tag{3.242}$$

By letting $N_{nb} := NP - N_b$, the particles and the associated discrete field variables can be separated into two groups where the subscript "b" denotes the boundary terms and "nb" denotes the non-boundary terms. Therefore, $g_{iI}(t) := g_i(\boldsymbol{x}_I, t)$, $\quad I = 1, \cdots\cdots, N_b$, and meshfree interpolation field can be written as:

$$u^h(\boldsymbol{x}) = \sum_{I=1}^{NP} N_I(\boldsymbol{x}) u_I = \sum_{I=1}^{N_b} N_I^b(\boldsymbol{x}) u_I^b + \sum_{I=1}^{N_{nb}} N_I^{nb}(\boldsymbol{x}) u_I^{nb}$$

$$= \mathbf{N}^b(\boldsymbol{x}) \mathbf{u}^b + \mathbf{N}^{nb}(\boldsymbol{x}) \mathbf{u}^{nb} \; ; \tag{3.243}$$

where

$$\mathbf{N}^b(\boldsymbol{x}) := \{N_1^b(\boldsymbol{x}), \cdots, N_{N_b}^b(\boldsymbol{x})\} \;, \quad \mathbf{u}^b := \{u_1^b, \cdots, u_{N_b}^b\}^T \; ; \tag{3.244}$$

$$\mathbf{N}^{nb}(\boldsymbol{x}) := \{N_1^{nb}(\boldsymbol{x}), \cdots, N_{N_{nb}}^{nb}(\boldsymbol{x})\} \;, \quad \mathbf{u}^{nb} := \{u_1^{nb}, \cdots, u_{N_{nb}}^{nb}\}^T \;. \tag{3.245}$$

Define

$$\mathbf{D}^b := \begin{pmatrix} N_1^b(\boldsymbol{x}_1) & \cdots & N_I^b(\boldsymbol{x}_1) & \cdots & N_{N_b}^b(\boldsymbol{x}_1) \\ \vdots & & \vdots & & \vdots \\ N_1^b(\boldsymbol{x}_I) & \cdots & N_I^b(\boldsymbol{x}_I) & \cdots & N_{N_b}^b(\boldsymbol{x}_I) \\ \vdots & & \vdots & & \vdots \\ N_1^b(\boldsymbol{x}_{N_b}) & \cdots & N_I^b(\boldsymbol{x}_{N_b}) & \cdots & N_{N_b}^b(\boldsymbol{x}_{N_b}) \end{pmatrix}^{N_b \times N_b} \; ; \tag{3.246}$$

$$\mathbf{D}^{nb} := \begin{pmatrix} N_1^{nb}(\boldsymbol{x}_1) & \cdots & N_I^{nb}(\boldsymbol{x}_1) & \cdots & N_{N_{nb}}^{nb}(\boldsymbol{x}_1) \\ \vdots & & \vdots & & \vdots \\ N_1^{nb}(\boldsymbol{x}_I) & \cdots & N_I^{nb}(\boldsymbol{x}_I) & \cdots & N_{N_{nb}}^{nb}(\boldsymbol{x}_I) \\ \vdots & & \vdots & & \vdots \\ N_1^{nb}(\boldsymbol{x}_{N_b}) & \cdots & N_I^{nb}(\boldsymbol{x}_{N_b}) & \cdots & N_{N_{nb}}^{nb}(\boldsymbol{x}_{N_b}) \end{pmatrix}^{N_b \times N_{nb}} \;. \tag{3.247}$$

Thus the discrete essential conditions, Eq. (3.253), may read as follows:

$$\mathbf{D}^b \mathbf{u}^b = \mathbf{g} - \mathbf{D}^{nb} \mathbf{u}^{nb} \;, \tag{3.248}$$

$$\mathbf{D}^b \delta\mathbf{u}^b = -\mathbf{D}^{nb} \delta\mathbf{u}^{nb} \;, \tag{3.249}$$

where $\mathbf{g} := \{g_1, \cdots, g_{N_b}\}^T$. Inverting matrix $\mathbf{D}^b$:

$$\mathbf{u}^b = (\mathbf{D}^b)^{-1} \mathbf{g} - (\mathbf{D}^b)^{-1} \mathbf{D}^{nb} \mathbf{u}^{nb} \; ; \tag{3.250}$$

$$\delta\mathbf{u}^b = -(\mathbf{D}^b)^{-1} \mathbf{D}^{nb} \delta\mathbf{u}^{nb} \; ; \tag{3.251}$$

Substituting Eq. (3.250) back into Eq. (3.243) yields:

$$u^h(\boldsymbol{x}) = \sum_{I=1}^{NP} N_I(\boldsymbol{x}) u_I = \mathbf{N}^b(\boldsymbol{x}) (\mathbf{D}^b)^{-1} \mathbf{g}$$

$$+ \left(\mathbf{N}^{nb}(\boldsymbol{x}) - \mathbf{N}^b(\boldsymbol{x})(\mathbf{D}^b)^{-1} \mathbf{D}^{nb}\right) \mathbf{u}^{nb} \;. \tag{3.252}$$

Obviously, for $\boldsymbol{x}_I \in \Gamma_u$, $I = 1, \cdots, N_b$:

$$u^h(\boldsymbol{x}_I) = g_I \; , \tag{3.253}$$

$$\delta u^h(\boldsymbol{x}_I) = 0 \; . \tag{3.254}$$

Eq. (3.252) can be also interpreted as a transformation of the shape functions:

$$u^e(\boldsymbol{x}) = \sum_{I=1}^{N_b} W_I^b(\boldsymbol{x}) u_I^b + \sum_{I=1}^{N_{nb}} W_I^{nb} u_I^{nb} = \mathbf{W}^b(\boldsymbol{x})\mathbf{g} + \mathbf{W}^{nb}(\boldsymbol{x})\mathbf{u}^{nb} \; ; \tag{3.255}$$

where $\mathbf{W}^b(\boldsymbol{x}) := \mathbf{N}^b(\boldsymbol{x})(\mathbf{D}^b)^{-1}$, and $\mathbf{W}^{nb}(\boldsymbol{x}) := \left[\mathbf{N}^{nb}(\boldsymbol{x}) - \mathbf{N}^b(\boldsymbol{x})(\mathbf{D}^b)^{-1}\mathbf{D}^{nb}\right]$. Here $\mathbf{W}^{nb}(\boldsymbol{x})$ can be viewed as modified, or new shape function, which takes zero value at boundary nodals, or boundary particles. It can be shown that $\forall \boldsymbol{x}_I \in \Gamma_u$

$$\mathbf{N}^b(\boldsymbol{x}_I)(\mathbf{D}^b)^{-1} = (\underbrace{0, \cdots, 0, 1, 0, \cdots, 0}_{I}) \tag{3.256}$$

$$(\underbrace{0, \cdots, 0, 1, 0, \cdots, 0}_{I})\mathbf{D}^{nb} = \mathbf{N}^{nb}(\boldsymbol{x}_I) \tag{3.257}$$

consequently

$$\mathbf{W}^{nb}(\boldsymbol{x}_I) = \left(\mathbf{N}^{nb}(\boldsymbol{x}_I) - \mathbf{N}^b(\boldsymbol{x}_I)(\mathbf{D}^b)^{-1}\mathbf{D}^{nb}\right)$$
$$= \left(0, 0, \cdots, 0\right) \; , \quad I = 1, 2, \cdots, N_b \tag{3.258}$$

Let

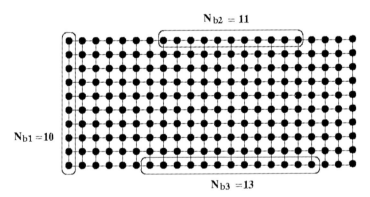

Fig. 3.12. Piecewise essential boundaries.

$$\mathbf{u}^h(\boldsymbol{x}) = \sum_{I \in \Lambda_{nb}} W_I^{nb}(\boldsymbol{x})\mathbf{u}_I + \sum_{I \in \Lambda_b} W_I^b(\boldsymbol{x})\mathbf{g}_I \tag{3.259}$$

$$\delta\mathbf{u}^h(\boldsymbol{x}) = \sum_{I \in \Lambda_{nb}} W_I^{nb}(\boldsymbol{x})\delta\mathbf{u}_I \tag{3.260}$$

Obviously, $\delta\mathbf{u} \neq 0, \forall \boldsymbol{x} \in \Gamma_u$. Thus the last term in (3.238) still can not be dropped. Substituting (3.260) into (3.238) yields the following algebraic equations,

$$\mathbf{K} \cdot \mathbf{U} = \mathbf{f} \tag{3.261}$$

where $\mathbf{U} = \{\mathbf{u}_1, \mathbf{u}_2, \cdots, \mathbf{u}_I, \cdots, \mathbf{u}_{NP}\}^T$. For 2D plane stress linear elasticity problem, one may have explicit expressions on both stiffness matrix, as well as force terms

$$\mathbf{K}_{IJ} = \int_{\Omega} \mathbf{B}_I^T \mathbf{D} \mathbf{B}_J d\Omega \tag{3.262}$$

$$\mathbf{f}_I = \int_{\Gamma_t} W_I^{nb} \bar{\mathbf{T}} dS + \int_{\Gamma_u} W_I^{nb} \mathbf{R} dS + \int_{\Omega} W_I^{nb} \mathbf{b} d\Omega \tag{3.263}$$

and

$$\mathbf{B}_I = \begin{pmatrix} W_{I,x}^{nb}, & 0 \\ 0, & W_{I,y}^{nb} \\ W_{I,y}^{nb}, & W_{I,x}^{nb} \end{pmatrix} \tag{3.264}$$

Our claim is that

$$\int_{\Gamma_u} W_I^{nb} \mathbf{R} dS = \int_{\Gamma_u} R^i(\boldsymbol{x})\{\mathbf{N}^{nb}(\boldsymbol{x}) - \mathbf{N}^b(\boldsymbol{x})(\mathbf{D}^b)^{-1}\mathbf{D}^{nb}\}\delta\mathbf{u}_i^{nb} d\Gamma \approx 0 \tag{3.265}$$

In 2D case, this fact can be shown by considering a special case: Assume that the essential boundary is a straight line segment $[a,b]$, and there are N_b particles distributed evenly on the segment, and $h = |b - a|/(N_b - 1)$. Based on trapezoidal rule, or Simpson Rule, the following estimate can then be reached immediately,

$$\left| \int_{\Gamma_u} \mathbf{R}(x) W_I^{nb}(x) d\Gamma \right| \leq \begin{cases} \sup\limits_{x \in [a,b]} \{|\mathbf{R}(x)|\} \dfrac{(b-a)h^2}{12} \left| W_I^{nb(2)}(\zeta) \right|, & a < \zeta < b \\[4mm] \sup\limits_{x \in [a,b]} \{|\mathbf{R}(x)|\} \dfrac{(b-a)h^4}{180} \left| W_I^{nb(4)}(\zeta) \right|, & a < \zeta < b \end{cases} \tag{3.266}$$

This is because we deliberately choose the sample points of trapezoidal rule, or Simpson Rule on Γ_u as the boundary particles, $I = 1, 2, \cdots, N_b$. Since at every boundary particle point, $W_I^{nb}(\boldsymbol{x}_I) = 0$, $I = 1, 2, \cdots, N_b$, and therefore the discrete summation $\sum\limits_{I=1}^{N_b} \mathbf{R}(x_I) W_I^{nb}(x_I) \Delta x_I = 0$, here Δx_I are the integration weight. This estimate can be further improved, provided that the window function is "very smooth". In other words, Eq. (3.266) suggests that if the window function is smooth enough, the magnitude of the error coming from the approximated essential boundary condition (3.253) can be made as the same, or even less than that of the interpolation error. This proves our claim. So far in practice, the transform method is the most efficient, and most popular approach. However, after the transformation, the support size as well as shape of new shape functions have changed due to the influence from the boundary particles. Thus, special consideration has to be taken when one retrieves the stored shape functions using the support size criteria. However, such problem does NOT exist, if the original shape functions are transformed at each time step. In Fig. 3.13, the change of the support for an interior node J is illustrated. Because of the influence from the boundary node I, the support of the node J has changed from its original shape (the shaded area in the top picture) to its final one (the shaded area in the bottom picture).

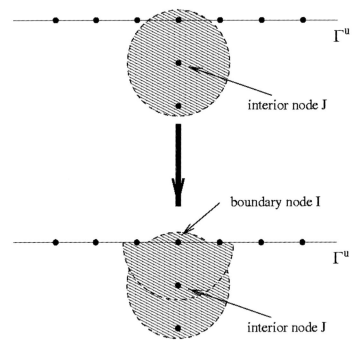

Fig. 3.13. Support change for an interior node after shape function transformation.

3.4.5 Boundary Singular Kernel Method

The idea of using singular kernel function to enforce the Kronecker delta property should be attributed to Lancaster & Salkauskas,[255] which they called the *interpolating moving least square interpolant*. Some authors later used it in computations, e.g. Kaljevic & Saigal[235] and Chen & Wang.[104] The idea is quite simple. Take a set of positive shape function $\{\Phi_h(\boldsymbol{x} - \boldsymbol{x}_I)\}_{I=1}^N$. Suppose $\boldsymbol{x}_J$ is on the boundary Γ_u, we modify the shape function basis as,

$$
\tilde{\Phi}_h(\boldsymbol{x} - \boldsymbol{x}_I) = \begin{cases} \dfrac{\Phi_h(\boldsymbol{x} - \boldsymbol{x}_I)}{|\boldsymbol{x} - \boldsymbol{x}_I|^2}, & \forall \ I \in \Gamma_u \\[2mm] \Phi_h(\boldsymbol{x} - \boldsymbol{x}_I), & \forall \ I \notin \Gamma_u \end{cases} \tag{3.267}
$$

and then build a new shepard basis on $\{\Phi_h(\boldsymbol{x} - \boldsymbol{x}_I)\}$ as

$$
\Psi_h(\boldsymbol{x} - \boldsymbol{x}_I) = \frac{\tilde{\Phi}_h(\boldsymbol{x} - \boldsymbol{x}_I)}{\sum_I \tilde{\Phi}_h(\boldsymbol{x} - \boldsymbol{x}_I)} \tag{3.268}
$$

one may verify that for the boundary nodes $\boldsymbol{x}_J$, $\Psi_h(\boldsymbol{x}_I - \boldsymbol{x}_J) = \delta_{IJ}$. In real computations, the procedure works in certain range of dilation parameter, h, but when h is either too large or too small, the convergence rate of interpolation error drops rapidly.[104]

3.4.6 Coupled Finite Element and Particle Approach

Another approach to enforce Dirichlet boundary condition is to couple finite element with particles close to the boundary and necklace the particle domain with a FEM boundary layer and apply essential boundary conditions to the finite element nodes (Krongauz & Belytschko[248] and Liu et al.[306]). In this approach, all the boundary and its neighborhood are meshed with finite element nodal points and there is a buffer zone between finite element zone and particle zone, which is connected with the so-called ramp functions. Denote the finite element basis as $\{N_i(x)\}$, particle basis as $\{\Phi_i(x)\}$, and ramp function as $R(x)$. The interpolation function in the buffer zone is the combination of FEM and particle interpolant

$$
\tilde{\Phi}_i(x) = \begin{cases} (1 - R(x))\Phi_i(x) + R(x)N_i(x) & x \in \Omega_{fem} \\[2mm] \Phi(x) & x \in \Omega_p \end{cases} \tag{3.269}
$$

where the ramp function is chosen as $R(x) = \sum_i N_i(x), \ x_i \in \partial\Omega_{fem}$.

This approach is recently used again by Liu & Gu in a meshfree local Petrov-Galerkin (MLPG) implementation.[283]

Although the method works well, it compromises the essential nature, and subsequently the advantages of the meshfree method, and one may be wondering why one chooses to use the method in the first place. Indeed, some numerical evidences show that, after FEM-particle coupling, the numerical algorithm loses its advantages as being meshfree method. For example, in shear band simulations, the mesh alignment sensitivity due to the finite element mesh around the boundary ruins the entire numerical simulation.

To enforce the Dirichlet boundary condition while still retaining the advantage of particle method, recently, a so-called hierarchical enrichment of technique is developed to enforce the essential boundary condition,[443] which is a further development of the work.[306] The idea is as follows. Around the boundary, one has a layer of finite element nodes, and all the nodes on the boundary are finite element nodes. Right within the boundary the particles are mixed with the finite element nodes, and there is no buffer zone. Denote the finite element shape function as $N_I(\boldsymbol{x})$ $I \in B$; and denote meshfree shape function as $\Phi_I(\boldsymbol{x}), I \in A$. One can view that particle discretization as an enrichment of finite element discretization.

$$u^h(\boldsymbol{x}) = \sum_{I \in B} N_I(\boldsymbol{x}) a_I + \sum_{I \in A} \tilde{\Phi}_I(\boldsymbol{x}) d_I \tag{3.270}$$

where $\tilde{\Phi}_I(\boldsymbol{x})$ is complementary to the finite element basis, i.e.

$$\tilde{\Phi}_I(\boldsymbol{x}) = \Phi_I(\boldsymbol{x}) - \sum_{J \in B} N_J(\boldsymbol{x}) \Phi_I(\boldsymbol{x}_J) \tag{3.271}$$

It is easy to verify that for a boundary particle $\boldsymbol{x}_I$, $I \in B$, $u^h(\boldsymbol{x}_I) = a_I$. Thus Dirichlet boundary condition can be specified directly through the coefficients $\{a_I\}_{I \in B}$. Since particles can be made very close to the boundary, as long as they are not the boundary nodes.

It is worth mentioning that even though meshfree interpolants have no difficulties in enforcing natural boundary conditions, the implementation of enforcing natural boundary conditions in meshfree setting is different from those in FEM setting. In finite element procedure, one only need to calculate a surface, or curve line integral in evaluating traction boundary conditions; whereas in meshfree setting, one has to take into account of the influences from the interior particles as well, though this is seldom mentioned in the literature.

3.5 H-P Clouds Method

Introducing p-enrichment in a meshfree discretization, Duarte and Oden[153, 154] attach a sequence of Legendre polynomials with moving least square interpolant to construct a first p-version meshfree interpolant, which they named as h-p Clouds.

The fundamental idea in the h-p cloud method is the construction of a hierarchical basis using the partition of unity $\mathcal{G}_N = \{\Phi_I^\ell(x)\}$. These class of functions can be constructed at a cost comparable to the computation of finite element shape functions and has the property that, for a proper choice of the $\mathbf{P}$ vector, we can ensure $\mathcal{P}_p \subset span\{\mathcal{F}_N^{\ell,p}\}$ where $\mathcal{P}_p$ demotes the space of polynomial of degree less to equal to p.

Let $\mathcal{L}_p$ denote a set of tensor-product complete polynomials L_{ijk} in $\mathbb{R}^3$,

$$L_{ijk}(\mathbf{x}) = L_i(x_1)L_j(x_2)L_k(x_3), \quad 0 \le i,j,k \le p \tag{3.272}$$

where L_i is a Legendre polynomial of degree i in $\mathbb{R}$. Let $\mathcal{G}_N^\ell := \{\Phi_I^\ell(x)\}_{I \in \Lambda}$ denotes a partition of unity. The $\mathcal{G}_N^\ell$ is called $\mathcal{L}_\ell$ reducible, if it can reproduce any element $L_{ijk} \in \mathcal{L}_\ell$, that is

$$L_{ijk}(\mathbf{x}) = \sum_{I \in \Lambda} L_{ijk}(\mathbf{x}_I)\Phi_I^\ell(x) \tag{3.273}$$

The so-called h-p Clouds is to add, hierarchically, appropriate basis element to the original partition of unity, $\{\Phi_I^\ell\}_{I \in \Lambda}$ such that the resulting basis can reproduce polynomial of degree $p > \ell$.

One hierarchical family constructed by Duarte and Oden is called $\mathcal{F}_N^{\ell,p}$, whose structure can be expressed as

$$\mathcal{F}_N^{\ell,p} = \Big(\{\Phi_I^\ell(x)\} \bigcup \{\Phi_I^\ell(x)L_{ijk}(x)\} : I \in \Lambda; 0 \le i,j,k \le p$$
$$i \text{ or } j \text{ or } k > \ell; \ p \ge \ell \Big) \tag{3.274}$$

Duarte and Oden[153] showed that $\mathcal{F}_N^{\ell,p}$ can reproduce $L_{ijk} \in \mathcal{L}_p$. Note that the point here is $p > \ell$!.

To prove $\mathcal{F}_N^{\ell,p}$ can reproduce any $L_{ijk} \in \mathcal{L}_p$. One must show that any $L_{ijk} \in Lcal_p$ can be represented by a linear combination of basis functions in $\mathcal{F}_N^{\ell,p}$, i.e.

$$L_{rst}(\mathbf{x}) = \sum_{I \in \Lambda} \Big(a_I \Phi_I^\ell(x) + \sum_{\substack{0 \le i,j,k \le p , \\ i \text{ or } j \text{ or } k > \ell}} b_{Iijk}\Phi_I^\ell(x)L_{ijk}(x) \Big) \tag{3.275}$$

If $r,s,t < \ell$, then choose

$$a_I = L_{rst}(\mathbf{x}_I), \ I \in \Lambda$$
$$b_{Iijk} = 0$$

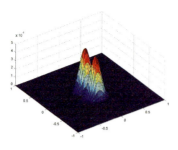

(a) hp-Clouds shape function,
$\Psi_I(\boldsymbol{x})\boldsymbol{x}^2$;

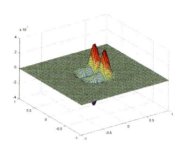

(b) hp-Clouds shape function,
$\Psi_I(\boldsymbol{x})\boldsymbol{x}^2\mathbf{y}$;

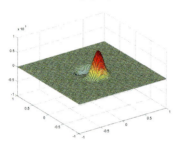

(c) hp-Clouds shape function,
$\Psi_I(\boldsymbol{x})\boldsymbol{x}^3$;

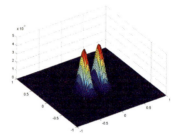

(d) hp-Clouds shape function
$\Psi_I(\boldsymbol{x})\mathbf{y}^2$;

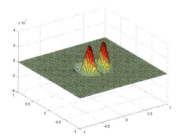

(a) hp-Clouds shape function,
$\Psi_I(\boldsymbol{x})\mathbf{y}^2\mathbf{x}$;

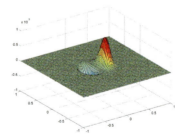

(b) hp-Clounds shape function,
$\Psi_I(\boldsymbol{x})\mathbf{y}^3$;

Fig. 3.14. Hp-cloud shape functions.

since $\{\Phi_I^\ell\}$ are $\mathcal{L}_\ell$ reducible.

If i, or j, or $k > \ell$, then take

$$a_I = 0$$

$$b_{Iijk} = \begin{cases} 1 & \text{if } i = r, j = s, k = t \\ \\ 0 & \text{otherwise} \end{cases} \quad , \quad I \in \Lambda \qquad (3.276)$$

Therefore,

$$\sum_{I \in \Lambda} \left(a_I \Phi_I^\ell(\boldsymbol{x}) + \sum_{\substack{0 \le i,j,k \le p \ , \\ i \text{ or } j \text{ or } k > \ell}} b_{Iijk} \Phi_I^\ell L_{ijk}(\boldsymbol{x}) \right) =$$

$$\sum_{I \in \Lambda} \left(b_{Irst} \Phi_I^\ell(\boldsymbol{x}) \right) = L_{rst}(\boldsymbol{x}) \sum_{I \in \Lambda} \Phi_I^\ell(\boldsymbol{x}) = L_{rst}(\boldsymbol{x}) \qquad (3.277)$$

In one dimensional case, the h-p Clouds hierarchical interpolation takes form

$$u^h(x) = \sum_{I \in \Lambda} \Phi_I^{n+1}(x) L_{IJ}(x) b_J$$

$$= \sum_{I \in \Lambda} \Phi_I^{n+1}(x) \left(u_I L_0 + \sum_{i=1}^{\ell} b_{iI} L_i(x) \right) \qquad (3.278)$$

where $\Phi_I^{n+1}(x)$ is the $n+1$ order moving least square interpolant. In general, $L_i(x)$ may be regarded as the Taylor expansion of $u(x)$ at point x_I. The reason Duarte and Oden choose Legendre polynomial is because of their better conditioning properties; a common procedure has been used in p-version finite element.[400]

This philosophy is similar to the philosophy of a more general framework — the so-called *partition of unity method*, which is proposed by Babuška and Melenk.[23, 323]

In principle, finite element interpolants are partition of unity, and almost all the meshfree interpolants are partition of unity. It would be an appropriate terminology to unify a general class of Galerkin interpolants under the category of partition of unity.

Nevertheless, Babuška's partition of unity method is a special technique to construct meshfree or mesh-based interpolant tailored to solve the differential equation that is under consideration.

3.6 The Partition of Unity Method (PUM)

The essense of the method is to take a partition of unity and multiply it with other independent basis that one deems "meritorious", or simply desirable for solving the specific partial differential equations. The method offers flexibility and leverage in computations, especially when the users have some prior knowledge about the problem they are solving.

Definition 3.6.1. *Let $\Omega \subset \mathbb{R}^d$ be an open set, $\{\Omega_I\}$ be an open cover of Ω satisfying a pointwise overlap condition,*

$$\exists M \in \mathbb{N} \quad \forall \, \boldsymbol{x} \in \Omega \quad card\{I \mid \boldsymbol{x} \in \Omega_I\} \le M \qquad (3.279)$$

Let $\{\Phi_I\}$ be a *Lipschitz partition of unity subordinate to the cover* $\{\Omega_I\}$ *satisfying*

$$supp\{\Phi_I\} \subset closure(\Omega_I), \quad \forall I \in \Lambda \tag{3.280}$$

$$\sum_{I \in \Lambda} \Phi_I = 1 \quad \forall \; x \; \in \Omega \tag{3.281}$$

$$\|\Phi\|_{L^\infty(\mathbb{R}^d)} \le C_\infty \tag{3.282}$$

$$\|\nabla \Phi_I\|_{L^\infty(\mathbb{R}^d)} \le \frac{C_G}{diam\,\Omega_I} \tag{3.283}$$

where C_∞, C_G *are two constants. Then* $\{\Phi_I\}$ *is called a* (M, C_∞, C_G) *partition of unity subordinate to the cover* $\{\Omega_I\}$. *The partition of unity* $\{\Phi_i\}$ *is said to be of degree* $m \in \mathbb{N}_0$ *if* $\{\Phi_I\} \subset C^m(\mathbb{R}^d)$. *The covering* $\{\Omega_I\}$ *are called patches.*

3.6.1 Examples of Partition of Unity

In the following, several examples of (M, C_∞, C_G) partition of unity of one dimensional are described, which are subordinated to a covering that covers the domain $(0, 1)$.

Example 3.6.1. The usual piecewise linear hat-function forms a partition of unity. Let

$$\Phi^1(x) = \begin{cases} 1 + \dfrac{x}{h} & \forall x \in (-h, 0) \\[2mm] 1 - \dfrac{x}{h} & \forall x \in (0, h) \\[2mm] 0 & elsewhere \end{cases} \tag{3.284}$$

and define the partition of unity by $\Phi^1_J(x) = \Phi^1(x - x_J), \; J \in \Lambda$.

where $x_J = \pm Jh$ and $J = 0, 1, 2, \cdots,$.

Example 3.6.2. Functions that are identically equal to one on a subset of their support can also form a partition of unity.

$$\Phi^2(x) = \begin{cases} \dfrac{3}{2} + 2\dfrac{x}{h} & \forall x \in (-\dfrac{3}{4}h, -\dfrac{h}{4}] \\[3mm] 1 & \forall x \in (-\dfrac{h}{4}, \dfrac{h}{4}) \\[3mm] \dfrac{3}{2} - 2\dfrac{x}{h} & \forall x \in (\dfrac{h}{4}, \dfrac{3}{4}h] \\[3mm] 0 & elsewhere \end{cases} \tag{3.285}$$

and define the partition of unity by $\Phi_J^2(x) = \Phi^2(x - x_J)$, $J \in \Lambda$. where $x_J = \pm Jh$ and $J = 0, 1, 2, \cdots$,.

The above two examples of partition of unity are piece-wise continuous (C^0). Partition of unity with higher order continuity can be constructed as well. The following is a piecewise C^1 polynomial example.

Example 3.6.3. Define

$$
\Phi^3(x) = \begin{cases} (x+h)^2(h-2x), & \forall\, x \in (-h, 0] \\[2mm] (x-h)^2(h+2x), & \forall\, x \in (0, h) \\[2mm] 0, & elsewhere \end{cases} \tag{3.286}
$$

The individual members of the partition of unity is defined by $\Phi_J^3(x) = \Phi^4(x - x_J)$ on the patch Ω_J.

3.6.2 Examples of PUM Interpolants

Before constructing problem dependent PUM interpolants, we first describe the construction process.

Definition 3.6.2. *Let $\{\Omega_I\}_{I \in \Lambda}$ be an open cover of $\Omega \subset \mathbb{R}^d$ and let $\{\Phi_I\}$ be a (M, C_∞, C_G) partition of unity subordinate to $\{\Omega_I\}$. Let $V_I \subset H^1\left(\Omega_I \cap \Omega\right)$ be a given function space with the basis $V_I := \{v_{IJ}\}_{J \in V_I}$. Then the space*

$$
V := \sum_{I \in \Lambda} \Phi_I V_I = \left\{ \sum_{I \in \Lambda} \sum_{J \in V_I} \Phi_I v_{IJ} \right\} \subset H^1(\Omega) \tag{3.287}
$$

is called the PUM space. The PUM space V is said to be of degree $m \in \mathbb{N}$ if $V \in C^m(\Omega)$. The space V_I are referred to as the local approximation spaces.

Consider an one-dimensional example,

$$
\begin{aligned} -u'' + k^2 u &= f, \quad \forall\, x \in (0,1) \\ u(0) &= 0 \\ u'(1) &= g \end{aligned} \tag{3.288}
$$

where $f \in C^2[0,1]$ and $g \in \mathbb{R}$.

Let $n \in \mathbb{N}$ $h = \dfrac{1}{n}$ and define $x_J = Jh$, $J = 0, 1, \cdots, n$. Define $x_{-1} = -h$ and $x_{n+1} = 1 + h$ and set the patches $\Omega_J = (x_{J-1}, x_{J+1})$, $J = 0, 1, \cdots, n$. On each path Ω_J, We defined a local space that approximate the solution of problem (3.288).

$$
\begin{aligned} V_J &= \mathrm{span}\{1, \sinh kx, \cosh kx\} \quad \forall\, x \in \Omega_J \cap \Omega, \quad J = 1, 2, \cdots, \\ V_0 &= \mathrm{span}\{\sinh kx, 1 - \cosh kx\} \quad \forall\, x \in \Omega_0 \cap \Omega \end{aligned} \tag{3.289}
$$

Choosing $\{\Phi_I^1(x)\}$ as the partition of unity, the resulting PUM interpolation space is

$$V^{PU} = span\ \big\{\Phi_J^1(x), \Phi_J^1(x)\sinh(kx), \Phi_J^1(x)\cosh(kx), \tag{3.290}$$

$$\Phi_0^1(x)\sinh(kx), \Phi_0^1(x)(1 - \cosh kx), \quad J \in \Lambda\big\} \tag{3.291}$$

In other example, Babuška and Melenk[323] used the following PUM interpolation

$$u^h(x) = \sum_{I \in \Lambda} \Phi_I(x)(a_{0I} + a_{1I}x + a_{2I}y + b_{1I}\sin(nx) + b_{2I}\cos(nx)) \tag{3.292}$$

to solve Helmholtz equation. Dolbow et al.[151] used the following interpolant to simulate strong discontinuity, i.e. the crack surfaces,

$$\mathbf{u}^h(\boldsymbol{x}) = \sum_I N_I(\boldsymbol{x})\Big[\mathbf{u}_I + H(\boldsymbol{x})\boldsymbol{b}_I + \sum_J \mathbf{c}_{IJL}F_L(\boldsymbol{x})\Big] \tag{3.293}$$

where $H(\boldsymbol{x})$ is the Heaviside function. If $\Phi_I(x)$ is a meshfree interpolant, then the method is a meshfree method; if $\Phi_I(x)$ is a finite elment interpolant, the method is called PUFEM, which is the acronym of *partition of unity finite element method*. Recently, Wagner et al.[442] used a similar version of PUFEM, which they termed as X-FEM, to simulate rigid particle movement in a Stokes flow. By attaching discontinuous function to a partition of unity, they can represent particle shape accurately, and particle surface need not to conform to the finite element boundary, so that moving particles can be simulated without remeshing. The so-called X-FEM technique is also used by Daux et al.[134] to model cracks, especially the crack with arbitrary branch, or intersecting cracks.

In fact, a widely advertised example is that one can build a h-p clouds on the simplest meshless partition of unity — Shepard interpolant, i.e. one can pile up higher order polynomials to a shepard interpoalnt. By doing so, one does not need the matrix inversion when constructing higher order meshfree shape function, but one may still enjoy good interpolation convergence.

3.7 Meshfree Quadrature and Finite Sphere Method

Most meshfree Galerkin methods, at least the early versions, still use background cell, or background grid to deploy quadrature points in order to integrate the weak form. The background cell, or "background mesh" is different from a finite element mesh. It is a rather loose data structure, which does not need to be structured and can be any non-overalpping domain decomposition, and hence is easy to be refined.

Nevertheless, there is still a "ghost mesh" present. The search for bona fide meshfree methods continues. Moreover, how to place such background cell, or how to place quadrature points in the domain of interests will directly influence the accuracy as well as the invertibility of stiffness matrix. The main concern of meshfree patch-test[40, 45, 149] is in fact the stability of quadrature integration. MLS interpolant is a partition of unity and the linear completeness or interpolation consistency is a priori, there is neither incompatibility nor rigid body mode left to be tested unlike the finite element shape function. However, if there are not enough quadrature points in a compact support, or quadrature points are not evenly distributed, spurious modes may occur.

Up to today, this is one of the two major shortcomings (the cost of meshfree methods is another one) left when meshfree methods are compared with finite element methods. Dolbow & Belytschko[150] studied Gauss quadrature integration error with respect to different set-up of background cells as well as local quadrature point distributions (distribution within the cell). They found that if background cell does not match with the compact support of the meshfree shape function, considerable integration error will rise. One may want to relate to this predicament with the similar foes in SPH computations. Since SPH is a strong form collocation type of approximation, there is no quadrature integration involved. It, nevertheless, can be viewed as a nodal point integration scheme, which, as mentioned above, may cause numerical pathologies, such as tensile instability, zero-energy mode, etc.

There is a simple, but effective replacement for backgound cell quadrature. That is the local, self-similar support integration. The idea is quite simple. Assume that meshfree shape functions are compactly supported, and the support for each and every particle is similar in shape, say circular region in 2D, a sphere in 3D.

The key idea here is to form a *local weak formulation*. To illustrate the idea, we consider linear elastostatics model problem,

$$\nabla \cdot \boldsymbol{\sigma} + \mathbf{b} = 0, \quad \forall \, \boldsymbol{x} \in \Omega \tag{3.294}$$

and

$$\mathbf{u} = \bar{\mathbf{u}}, \quad \forall \, \boldsymbol{x} \in \Gamma_u \tag{3.295}$$

$$\mathbf{n} \cdot \boldsymbol{\sigma} = \bar{\mathbf{t}}, \quad \forall \, \boldsymbol{x} \in \Gamma_t \tag{3.296}$$

where Ω is the problem domain, Γ_u is the essential boundary, and Γ_t is the traction boundary.

A global Galerkin weak form may be formed as

$$\int_{\Omega} \boldsymbol{\epsilon}_v : \boldsymbol{\sigma} d\Omega + \alpha \int_{\Gamma_u} \mathbf{v} \cdot \mathbf{u} d\Gamma - \int_{\Gamma_u} \mathbf{v} \cdot \mathbf{t} d\Gamma = \int_{\Gamma_t} \mathbf{v} \cdot \bar{\mathbf{t}} d\Gamma$$

$$+ \alpha \int_{\Gamma_u} \mathbf{v} \cdot \bar{\mathbf{u}} d\Gamma + \int_{\Omega} \mathbf{v} \cdot \mathbf{b} d\Omega \tag{3.297}$$

where α is a penalty parameter.

Consequently, the stiffness matrix obtained from the global weak form is as follows

$$\mathbf{K}_{IJ} = \int_{\Omega} (\mathbf{B}_I)^T \mathbf{D} \mathbf{B}_J d\Omega + \alpha \int_{\Gamma_u} \mathbf{V}_I N_J^\varrho d\Gamma - \int_{\Gamma_u} \mathbf{V}_I \mathbf{N} \mathbf{D} \mathbf{B}_J d\Gamma \quad (3.298)$$

where in two-dimensional space

$$\mathbf{V}_I = \begin{bmatrix} N_I^\varrho & 0 \\ 0 & N_I^\varrho \end{bmatrix}, \quad \hat{\mathbf{u}}_J = \begin{bmatrix} \hat{u}_{1J} \\ \hat{u}_{2J} \end{bmatrix}$$

$$\mathbf{B}_I = \begin{bmatrix} N_{I,1}^\varrho & 0 \\ 0 & N_{I,2}^\varrho \\ N_{I,2}^\varrho & N_{I,1}^\varrho \end{bmatrix}$$

$$\mathbf{N} = \begin{bmatrix} n_1 & 0 & n_2 \\ 0 & n_2 & n_1 \end{bmatrix}$$

$$\mathbf{D} = \frac{\bar{E}}{1 - \bar{\nu}^2} \begin{bmatrix} 1 & \bar{\nu} & 0 \\ \bar{\nu} & 1 & 0 \\ 0 & 0 & (1 - \bar{\nu}/2) \end{bmatrix} \quad (3.299)$$

where

$$\bar{E} = \begin{cases} E \\ \dfrac{E}{1 - \nu^2} \end{cases} \quad \text{and} \quad \bar{\nu} = \begin{cases} \nu & \text{for plane stress} \\ \dfrac{\nu}{1 - \nu} & \text{for plane strain} \end{cases} \quad (3.300)$$

If both trial and test functions have the same shape of compact support (e.g. a circular region in 2D, and a sphere in 3D), the above integration can be rewritten with respect to the subdomain of a test function, $\Omega_I \cap \Omega$ for $N_I^\varrho(\boldsymbol{x})$,

$$\mathbf{K}_{IJ} = \int_{\Omega_I \cap \Omega} (\mathbf{B}_I)^T \mathbf{D} \mathbf{B}_J d\Omega + \alpha \int_{\Gamma_{Iu}} \mathbf{V}_I N_J^\varrho d\Gamma - \int_{\Gamma_{Iu}} \mathbf{V}_I \mathbf{N} \mathbf{D} \mathbf{B}_J d\Gamma \quad (3.301)$$

where Ω_I is the support of particle I, and $\Gamma_{Iu} = \partial\Omega_I \cap \Gamma_u$

Because all shape functions are compactly supported, outside $\Omega_I \cap \Omega$, meshfree interpolant, Φ_I, and its derivatives will be automatically zero, the integrals in the rest of domain, i.e. Ω/Ω_I vanish. Eq. (3.298) is identical with Eq. (3.301). To evaluate K_{IJ}, one only need to carry out integration in effective domain $\Omega_I \cap \Omega$ and Γ_{Iu}. On the other hand, Eq. (3.301) may be derived from a local weak formulation,

$$\int_{\Omega_I \cap \Omega} \boldsymbol{\epsilon}_v : \boldsymbol{\sigma} d\Omega + \alpha \int_{\Gamma_{Iu}} \mathbf{v} \cdot \mathbf{u} d\Gamma - \int_{\Gamma_{Iu}} \mathbf{v} \cdot \mathbf{t} d\Gamma = \int_{\Gamma_{It}} \mathbf{v} \cdot \bar{\mathbf{t}} d\Gamma$$

$$+ \alpha \int_{\Gamma_{Iu}} \mathbf{v} \cdot \bar{\mathbf{u}} d\Gamma + \int_{\Omega_I \cap \Omega} \mathbf{v} \cdot \mathbf{b} d\Omega \quad (3.302)$$

where $\Gamma_{It} := \Gamma_t \cap \partial\Omega$.

In fact, the finite element weak formulation can be also viewed as a local formulation. However, in finite element discretization, local elements do not

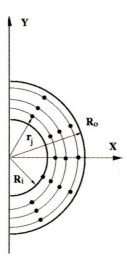

Fig. 3.15. A semi-annular on which the cubature rule is described.

overlap, whereas in meshfree discretization, local supports do overlap, though both of them are equivalent to the corresponding global weak formulations.

Since every Ω_I , $I = 1, ..., n$ have the same shape, once a quadrature rule is fixed for a compact support, it will be the same for other compact supports as well in computer implementation. Therefore, computer implementation of numerical quadrature can be easily modulated. The quadrature rule is then set for each and every compact support. Consequently, we can integrate the weak form locally from one compact support to another compact support by the same integration subroutine. Note that there is a difference between the global domain quadrature integration and local domain quadrature integration, because compact supports of meshfree interpolants are overlapped with each other the overlapped region will be integrated several times. Nonetheless, mathematically, the global weak formulation is the equivalent to the local weak formulation.

By using local quadrature, we are free from background integration cell, and hence we are free from any implicit mesh.

It is worth mentioning that since the supports of most of meshfree interpolants are n-dimensional spheres. The conventional Gauss quadrature rule may not be suitable in numerical integration. A so-called cubature rule is suitable for numerical integration in an annular region. It is recently documented and implemented in details in a meshfree Galerkin method known as the Finite Sphere Method by De and Bathe.[136]

3.7.1 Cubature on Annular Sectors in $\mathbb{R}^2$

A special case of the cubature rules discussed in reference[136] is documented here, which is in fact due to Pierce (1957).

Consider a semi-annular sector $\Omega \in \mathbb{R}^2$. A polynomial function with degree $k = 4m + 3$ is defined the semi-annular sector, $f(x_1, x_2)$, $\forall (x_1, x_2) \in \Omega$. It can be exactly integrated based on the following cubature rules, i.e.

$$\iint_\Omega f(x, y)d\Omega = \sum_{i=1}^{4(m+1)} \sum_{j=1}^{m+1} C_j f(r_j \cos\theta_i, r_j \sin\theta_i) \tag{3.303}$$

The quadrature points, $(r_j \cos\theta_i, r_j \sin\theta_i)$, are located on

$$\theta_i = \frac{(i\pi)}{2(m+1)} \tag{3.304}$$

$$P_{m+1}(r_j^2) = 0 \tag{3.305}$$

where $P_{m+1}(r^2)$ is the Legendre polynomial in r^2 of degree $m+1$, orthogonalized on $[R_i^2, R_o^2]$. The quadrature weight is determined by the formula

$$C_j = \frac{1}{(4(m+1)P'_{m+1}(r_j^2))} \int_{R_i^2}^{R_o^2} \frac{P_{m+1}(r^2)}{(r^2 - r_j^2)} dr^2 \tag{3.306}$$

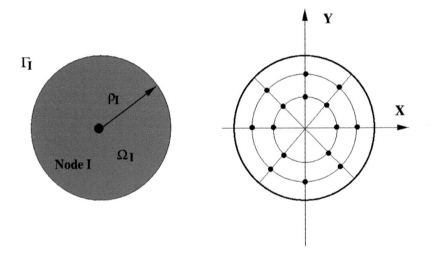

Fig. 3.16. An interior disk on which cubature rule is described.

3.8 Meshfree Local Petrov-Galerkin (MLPG) Method

The above approach is generalized by Atluri and his co-workers, and they labeled it as the so-called local Petrov-Galerkin formulation (MLPG).

As mentioned above, there is no real substance in the term *local*. A local Galerkin weak formulation is different from a Galerkin global weak formulation, if only the covering generated by the supports of test function can not cover the whole domain, otherwise the two weak formulations are equivalent. Nevertheless, local quadrature scheme is usually used, which allows local integration patch overlapping.

A distinct feature of MLPG is hinted in the term *Petrov-Galerkin*. The term Petrov-Galerkin here refers to the fact that in MLPG, the trial function and test function are different. Even though they may have the same analytical expression, they may still have different support sizes, and subsequently cover different numbers of particles. This indicates that they are in principle different functions even if the same moving least square procedure is adopted to construct both trial and test function. Denote Ω_I^{te} as the support for the I-th test function and Ω_I^{tr} as the support of I-th trial function. $\Omega_I^{te} \neq \Omega_I^{tr}$ in general. Let

$$u_I^h(\boldsymbol{x}) = \sum_{I \in \Lambda} \Phi_{\varrho I}^{tr}(\boldsymbol{x})\hat{u}_I$$

$$v_I^h(\boldsymbol{x}) = \sum_{I \in \Lambda} \Phi_{\varrho I}^{te}(\boldsymbol{x})\hat{v}_I \qquad (3.307)$$

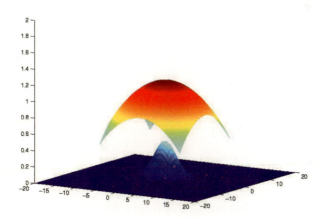

Fig. 3.17. Illustration of meshfree trial and test functions.

Consider the same elastostatics problem. The weak form with respect to the I-th test function is

$$\int_{\Omega_I^{te} \cap \Omega} \boldsymbol{\epsilon}_v : \boldsymbol{\sigma} d\Omega + \alpha \int_{\Gamma_u \cap \partial \Omega_I^{te}} \mathbf{v} \cdot \mathbf{u} d\Gamma - \int_{\Gamma_u \cap \partial \Omega_I^{te}} \mathbf{v} \cdot \mathbf{t} d\Gamma$$

$$= \int_{\Gamma_t \cap \partial \Omega_I^{te}} \mathbf{v} \cdot \bar{\mathbf{t}} d\Gamma + \alpha \int_{\Gamma_u \cap \partial \Omega_I^{te}} \mathbf{v} \cdot \bar{\mathbf{u}} d\Gamma + \int_{\Omega_I^{te} \cap \Omega} \mathbf{v} \cdot \mathbf{b} d\Omega \qquad (3.308)$$

Consequently, one may derive N local Petrov-Galerkin weak forms, each of them around a distinct particle I,

$$\sum_{J \in \Lambda} \mathbf{K}_{IJ} \mathbf{d}_J = \mathbf{f}_I, \quad I \in \Lambda \qquad (3.309)$$

Here the trial function's support is Ω_I^{tr} whereas the I-th test function's support is denoted as Ω_I^{te}.

Note that the integration of weak form is local, which means that no background cell is needed,

$$\mathbf{K}_{IJ} = \int_{\Omega_I^{te} \cap \Omega} (\mathbf{B}_{vI})^T \mathbf{D} \mathbf{B}_J d\Omega + \alpha \int_{\Gamma_u \cap \partial \Omega_I^{te}} \mathbf{V}_I N_J^\varrho d\Gamma$$

$$- \int_{\Gamma_u \cap \partial \Omega_I^{te}} \mathbf{V}_I \mathbf{N} \mathbf{D} \mathbf{B}_J d\Gamma \qquad (3.310)$$

$$\mathbf{f}_I = \int_{\Gamma_t \cap \partial \Omega_I^{te}} \mathbf{V}_I \bar{\mathbf{t}} d\Gamma + \int_{\Gamma_u \cap \partial \Omega_I^{te}} \mathbf{V}_I \bar{\mathbf{u}} d\Gamma + \int_{\Omega_I^{te} \cap \Omega} \mathbf{V}_I \mathbf{b} d\Omega \qquad (3.311)$$

where

$$\mathbf{V}_I = \begin{bmatrix} N_I^{\varrho(te)} & 0 \\ 0 & N_I^{\varrho(te)} \end{bmatrix},$$

$$\mathbf{B}_{vI} = \begin{bmatrix} N_{I,1}^{\varrho(te)} & 0 \\ 0 & N_{I,2}^{\varrho(te)} \\ N_{I,2}^{\varrho(te)} & N_{I,1}^{\varrho(te)} \end{bmatrix},$$

$$\mathbf{B}_J = \begin{bmatrix} N_{J,1}^{\varrho(tr)} & 0 \\ 0 & N_{J,2}^{\varrho(tr)} \\ N_{J,2}^{\varrho(tr)} & N_{J,1}^{\varrho(tr)} \end{bmatrix}.$$

Since trial and test functions are different, it will result unsymmetric stiffness matrix in general. If $\Omega_I^{tr} = \Omega_I^{te}$, and the trial function be the same as the weighting function. Then the above Petrov-Galerkin formulation returns to the conventional Bubnov-Galerkin formulation. In that case, it recovers the local symmetric Galerkin weak formulation we presented in the previous section.

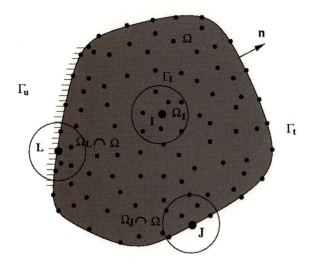

Fig. 3.18. An illustration on local Galerkin discretization.

3.9 Finite Point Method

If in meshfree local Petrov-Galerkin formulation, the covering generated by the supports of the test functions can not cover the whole domain, the local Galerkin weak formulation will be substantially different from global Galerkin weak formulation.

An extreme case is the so-called *finite point method*,[359, 360] in which Dirac delta function is chosen as the test function. In this case, no weak formulation is resulted. The strong form residual form is being collocated into a set of algebraic equations.

Let us consider a scalar function problem governed by a differential equation,

$$A(u) = b , \quad x \in \Omega \tag{3.312}$$

with Neumann boundary conditions

$$B(u) = t , \quad x \in \Gamma_t \tag{3.313}$$

and Dirichlet (essential) boundary conditions

$$u - u_p = 0, \quad x \in \Gamma_u \tag{3.314}$$

to be satisfied in a domain Ω with boundary $\partial\Omega \bigcup \Gamma_t \cup \Gamma_u$. In the above, A and B are appropriate differential operators, u is the problem unknown and b and t represent external forces or sources acting over the domain Ω and along the boundary Γ_t, respectively; u_p is the prescribed value of u over the boundary Γ_t.

Consider the following Petrov-Galerkin weighted residual formulation,

$$
\int_{\Omega} W_I[A(u^\varrho) - b]d\Omega + \int_{\Gamma_t} \bar{W}_I[B(u^\varrho) - t]d\Gamma
$$
$$
+ \int_{\Gamma_u} \tilde{W}_I[u^\varrho - u_p]d\Gamma = 0 , \quad I \in \Lambda \tag{3.315}
$$

where $W_I, \bar{W}_I$, and $\tilde{W}_I$ are the test functions in interior of the domain, on natural boundary, or on essential boundary; and $u^{ha}(\boldsymbol{x})$ is the meshfree approximation,

$$
u^\varrho(\boldsymbol{x}) = \sum_{I \in \Lambda} N_I^\varrho(\boldsymbol{x})u_I \tag{3.316}
$$

In both Onate et al.'s finite point method[359, 360] and Aluru's collocated RKPM,[10] the test functions are chosen as

$$
W_I := \delta_\Omega(\boldsymbol{x} - \boldsymbol{x}_I) \tag{3.317}
$$
$$
\bar{W}_I := \delta_{\Gamma_t}(\boldsymbol{x} - \boldsymbol{x}_I) \tag{3.318}
$$
$$
\tilde{W}_I := \delta_{\Gamma_u}(\boldsymbol{x} - \boldsymbol{x}_I) \tag{3.319}
$$

The above Petrov-Galerkin collocation formulation can be written in a local form as well. Let $\{\Omega_I\}_{I \in \Lambda}$ be the set of compact supports for trail function, and define bilinear form

$$
(f, g)_\Omega := \int_\Omega fg d\Omega \tag{3.320}
$$

A local Petrov-Galerkin collocation formula may be written as

$$
\Big(\delta_\Omega(\boldsymbol{x} - \boldsymbol{x}_I), A(u^h(\boldsymbol{x}))\Big)_{\Omega_I \cap \Omega} + \Big(\delta_{\Gamma_t}(\boldsymbol{x} - \boldsymbol{x}_I), B(u^h(\boldsymbol{x}))\Big)_{\partial\Omega_I \cap \Gamma_t}
$$
$$
\Big(\delta_{\Gamma_u}(\boldsymbol{x} - \boldsymbol{x}_I), u^h(\boldsymbol{x})\Big)_{\partial\Omega_I \cap \Gamma_u} = \Big(\delta_\Omega(\boldsymbol{x} - \boldsymbol{x}_I), b(\boldsymbol{x})\Big)_{\Omega_I \cap \Omega}
$$
$$
+ \Big(\delta_{\Gamma_t}(\boldsymbol{x} - \boldsymbol{x}_I), t(\boldsymbol{x})\Big)_{\partial\Omega_I \cap \Gamma_t} + \Big(\delta_{\Gamma_u}(\boldsymbol{x} - \boldsymbol{x}_I), u_p(\boldsymbol{x})\Big)_{\partial\Omega_I \cap \Gamma_u},
$$
$$
I \in \Lambda \tag{3.321}
$$
$$
\Big(\delta_{\Gamma_u}(\boldsymbol{x} - \boldsymbol{x}_I), u^h(\boldsymbol{x})\Big)_{\partial\Omega_I \cap \Gamma_u} = \Big(\delta_{\Gamma_u}(\boldsymbol{x} - \boldsymbol{x}_I), u_p(\boldsymbol{x})\Big)_{\partial\Omega_I \cap \Gamma_u},
$$
$$
I \in \Lambda \tag{3.322}
$$

Note the difference between the global collocation (3.315) and local collocation (3.321) and (3.322).

Finally, we obtain a set of algebraic equations,

$$
A(u^\varrho)_I - b_I = 0 \tag{3.323}
$$
$$
B(u^\varrho)_I - t_I = 0 \tag{3.324}
$$
$$
u_I^\varrho - u_{pI} = 0 \tag{3.325}
$$

which lead to a system of equations

$$\mathbf{K}\mathbf{u}^\varrho = \mathbf{f} \tag{3.326}$$

with $\mathbf{u}^\varrho := [u_1^\varrho, u_2^\varrho, \cdots, u_{NP}^\varrho]$, $K_{IJ} = [A(N_J)]_I + [B(N_J)]_I$ where the symmetry of the coefficient matrix $\mathbf{K}$ is not generally achieved, and $\mathbf{f}$ is a vector containing the contributions from the source terms b, t , and prescribed values u_p.

Distinguished from other collocation methods, the unique merite of meshfree collocation is its excellent stability property. Due to the large support of the trial function, a single meshfree collocation equation involves much more particles than the usual finite difference method as well as finite volume method. It thus provides stabilization effect to convective term in simulations of advective-diffusive transport and fluid flow problems in general [359, 360].

3.10 Meshfree Local Boundary Integral Equation

This local quadrature idea is extended by Atluri and his colleagues to form other meshfree formulations.[16, 17, 19, 471, 472] Another meshfree formulation formed by Atluri et al. is the so-called local boundary integral equation (LBIE).

Consider a boundary value problem of Poisson's equation,

$$\nabla^2 u = p(x), \quad x \in \Omega \tag{3.327}$$

$$u = \bar{u}, \quad x \in \Gamma_u \tag{3.328}$$

$$\frac{\partial u}{\partial n} = q, \quad \Gamma_q \tag{3.329}$$

One may establish a boundary integral equation for a chosen subdomain Ω_s (Note that Ω_s has nothing to do with a particle's compact support),

$$\alpha u(\boldsymbol{y}) = -\int_{\Omega_s} u(\boldsymbol{x}) \frac{\partial \tilde{u}^*}{\partial n}(\boldsymbol{x}, \boldsymbol{y}) d\Gamma + \int_{\Gamma_s} \frac{\partial u}{\partial n}(\boldsymbol{x}) \tilde{u}^*(\boldsymbol{x}, \boldsymbol{y}) d\Gamma$$
$$- \int_{\partial \Gamma_s} \tilde{u}^*(\boldsymbol{x}, \boldsymbol{y}) p(\boldsymbol{x}) d\Omega \tag{3.330}$$

where $\tilde{u}^*$ is the Green's function

$$\tilde{u}^*(\boldsymbol{x}, \boldsymbol{y}) = \frac{1}{2\pi} \ln \frac{r_0}{r} \tag{3.331}$$

For each particle in the domain Ω, one can form one such local boundary integral equation, and let $u^h(\boldsymbol{x}) = \sum_i \phi_i(\boldsymbol{x}) d_i$, one may obtain the following algebraic equations

$$\alpha_i u_i = \sum_{j=1}^{N} K_{ij}^* d_j + f_i^* \tag{3.332}$$

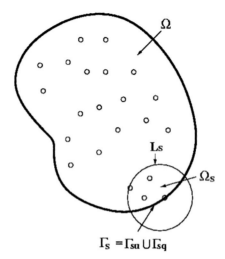

$$\Gamma_S = \Gamma_{su} \cup \Gamma_{sq}$$

Fig. 3.19. Local meshfree-Galerkin illustration ($\partial \Omega_s = L_s \cup \Gamma_s$).

and

$$K_{ij}^* = \int_{\Gamma_{su}} \tilde{u}^*(\boldsymbol{x}, \boldsymbol{y}_i) \frac{\partial \phi_j}{\partial n} d\Gamma - \int_{\Gamma_{sq}} \phi_j \frac{\partial \tilde{u}^*}{\partial n}(\boldsymbol{x}. \boldsymbol{y}_i) d\Gamma$$

$$- \int_{L_s} \phi_j \frac{\partial \tilde{u}^*}{\partial n}(\boldsymbol{x}, \boldsymbol{y}_i) d\Gamma \qquad (3.333)$$

$$f_i^* = \int_{\Gamma_{sq}} \tilde{u}^*(\boldsymbol{x}, \boldsymbol{y}_i) \bar{q} d\Gamma - \int_{\Gamma_{su}} \bar{u} \frac{\partial \tilde{u}^*}{\partial n}(\boldsymbol{x}, \boldsymbol{y}) d\Gamma$$

$$- \int_{\Omega_s} \tilde{u}^*(\boldsymbol{x}, \boldsymbol{y}_i) p(\boldsymbol{x}) d\Omega \qquad (3.334)$$

Those local boundary integrals and local domain integrals can be integrated by fixed quadrature rules. An obvious advantage of this formulation is that it does not need to enforce essential boundary condition. Nevertheless, this formulation relies on Green's function, and it is limited in a handful of linear problems.

3.11 Meshfree Quadrature and Nodal Integration

How to integrate meshfree Galerkin weak formulation is a critical issue in computations. Most meshfree Galerkin methods integrate Galerkin weak formulation by using Gauss quadrature employed in a non-overlapping (subdivision) background cell. This is a very efficient procedure. Nevertheless, it incurred three criticisms: (1) MLS based meshfree interpolant is irrational

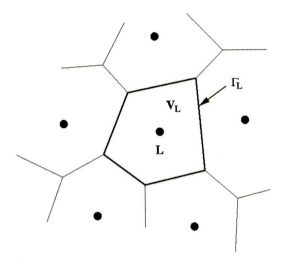

Fig. 3.20. Geometry definition of a representative nodal domain.

function, exact integration is almost impossible; (2) to maintain given numerical accuracy the deployment of Gauss quadrature points has to depend on particle distribution, this requires a complicated background cell structure; in other words, this requires a background mesh.

Because of this reason, the search for "truly meshfree method" continues. In order to completely eliminate Gauss quadrature, Chen et al.[103] proposed a so-called stabilized conforming nodal integration procedure for meshfree Galerkin methods.

The basic question is: how to find an accurate and stable nodal integration scheme ? The next question is: what is the criterion to control accuracy ? From a physical modeling standpoint, conservation properties ensure physically correct solutions. As pointed out previously, unlike finite element methods and finite volume methods, meshfree Galerkin methods, in specific, RKPM or SPH have only global conservative properties, they lack local conservative properties.

It is shown in their study[105] that a direct integration introduces numerical instability due to rank deficiency in the stiffness matrix. To stabilize the nodal integration, they proposed a so-called smoothing stabilization technique. The basic proposition of Chen et al.[103] is: for meshfree solution of a nodally integrated weak form to be stable and convergent, it has to be locally conservative. For example, the linear momentum, or force has to be balanced at point I, i.e.

$$\mathbf{f}_I^{int} = 0, \qquad \boldsymbol{x}_I \text{ is an interior point} \tag{3.335}$$

$$\mathbf{f}_I^{int} = \mathbf{f}_I^{ext}, \quad \boldsymbol{x}_I \text{ is close to boundary} \tag{3.336}$$

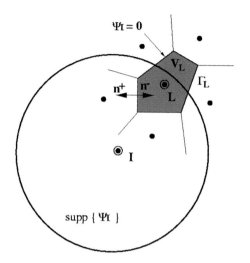

Fig. 3.21. Illustration of stabilized nodal integration.

The procedure is: one first smoothes strain in a chosen neighborhood of the particle I, say Ω_I, to replace the strain at point I with an average strain, provided a general triangulation is possible,

$$\bar{\epsilon}^h_{ij}(\boldsymbol{x}_I) = \int_{\Omega} \epsilon^h_{ij}(\boldsymbol{x})\Phi(\boldsymbol{x} - \boldsymbol{x}_I)d\Omega \tag{3.337}$$

where $\Phi(\boldsymbol{x})$ is the characteristic function of the domain Ω_I,

$$\Phi(\boldsymbol{x} - \boldsymbol{x}_I) = \begin{cases} \dfrac{1}{A_I} & \boldsymbol{x} \in \Omega_I \\ 0 & \boldsymbol{x} \notin \Omega_I \end{cases} \tag{3.338}$$

where $A_I = meas(\Omega_I)$. Note that here Ω_I is not the compact support of the particle I $(supp(\Psi_I))$, it is the Voronoi cell that contains the particle I, as illustrated in Fig. 3.20.

The divergence theorem is then used to replace the area, or volume integration around particle I by a contour integration of the Voronoi cell boundary.

$$\bar{\epsilon}^h_{ij}(\boldsymbol{x}_I) = \frac{1}{2A_I} \int_{\Omega_I} \left(\frac{\partial u^h_i}{\partial x_j} + \frac{\partial u^h_j}{\partial x_i} \right) d\Omega = \frac{1}{2A_I} \int_{\Gamma_I} (u^h_i n_j + u^h_j n_i)dS \tag{3.339}$$

Substituting RKPM interpolation field

$$u^h(\boldsymbol{x}_I) = \sum_{I \in G} K_{\varrho I}(\boldsymbol{x})\mathbf{d}_I \tag{3.340}$$

into (3.339) yields

$$\bar{\epsilon}^h(\boldsymbol{x}_L) = \sum_{I \in G_L} \bar{\mathbf{B}}_I(\mathbf{x}_L)\mathbf{d}_I \tag{3.341}$$

where

$$\bar{\epsilon}^h = [\bar{\epsilon}_{11}^h, \bar{\epsilon}_{22}^h, \bar{\epsilon}_{12}^h]^T, \quad \mathbf{d}^T = [d_{1I}, d_{2I}]^T \tag{3.342}$$

$$\bar{\mathbf{B}}_I(\boldsymbol{x}_L) = \begin{bmatrix} \bar{b}_{I1}(\boldsymbol{x}_L), & 0 \\ 0, & \bar{b}_{I2}(\boldsymbol{x}_L) \\ \bar{b}_{I2}(\boldsymbol{x}_L), & \bar{b}_{I1}(\boldsymbol{x}_L) \end{bmatrix} \tag{3.343}$$

and

$$\bar{b}_{Ii}(\boldsymbol{x}_L) = \frac{1}{A_L} \int_{\Gamma_L} \mathcal{K}_{\varrho I}(\boldsymbol{x}) n_i(\boldsymbol{x}) d\Gamma \tag{3.344}$$

By doing this, one can automatically ensure the local conservative of linear momentum or balance of force at point I (3.335) and (3.336), because it can be shown that for all interior nodes $\{I : supp(\Phi_I) \cap \Gamma = \emptyset\}$,

$$\mathbf{f}_I^{int} = \int_\Omega \mathbf{B}_I^T \sigma d\Omega = \int_\Omega \mathbf{B}_I^T d\Omega \sigma = \sum_L \mathbf{B}_I(\boldsymbol{x}_L) A_L \sigma = 0 \tag{3.345}$$

if the stress inside the domain is constant. This stems from the fact that

$$\sum_L \mathbf{B}_I(\boldsymbol{x}_L) A_L = \begin{bmatrix} \sum_L \bar{b}_{I1}(\boldsymbol{x}_L), & 0 \\ 0, & \sum_L \bar{b}_{I2}(\boldsymbol{x}_L) \\ \sum_L \bar{b}_{I2}(\boldsymbol{x}_L), & \sum_L \bar{b}_{I1}(\boldsymbol{x}_L) \end{bmatrix} = 0 \tag{3.346}$$

because each component in the matrix of Eq. (3.346) must vanish,

$$\sum_L b_{Ii}(\boldsymbol{x}_L) A_L = \sum_L \int_{\Gamma_L} \Phi_I(\boldsymbol{x}) n_i(\boldsymbol{x}) d\Gamma = 0 . \tag{3.347}$$

For Γ_L is completly or partially inside $supp(\Phi_I)$, each segment of Γ_L inside $supp(\Phi_I)$ is shared by two nodal domains with opposite surface normals on each side of the domain as shown in Fig. 3.21. The condition $\mathbf{n}^+(\boldsymbol{x}) = -\mathbf{n}^-(\boldsymbol{x})$ for $\boldsymbol{x} \in \Gamma_L$, $\boldsymbol{x} \in supp(\Phi_I)$ leads to a vanishing summation $\sum_L \int_{\Gamma_L} \Phi_I(\boldsymbol{x}) n_i d\Gamma$ in Eq. (3.347). For Γ_L that is completely or partially outside $supp(\Phi_I)$, $\Phi_I(\boldsymbol{x}) = 0$ for $\boldsymbol{x} \in \Gamma_L$, $\boldsymbol{x} \notin supp(\Phi_I)$, and as such the vanishing $supp(\Phi_I)$, $\Phi_I(\boldsymbol{x}) = 0$ remains. The virtue of this technique is that it completely elimi-nates Gauss quadrature integration, which is especially attractive in inelastic large deformation calculation with a Lagrangian formulation.

4. Approximation Theory of Meshfree Interpolants

The approximation theory of meshfree methods consists of several key elements. First how to distribute particles to represent a continua and to build a valid meshfree discretization; second how to measure the quality of meshfree interpolation, third, how to measure the convergence of a meshfree Galerkin weak form. Overall, we are dealing with the theoretical foundation or mathematical structure of meshfree interpolation.

Around 1995, three papers were published dealing with the subject of mathematical structure and convergence of meshfree methods: Duarte and Oden [1996], Li and Liu [1996], Liu, Li, and Belytschko [1997], and Babuska and Melenk [1997], which laid a theoretical foundation for meshfree methods. Few years later, Li and Liu[266, 267] Han and Meng[190] Huerta and Fernandex-Mendez,[217] De and Bath,[136] and more recently, Babuska, Banerjee, and Osborn [2003] made further contribution to this subject.

4.1 Requirements and Properties of Meshfree Discretization

Even though there is no mesh involved in meshfree discretization, the particle distribution of a meshfree discretization has to satisfy certain requirements to ensure a valid computation or a sensible simulation. Before detailing the requirements of meshfree discretization, a few definitions are in order.

In this book, the domain of influence for a point, $x \in \bar{\Omega}$, is defined as a subdomain

$$V_{x} = \left(\max_{\{\varrho_I | N_I(x) \neq 0, \ \forall I \in \Lambda\}} \{y \mid |y - x| < \varrho_I\} \right) \bigcap \Omega \tag{4.1}$$

The concept of the domain of influence is different from the support of a particle. First of all, every point in Ω has its domain of influence, no matter if it is a particle or just a sampling point. For each particle, it associates with an interpolation function, and hence a compact support. The compact support for a particle is defined as

$$\Omega_I := \{x \mid |x - x_I| < \varrho_I\} \bigcap \Omega . \tag{4.2}$$

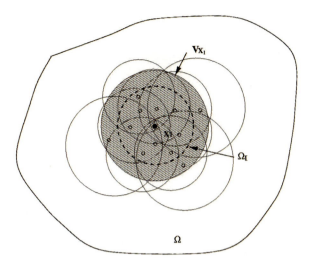

Fig. 4.1. Comparison between domain of influence and support size.

where ϱ_I is the radius of the compact support. Second, even for a particle, its compact support is not the same as its domain of influence. If the surounding particles have larger support sizes, the domain of influence at $\boldsymbol{x}_I$, $V_{\boldsymbol{x}_I}$, will be larger than its support, i.e. $\Omega_I \subset V_{\boldsymbol{x}_I}$. Fig. (4.1) illustrates the difference between the two.

4.1.1 Regularity of Particle Distributions

In,[348] Nayroles et al. found that the rank and the conditioning of the moment matrix, $\mathcal{M}(\boldsymbol{x})$, depend on the relative locations of the particles residing within the domain of influence of point $\boldsymbol{x}$. For instance, in 2D case, $\mathcal{M}(\boldsymbol{x})$ would be singular, if the particles inside the domain of influence are aligned along a straight line or less than three points. On the other hand, the minimum number of particles within a domain of influence also depend on the rank of basis vector $\mathbf{P}(\boldsymbol{x})$ as well. A necessary condition to form a non-singular moment matrix, $\mathcal{M}(\boldsymbol{x})$, is given by Nayroles et al. : *there exist at least $\ell + 1$ particles in $V_{\boldsymbol{x}}$, if the basis vector $\mathbf{P}(\boldsymbol{x})$ has $\ell + 1$ independent components.*

Let's consider a simply connected region, $\bar{\Omega} \subset \mathbb{R}^3$, in which there is a particle distribution. By particle distribution we mean that

Definition 4.1.1. *(Particle Distribution) Each particle inside or on the boundary of Ω is assigned with a parametric dilation vector $\boldsymbol{\varrho}_I$, $I = 1, \cdots\cdots, NP$. With the parametric dilation vector $\boldsymbol{\varrho}_I$, one can construct a compact support around each particle. Suppose the particle I occupies the position $\boldsymbol{x}_I$ in reference coordinate system, the compact support can be constructed as a local*

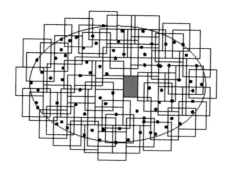

Fig. 4.2. *Inadmissible particle distribution I: shaded area is not covered by any nodal support.*

"sphere"

$$S_I := \{ \boldsymbol{x} \mid |\boldsymbol{x} - \boldsymbol{x}_I| \leq |\boldsymbol{\varrho}_I| \} \tag{4.3}$$

or, a parallelogram box

$$S_I := \{ \boldsymbol{x} \mid |x_i - x_{Ii}| \leq \varrho_{Ii} \, , \; \varrho_{Ii} > 0, \;\; 1 \leq i \leq 3 \} \tag{4.4}$$

Not all particle distributions can be used in numerical computation. The valid particle distribution is referred to as the *"admissible particle distribution"*. Admissibility of the particle distribution depends on computation feasibility. To define admissible particle distribution, we first define how to measure the density of particle distribution. There are many different ways to measure the density of particle distribution. In this book, it is assumed that first the particle distribution is quasi-uniform, which means that the particle distribution is statistically isotropic, and homogenuous, and second in every domain of influence the number of particles inside in each support or domain of influence is always finite. More precisely, there exist an upper bound and lower bound on the number of particles inside a support. Based on this assumption, the density of a particle distribution can be measured by the radius of a compact support, which is why the radius of a compact support is often called as dilation parameter. This is convenient in refinement procedures, because the radius of the support size becomes the numerical length scale, and it is easy to control.

Definition 4.1.2. *(Particle Density Index) The density of a particle distribution is measured by the indices that are associated with the dilation parameters of all particles,*

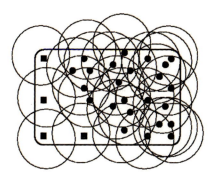

Fig. 4.3. *Inadmissible particle distribution II.*

$$\varrho_{max} := \max\{|\boldsymbol{\varrho}_I|, I \in \Lambda\} \tag{4.5}$$

$$\varrho_{min} := \min\{|\boldsymbol{\varrho}_I|, I \in \Lambda\} \tag{4.6}$$

For simplicity, in this book, the presentation is restricted to the case particle distribution is quasi-uniform, which implies that the dilation parameters, or the support sizes variation is quasi-uniform, i.e. there exist two constants $c_1, c_2 \in (0, \infty)$ such that

$$c_1 \leq \frac{\varrho_I}{\varrho_J} \leq c_2 , \quad \forall\, I, J, \in \Lambda \tag{4.7}$$

For such particle distributions, there exist a parameter $\varrho_{min} \leq \varrho \leq \varrho_{max}$, and constants, $\tilde{c}_1, \tilde{c}_2 > 0$, such that

$$\tilde{c}_1 \leq \frac{\varrho_I}{\varrho} \leq \tilde{c}_2, \quad \forall\, I \in \Lambda \tag{4.8}$$

Obviously, based on our assumption, the smaller ϱ is, the denser the particle distribution will be. Therefore, a particle refinement process can be defined as a decreasing process of the dilation parameter, provided that in this process the new particle distribution remains an admissible particle distribution.

Definition 4.1.3. *(Admissible Particle Distribution)*
An admissible particle distribution is a particle distribution that satisfies the following conditions
1. Every particle of a distribution associates with a compact support

$$S_I := \left\{ \boldsymbol{x} \,\middle|\, |\boldsymbol{x} - \boldsymbol{x}_I| \leq \varrho \right\} \quad , \tag{4.9}$$

and the union of all the compact support S_I

$$S := \bigcup_{I \in \Lambda} S_I \tag{4.10}$$

generates a covering for the domain $\bar{\Omega}$ in which the particles reside, viz.

$$\bar{\Omega} \subseteq S \quad . \tag{4.11}$$

2. $\forall \, \bar{\mathbf{x}} \in \bar{\Omega}$, *there exists a ball*

$$\mathcal{B}(\boldsymbol{x}) := \left\{ \boldsymbol{x} \mid |\boldsymbol{x} - \bar{\mathbf{x}}| \leq \varrho \right\} \tag{4.12}$$

so that the number of particles in the $\mathcal{B}$, N_p, satisfies the condition

$$N_{min} \leq N_p \leq N_{max} \tag{4.13}$$

where both $0 < N_{min} < N_{max} < \infty$ are a priori assigned numbers, such that it is necessary to fulfill the following requirements:

- *the moment matrix is finite dimension;*
- *the moment matrix is invertible;*
- *the moment matrix is well conditioned;*

3. *The particle distribution pattern should be non-degenerate, which refers to that, in 1-D case, there are at least two particles in $\mathcal{B}(\boldsymbol{x})$, and the difference of the two particles should be nonzero. Similarly, in 2-D case, there are at least three particles in $\mathcal{B}(\boldsymbol{x})$, and the particles form a triangular element with nonzero area. Generally speaking, for $\mathcal{B}(\boldsymbol{x}) \in \mathbb{R}^d$, there are at least $d + 1$ particles in $\mathcal{B}(\boldsymbol{x})$, and their position vectors form a non-degenerate d-th rank "simplex" element, such that it is necessary for the moment matrix to be inverted.*

It may be noted that first though the above conditions are required for the domain of influence of any point, $\boldsymbol{x} \in \Omega$, in practice, if they are enforced in each compact support of the particle, $I \in \Lambda$, they seem to be automatically satisfied for the domain of influence of any point, $\boldsymbol{x} \in \Omega$. This is in fact an unproved proposition, which is most likely true and it can be proved; second, both conditions (2) and (3) are restrictions on particle number and distribution pattern in a domain of influence, and both of them are necessary conditions for invertibility of the moment matrix $\mathbf{M}$; and the second condition may be expressed as

$$N_{min} \leq card\left\{ J \mid J \in \Lambda \text{ and } \boldsymbol{x}_J \in V_{\boldsymbol{x}} \right\} \leq N_{max} \tag{4.14}$$

which is partially *the pointwise overlap condition* referred by Babuska et al. A non-admissible particle distribution is shown in Figure (4.1.1), which fails to satisfy the condition **1**. In Figure (4.1.1), one can see that in the middle

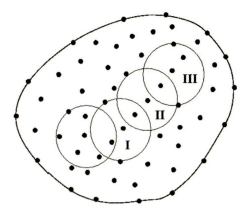

Fig. 4.4. *A degenerate 2-D particle distribution.*

of the area there is a small dark region which is not covered by the compact support of any particles. As a matter of fact, condition **2** implies condition **1**. Moreover, condition (4.13) is closely related to computational feasibility. First an upper bound on particle number in a compact support guarantees the finite dimension of moment matrix, second the condition number of the moment matrix increases as the number of particles inside a support increases, and third as the number of shape functions covering a local area increases, the shape functions tends to be more and more linearly dependent in the local area, which is also related to the second issue. This condition is restated as the following hypothesis ([190]).

Hypothesis 4.1.1 *(H) There is a constant I_0 that is independent with NP such that for any $x \in \bar{\Omega}$, there are at most I_0 of x_I satisfying the relation $\|x - x_I\| < \varrho_I$, i.e. each point in $\bar{\Omega}$ is covered by at most I_0 shape functions.*

The condition $N_p \geq N_{min}$ guarantees the stability condition of the shape function, or regularity of the moment matrix $\mathbf{M}$. The second part of inequality guarantees the bandedness of the resulting stiffness matrix. To estimate the lower bound, N_{min}, of the particle number in the domain of influence of any point, $x \in \Omega$, we have the following proposition:

Proposition 4.1.1. *For any $x \in \Omega$, a necessary condition for $\mathbf{M}(x)$ to be invertible is that x is covered by at least $N_p = dim\mathcal{P}_p$ shape functions.*

The reasoning used in the following proof is largely due to Duarte and Oden [1996].
   Proof:
   Let x be an arbitrary point in $\bar{\Omega}$ and assume that there are particles with indices, $I_1, I_2, \cdots, I_k$ within its domain of influence, i.e.

$$\boldsymbol{x}_{I_i} \in V(\boldsymbol{x}), \quad \forall, \ i = 1, 2, \cdots, k \tag{4.15}$$

or

$$\boldsymbol{x} \in \bigcap_{i=1}^{k} supp(\varPhi_{I_i}) \tag{4.16}$$

The discrete moment is

$$\mathbf{M}_{\varrho}^{h}(\boldsymbol{x}) = \sum_{i=1}^{k} \mathbf{P}\left(\frac{\boldsymbol{x} - \boldsymbol{x}_{I_i}}{\varrho}\right) \mathbf{P}^{T}\left(\frac{\boldsymbol{x} - \boldsymbol{x}_{I_i}}{\varrho}\right) \varPhi_{\varrho}(\boldsymbol{x} - \boldsymbol{x}_{I_i}) \varDelta V_I \tag{4.17}$$

which can be written as

$$
\begin{bmatrix}
P_1\left(\frac{\boldsymbol{x}-\boldsymbol{x}_{I_1}}{\varrho}\right) & P_1\left(\frac{\boldsymbol{x}-\boldsymbol{x}_{I_2}}{\varrho}\right) & \cdots & P_1\left(\frac{\boldsymbol{x}-\boldsymbol{x}_{I_k}}{\varrho}\right) \\
P_2\left(\frac{\boldsymbol{x}-\boldsymbol{x}_{I_1}}{\varrho}\right) & P_2\left(\frac{\boldsymbol{x}-\boldsymbol{x}_{I_2}}{\varrho}\right) & \cdots & P_2\left(\frac{\boldsymbol{x}-\boldsymbol{x}_{I_k}}{\varrho}\right) \\
\vdots & \vdots & \ddots & \vdots \\
P_{n_p}\left(\frac{\boldsymbol{x}-\boldsymbol{x}_{I_1}}{\varrho}\right) & P_{n_p}\left(\frac{\boldsymbol{x}-\boldsymbol{x}_{I_2}}{\varrho}\right) & \cdots & P_{n_p}\left(\frac{\boldsymbol{x}-\boldsymbol{x}_{I_k}}{\varrho}\right)
\end{bmatrix}
$$

$$
* \begin{bmatrix}
\varPhi_{\varrho}\left(\frac{\boldsymbol{x}-\boldsymbol{x}_{I_1}}{\varrho}\right)\varDelta_{I_1} & 0 & \cdots & 0 \\
0 & \varPhi_{\varrho}\left(\frac{\boldsymbol{x}-\boldsymbol{x}_{I_2}}{\varrho}\right)\varDelta_{I_2} & \cdots & 0 \\
\vdots & \vdots & \ddots & \vdots \\
0 & 0 & \cdots & \varPhi_{\varrho}\left(\frac{\boldsymbol{x}-\boldsymbol{x}_{I_k}}{\varrho}\right)\varDelta_{I_k}
\end{bmatrix}
$$

$$
* \begin{bmatrix}
P_1\left(\frac{\boldsymbol{x}-\boldsymbol{x}_{I_1}}{\varrho}\right) & P_2\left(\frac{\boldsymbol{x}-\boldsymbol{x}_{I_2}}{\varrho}\right) & \cdots & P_{n_p}\left(\frac{\boldsymbol{x}-\boldsymbol{x}_{I_k}}{\varrho}\right) \\
P_1\left(\frac{\boldsymbol{x}-\boldsymbol{x}_{I_1}}{\varrho}\right) & P_2\left(\frac{\boldsymbol{x}-\boldsymbol{x}_{I_2}}{\varrho}\right) & \cdots & P_{n_p}\left(\frac{\boldsymbol{x}-\boldsymbol{x}_{I_k}}{\varrho}\right) \\
\vdots & \vdots & \ddots & \vdots \\
P_1\left(\frac{\boldsymbol{x}-\boldsymbol{x}_{I_1}}{\varrho}\right) & P_2\left(\frac{\boldsymbol{x}-\boldsymbol{x}_{I_2}}{\varrho}\right) & \cdots & P_{n_p}\left(\frac{\boldsymbol{x}-\boldsymbol{x}_{I_k}}{\varrho}\right)
\end{bmatrix} := \mathbf{P}\mathbf{W}\mathbf{P}^{T} \tag{4.18}
$$

The rank of $\mathbf{P}$, which will be consequently the rank of $\mathbf{M}_{\varrho}^{h}(\boldsymbol{x})$ as well, will be most equal to $N_k := card\left\{\boldsymbol{x}_J \mid \boldsymbol{x}_J \in V_{\boldsymbol{x}}\right\}$. For $\mathbf{M}_{\varrho}^{h}(\boldsymbol{x})$ to be positive definite, it requires that $N_k \geq N_p$.

In one dimensional case, we have $n_p = dim \mathcal{P}_p = p + 1$. Suppose in the domain of influence of x, $V(x)$, there are $p + 1$ particles, $x_{I_1}, x_{I_2}, \cdots, x_{I_{p+1}}$. The moment matrix is

$$\mathbf{M}(x) = \sum_{I \in \Lambda} \varPhi_{\varrho}\left(\frac{x - x_I}{\varrho}\right) \mathbf{P}\left(\frac{x - x_I}{\varrho}\right) \mathbf{P}\left(\frac{x - x_I}{\varrho}\right)^{T} \varDelta x_I \tag{4.19}$$

where $\mathbf{P}^T(x) = (1, x, \cdots, x^p)$. Then the matrix $\mathbf{P}(x)$ becomes

$$
\begin{pmatrix}
1 & \left(\dfrac{x - x_{I_1}}{\varrho}\right) & \cdots\cdots & \left(\dfrac{x - x_{I_1}}{\varrho}\right)^p \\
1 & \left(\dfrac{x - x_{I_2}}{\varrho}\right) & \cdots\cdots & \left(\dfrac{x - x_{I_2}}{\varrho}\right)^p \\
\vdots & \vdots & \ddots & \vdots \\
1 & \left(\dfrac{x - x_{I_{p+1}}}{\varrho}\right) & \cdots\cdots & \left(\dfrac{x - x_{I_{p+1}}}{\varrho}\right)^p
\end{pmatrix}
\tag{4.20}
$$

The determinant of the coefficient matrix is equal to

$$
\prod_{1 \le \ell < k \le p+1} \left[\left(\frac{x - x_{I_k}}{\varrho}\right) - \left(\frac{x - x_{I_\ell}}{\varrho}\right)\right] = \prod_{1 \le \ell < k \le p+1} \left(\frac{x_{I_\ell} - x_{I_k}}{\varrho}\right)
\tag{4.21}
$$

which is nonzero since the particles are distinct. Therefore, the discrete moment matrix $M(x)$ is invertible if and only if x is covered by at least $p + 1$ shape functions. ♣

Definition 4.1.4. *(Han and Meng) A family of particle distribution $\{\{x_I\}_{I \in \Lambda}\}$ is said to be (ϱ, p)-regular, if there exists a constant L_0 such that*

$$
\max_{\boldsymbol{x} \in \bar{\Omega}} \|\mathbf{M}(\boldsymbol{x})^{-1}\| \le L_0
\tag{4.22}
$$

for all the particle distributions in the family. Note that the moment matrix is defined in (9.32) and p is the highest order polynomial in basis vector $\mathbf{P}(\boldsymbol{x})$.

This definition links the invertibility of the moment matrix with regularity of particle distribution. To precisely describe the regularity of a particle distribution, Han and Meng proved the following two theorems for admissible particle distributions in one-dimensional case and multiple dimensional case $(p = 1)$.

Theorem 4.1.1. *(Admissible particle distribution in 1D)*
Assume that there exist two constants, $c_0 > 0$ and $\sigma_0 > 0$, such that for any $x \in [0, L]$, there are $I_1 < I_2 < \cdots < I_{p+1}$ with

$$
\min_{1 \le j \le p+1} \Phi\left(\frac{x - x_{I_j}}{\varrho_{I_j}}\right) \ge c_0 > 0
\tag{4.23}
$$

and

$$
\min_{j \ne k} \left|\frac{x_{I_j} - x_{I_k}}{\varrho}\right| \ge \sigma_0 > 0
\tag{4.24}
$$

Then the family of particle distributions $\{\{x_I\}_{I \in \Lambda}\}$ is (r, p)-regular, i.e. there exists a constant $L(c_0, \sigma_0)$ such that

$$
\max_{0 \le x \le L} \|\mathbf{M}^{-1}(x)\|_2 \le L(c_0, \sigma_0)
\tag{4.25}
$$

Few comments are in order. First the requirement (4.23) is a strengthened version of overlapping condition $N_{min} > p + 1$. Second, the condition (4.24) is essentially a requirement that within the domain of influence of any point $x \in [0, L]$ at least $p + 1$ particles do not coalesce as refinement proceeds or deformation goes. This will prevent a regular particle distribution being degenerate. Now we prove the theorem.

Proof;

For a real, symmetric, and positive semi-definite matrix, $A \in \mathbb{R}^{(p+1) \times (p+1)}$, one may arrange its eigenvalues in increasing order:

$$(0 \leq) \lambda_1(A) \leq \lambda_2(A) \leq \cdots \leq \lambda_{p+1}(A) \tag{4.26}$$

and $\|A\|_2 = \lambda_p(A)$.

If $\lambda_1(A) > 0$ (A is positive definite), the eigenvalue of A^{-1} are

$$(0 <) \lambda_{p+1}(A)^{-1} \leq \lambda_p \leq \cdots \leq \lambda_1(A)^{-1} \tag{4.27}$$

and $\|A^{-1}\|_2 = \lambda_1(A)^{-1}$. If A and B are both real and symmetric and $A - B$ is also positive semi-definite,

$$\lambda_k(A) \geq \lambda_k(B), \quad k = 1, 2, \cdots, p + 1 \tag{4.28}$$

Under the given assumptions, the moment matrix, $\mathbf{M}(x)$ is symmetric and positive definite. Thereby,

$$\|\mathbf{M}(x)^{-1}\|_2 = \lambda_0(\mathbf{M}(x))^{-1} \tag{4.29}$$

Denote

$$z_I := \frac{x - x_I}{\varrho} \tag{4.30}$$

and matrix

$$H(z_{I_1}, \cdots, z_{I_{p+1}}) = \left(\mathbf{P}(z_{I_1}), \cdots, \mathbf{P}(z_{I_{p+1}}) \right) = \begin{bmatrix} 1 & 1 & \cdots & 1 \\ z_{I_1} & z_{I_2} & \cdots & z_{I_{p+1}} \\ \vdots & \vdots & \ddots & \vdots \\ z_{I_1}^p & z_{I_2}^p & \cdots & z_{I_{p+1}}^p \end{bmatrix} \tag{4.31}$$

Consider (4.23).

$$\mathbf{M}(x) - c_0 \sum_{j=1}^{p+1} \mathbf{P}(z_{I_j}) \mathbf{P}(z_{I_j})^T = \sum_{j=1}^{p+1} \left(\Phi(z_{I_j}) - c_0 \right) \mathbf{P}(z_{I_j}) \mathbf{P}(z_{I_j})^T$$

$$+ \sum_{I \neq I_j, 1 \leq j \leq p+1} \Phi(z_I) \mathbf{P}(z_I) \mathbf{P}(z_I)^T \geq 0 \tag{4.32}$$

is positive semi-definite. Then

$$\lambda_1(\mathbf{M}(x)) \geq \left(c_0 \sum_{j=1}^{p+1} \mathbf{P}(z_{I_j})\mathbf{P}(z_{I_j})^T\right) = c_0\lambda_1\left(H(z_{I_1},\cdots,z_{I_{p+1}})H(z_{I_1},\cdots,z_{I_{p+1}})^T\right)$$

$$(4.33)$$

Thereby,

$$\|\mathbf{M}(x)^{-1}\|_2 \leq c_0^{-1}\left\|\left(H(z_{I_1},\cdots,z_{I_{p+1}})H(z_{I_1},\cdots,z_{I_{p+1}})^T\right)^{-1}\right\|_2$$

$$= c_0^{-1}\|H(z_{I_1},\cdots,z_{I_{p+1}})^{-1}\|_2^2 \qquad (4.34)$$

Consider the formula,

$$det H(z_{I_1},\cdots,z_{I_{p+1}}) = \prod_{1\leq\ell<k\leq p+1}(z_{I_k}-z_{I_j}) \qquad (4.35)$$

and the condition (4.24)

$$\min_{j\neq k}|z_{I_j}-z_{I_k}| \geq \sigma_0 > 0. \qquad (4.36)$$

Therefore $\|H^{-1}(z_{I_1},\cdots,z_{I_{p+1}})\|_2$ is uniformly bounded and hence $\|\mathbf{M}(x)\|_2$ is uniformly bounded.

This result can be generalized to multiple dimensions for case $p = 1$.

Theorem 4.1.2. *(Admissible particle distribution in multiple dimension)*
A family of particle distributions $\{\{\mathbf{x}_I\}_{I\in\Lambda}$ in $\mathbb{R}^d$ is $(\varrho, 1)$-regular if there exist two constants $c_0, \tilde{c}_0 > 0$ such that for any $\mathbf{x} \in \bar{\Omega}$, there are at least $d+1$ particles, $\mathbf{x}_{I_0},\cdots,\mathbf{x}_{I_d}$ satisfying

$$\min_{1\leq j\leq d+1} \Phi\left(\frac{\mathbf{x}-\mathbf{x}_{I_j}}{\varrho}\right) \geq c_0 > 0 \qquad (4.37)$$

and the d-simplex with the vertices, $\mathbf{x}_{I_1},\cdots,\mathbf{x}_{I_{d+1}}$, has a volume large than $\tilde{c}_0\varrho^d$.

The proof of Theorem 4.1.2 is similar to the proof of Theorem 4.1.1, it is left to the readers for exercise. Here we only outline the key steps. Let

$$\mathbf{z}_{I_j} = \frac{\mathbf{x}-\mathbf{x}_{I_j}}{\varrho} = (z_{I_{j,1}},\cdots,z_{I_{j,d}})^T \qquad (4.38)$$

and

$$\mathbf{P}(\mathbf{z}_{I_j}) = (1, z_{I_{j,1}},\cdots,z_{I_{j,d}})^T \qquad (4.39)$$

By condition (4.37), one may deduce that $\mathbf{M}(\mathbf{x}) - c_0\sum_{j=1}^d \mathbf{P}(\mathbf{z}_{I_j})\mathbf{P}^T(\mathbf{z}_{I_j})$ is positive semi-definite. Hence $\|\mathbf{M}(\mathbf{x})^{-1}\|_2 \leq c_0^{-1}\|H(\mathbf{z}_{I_1},\cdots,\mathbf{z}_{I_d})^{-1}\|_2^2$ where

$$H(\boldsymbol{z}_{I_1}, \cdots, \boldsymbol{z}_{I_d}) = \begin{bmatrix} 1 & 1 & \cdots & 1 \\ z_{I_1,1} & z_{I_2,1} & \cdots & z_{I_d,1} \\ \vdots & \vdots & \ddots & \vdots \\ z_{I_1,d} & z_{I_2,d} & \cdots & z_{I_d,d} \end{bmatrix} \tag{4.40}$$

Substituting (4.38) into (4.40), one may find that

$$det\{H(\boldsymbol{z}_{I_1}, \cdots, \boldsymbol{z}_{I_d})\} = \frac{(-1)^d}{\varrho^d} \begin{bmatrix} 1 & 1 & \cdots & 1 \\ x_{I_1,1} & x_{I_2,1} & \cdots & x_{I_d,1} \\ \vdots & \vdots & \ddots & \vdots \\ x_{I_1,d} & x_{I_2,d} & \cdots & x_{I_d,d} \end{bmatrix} \tag{4.41}$$

So the determinant is nonzero if and only if the points $\boldsymbol{x}_{I_1}, \cdots, \boldsymbol{x}_{I_d}$ are the vertices of a nondegenerate d-simplex, which is equivalent to the condition $|detH(\boldsymbol{z}_{I_1}, \cdots, \boldsymbol{z}_{I_d})| \geq \tilde{c}_0 > 0$, or

$$\begin{vmatrix} 1 & 1 & \cdots & 1 \\ x_{I_1,1} & x_{I_2,1} & \cdots & x_{I_d,1} \\ \vdots & \vdots & \ddots & \vdots \\ x_{I_1,d} & x_{I_2,d} & \cdots & x_{I_d,d} \end{vmatrix} \geq \tilde{c}_0 \varrho^d \tag{4.42}$$

That is, the d-simplex with the vertices, $\boldsymbol{x}_{I_1}, \cdots, \boldsymbol{x}_{I_d}$, has a volume larger than $\tilde{c}_0 \varrho^d$ for a constant $\tilde{c}_0 > 0$. This condition will ensure the norm of the matrix $\|H^{-1}(\boldsymbol{z}_{I_1}, \cdots, \boldsymbol{z}_{I_d})\|_2$ is uniformly bounded, and hence the norm of the inverse matrix $\|\mathbf{M}^{-1}(\boldsymbol{x})\|_2$ is uniformly bounded.

4.1.2 Bounds on Shape Functions and Their Derivatives

In convergence study, it is important to estimate asymptotic behaviors of meshfree shape functions and their derivatives with respect to dilation parameter ϱ. Rigorous estimations on bounds of meshfree shape function and their derivatives are given by Han and Meng, which is huristically motivated by the original work of Liu, Li, and Belytschko [1997]. The essense of the above analyses is recaptured in the followings

Define meshfree shape function,

$$\Psi_I^\varrho(\boldsymbol{x}) := \mathbf{P}^T \left(\frac{\boldsymbol{x} - \boldsymbol{x}_I}{\varrho} \right) \mathbf{b}(\boldsymbol{x}) \Phi_\varrho(\boldsymbol{x} - \boldsymbol{x}_I) \Delta V_I \tag{4.43}$$

where the vector $\mathbf{b}(\boldsymbol{x})$ is determined by

$$\mathbf{M}\boldsymbol{b}(\boldsymbol{x}) = \mathbf{P}(0) = \begin{pmatrix} 1 \\ 0 \\ 0 \\ \vdots \\ \vdots \\ 0 \end{pmatrix}. \quad \Rightarrow \quad \boldsymbol{b}(\boldsymbol{x}) = \mathbf{M}^{-1}(\boldsymbol{x})\mathbf{P}(0) \tag{4.44}$$

For a (ϱ, p)-regular particle distribution, $\mathbf{M}^{-1}(\boldsymbol{x})$ is uniformly bounded. Therefore, $\mathbf{b}(\boldsymbol{x})$ is uniformly bounded. We immediately have the following result.

Theorem 4.1.3. *Assume that $\Phi \in C_c^k(\mathbb{R}^d)$ and the particle distributions are (ϱ, p)-regular, then there exists a constant $C < \infty$ such that*

$$\max_{I \in \Lambda} \|\Psi_I^\varrho(\boldsymbol{x})\|_\infty \leq C \tag{4.45}$$

♣

Moreover, by definition, it is easy to deduce that the correction function

$$C_\varrho(\mathbf{x} - \mathbf{x}_I; \mathbf{x}) = \mathbf{P}^T\left(\frac{\boldsymbol{x} - \boldsymbol{x}_I}{\varrho}\right)\boldsymbol{b}(\boldsymbol{x}) \in C_c^k(\mathbb{R}^d) \tag{4.46}$$

is also uniformly bounded.

Let $\boldsymbol{z} = \dfrac{\boldsymbol{x}}{\varrho}$. The moment matrix

$$\mathbf{M}(\boldsymbol{z}) = \sum_{I \in \Lambda} \mathbf{P}(\boldsymbol{z} - \boldsymbol{z}_I)\mathbf{P}^T(\boldsymbol{z} - \boldsymbol{z}_I)\Phi(\boldsymbol{z} - \boldsymbol{z}_I)\frac{w_I}{\varrho^d} \tag{4.47}$$

is then a homogeneous function of $\boldsymbol{z}$. By chain rule,

$$D_{\boldsymbol{x}}^\beta \mathbf{M}(\boldsymbol{z}) = D_{\boldsymbol{z}}^\beta \mathbf{M}(\boldsymbol{z})\varrho^{-|\beta|} \sim \mathcal{O}(\varrho^{-|\beta|}) \tag{4.48}$$

For any multi-index β with $|\beta| = 1$, differentiation of the moment equation

$$\mathbf{M}(\boldsymbol{x})\mathbf{b}(\boldsymbol{x}) = \mathbf{P}(0) \tag{4.49}$$

yields

$$\mathbf{M}(\boldsymbol{x})D_{\boldsymbol{x}}^\beta \mathbf{b}(\boldsymbol{x}) = -D_{\boldsymbol{x}}^\beta \mathbf{M}(\boldsymbol{x})\mathbf{b}(\boldsymbol{x}) \tag{4.50}$$

Since $\mathbf{M}^{-1}(\boldsymbol{x})$ is uniformly bounded and $D_{\boldsymbol{x}}^\beta \mathbf{M}(\boldsymbol{x}) \sim \mathcal{O}(\varrho^{-1})$, it is easy to show

$$\max_{\beta:|\beta|=1} \|D_{\boldsymbol{x}}^\beta \mathbf{b}\|_\infty \sim \mathcal{O}(\varrho^{-1}) \tag{4.51}$$

In general, under the assumption that $\Phi \in C_0^k(\mathbb{R}^d)$, it can be shown inductively that

$$\max_{\beta:|\beta|\leq k} \|D_{\boldsymbol{x}}^\beta \mathbf{b}\|_\infty \leq \frac{c}{\varrho^k} . \tag{4.52}$$

Hence, it may be shown that

Theorem 4.1.4. *Assume that particle distributions are (ϱ, p)-regular and the generating function Φ is k-time continuously differentiable. Then*

$$\max_{I \in \Lambda} \max_{\beta:|\beta|=\ell} \|D_{\boldsymbol{x}}^\beta \Psi_I^\varrho(\boldsymbol{x})\|_\infty \leq \frac{c}{\varrho^\ell}, \quad 0 \leq \ell \leq k . \tag{4.53}$$

♣

4.2 Completeness and Consistency of Meshfree Interpolants

In finite element analysis, in order to meet convergence requirements, finite element shape functions have to satisfy the completeness condition. Since in early days most FEM computations were applications in structural mechanics, it is often said that FEM shape functions have to be able to represent rigid body motion and a constant strain field exactly if the structural mechanics terminlogy is used.

For non-conforming elements, due to lack of analytic estimates, this completeness requirement is often tested through the so-called patch test. Following the same idea, Belytschko et al. [1994] have systematically conducted various patch tests for moving least square interpolant. It turns out that all the MLS interpolants have passed the patch test without exception, this is because that the meshfree interpolant has the built-in consistency property. Today the meshfree patch test is solely used to examine the adequacy of numerical quadrature with slightly different purpose for non-conforming FEM elements. What interests us most here is the meshfree shape function's built-in consistency property.

It was shown by both Nayroles et al. [1992] and Liu et al. [1995a] that for the moving least square interpolant, the completeness of the interpolant shape function is automatic. As a matter of fact, one can show that the reproducing kernel formula can reproduce any type of polynomials, or other independent basis for that matter, based on initial design.

It was found in[264, 303] that the polynomial reproducing ability comes from a so-called p-consistency structure, which is the reminisence of moment conditions in constructing wavelet functions,[129] except that now these conditions are being enforced on a set of randomly distributed particles. The first moment condition, or the zero-th order moment condition, is

$$\sum_{I \in \Lambda} \Psi_I^{\varrho}(\mathbf{x}) = 1 \ . \tag{4.54}$$

When a meshfree interpolation basis satisfies this condition, it is called as a meshfree partition of unity.

Furthermore, the meshfree partition of unity that satisfies certain consistency conditions yields additional differential consistency conditions. The consistency conditions, or the moment conditions, combining with the corresponding differential consistency conditions are the essential mathematical structure of meshfree interpolant, and they are special characteristics of meshfree interpolants, and represent a unique interpolation framework, which is central to the estimation of meshfree interpolation error, local refinement and enrichment of interpolation basis, etc.

4.2.1 p-th Order Consistency Condition

Previously, we have shown the polynomial reproducing property of RKPM interpolant in one dimensional case. Now we consider the general polynomial reproducing property of RKPM interpolant in $\mathbb{R}^d$.

Lemma 4.2.1. *The reproducing kernel particle interpolant basis,* $\{\Psi_I^\varrho(\mathbf{x})\}$, *generated by a complete p-th order,* N_p*-term polynomial basis,* $\mathbf{P}^T(\mathbf{x}) = \{P_1(\mathbf{x}), P_2(\mathbf{x}), \cdots, P_{N_p}(\mathbf{x})\}$, *satisfies the following p-th order consistency condition:*

$$\sum_{I \in \Lambda} P_k(\frac{\mathbf{x} - \mathbf{x}_I}{\varrho})\Psi_I^\varrho(\mathbf{x}) = \delta_{k1}, \qquad 1 \le k \le N_p. \qquad (4.55)$$

where $N_p = \begin{pmatrix} p+d \\ d \end{pmatrix}$.

Proof:
By definition, it is straightforward,

$$\sum_{I \in \Lambda} \Psi_I^\varrho(\mathbf{x}) P_k(\frac{\mathbf{x} - \mathbf{x}_I}{\varrho})$$

$$= \mathbf{P}^T(0)\mathbf{M}^{-1}(\mathbf{x}) \sum_{I \in \Lambda} \mathbf{P}(\frac{\mathbf{x} - \mathbf{x}_I}{\varrho}) P_k(\frac{\mathbf{x} - \mathbf{x}_I}{\varrho}) \Phi_\varrho(\mathbf{x} - \mathbf{x}_I) \Delta V_I$$

$$= \mathbf{P}(0)^T \mathbf{M}^{-1}(\mathbf{x}) \sum_{I \in \Lambda} \begin{pmatrix} P_1 P_k \\ P_2 P_k \\ \vdots \\ \vdots \\ P_\ell P_k \end{pmatrix} \Phi_\varrho(\mathbf{x} - \mathbf{x}_I) \Delta V_I$$

$$= \frac{1}{D_\ell}(A_{11}, -A_{12}, \cdots, (-1)^{1+\ell}A_{1\ell}) \begin{pmatrix} M_{1k} \\ M_{2k} \\ \cdot \\ \cdot \\ \cdot \\ M_{\ell k} \end{pmatrix} = \delta_{1k} \qquad . \qquad (4.56)$$

The *Laplace* theorem is used in the last step. ♣
A direct consequence of Lemma (4.2.1) is the following.

Corollary 4.2.1 (p-th order consistency condition). *For a multi-index* α, $1 \le |\alpha| \le p$, *the* α*-order moments of the RKPM shape function are constants, i.e.*

$$\sum_{I \in \Lambda} (\boldsymbol{x} - \boldsymbol{x}_I)^{\alpha} \Psi_I^{\varrho}(\boldsymbol{x}) = \delta_{\alpha 0} \ ;$$

(4.57)

or

$$\sum_{I \in \Lambda} \boldsymbol{x}_I^{\alpha} \Psi_I^{\varrho}(\boldsymbol{x}) = \boldsymbol{x}^{\alpha}$$

(4.58)

Proof:

We first show (4.57). By construction, for a complete p-th order MLSRK approximation, there is a one-to-one correspondence between the polynomial basis $P_i(\boldsymbol{z})$ and the polynomial function $\boldsymbol{z}^{\alpha}$. Therefore, for $0 \leq |\alpha| \leq p$, $\exists \ 1 \leq k \leq N_p$ such that

$$P_k(\frac{\boldsymbol{x} - \boldsymbol{x}_I}{\varrho}) = (\frac{\boldsymbol{x} - \boldsymbol{x}_I}{\varrho})^{\alpha}$$
$$= (\frac{x_{1I} - x_1}{\varrho_1})^{\alpha_1} (\frac{x_2 - x_{2I}}{\varrho_2})^{\alpha_2} \cdots (\frac{x_n - x_{nI}}{\varrho_n})^{\alpha_n}$$

(4.59)

Then, for $0 \leq |\alpha| \leq p$,

$$\sum_{I \in \Lambda} \left(\frac{\boldsymbol{x} - \boldsymbol{x}_I}{\varrho}\right)^{\alpha} \Psi_I^{\varrho}(\boldsymbol{x}) = \sum_{I \in \Lambda} P_k(\frac{\boldsymbol{x} - \boldsymbol{x}_I}{\varrho}) \Psi_I^{\varrho}(\boldsymbol{x}) = \delta_{k1} = \delta_{\alpha 0}$$

(4.60)

which implies

$$\sum_{I \in \Lambda} (\boldsymbol{x} - \boldsymbol{x}_I)^{\alpha} \Psi_I^{\varrho}(\boldsymbol{x}) = \sum_{I \in \Lambda} (\boldsymbol{x} - \boldsymbol{x}_I)^{\alpha} = \delta_{\alpha 0}$$

(4.61)

Now, we show (4.58).

$$\sum_{I \in \Lambda} \boldsymbol{x}_I^{\alpha} \Psi_I^{\varrho}(\boldsymbol{x}) = \sum_{I \in \Lambda} (\boldsymbol{x}_I - \boldsymbol{x} + \boldsymbol{x})^{\alpha} \Psi_I^{\varrho}(\boldsymbol{x})$$
$$= \sum_{I \in \Lambda} \left(\sum_{\beta \leq \alpha} \binom{\alpha}{\beta} (\boldsymbol{x}_I - \boldsymbol{x})^{\alpha - \beta} \boldsymbol{x}^{\beta}\right) \Psi_I^{\varrho}(\boldsymbol{x})$$
$$= \sum_{\beta \leq \alpha} \binom{\alpha}{\beta} \boldsymbol{x}^{\beta} \sum_{I \in \Lambda} (\boldsymbol{x}_I - \boldsymbol{x})^{\alpha - \beta} \Psi_I^{\varrho}(\boldsymbol{x})$$
$$= \sum_{\beta \leq \alpha} (-1)^{|\alpha - \beta|} \binom{\alpha}{\beta} \boldsymbol{x}^{\beta} \delta_{(\alpha - \beta)0} = \boldsymbol{x}^{\alpha}$$

(4.62)

♣

4.2.2 Differential Consistency Conditions

Lemma (4.2.1) leads a profound consequence for the moments of the derivatives of the meshfree shape function. It plays a key role in the convergence study of the MLSRK approximation.

Lemma 4.2.2 (p-th order differential consistency condition).
   Let $\Phi \in C^k(\mathbb{R}^d)$; for $\Omega \subset \mathbb{R}^d$, the shape function basis generated by the complete p-order, N_p-component polynominal basis, satisfies the following conditions, $\forall\ |\alpha| < p, |\beta| \leq k$,

$$\sum_{I \in \Lambda} (\boldsymbol{x} - \boldsymbol{x}_I)^\alpha D_{\boldsymbol{x}}^\beta \Psi_I^\varrho(\boldsymbol{x}) = (-1)^{|\beta|} \alpha! \delta_{\alpha\beta}$$

$$(4.63)$$

or equivalently,

$$\sum_{I \in \Lambda} \boldsymbol{x}_I^\alpha D_{\boldsymbol{x}}^\beta \Psi_I^\varrho(\boldsymbol{x}) = \frac{\alpha!}{(\alpha - \beta)!} \boldsymbol{x}^{\alpha-\beta}$$

$$(4.64)$$

Proof:
We first prove (4.63). The proof proceeds by induction.
1. $|\beta| = 0, |\alpha| \leq m$; by Lemma (4.2.1),

$$\sum_{I \in \Lambda} (\boldsymbol{x} - \boldsymbol{x}_I)^\alpha \Psi_I^\varrho(\boldsymbol{x}) = \delta_{\alpha 0} = \alpha! \delta_{\alpha 0}\ .$$

$$(4.65)$$

Thus, Eq. (4.63) holds for $\beta = 0$.
2. $|\beta| = 1, |\alpha| \leq m$; without loss of generality, one can assume that

$$\beta := (0, 0, \cdots, \beta_j, 0, \cdots, 0)\ \text{ and }\ \beta_j = 1$$
$$\Rightarrow\quad D_{\boldsymbol{x}}^\beta = \partial_{x_j}$$

$$(4.66)$$

Hence,

$$D_{\boldsymbol{x}}^\beta \left\{ \sum_{I \in \Lambda} (\boldsymbol{x} - \boldsymbol{x}_I)^\alpha \Psi_I^\varrho(\boldsymbol{x}) \right\} = \partial_{x_j} \left(\sum_{I \in \Lambda} (\boldsymbol{x} - \boldsymbol{x}_I)^\alpha \Psi_I^\varrho(\boldsymbol{x}) \right)$$
$$= \alpha_j \sum_{I \in \Lambda} (x_1 - x_{1I})^{\alpha_1} \cdots (x_j - x_{jI})^{\alpha_j - 1} \cdots (x_n - x_{nI})^{\alpha_n} \Psi_I^\varrho(\boldsymbol{x})$$
$$+ \sum_{I \in \Lambda} (\boldsymbol{x} - \boldsymbol{x}_I)^\alpha \partial_{x_j} \Psi_I^\varrho(\boldsymbol{x}) = 0$$

$$(4.67)$$

This leads to

$$\sum_{I \in \Lambda} (\boldsymbol{x} - \boldsymbol{x}_I)^\alpha D_{\boldsymbol{x}}^\beta \Psi_I^\varrho(\boldsymbol{x}) = -\alpha_1! \cdots [(\alpha_j - 1)! \alpha_j] \cdots \alpha_n! \delta_{\alpha_1 0} \cdots \delta_{(\alpha_j - 1)0} \cdots \delta_{\alpha_n 0}$$

$$= -\alpha_1! \cdots \alpha_j! \cdots \alpha_n! \delta_{\alpha_1 0} \cdots \delta_{\alpha_j 1} \cdots \delta_{\alpha_n 0}$$

$$= (-1)^{|\beta|} \alpha! \delta_{\alpha\beta} \tag{4.68}$$

Hence, Eq.(4.63) is true for $|\beta| = 0, 1$. Assume that (4.63) holds for $|\beta| \leq k - 1$ and $|\alpha| \leq p$, we need to show that it is true for the case $|\beta'| = k$, and $|\alpha| \leq p$. By assumption

$$\sum_{I \in \Lambda} (\boldsymbol{x} - \boldsymbol{x}_I)^\alpha D_{\boldsymbol{x}}^\beta \Psi_I^\varrho(\boldsymbol{x}) = (-1)^{|\beta|} \alpha! \delta_{\alpha\beta} \ , \tag{4.69}$$

Let

$$\gamma := (0, 0, \cdots, \gamma_j, \cdots, 0) \ , \qquad \gamma_j = 1 \ . \tag{4.70}$$

Then

$$D_{\boldsymbol{x}}^\gamma = \partial_{x_j} \quad \Rightarrow \quad D_{\boldsymbol{x}}^\gamma D_{\boldsymbol{x}}^\beta = D_{\boldsymbol{x}}^{\beta'} \tag{4.71}$$

where $|\beta'| = k$.

Differentiation Eq.(4.69) yields

$$\alpha_j \sum_{I \in \Lambda} (x_1 - x_{1I})^{\alpha_1} (x_2 - x_{2I})^{\alpha_2} \cdots (x_j - x_{jI})^{\alpha_j - 1} \cdots (x_j - x_{jI})^{\alpha_n} D_{\boldsymbol{x}}^\beta \Psi_I^\varrho(\boldsymbol{x})$$

$$+ \sum_{I \in \Lambda} (\boldsymbol{x} - \boldsymbol{x}_I)^\alpha D_{\boldsymbol{x}}^\gamma D_{\boldsymbol{x}}^\beta \Psi_I^\varrho(\boldsymbol{x}) = 0 \tag{4.72}$$

It follows that

$$\sum_{I \in \Lambda} (\boldsymbol{x} - \boldsymbol{x}_I)^\alpha D_{\boldsymbol{x}}^{\beta'} \Psi_I^\varrho(\boldsymbol{x})$$

$$= \alpha_j \sum_{I \in \Lambda} (x_1 - x_{1I})^{\alpha_1} (x_2 - x_{2I})^{\alpha_2} \cdots (x_j - x_{jI})^{\alpha_j - 1} \cdots (x_j - x_{jI})^{\alpha_n} D_{\boldsymbol{x}}^\beta \Psi_I^\varrho(\boldsymbol{x})$$

$$= -(-1)^{|\beta|} \alpha_j \left(\alpha_1! \alpha_2! \cdots (\alpha_j - 1)! \cdots \alpha_n! \right) \delta_{\alpha_1 \beta_1} \delta_{\alpha_2 \beta_2} \cdots \delta_{\alpha_{j-1} \beta_j} \cdots \delta_{\alpha_n \beta_n}$$

$$= (-1)^{|\beta'|} \left(\alpha_1! \alpha_2! \cdots \alpha_j! \cdots \alpha_n! \right) \delta_{\alpha_1 \beta_1} \delta_{\alpha_2 \beta_2} \cdots \delta_{\alpha_j \beta_{j+1}} \cdots \delta_{\alpha_n \beta_n}$$

$$= (-1)^{|\beta'|} \alpha! \delta_{\alpha\beta'} \tag{4.73}$$

We now show (4.2.2).

4.3 Meshfree Interpolation Error Estimate

In this section, meshfree interpolation estimate is discussed, which is essential to the convergence of the related meshfree Galerkin method and its error estimation. In most mathematical literature, a reproducing kernel formulation is often defined in a Hilbert space. Therefore, it is pertinent to only discuss the RKPM interpolation estimation in Hilbert space. Nevertheless, the general meshfree interpolation estimate in Sobolev space has been extensively studied in the literature. Therefore, relevent interpolation estimates in Sobolev space are also briefly discussed.

4.3.1 Local Interpolation Estimate

In this book, we only consider the case of regular, quasi-uniform particle distribution, which also implies that the compact support of the meshfree shape functions has quasi-uniform support size variation. Denote meshfree RKPM shape function as

$$\Psi_I^\varrho(\boldsymbol{x}) = \Phi_\varrho^h(\boldsymbol{x} - \boldsymbol{x}_I)\mathbf{P}^T\left(\frac{\boldsymbol{x} - \boldsymbol{x}_I}{\varrho}\right)\mathbf{b}(\mathbf{x})\Delta V_I \tag{4.76}$$

Define the RKPM interpolation

$$\mathcal{R}_{\varrho,h}^p u(\boldsymbol{x}) := \sum_{I \in \Lambda} \Psi_I^\varrho(\boldsymbol{x})u_I \tag{4.77}$$

where p denotes the order of the polynomial basis of $\mathbf{P}$ vector, ϱ denotes the scaling dilation parameter, and subscript, h, emphasizes on discrete interpolation.

The goal here is to estimate the interpolation error $u(\boldsymbol{x}) - \mathcal{R}_{\varrho,h}^p u(\boldsymbol{x})$ in Hilbert space. To do so, we first introduce a few notations. More detailed information may be found in standard finite element text books, e.g. Brenner and Scott.[84]

Let B be a ball. Then a domain Ω_1 is said to be star-shaped with respect to B if for any $\boldsymbol{x} \in \Omega_1$ the closed convex hull of $\{\boldsymbol{x}\} \bigcup B$ is a subset of Ω_1. The chunkiness parameter of Ω_1 is defined as

$$P_c = \frac{diam\{\Omega_1\}}{\varrho_{max}} \tag{4.78}$$

where $\varrho_{max} = sup\{\varrho : \Omega_1 \text{ is star−shaped with respect to a ball of radius } \varrho\}$. Define open sets

$$\Omega_J := \left\{\boldsymbol{x} \,\middle|\, \|\boldsymbol{x} - \boldsymbol{x}_J\| < \varrho_J + \varrho_{max}\right\} \tag{4.79}$$

and define the index set

$$S_J = \left\{I \,\middle|\, dist(\boldsymbol{x}_I, B_J) < \varrho_I\right\} \tag{4.80}$$

$$\sum_{I\in\Lambda} x_I^\alpha D_x^\beta \Psi_I^\varrho(x) = \sum_{I\in\Lambda}(x_I - x + x)^\alpha D_x^\beta \Psi_I^\varrho(x)$$

$$= \sum_{I\in\Lambda}\sum_{\gamma\leq\alpha}\binom{\alpha}{\gamma}(x_I - x)^{\alpha-\gamma} x^\gamma D_x^\beta \Psi_I^\varrho(x)$$

$$= \sum_{\gamma\leq\alpha}\binom{\alpha}{\gamma} x^\gamma \sum_{I\in\Lambda}(-1)^{|\alpha-\gamma|}(x - x_I)^{\alpha-\gamma} D_x^\beta \Psi_I^\varrho(x)$$

$$= \sum_{\gamma\leq\alpha}(-1)^{2|\alpha-\gamma|}\binom{\alpha}{\gamma} x^\gamma (\alpha-\gamma)! \delta_{(\alpha-\gamma)\beta} \tag{4.74}$$

There can be only one term left; i.e. the term $\beta = \alpha - \gamma$.
It then follows immediately that

$$\sum_{I\in\Lambda} x_I^\alpha D_x^\beta \Psi_I^\varrho(x) = \binom{\alpha}{\alpha-\beta}\beta! x^{\alpha-\beta} = \frac{\alpha!}{(\alpha-\beta)!} x^{\alpha-\beta} \tag{4.75}$$

♣

Theorem 4.3.1. *Assume that the particle distribution are (ϱ, p)-regular, $\Phi(\boldsymbol{x}) \in C_0^k(\mathbb{R}^d)$, and $u(\boldsymbol{x}) \in H^{p+1}(\Omega)$, where Ω is a bounded open set in $\mathbb{R}^d$. Suppose the boundary $\partial\Omega$ is Lipschitz continuous. Then the following interpolation estimates hold*

$$\left\| u - \mathcal{R}_{\varrho,h}^p u \right\|_{H^\ell(\Omega)} \leq C_p \varrho^{p+1-\ell} \|u\|_{H^{p+1}(\Omega)}, \quad \forall\, 0 \leq \ell \leq p \qquad (4.81)$$

in particularly, for $\ell = 0$

$$\left\| u - \mathcal{R}_{\varrho,h}^p u \right\|_{L^2(\Omega)} \leq C_0 \varrho^{p+1} \|u\|_{H^{p+1}(\Omega)} \qquad (4.82)$$

where C_p, C_0 are constants, which are independent with the dilation parameter ϱ.

Proof:
It suffices to show

$$|u - \mathcal{R}_{\varrho,h}^p u|_{H^k(\Omega)} \leq C \varrho^{p+1-k} |u|_{H^{m+1}(\Omega)} \ . \qquad (4.83)$$

By definition

$$\mathcal{R}_{\varrho,h}^p u(\boldsymbol{x}) = \sum_{I \in \Lambda} u(\boldsymbol{x}_I) \Psi_I^\varrho(\boldsymbol{x}) \ , \qquad (4.84)$$

Taking the derivative of (4.84) yields

$$D_{\boldsymbol{x}}^\beta \mathcal{R}_{\varrho,h}^p u(\boldsymbol{x}) = \sum_{I \in \Lambda} u(\boldsymbol{x}_I) D_{\boldsymbol{x}}^\beta \Psi_I^\varrho(\boldsymbol{x}) \qquad (4.85)$$

By Taylor expansion (e.g. Brenner and Scott[84]),

$$u(\boldsymbol{x}_I) = \sum_{|\alpha| \leq p} \frac{1}{\alpha!} (\boldsymbol{x}_I - \boldsymbol{x})^\alpha D_{\boldsymbol{x}}^\alpha u(\boldsymbol{x})$$

$$+ \sum_{|\alpha| = p+1} \frac{p+1}{\alpha!} \int_0^1 (1-\theta)^p D_{\boldsymbol{x}}^\alpha u(\boldsymbol{x} + \theta(\boldsymbol{x}_I - \boldsymbol{x})) d\theta (\boldsymbol{x}_I - \boldsymbol{x})^\alpha \quad (4.86)$$

where $0 < \theta < 1$. To simplify notation, we denote $p(\alpha) = (p+1)/\alpha!$ in the rest of proof.

Substituting (4.86) back to (4.85) yields

$$
D_{\boldsymbol{x}}^{\beta}\left(\mathcal{R}_{\varrho,h}^{p}u(\boldsymbol{x})\right) = \sum_{I \in \Lambda}\left\{\sum_{|\alpha| \leq p}\frac{(\boldsymbol{x}_I - \boldsymbol{x})^{\alpha}}{\alpha!}D_{\boldsymbol{x}}^{\alpha}u(\boldsymbol{x})\right.
$$
$$
\left. + \sum_{|\alpha|=p+1}p(\alpha)(\boldsymbol{x}_I - \boldsymbol{x})^{\alpha}\int_0^1(1-\theta)^p D_{\boldsymbol{x}}^{\alpha}u(\boldsymbol{x}+\theta(\boldsymbol{x}_I - \boldsymbol{x}))d\theta\right\} \cdot D_{\boldsymbol{x}}^{\beta}\Psi_I^{\varrho}(\boldsymbol{x})
$$
$$
= \sum_{|\alpha| \leq p}\frac{1}{\alpha!}D_{\boldsymbol{x}}^{\alpha}u(\boldsymbol{x})\left(\sum_{I \in \Lambda}(\boldsymbol{x}_I - \boldsymbol{x})^{\alpha}D_{\boldsymbol{x}}^{\beta}\Psi_I^{\varrho}(\boldsymbol{x})\right)
$$
$$
+ \sum_{I \in \Lambda}\left\{\sum_{|\alpha|=p+1}p(\alpha)(\boldsymbol{x}_I - \boldsymbol{x})^{\alpha}\int_0^1(1-\theta)^p D_{\boldsymbol{x}}^{\alpha}u(\boldsymbol{x}+\theta(\boldsymbol{x}_I - \boldsymbol{x}))d\theta\right\}
$$
$$
\cdot D_{\boldsymbol{x}}^{\beta}\Psi_I^{\varrho}(\boldsymbol{x})
$$
$$
= \sum_{|\alpha| \leq p}\frac{1}{\alpha!}D_{\boldsymbol{x}}^{\alpha}u(\boldsymbol{x})\alpha!\delta_{\alpha\beta}\qquad\qquad \leftarrow \text{by}\quad \text{Lemma}(4.2.2)
$$
$$
+ \sum_{I \in \Lambda}\left\{\sum_{|\alpha|=p+1}p(\alpha)(\boldsymbol{x}_I - \boldsymbol{x})^{\alpha}\int_0^1(1-\theta)^p D_{\boldsymbol{x}}^{\alpha}u(\boldsymbol{x}+\theta(\boldsymbol{x}_I - \boldsymbol{x}))d\theta\right\}
$$
$$
\cdot D_{\boldsymbol{x}}^{\beta}\Psi_I^{\varrho}(\boldsymbol{x}) \tag{4.87}
$$

If $\partial\Omega$ is smooth enough, one can carefully choose a particle distribution and dilation parameter ϱ such that each sub-domain $supp\{\Psi_I\} \cap \Omega$ is *star-shaped* with respect to $\boldsymbol{x}_I$, i.e. $\forall \boldsymbol{x} \in supp\{\Psi_I\} \cap \Omega$,

$$
\boldsymbol{x} + \theta(\boldsymbol{x}_I - \boldsymbol{x}) \in supp\{\Psi_I\} \cap \Omega \ . \tag{4.88}
$$

thus the above expansion always make sense. It should be noted that this is a rather loose condition, which does not require domain Ω to be convex.

Let $c(p) = \max_{|\alpha|=p+1}p(\alpha)$. It follows then

$$
|D_{\boldsymbol{x}}^{\beta}u - D_{\boldsymbol{x}}^{\beta}\mathcal{R}_{\varrho,h}^{p}u| \leq c(p)\sum_{I \in \Lambda}\left\{\sum_{|\alpha|=p+1}\int_0^1(1-\theta)^p|D_{\boldsymbol{x}}^{\alpha}u(\boldsymbol{x}+\theta(\boldsymbol{x}_I - \boldsymbol{x}))|d\theta\right\}
$$
$$
\cdot|\mathbf{x}_I - \mathbf{x}|^{p+1}|D_{\boldsymbol{x}}^{\beta}\Psi_I^{\varrho}(\boldsymbol{x})| \tag{4.89}
$$

If $\boldsymbol{x} \in supp\{\Psi_I^{\varrho}(\boldsymbol{x})\}$ then $\exists\, 0 < r < 1$, such that $|\boldsymbol{x}_I - \boldsymbol{x}| \leq r\varrho$; therefore

$$
|D_{\boldsymbol{x}}^{\beta}u - D_{\boldsymbol{x}}^{\beta}\mathcal{R}_{\varrho,h}^{p}u| \leq C_p\varrho^{p+1}\sum_{I \in \Lambda(\boldsymbol{x})}\left(\sum_{|\alpha|=p+1}\int_0^1(1-\theta)^p|D_{\boldsymbol{x}}^{\alpha}u(\boldsymbol{x}+\theta(\boldsymbol{x}_I - \boldsymbol{x}))|d\theta\right)
$$
$$
\cdot\chi_{supp\{\Psi_I(\boldsymbol{x})\}}|D_{\boldsymbol{x}}^{\beta}\Psi_I^{\varrho}(\boldsymbol{x})| \tag{4.90}
$$

where $\chi_{supp\{\Psi_I(\boldsymbol{x})\}}$ is the characteristic function of $supp\{\Psi_I(\mathbf{x})\}$,

$$\chi_{supp\{\Psi_I(\boldsymbol{x})\}} = \begin{cases} 1 \,, & \boldsymbol{x} \in supp\{\Psi_I(\boldsymbol{x})\} \\ 0 \,, & \boldsymbol{x} \notin supp\{\Psi_I(\boldsymbol{x})\} \end{cases} \tag{4.91}$$

and $\Lambda(\boldsymbol{x}) := \{I \in \Lambda \mid \boldsymbol{x} \in supp\{\Psi_I\} \cap \bar{\Omega}\}$.

By Theorem (4.1.4),

$$|D_{\boldsymbol{x}}^{\beta}\Psi_I(\mathbf{x})| \le c\varrho^{-|\beta|} \tag{4.92}$$

and Minkowski's inequality, one may have the following inequality,

$$\int_{\Omega} |D_{\boldsymbol{x}}^{\beta} u - D_{\boldsymbol{x}}^{\beta}\left(\mathcal{R}_{\varrho,h}^{p} u\right)|^2 d\Omega$$

$$\le C(p)\varrho^{2(p+1-|\beta|)} \sum_{|\alpha|=p+1} \int_{\Omega} \left[\int_0^1 (1-\theta)^p \sum_{I\in\Lambda(\boldsymbol{x})} |D_{\boldsymbol{x}}^{\alpha} u(\boldsymbol{x} + \theta(\boldsymbol{x}_I - \boldsymbol{x}))|\chi_{supp\{\Psi_I(\boldsymbol{x})\}} d\theta \right]^2 d\Omega$$

$$\le C(p)\varrho^{2(p+1-|\beta|)} \sum_{|\alpha|=p+1} \left[\int_0^1 \left(\int_{\Omega} (1-\theta)^{2p} \right. \right.$$

$$\left. \left. \sum_{I\in\Lambda(\boldsymbol{x})} |D_{\boldsymbol{x}}^{\alpha} u(\boldsymbol{x} + \theta(\boldsymbol{x}_I - \boldsymbol{x}))|^2 \chi_{supp\{\Psi_I(\boldsymbol{x})\}} d\Omega_{\boldsymbol{x}} \right)^{\frac{1}{2}} d\theta \right]^2 \tag{4.93}$$

Change of variable $\mathbf{y} = \mathbf{x} + \theta(\boldsymbol{x}_I - \boldsymbol{x})$:

$$B(\boldsymbol{x}_I, \varrho) = \{\boldsymbol{x} \mid |\mathbf{x}-\mathbf{x}_I| \le r\varrho\} \Rightarrow B(\boldsymbol{y}_I, (1-\theta)\varrho) = \{\boldsymbol{y} \mid |\mathbf{y}-\mathbf{y}_I| \le (1-\theta)r\varrho\}$$

and hence $d\Omega_{\boldsymbol{x}} = (1-\theta)^{-n} d\Omega_{\mathbf{y}}$ where n is the space dimension.

Therefore,

$$\int_{\{B(\boldsymbol{x},\varrho)\}} |D_{\boldsymbol{x}}^{\alpha}(\boldsymbol{x} + \theta(\boldsymbol{x}_I - \boldsymbol{x}))|^2 d\Omega_{\boldsymbol{x}} = \int_{\{B(\boldsymbol{y},(1-\theta)\varrho)\}} |D_{\boldsymbol{y}}^{\alpha}(\boldsymbol{y})|^2 \frac{d\Omega_{\boldsymbol{y}}}{(1-\theta)^n}$$

$$\le \int_{\{B(\boldsymbol{x}_I,\varrho)\}} |D_{\boldsymbol{x}}^{\alpha}(\boldsymbol{x})|^2 \frac{d\Omega_{\boldsymbol{x}}}{(1-\theta)^n} \tag{4.94}$$

since $\boldsymbol{x}_I = \boldsymbol{y}_I$ and $B(\boldsymbol{y}_I, (1-\theta)\varrho) \subset B(\boldsymbol{x}_I, \varrho)$.

Thereby, if $p \ge \dfrac{n}{2}$, from the pointwise overlap condition (4.13), for fixed $\boldsymbol{x}$, $card\{\Lambda(\boldsymbol{x})\} \le N_{max}$, one then has

$$\sum_{|\alpha|=p+1} \left[\int_0^1 \left(\int_{\Omega} (1-\theta)^{2p} \sum_{I\in\Lambda(\boldsymbol{x})} |D_{\boldsymbol{x}}^{\alpha} u(\boldsymbol{x} + \theta(\boldsymbol{x}_I - \boldsymbol{x}))|^2 \chi_{supp\{\Psi_I(\boldsymbol{x})\}} d\Omega_{\boldsymbol{x}} \right)^{\frac{1}{2}} d\theta \right]^2$$

$$\le C_1(p) N_{max} \sum_{|\alpha|=p+1} \|D_{\boldsymbol{x}}^{\alpha} u\|_{L^2(\Omega)}^2 \le C_2(p, N_{max}) |u|_{H^{p+1}(\Omega)}^2 \tag{4.95}$$

where N_{max} is the maximum number of RKPM shape functions in the domain of influence of an arbitrary point $\boldsymbol{x} \in \Omega$.

One can readily show that the following pointwise estimate holds

$$\|D_{\boldsymbol{x}}^{\beta}u - D_{\boldsymbol{x}}^{\beta}\mathcal{R}_{\varrho,h}^{m}u\|_{L^{2}(\Omega)}^{2} \leq C(p,\beta,N_{max})\varrho^{2(p+1-|\beta|)}|u|_{H^{p+1}(\Omega)}^{2} \qquad (4.96)$$

Hence, we conclude that $\exists\ 0 < C_k, C_k' < \infty,$

$$|u - \mathcal{R}_{\varrho,h}^{p}u|_{H^{k}(\Omega)} \leq C_k'\varrho^{p+1-k}|u|_{H^{p+1}(\Omega)} \qquad k = 0, 1, \cdots, p \qquad (4.97)$$

or

$$\|u - \mathcal{R}_{\varrho,h}^{p}u\|_{H^{k}(\Omega)} \leq C_k\varrho^{p+1-k}\|u\|_{H^{p+1}(\Omega)} \qquad k = 0, 1, \cdots, p. \qquad (4.98)$$

♣

4.4 Convergence of Meshfree Galerkin Procedures

The interpolation estimate obtained in the last section can be used in error estimation of numerical solutions obtained by using moving least square kernel Galerkin method. To illustrate the general procedure, error estimates for elliptic partial differential equations are considered in two different boundary conditions.

4.4.1 The Neumann Boundary Value Problem (BVP)

Since natural boundary condition problems require fewer restrictions on both trial functions and weighting functions, it is convenient to consider the following model problem first, a Neumann problem for the second order elliptic partial differential equation,

$$L(u) = -\nabla^2 u + u = f(\boldsymbol{x}), \quad \boldsymbol{x} \in \Omega$$

$$\frac{\partial u}{\partial n} = g(\boldsymbol{x}), \qquad\qquad \boldsymbol{x} \in \partial\Omega$$

$$\tag{4.99}$$

where f, g are assumed to satisfy sufficient regularity requirements.

Define a bilinear form,

$$a(u, v) := \int_\Omega \left(\nabla u \cdot \nabla v + u \cdot v \right) d\Omega \ . \tag{4.100}$$

It is obvious that $a(u, v)$ is coercive on $H_1(\Omega)$,

$$a(u, u) = \|u\|_{H^1(\Omega)}^2 \geq \gamma \|u\|_{H^1(\Omega)}^2, \quad \forall u \in H^1(\Omega) \tag{4.101}$$

where $0 < \gamma \leq 1$. Or we should say that $a(u, v)$ is a $H^1(\Omega)$-elliptic form. It is also straightforward to show that $a(u, v)$ is continuous, i.e. $\exists C > 0$ such that

$$a(u, v) \leq C\|u\|_{H^1(\Omega)}\|v\|_{H^1(\Omega)} \tag{4.102}$$

As a matter of fact, by Cauchy's inequality

$$a(u, v) \leq \sqrt{\int_\Omega (\nabla u)^2 d\Omega} \sqrt{\int_\Omega (\nabla v)^2 d\Omega} + \sqrt{\int_\Omega (u)^2 d\Omega} \sqrt{\int_\Omega (v)^2 d\Omega}$$

$$\leq \left(\sqrt{\int_\Omega (\nabla u)^2 d\Omega} + \sqrt{\int_\Omega (u)^2 d\Omega} \right) \left(\sqrt{\int_\Omega (\nabla v)^2 d\Omega} + \sqrt{\int_\Omega (v)^2 d\Omega} \right)$$

$$\leq 2 \left(\|u\|_{H^1(\Omega)} \|v\|_{H^1(\Omega)} \right) \tag{4.103}$$

In the last step, the arithmetic-geometric inequality is used.

Then, by the Lax-Milgram theorem, the original problem (4.99) is equivalent to the following variational weak formulation,

$$\begin{cases} \text{Find } u \in H^1(\Omega) \text{ such that } \forall v \in H^1(\Omega) \\ \int_{\Omega}\Big(\nabla u \cdot \nabla v + uv\Big)d\Omega = \int_{\Omega}fvd\Omega + \int_{\partial\Omega}gvdS \end{cases} \quad (4.104)$$

Introduce an admissible particle distribution on the domain of interests $\bar{\Omega}$, some of the particles lie on the boundary $\partial\Omega$. Construct meshfree shape function basis by following Eqs. (4.43) and (4.43). The meshfree interpolant space is

$$\mathcal{V}_{\varrho}^p(\Omega) := span\{\Psi_I^{\varrho} \mid I \in \Lambda; \quad supp\{\Psi_I^{\varrho}\} \cap \Omega \neq \emptyset\} \cap H^1(\Omega) \quad (4.105)$$

Here the superscript p indicates that the shape function is constructed by the complete p-th order polynomial. Clearly $\mathcal{V}_{\varrho}^p(\Omega) \subset C^p(\Omega) \subset H^1(\Omega)$ provided that $p \geq 1$.

Then we formulate the meshfree Galerkin problem I as

$$\text{MGP(I)} \quad \begin{cases} \text{Find } u^{\varrho} \in \mathcal{V}_{\varrho}^p(\Omega) \text{ such that } \forall v^{\varrho} \in \mathcal{V}_{\varrho}^p(\Omega) \\ \int_{\Omega}\Big(\nabla u^{\varrho} \cdot \nabla v^{\varrho} + u^{\varrho}v^{\varrho}\Big)d\Omega = \int_{\Omega}fv^{\varrho}d\Omega + \int_{\partial\Omega}gv^{\varrho}dS \end{cases} \quad (4.106)$$

For the numerical meshfree Galerkin solution u^{ϱ} of (4.106), we have the following error estimate, which is based on the celebrated Céa Lemma (Ciarlet [1978]).

Theorem 4.4.1. *Let $u \in C^{p+1}(\Omega)$; If u is the solution of Neumann problem (4.99), and $u^{\varrho} \in \mathcal{V}_{\varrho}^p(\Omega)$ is the solution of weak formulation (4.106), then $\exists\, C_k > 0$ such that*

$$\|u - u^{\varrho}\|_{H^k(\Omega)} \leq C_1\varrho^{p+1-k}\|u\|_{H^{p+1}(\Omega)}, \quad k = 0, 1 \quad (4.107)$$

where the constants, C_k, do not depend on dilation parameter ϱ.

Proof:
We first show (4.107). Since $v^{\varrho} \in \mathcal{V}_{\varrho}^p(\Omega) \subset H^1(\Omega)$,

$$a(u, v^{\varrho}) = \int_{\Omega} fv^{\varrho}d\Omega + \int_{\partial\Omega} gv^{\varrho}dS \quad (4.108)$$

$$a(u^{\varrho}, v^{\varrho}) = \int_{\Omega} fv^{\varrho}d\Omega + \int_{\partial\Omega} gv^{\varrho}dS \quad (4.109)$$

Subtraction (4.109) from (4.108) yields

$$a(u - u^{\varrho}, v^{\varrho}) = 0 , \qquad\qquad \forall v^{\varrho} \in \mathcal{V}_{\varrho}^p(\Omega) \quad (4.110)$$

thus

$$\|u - u^\varrho\|^2_{H^1(\Omega)} = a(u - u^\varrho, u - u^\varrho)$$
$$= a\big((u - u^\varrho), (u - v^\varrho) + (v^\varrho - u^\varrho)\big)$$
$$= a(u - u^\varrho, u - v^\varrho) + a(u - u^\varrho, v^\varrho - u^\varrho)$$
$$= a(u - u^\varrho, u - v^\varrho) \qquad \Leftarrow \; v^\varrho - u^\varrho \in \mathcal{V}^p_\varrho(\Omega)$$
$$\leq C\|u - u^\varrho\|_{H^1(\Omega)}\|u - v^\varrho\|_{H^1(\Omega)} \qquad \Leftarrow \; \text{by continuity of } a(u,v)$$

$$(4.111)$$

Thus $\forall v^\varrho \in \mathcal{V}^p_\varrho(\Omega)$,

$$\|u - u^\varrho\|_{H^1(\Omega)} \leq C \inf_{v^\varrho \in \mathcal{V}^p_\varrho(\Omega)} \|u - v^\varrho\|_{H^1(\Omega)}$$
$$= C \min_{v^\varrho \in \mathcal{V}^p_\varrho(\Omega)} \|u - v^\varrho\|_{H^1(\Omega)} \qquad (4.112)$$

Since v^ϱ is an arbitrary element in $\mathcal{V}^p_\varrho(\Omega)$, let

$$v^\varrho = \mathcal{R}^p_{\varrho,h} u \quad . \qquad (4.113)$$

Note that $u^\varrho \neq \mathcal{R}^p_{\varrho,h} u$! Consequently, by the interpolation estimate (4.81), we obtain

$$\|u - u^\varrho\|_{H^1(\Omega)} \leq C\|u - \mathcal{R}^p_{\varrho,h} u\|_{H^1(\Omega)}$$
$$\leq C_1 \varrho^p \|u\|_{H^{p+1}(\Omega)} \qquad (4.114)$$

Next, we show the case $k = 0$. The procedure is the standard duality technique known as Nitsche's trick. Let $e^\varrho = u - u^\varrho$ and consider the following auxiliary problem,

$$L(w) = e^\varrho \quad . \qquad (4.115)$$

The corresponding weak solution satisfies

$$a(w, v) = (e^\varrho, v) , \qquad \forall \, v \in \mathcal{V}^p_\varrho(\Omega) \qquad (4.116)$$

Choosing $v = e^\varrho$ and considering

$$a(v^\varrho, u - u^\varrho) = a(v^\varrho, e^\varrho) \qquad \forall \, v^\varrho \in \mathcal{V}^p_\varrho(\Omega) \qquad (4.117)$$

one has

$$\|e^\varrho\|_{L^2(\Omega)} = a(w, e^\varrho) = a(w - v^\varrho, e^\varrho) \qquad (4.118)$$

Let $v^\varrho = w^\varrho$. By Cauchy's inequality, the above expression can be bounded as follows

$$\|e^\varrho\|^2_{L^2(\Omega)} = a(w - w^\varrho, u - u^\varrho)$$
$$\leq C\|w - w^\varrho\|_{H^1(\Omega)}\|u - u^\varrho\|_{H^1(\Omega)} \qquad \Leftarrow \text{ by continuity of } a(u,v)$$
$$\leq CC' \varrho\|w\|_{H^2(\Omega)}\|u - u^\varrho\|_{H^1(\Omega)} \qquad \Leftarrow \text{ by (4.81)}$$

On the other hand, the continuous dependence of the solution on the data requires

$$\|w\|_{H^2(\Omega)} \leq C''\|e^\varrho\|_{L^2(\Omega)} \tag{4.119}$$

Thus,

$$\|e^\varrho\|^2_{L^2(\Omega)} \leq CC'C'' \varrho\|e^\varrho\|_{L^2(\Omega)}\|u - u^\varrho\|_{H^1(\Omega)} \tag{4.120}$$

which finally leads to

$$\|u - u^\varrho\|_{L^2(\Omega)} \leq CC'C'' \varrho\|u - u^\varrho\|_{H^1(\Omega)} \leq C_0\varrho^{p+1}\|u\|_{H^{p+1}(\Omega)} \tag{4.121}$$

♣

In most numerical computations, the interpolated function $f^\varrho(\boldsymbol{x})$, $g^\varrho(\boldsymbol{x})$ are used instead of exact input data function $f(\boldsymbol{x})$ and $g(\boldsymbol{x})$. In generally, meshfree interpolants are usually "non-interpolation" schemes; this makes numerical computations intriguing. In the actual computations, the following scheme is often implemented:

$$f^\varrho(\boldsymbol{x}) := \mathcal{R}^p_{\varrho,h}f(\boldsymbol{x}), \quad \boldsymbol{x} \in \Omega \tag{4.122}$$

$$g^\varrho(\boldsymbol{x}) := \mathcal{R}^p_{\varrho,h}\tilde{g}(\boldsymbol{x}), \quad \boldsymbol{x} \in \bar{\Omega} \tag{4.123}$$

where function $\tilde{g}(\boldsymbol{x}) \in H^1(\bar{\Omega}) \cap C^1(\Omega)$ or even more smooth, such that

$$\tilde{g}(\boldsymbol{x}) := \begin{cases} g\,, & \boldsymbol{x} \in \partial\Omega \\ \text{continuous function; } \boldsymbol{x} \in \Omega \end{cases} \tag{4.124}$$

In this manner, a second meshfree Galerkin variational formulation is set as follows

$$\text{MGP(II)} : \begin{cases} \text{Find } u^\varrho \in \mathcal{V}^p_\varrho(\Omega) \text{ such that } \forall\, v^\varrho \in \mathcal{V}^p_\varrho(\Omega) \\ \int_\Omega (\nabla u^\varrho \nabla v^\varrho + u^\varrho v^\varrho)d\Omega = \int_\Omega f^\varrho v^\varrho d\Omega + \int_{\partial\Omega} g^\varrho v^\varrho dS \end{cases} \tag{4.125}$$

Remark 4.4.1. In (4.125), the expression $\displaystyle\int_{\partial\Omega} g^\varrho v^\varrho dS$ should be interpreted as

$$\int_{\partial\Omega} g^\varrho(\varrho, \boldsymbol{y}, \boldsymbol{x}_I)v^\varrho(\boldsymbol{y}, \boldsymbol{x}_I)dS_{\boldsymbol{y}} \tag{4.126}$$

where $\boldsymbol{y} \in \partial\Omega$, but $\boldsymbol{x}_I \in \bar{\Omega}$. An elucidate description on implementation of natural boundary conditions of this sort is given by Pang.[367]

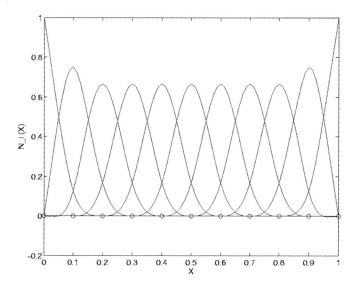

Fig. 4.5. *A finite-dimensional trial function basis that belongs to the space* $\mathcal{U}_\varrho^1([0,1])$.

4.4.2 The Dirichlet Boundary Value Problem

As mentioned previously, unlike finite element interpolation shape functions, most meshfree interpolation trial and test functions can not satisfy the essential boundary condition unless special care is taken (see Chapter 3).

Consequently, the error estimates of meshfree Galerkin formulations concerning Dirichlet boundary value problems are more involved.

However, it has been discovered that at least in one dimensional cases the correction function can be adjusted such that the shape function may satisfy the essential boundary condition by carefully choosing the dilation parameter. It is possible to construct a finite-dimensional trial function space $\mathcal{U}_\varrho^p$, such that

$$\mathcal{U}_\varrho^p(\Omega) := span\{\Psi_I^\varrho(\boldsymbol{x}) \mid supp\{\Psi_I^\varrho\} \cap \Omega \neq \emptyset; \Psi_I(\boldsymbol{x}_J) = \delta_{IJ}, \forall \boldsymbol{x}_J \in \partial\Omega\}$$

(4.127)

and it is feasible to construct a finite-dimensional test function space such that,

$$\mathcal{W}_{\varrho,0}^p(\Omega) := span\{\Psi_I^\varrho(\boldsymbol{x}) \mid supp\{\Psi_I^\varrho\} \cap \Omega \neq \emptyset; \Psi_I^\varrho(\boldsymbol{x}) = 0, \forall \boldsymbol{x} \in \partial\Omega\} .$$ (4.128)

where $I \in \Lambda$.

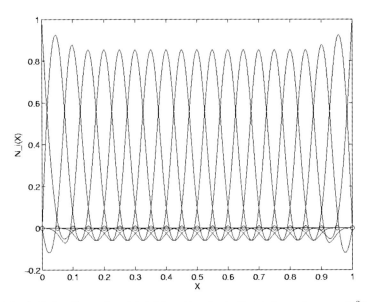

Fig. 4.6. *A finite-dimensional trial function basis that belongs to space $\mathcal{U}_\varrho^2([0,1])$.*

Two pictorial examples of such finite-dimensional trial function spaces are shown in Figs. 4.5 and 4.6. Some other examples of such finite-dimensional spaces may be constructed by proper choice of the dilation parameter. The family of shape functions in Fig. 4.5 are constructed based on cubic spline window function with linear generating polynomial, and the family of shape functions in Fig. 4.6 are constructed based on the fifth order spline window function with quadratic generating polynomial. In both cases, the particle distributions are uniform. The group of shape functions in Fig. 4.5 have the support radius $r\varrho = 2\Delta x$; for those shape functions in Fig. 4.6, the support radius is $r\varrho = 3\Delta x$. For the multiple dimensional problem, proper trial function base or weighting function base can be devised with care.

Consider the following model problem

$$-\nabla^2 u + u = f(\boldsymbol{x}) \qquad\qquad \boldsymbol{x} \in \Omega \qquad\qquad (4.129)$$

$$u = h(\boldsymbol{x}) \qquad\qquad \boldsymbol{x} \in \partial\Omega \qquad\qquad (4.130)$$

By employing the classic variational technique, a meshfree Galerkin formulation for 1D Dirichlet problem (4.129) and (4.130) can be set as follows

$$\text{MGP(III)}: \begin{cases} \text{Find } u^\varrho \in \mathcal{U}_\varrho^m(\Omega), \text{ such that } \forall w^\varrho \in \mathcal{W}_{\varrho,0}^m(\Omega) \\[2mm] \int_\Omega (\nabla u^\varrho \cdot \nabla w^\varrho + u^\varrho \cdot w^\varrho) d\Omega = \int_\Omega f w^\varrho d\Omega \end{cases} \qquad (4.131)$$

Following the similar procedure in previous sections, one can show that the following statement holds.

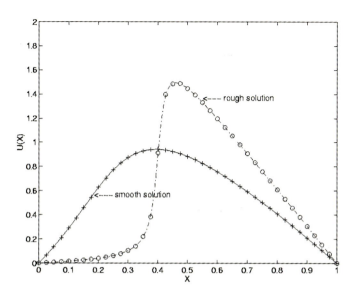

Fig. 4.7. The exact solutions and numerical solutions of the benchmark problem.

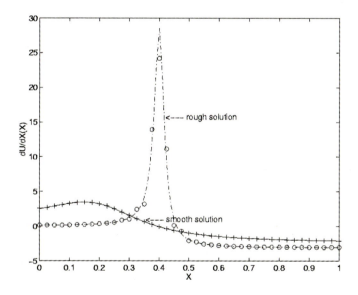

Fig. 4.8. The derivatives of exact solutions and numerical solutions.

Theorem 4.4.2. *Let $u \in C^{p+1}(\Omega)$. If u is the solution of Dirichlet problem (4.129)– (4.130), and u^{ϱ} is the solution of weak formulation (4.131), then $\exists C$ such that*

$$\|u - u^\varrho\|_{H^1(\Omega)} \leq C_{d1} \varrho^p \|u\|_{H^{p+1}(\Omega)} \qquad (4.132)$$

and

$$\|u - u^\varrho\|_{L^2(\Omega)} \leq C_{d0} \varrho^{p+1} \|u\|_{H^{p+1}(\Omega)} \qquad (4.133)$$

where the constants C_{d0}, C_{d1} do not depend on dilation parameter ϱ.

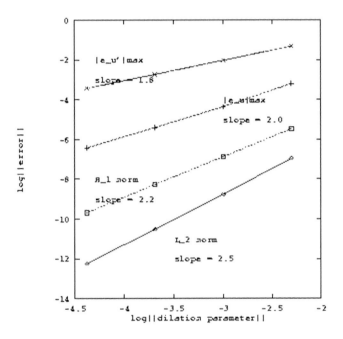

Fig. 4.9. The convergence rate for the smooth problem, $\alpha = 5.0, \bar{x} = 0.2$, for shape functions with basis of linear polynomial.

The above finite-dimensional trial and test function spaces can be constructed for non-uniform particle distribution as well. The construction for one dimensional as well as some two dimensional examples are provided in a paper by Gosz and Liu.[182]

4.4.3 Numerical Examples

In the following, several demonstrative problems have been solved by using a meshfree interpolant — the RKPM shape function. In order to compare

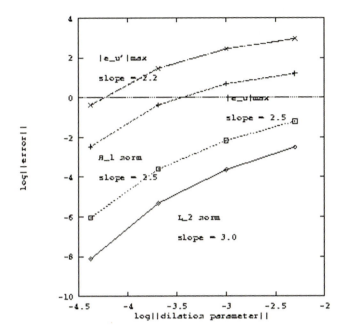

Fig. 4.10. The convergence rate for the rough problem, $\alpha = 5.0, \bar{x} = 0.2$, for shape functions with basis of linear polynomial.

with other numerical algorithms, here, a special benchmark problem is tested. This problem was originally proposed by Rachford and Wheeler [1974] to test the convergence property of the H^{-1}-Galerkin method, and was used again by Babuška, Oden, and Lee [1977] to test the mixed-hybrid finite element method. It is a two point boundary value problem,

$$\begin{cases} -u_{,xx} + u = f(x) & x \in (0,1) \\ u(0) = u(1) = 0 \end{cases} \tag{4.134}$$

where

$$f(x) = \frac{2\alpha(1 + \alpha^2(1 - \bar{x})(x - \bar{x}))}{(1 + \alpha^2(x - \bar{x})^2)^2}$$
$$+ (1 - x)(\arctan(\alpha(x - \bar{x})) + \arctan(\alpha\bar{x})) \tag{4.135}$$

The exact solution of Eq. (4.134) is

$$u(x) = (1 - x)(\arctan(\alpha(x - \bar{x})) + \arctan(\alpha\bar{x})) \tag{4.136}$$

According to the designed feature, the solution (4.136) changes its roughness as the parameter α varies. When α is relatively small, the solution (4.136)

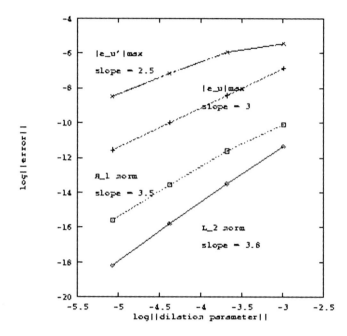

Fig. 4.11. The convergence rate for the smooth problem, $\alpha = 50.0, \bar{x} = 0.4$, while using shape functions with the base of quadratic polynomial.

is smooth; as α becomes large, there will be a sharp knee arising close to the location $x = \bar{x}$. Thus, it provides quite a challenging test for numerical computations.

Following the choice of Babuška et al. [1977], the two representative parameter groups chosen are as follows,

$$\text{the smooth solution :} \begin{cases} \alpha = 5.0 \\ \bar{x} = 0.2 \end{cases} ; \qquad (4.137)$$

$$\text{the rough solution :} \begin{cases} \alpha = 50.0 \\ \bar{x} = 0.40 \end{cases} . \qquad (4.138)$$

In Fig. 4.7, the exact solutions— both smooth and rough are plotted in comparison with numerical results. One can see that the numerical solutions agree with exact solutions fairly well in both cases—the smooth solution as well as the rough solution. In Fig. 4.8, the comparison between exact solution and numerical solution is made for the first order derivatives.

As mentioned above, two types of shape functions have been used in numerical computation: the shape functions generated by cubic spline window

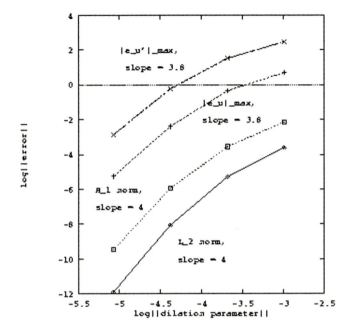

Fig. 4.12. The convergence rate for the rough problem, $\alpha = 50.0, \bar{x} = 0.4$, while using shape functions with base of quadratic polynomial.

function and those generated by fifth order spline window function, i.e. the shape function families shown in both Figs. 4.5 and 4.6. The computation is carried out for four different particle distributions: 11 particles, 21 particles, 41 particles, and 81 particles. The results shown in Figs. 4.7 and 4.8 are obtained by using the first group of shape function with 41 particles uniformly distributed in the domain.

Based on numerical results, the convergence rate of the algorithm is also shown with respect to different norms: L_2 norm, H_1 norm, and $|\cdot|_{max}$ norm. For the shape functions generated by cubic spline window function, the corresponding convergence results are plotted in both Figs. 4.9 and 4.10, and the convergence results for the shape function based on fifth order spline window function are displayed in Figs. 4.11 and (4.12).

As mentioned above, the shape function family in Fig. 4.5 is generated by linear polynomials, i.e. $m = 1$, and the shape function family in Fig. 4.6 is generated by quadratic polynomials, i.e. $m = 2$. Based on the Theorem (4.132), the convergence rates with respect to L_2 norm are 2 and 3 respectively. The numerical results shown in Figs. 4.9-4.12 seem to be better than this estimate. Nevertheless, when the particle density increases, the theoretical bound will become evident. On the other hand, one may observe that there seems to be a tendency that H_1 error norm converges almost as fast as the L_2 error norm.

One may also notice an interesting fact that both L_2 norm and H_1 norm have faster convergence rate than that of the maximum norm, which is totally in contrast with the conventional finite element method; the regular finite element method has an opposite tendency that maximum norm always have a faster convergence rate than H norms.

4.5 Approximation Theory of Meshfree Wavelet Functions

Meshfree hierarchical partition of unity, or meshfree wavelet shape functions, have been introduced in Chapter 3. In this section, a detailed approximation theory is presented for meshfree wavelet functions.

Denote,

$$\Psi_I^{[\alpha]}(\boldsymbol{x}) := \mathbf{P}^T\left(\frac{\boldsymbol{x} - \boldsymbol{x}_I}{\varrho}\right)\mathbf{b}^{[\alpha]}(\boldsymbol{x})\Phi_\varrho(\boldsymbol{x} - \boldsymbol{x}_I)\Delta V_I, \quad 0 \le |\alpha| \le p \quad (4.139)$$

where

$$\mathbf{P}^T(\mathbf{x}) = (1, \mathbf{x}, \cdots, \mathbf{x}^\alpha, \cdots), \tag{4.140}$$

$$\mathbf{b}^{[\alpha]}(\boldsymbol{x}) = \mathbf{M}^{-1}(\boldsymbol{x})\mathbf{P}^{[\alpha]}(0), \tag{4.141}$$

$$\mathbf{P}^{[\alpha]}(0) := \frac{1}{\alpha!}D_{\boldsymbol{x}}^\alpha\mathbf{P}(\boldsymbol{x})\Big|_{\boldsymbol{x}=0} \tag{4.142}$$

When $\alpha = 0$, $\Psi^{[0]}(\boldsymbol{x})$ is the regular RKPM interpolant that has been analyzed in the previous sections. When $\alpha \ne 0$, $\Psi^{[\alpha]}(\boldsymbol{x})$ is the so-called meshfree wavelet functions (see[266, 267]).

Assume $\alpha_1 = 0, |\alpha_2| = 1, \cdots, |\alpha_{N_p}| = p$ and $N_p := dim\mathcal{P} = \dfrac{(p+d)!}{p!d!}$. where d is space dimension. For $d = 1$, $\alpha_i = i+1$ and $\alpha_p = p+1$. We then have

$$\left(\mathbf{P}^{[\alpha_i]}(0)\right)^T = (0, \cdots, 0, \underbrace{1, 0, \cdots, 0}_{p-i}) \tag{4.143}$$

and

$$\left(\mathbf{P}^{[\alpha_1]}(0), \mathbf{P}^{[\alpha_2]}(0), \cdots, \mathbf{P}^{[\alpha_{N_p}]}(0)\right) = \mathbf{I}. \tag{4.144}$$

A useful identity may then be derived from Eq. (4.139)

$$\mathbf{M}(\boldsymbol{x})\begin{pmatrix} \Psi_I^{[\alpha_1]}(\boldsymbol{x}) \\ \Psi_I^{[\alpha_2]}(\boldsymbol{x}) \\ \vdots \\ \Psi_I^{[\alpha_{N_p}]}(\boldsymbol{x}) \end{pmatrix} = \mathbf{M}(\boldsymbol{x})\mathbf{I}\mathbf{M}^{-1}(\boldsymbol{x})\mathbf{P}\left(\frac{\boldsymbol{x} - \boldsymbol{x}_I}{\varrho}\right)\Phi_\varrho(\boldsymbol{x} - \boldsymbol{x}_I)\Delta V_I$$

$$= \mathbf{P}\left(\frac{\boldsymbol{x} - \boldsymbol{x}_I}{\varrho}\right)\Phi_\varrho(\boldsymbol{x} - \boldsymbol{x}_I)\Delta V_I \tag{4.145}$$

4.5.1 The Generalized Consistency Conditions

The following generalized consistency conditions are the intrinsic properties of the above meshfree wavelet functions.

For the wavelet interpolation bases generated by a complete p-th order polynomial basis, the following generalized consistency conditions, or moment equations hold by definition,

$$\sum_{I \in \Lambda} \left(\frac{x - x_I}{\varrho} \right)^{\alpha} \Psi_I^{[\beta]}(x) = \delta_{\alpha\beta}, \quad |\alpha|, |\beta| \leq p \tag{4.146}$$

Subsequently, they yield the generalized reproducing kernel condition

$$\sum_{I \in \Lambda} x_I^{\alpha} \Psi_I^{[\beta]}(x) = \sum_{I \in \Lambda} (x + (x_I - x))^{\alpha} \Psi_I^{[\beta]}(x)$$

$$= \sum_{I \in \Lambda} \binom{\alpha}{\gamma} (-1)^{\gamma} \varrho^{\gamma} x^{\alpha-\gamma} \left(\frac{x - x_I}{\varrho} \right)^{\gamma} \Psi_I^{[\beta]}(x)$$

$$= \binom{\alpha}{\gamma} (-1)^{\gamma} \varrho^{\gamma} x^{\alpha-\gamma} \delta_{\gamma\beta}$$

$$= (-1)^{|\beta|} \frac{\alpha! \varrho^{\beta}}{(\alpha-\beta)!\beta!} x^{\alpha-\beta}, \quad |\alpha|, |\beta| \leq p \tag{4.147}$$

which is the reminiscence of the differential consistency conditions for the regular RKPM shape basis $\{\Psi_I^{[0]}(x)\}$, e.g. Eqs. (4.63) and (4.2.2).
Let,

$$N_I^{[\beta]}(x) = \frac{(-1)^{\beta} \beta!}{\varrho^{\beta}} \Psi_I^{[\beta]}(x) . \tag{4.148}$$

We recover exactly Eqs. (4.63) and (4.2.2),

$$\sum_{I \in \Lambda} \left(\frac{x - x_I}{\varrho} \right)^{\alpha} N_I^{[\beta]}(x) = (-1)^{\beta} \frac{\beta!}{\varrho^{\beta}} \delta_{\alpha\beta}, \quad |\alpha|, |\beta| \leq p \tag{4.149}$$

$$\sum_{I \in \Lambda} x_I^{\alpha} N_I^{[\beta]}(x) = \frac{\beta!}{(\alpha-\beta)!} x^{\alpha-\beta}, \quad |\alpha|, |\beta| \leq p \tag{4.150}$$

The above generalized consistency conditions may be viewed as special cases of the differential consistency conditions.

Lemma 4.5.1. *For the β-th order meshfree wavelet interpolant generated by the p-order polynomial basis $\{\Psi_I^{(\beta)}(x)\}$, the following differential consistency conditions hold*

$$\sum_{I \in \Lambda} \left(\frac{x - x_I}{\varrho} \right)^{\alpha} D_x^{\gamma} \Psi_I^{[\beta]}(x) = (-1)^{|\gamma|} \frac{\alpha! \varrho^{-|\gamma|}}{\beta!} \delta_{\alpha(\beta+\gamma)} , \quad |\alpha|, |\beta|, |\gamma| \leq p \tag{4.151}$$

Proof:
    The proof is by induction on γ. First, assume $|\gamma| = 0$ and then by (4.146)

$$\sum_{I\in\Lambda}\left(\frac{\boldsymbol{x}-\boldsymbol{x}_I}{\varrho}\right)^{\alpha}\Psi^{[\beta]}(\boldsymbol{x})=\delta_{\alpha\beta} \tag{4.152}$$

Eq. (4.151) holds.

Second, assume that (4.151) holds for $0\le|\gamma|\le p-1$, namely,

$$\sum_{I\in\Lambda}\left(\frac{\boldsymbol{x}_I-\boldsymbol{x}}{\rho}\right)^{\alpha}D_x^{\gamma}\Psi^{[\beta]}(\mathbf{x})=(-1)^{\gamma}\frac{\alpha!\varrho^{-\gamma}}{\beta!}\delta_{\alpha(\beta+\gamma)} \ . \tag{4.153}$$

We need to show that (4.151) holds for $0\le|\gamma'|\le p$. Let $\gamma'=\gamma+\eta$, $\eta=(\eta_1,\eta_2,\cdots,\eta_n)$, $|\eta|=1$, $0\le|\gamma'|\le p$. Since $|\eta|=1$, differentiate (4.153) and then by the chain rule,

$$\sum_{I\in\Lambda}\left\{D_x^{\eta}\left(\frac{\mathbf{x}-\mathbf{x}_I}{\varrho}\right)^{\alpha}D_x^{\gamma}\Psi^{[\beta]}(\mathbf{x})\right.$$
$$\left.+\left(\frac{\mathbf{x}-\mathbf{x}_I}{\varrho}\right)^{\alpha}D_x^{\gamma+\eta}\Psi^{[\beta]}(\mathbf{x})\right\}=0 \tag{4.154}$$

It can be shown that

$$D_x^{\eta}\left(\frac{x-x_I}{\rho}\right)^{\alpha}=\frac{\alpha_1!\alpha_2!\cdots\alpha_n!}{(\alpha_1-\eta_1)!(\alpha_2-\eta_2)!\cdots(\alpha_n-\eta_n)!}\varrho^{-\eta_1}\cdots\varrho^{-\eta_n}$$
$$\cdot\left(\frac{x_1-x_{1I}}{\rho}\right)^{\alpha_1-\eta_1}\left(\frac{x_2-x_{2I}}{\rho}\right)^{\alpha_2-\eta_2}\cdots\left(\frac{x_n-x_{nI}}{\rho}\right)^{\alpha_n-\eta_n}$$
$$=\frac{\rho^{-\eta}\alpha!}{(\alpha-\eta)!}\left(\frac{x-x_I}{\rho}\right)^{\alpha-\eta} \tag{4.155}$$

Thereby, Eq.(4.154) yields

$$\sum_{I\in\Lambda}\left(\frac{x-x_I}{\rho}\right)^{\alpha}D_x^{\gamma'}\Psi^{[\beta]}(\mathbf{x})$$
$$=\frac{(-1)^{|\eta|}\varrho^{-\eta}\alpha!}{(\alpha-\eta)!}\sum_{I\in\Lambda}\left(\frac{x_I-x}{\rho}\right)^{\alpha-\eta}D_x^{\gamma}\Psi_I^{[\beta]}(\boldsymbol{x})\Delta V_I$$
$$=(-1)^{|\eta|}\varrho^{-\eta}\frac{\alpha!}{(\alpha-\eta)!}(-1)^{|\gamma|}\varrho^{-\gamma}\frac{(\alpha-\eta)!}{(\alpha-\eta-\gamma)!}\delta_{(\alpha-\eta)(\beta+\gamma)}$$
$$=\frac{(-1)^{|\gamma'|}\varrho^{-\gamma'}\alpha!}{(\alpha-\gamma')!}\delta_{(\alpha-\eta)(\beta+\gamma)}=\frac{(-1)^{|\gamma'|}\varrho^{-\gamma'}\alpha!}{(\alpha-\gamma')!}\delta_{\alpha(\beta+\gamma')} \tag{4.156}$$

In the second step, (4.153) is used, and in the last step, the identity, $\delta_{(\alpha-\eta)(\beta+\gamma)}=\delta_{\alpha(\beta+\gamma')}$, is used.   ♣

To understand the implication of such conditions, we have a one-dimensional example to illustrate the differential consistency conditions for meshfree wavelet interpolant,

Table 4.1. Differential Consistency Conditions For $\Psi_I^{[0]}(x)$

$M_3^{[0](0)} = 0$	$M_3^{[0](1)} = 0$	$M_3^{[0](2)} = 0$	$M_3^{[0](3)} = -3!\varrho^{-3}$
$M_2^{[0](0)} = 0$	$M_2^{[0](1)} = 0$	$M_2^{[0](2)} = 2!\varrho^{-2}$	$M_2^{[0](3)} = 0$
$M_1^{[0](0)} = 0$	$M_1^{[0](1)} = -1!\varrho^{-1}$	$M_1^{[0](2)} = 0$	$M_1^{[0](3)} = 0$
$M_0^{[0]} = 0!$	$M_0^{[0](1)} = 0$	$M_0^{[0](2)} = 0$	$M_0^{[0](3)} = 0$

Table 4.2. Differential Consistency Conditions For The 1st Wavelet $\Psi_I^{[1]}(x)$

$M_3^{[1](0)} = 0$	$M_3^{[1](1)} = 0$	$M_3^{[1](2)} = 3!\varrho^{-2}$	$M_3^{[1](3)} = 0$
$M_2^{[1](0)} = 0$	$M_2^{[1](1)} = -2!\varrho^{-1}$	$M_2^{[1](2)} = 0$	$M_2^{[1](3)} = 0$
$M_1^{[1](0)} = 1$	$M_1^{[1](1)} = 0$	$M_1^{[1](2)} = 0$	$M_1^{[1](3)} = 0$
$M_0^{[1](0)} = 0$	$M_0^{[1](1)} = 0$	$M_0^{[1](2)} = 0$	$M_0^{[1](3)} = 0$

Table 4.3. Differential Consistency Conditions For The 2nd Wavelet $\Psi_I^{[2]}(x)$

$M_3^{[2](0)} = 0$	$M_3^{[2](1)} = -3!/2!\varrho^{-1}$	$M_3^{[2](2)} = 0$	$M_3^{[2](3)} = 0$
$M_2^{[2](0)} = 1$	$M_2^{[2](1)} = 0$	$M_2^{[2](2)} = 0$	$M_2^{[2](3)} = 0$
$M_1^{[2](0)} = 0$	$M_1^{[2](1)} = 0$	$M_1^{[2](2)} = 0$	$M_1^{[2](3)} = 0$
$M_0^{[2](0)} = 0$	$M_0^{[2](1)} = 0$	$M_0^{[2](2)} = 0$	$M_0^{[2](3)} = 0$

Table 4.4. Differential Consistency Conditions For The 3rd Wavelet $\Psi_I^{[3]}(x)$

$M_3^{[3](0)} = 1$	$M_3^{[3](1)} = 0$	$M_3^{[3](2)} = 0$	$M_3^{[3](3)} = 0$
$M_2^{[3](0)} = 0$	$M_2^{[3](1)} = 0$	$M_2^{[3](2)} = 0$	$M_2^{[3](3)} = 0$
$M_1^{[3](0)} = 0$	$M_1^{[3](1)} = 0$	$M_1^{[3](2)} = 0$	$M_1^{[3](3)} = 0$
$M_0^{[3](0)} = 0$	$M_0^{[3](1)} = 0$	$M_0^{[3](2)} = 0$	$M_0^{[3](3)} = 0$

$$M_\alpha^{[\beta]} := \sum_{I \in \Lambda} \left(\frac{x - x_I}{\varrho} \right)^\alpha \Psi_I^{[\beta]}(x) \tag{4.157}$$

$$M_\alpha^{[\beta](\gamma)} := \sum_{I \in \Lambda} \left(\frac{x - x_I}{\varrho} \right)^\alpha D_x^\gamma \Psi^{[\beta]}(x) \ . \tag{4.158}$$

The differential consistency conditions for the fundamental kernel and the wavelet kernels can be interpreted as the following moment identities:

$$M_\alpha^{[0](\gamma)} = \delta_{\alpha\gamma} \ ; \qquad M_\alpha^{[\beta](\gamma)} = (-1)^{|\beta|} \varrho^{-\gamma} \frac{\alpha!}{\beta!} \delta_{(\alpha-\gamma)\beta} \tag{4.159}$$

Tables (4.1)–(4.4) display graphically the differential consistency conditions of the hierarchical partition of unity in Example (3.2.1). For the fundamental kernel, all the non-zero entries lie on the main diagonal line of the table; for the 1st order wavelet kernel, all the non-zero entries lie on the first sub-diagonal line; and for the 2nd order wavelet, the non-zero entries move to the second sub-diagonal line, and the pattern continues until the 3rd order wavelet.

4.5.2 Interpolation Estimate

Since interpolant $N_I^{[\beta]}(\boldsymbol{x})$ shares the same differential consistency conditions with the β-th derivative of fundamental meshfree interpolant, $D_x^\beta \Psi_I^{[0]}(\boldsymbol{x})$. We propose the following meshfree interpolation scheme,

$$\mathcal{R}_\varrho^{p[\beta]} u(\boldsymbol{x}) = \sum_{I \in \Lambda} (-1)^{|\beta|} \frac{\beta!}{\varrho^\beta} \Psi_I^{[\beta]}(\boldsymbol{x}) u(\boldsymbol{x}_I) \Delta V_I$$

$$= \sum_{I \in \Lambda} N_I^{[\beta]}(\boldsymbol{x}) u_I \Delta V_I \tag{4.160}$$

The main result of this section is the following interpolation estimate for the β-th order meshfree interpolant.

Theorem 4.5.1 (Local Estimate). *Assume $u \in H^{m+1}(\Omega) \cap C^0(\Omega)$ and $\phi \in C_0^m(\Omega) \cap H^{m+1}(\Omega)$ [1] and $2m \geq d$, where d is the dimension of the space. For given bounded domain Ω, there is a meshless hierarchical discretization $\{\mathcal{D}, \mathcal{F}_\varrho, \mathcal{H}_m\}$. Then $\forall \omega_I \in \mathcal{F}_\varrho$ and $\mathcal{R}_{\varrho,h}^{m[\beta]} u \in \text{span}\{\mathcal{H}_m\}$ the following interpolation error estimate holds*

$$\|D_x^\beta u(\boldsymbol{x}) - \mathcal{R}_{\varrho,h}^{m[\beta]} u(\boldsymbol{x})\|_{H^\beta(\omega_I \cap \Omega)} \leq C_I \varrho^{m+1-|\beta|} \|u\|_{H^{m+1}(\omega_I \cap \Omega)} ,$$

$$\forall \ 0 \leq |\beta| \leq m \tag{4.161}$$

where $\mathcal{R}_{\varrho,h}^{m[\beta]} u(x)$ is given in Eq. (3.174).

Proof:

We only need to show that $\exists \, C$ such that for fixed $I \in \Lambda$,

$$\left| D_x^\beta u(x) - \mathcal{R}_{\varrho,h}^{m[\beta]} u(x) \right|_{L^2(\omega_I \cap \Omega)} \leq C \varrho^{m+1-|\beta|} |u|_{H^{m+1}(\omega_I \cap \Omega)} ,$$

$$\forall \ 0 \leq |\beta|, \leq m \tag{4.162}$$

By Taylor's expansion, for $x \in \omega_I \cap \Omega$, one has

$$\left(D_x^\beta u - \mathcal{R}_{\varrho,h}^{m[\beta]} u \right) = D_x^\beta u(x) - \frac{(-1)^{|\beta|} \beta!}{\varrho^\beta} \sum_{J \in \Lambda_I} \Psi_J^{[\beta]}(x) u(x_J)$$

$$= D_x^\beta u(x) - \frac{(-1)^{|\beta|} \beta!}{\varrho^\beta} \sum_{J \in \Lambda_I} \Psi_J^{[\beta]}(x)$$

$$\cdot \left(\sum_{|\alpha| < m+1} \frac{1}{\alpha!} (x_J - x)^\alpha D^\alpha u(x) + \sum_{|\alpha| = m+1} \left(\frac{m+1}{\alpha!} \right) (x_J - x)^\alpha \right.$$

$$\left. \cdot \left(\int_0^1 s^m D^\alpha u(x_J + s(x - x_J)) ds \right) \right) \tag{4.163}$$

---

[1] This restriction is imposed for the sake of an easy proof; it may be relaxed.

Applying the consistency condition (4.146) to (4.163), one may find that

$$\left| D_x^\beta u - \mathcal{R}_{\varrho,h}^{m[\beta]} u \right| = \left| D_x^\beta u(x) - (-1)^{|\alpha|-|\beta|} \varrho^{\alpha-\beta} \frac{\beta!}{\alpha!} \delta_{\alpha\beta} D_x^\beta u(x) \right.$$

$$- \frac{\beta!(-1)^{-\beta}}{\varrho^\beta} \sum_{J\in\Lambda_I} \sum_{|\gamma|=m+1} \left(\frac{m+1}{\gamma!} \right) (x_J - x)^\gamma \Psi_J^{[\beta]}(x)$$

$$\left. \cdot \left(\int_0^1 s^m D^\gamma u(x_J + s(x - x_J)) ds \right) \right|$$

$$\leq C(\beta, m) \varrho^{-|\beta|} \left| \sum_{J\in\Lambda_I} \sum_{|\alpha|=m+1} \frac{(x_J - x)^\alpha}{\alpha!} \Psi_J^{[\beta]}(x) \right.$$

$$\left. \cdot \left(\int_0^1 s^m D^\alpha u(x_J + s(x - x_J)) ds \right) \right| \qquad (4.164)$$

Let $\zeta = x/\rho$, $\zeta_J = x_J/\rho$ and $E_{\omega_I}^{\gamma,m}(x) := \left| D^\beta u - \mathcal{R}_{\varrho,h}^{m[\beta]} u \right|$. Considering the fact that $\Psi_J^{[\beta]}(\zeta)$ is compactly supported and its support size equals $diam\{\omega_J\}$, then identically,

$$\Psi_J^{[\beta]}(\zeta) = \Psi_J^{[\beta]}(\zeta)\chi_J(\zeta)$$

where $\chi_J(\zeta)$ is the characteristic function of ω_J, i.e.

$$\chi_J(\zeta) = \begin{cases} 1, & |\zeta_J - \zeta| \leq a_J \\ 0, & |\zeta_J - \zeta| > a_J \end{cases}$$

Repeat using Cauchy-Schwarz inequality yields

$$E_{\omega_I}^{\gamma,m}(x) \leq C(\beta,m)\rho^{m+1-|\beta|} \left| \sum_{J\in\Lambda_I} \sum_{|\alpha|=m+1} \frac{(\zeta_J-\zeta)^\alpha}{\alpha!} \left(\Psi_J^{[\beta]}(\zeta)\right)\right.$$

$$\left. \cdot \left(\int_0^1 s^m D^\alpha u\big(\varrho[\zeta_J + s(\zeta-\zeta_J)]\big)ds\right) \chi_J(\zeta) \right|$$

$$\leq C(\beta,m)\varrho^{m+1-|\beta|} \left| \int_0^1 \sum_{J\in\Lambda_I} \left\{ \left[\sum_{|\alpha|=m+1} \left(\frac{(\zeta_J-\zeta)^\alpha}{\alpha!}\Psi_J^{[\beta]}(\zeta)\right)^2\right]^{1/2} \right.\right.$$

$$\left.\left. \cdot \left[\sum_{|\alpha|=m+1} \left(s^m D^\alpha u\big(\varrho[\zeta_J + s(\zeta-\zeta_J)]\big)\right)^2\right]^{1/2} \right\} \chi_J(\zeta)ds \right|$$

$$\leq C(\beta,m)\rho^{m+1-|\beta|} \left| \int_0^1 \left\{ \left[\sum_{J\in\Lambda_I}\chi_J(\zeta)\right.\right.\right.$$

$$\sum_{|\alpha|=m+1}\left(\frac{(\zeta_J-\zeta)^\alpha}{\alpha!}\Psi_J^{[\beta]}(\zeta)\right)^2\Big]^{1/2}\left[\sum_{J\in\Lambda_I}\chi_J(\zeta)\right.$$

$$\left.\left.\left. \cdot \sum_{|\alpha|=m+1}\left(s^m D^\alpha u\big(\rho[\zeta_J + s(\zeta-\zeta_J)]\big)\right)^2\right]^{1/2}\right\}ds\right| \qquad (4.165)$$

Since $\Psi_J^{[\beta]}(\zeta)$ is bounded, i.e. $\exists C > 0$ such that

$$sup_{x\in\omega_I}\Psi_I^{[\beta]}(x) \leq C \qquad (4.166)$$

Moreover, $\chi_J(\zeta)\left|\zeta_J-\zeta\right| \leq a_J \leq C_d$,

$$E_{\omega_I}^{\beta,m}(x) \leq C(\alpha,\beta,\gamma,C_d,m)\rho^{m+1-|\beta|} \left| \int_0^1 \left[\sum_{J\in\Lambda_I}\chi_J(\zeta)\right.\right.$$

$$\left.\left. \cdot \sum_{|\alpha|=m+1}\left(s^m D^\alpha u\big(\varrho[\zeta_J + s(\zeta-\zeta_J)]\big)\right)\right]ds\right| \qquad (4.167)$$

To estimate L_2 norm of the error $E_{\omega_I}^{\beta,m}$,

$$\left| D_x^\beta u - \mathcal{R}_{\varrho,h}^{m[\beta]}u \right|_{L^2(\omega_I\cap\Omega)} \leq C(\alpha,\beta,\gamma,C_d,m)\varrho^{m+1-|\beta|}$$

$$\cdot \left\{ \sum_{J\in\Lambda_I}\int_0^1\int_{\omega_I\cap\omega_J\cap\Omega}\sum_{|\alpha|=m+1}s^{2m}\left(D^\alpha u(x_J + s(x-x_J))\right)^2 d\Omega_x ds \right\}^{1/2}$$

Change variables $z = x_J + s(x-x_J)$, and $\Omega_x ds = s^{-n}d\Omega_z ds$. The new integration domain for each $J\in\Lambda_I$ is

$$A_J(z,s) = \Big\{(z,s)\ \Big|\ s\in(0,1],\ \ \tilde{\omega}_I\cap\tilde{\omega}_J\cap\tilde{\Omega}\Big\} \qquad (4.168)$$

where $\forall\, J \in \Lambda_I$ and $\tilde{\omega}_J := \{z \mid \frac{1}{s}|z - z_J| \le a_J\rho,\ 0 < s \le 1\}$.

Consider the identity

$$\int_{B_I(\boldsymbol{x},\varrho)} \left| D_{\boldsymbol{x}}^\alpha u(\boldsymbol{x} + \theta(\boldsymbol{x}_I - \boldsymbol{x}))\chi_I \right|^2 d\Omega = \int_{B_I(\boldsymbol{y},(1-\theta)\varrho)} |D_{\boldsymbol{y}}^\alpha(\boldsymbol{y})|^2 \frac{d\Omega_{\boldsymbol{y}}}{(1-\theta)^n}$$

$$\le \int_{B_I(\boldsymbol{x},\varrho)} |D_{\boldsymbol{x}}^\alpha(\boldsymbol{x})|^2 \frac{d\Omega_{\boldsymbol{x}}}{(1-\theta)^n}\ .$$

Since $s \le 1$ and $z_J = x_J$, one has $\tilde{\omega}_J \subset \omega_J\ \forall\, J \in \Lambda_I$, and

$$\left\| D_x^\beta u - \mathcal{R}_{\varrho,h}^{m[\beta]} u \right\|_{L^2(\omega_I \cap \Omega)} \le C(\alpha,\beta,\gamma,C_d,m)\varrho^{m+1-|\beta|}$$

$$\cdot \left\{ \sum_{J \in \Lambda_I} \int_0^1 \int_{\omega_I \cap \omega_J \cap \Omega} \sum_{|\alpha|=m+1} s^{2m-n} \left(D^\alpha u(z)\right)^2 d\Omega_z ds \right\}^{1/2}$$

By the assumption $2m - n \ge 0$, Fubini's theorem, and the stability condition,

$$\left\| D_x^\beta u - \mathcal{R}_{\varrho,h}^{m[\beta]} u \right\|_{L^2(\omega_I \cap \Omega)} \le C(\alpha,\beta,\gamma,C_d,m,n,N_{max})\varrho^{m+1-|\gamma|}|u|_{H^{m+1}(\omega_I \cap \Omega)}$$

$$(4.169)$$

♣

Note the fact that the constant C in Eq.(4.169) is a function of C_d implies that C does not depend on a_I.

Theorem 4.5.2 (Global Estimate). *For $u \in H^{m+1}(\Omega) \cap C^0(\Omega)$, $\phi \in C_0^m(\Omega) \cap H^{m+1}(\Omega)$, the global discretization, $\{\mathcal{D}, \mathcal{F}_\rho, \mathcal{H}_m\}$, yields following estimate*

$$\|D_x^\beta u - \mathcal{R}_{\varrho,h}^{m[\beta]} u\|_{L^2(\Omega)} \le C\rho^{m+1-|\gamma|}\|u\|_{H^{m+1}(\Omega)}, \quad 0 \le |\beta| \le m \quad (4.170)$$

Proof

Again, we only need to show following semi-norm estimate

$$\left\| D_x^\beta u - \mathcal{R}_{\varrho,h}^{m[\beta]} u \right\|_{L^2(\Omega)} \le C\rho^{m+1-|\beta|}|u|_{H^{m+1}(\Omega)} \quad (4.171)$$

By (4.169) $\exists\, 0 < C_0 < \infty$ such that

$$\left\| D_x^\beta u - \mathcal{R}_{\varrho,h}^{m[\beta]} u \right\|_{L^2(\Omega)}^2 \le \sum_{I \in \Lambda} \left\| D_x^\beta u - \mathcal{R}_{\varrho,h}^{m[\beta]} u \right\|_{L^2(\omega_I \cap \Omega)}^2$$

$$\le C_0^2 \varrho^{2(m+1-|\beta|)} \sum_{I \in \Lambda} |u|_{H^{m+1}(\omega_I \cap \Omega)}^2 \quad (4.172)$$

where C_0 can be chosen as the constant C in (4.169).

The key technical ingredient of the global estimate is the following fact: there exists an auxiliary, virtual background cell discretization, $\{\mathring{\omega}_I\}_{I \in \Lambda}$, that has the properties:

$$(\mathbf{1}): \qquad x_I \in \overset{\circ}{\omega}_I \cap \Omega \,, \tag{4.173}$$

$$(\mathbf{2}): \qquad \overset{\circ}{\omega}_I \subset \omega_I \,, \tag{4.174}$$

$$(\mathbf{3}): \qquad \bigcup_{I \in \Lambda} \overset{\circ}{\omega}_I \cap \Omega = \Omega, \tag{4.175}$$

in which

$$int\{\overset{\circ}{\omega}_I\} \cap int\{\overset{\circ}{\omega}_J\} = \begin{cases} int\{\overset{\circ}{\omega}_I\} \,, & I = J \\ \emptyset \,, & I \neq J \end{cases} \tag{4.176}$$

such that $\forall\, I \in \Lambda$

$$\omega_I \cap \Omega \subset \bigcup_{J \in \Lambda_I} \overset{\circ}{\omega}_J \cap \Omega \tag{4.177}$$

We show that the claim is true by contradictory argument.

Suppose there is no such virtual cell discretization (4.173)–(4.176) that satisfies the condition (4.177). Then, $\exists I \in \Lambda$ and $x \in \Omega$ such that

$$x \in \omega_I \cap \Omega \,, \quad \text{but } x \notin \bigcup_{J \in \Lambda_I} \overset{\circ}{\omega}_J \cap \Omega$$

It is obvious that $x \notin \bigcup_{J \in \Lambda \backslash \Lambda_I} \overset{\circ}{\omega}_J \cap \Omega$, which leads to the contradiction $x \notin \Omega$ because of condition (4.175).

Hence, the overlapping condition (3.167) suggests that

$$\sum_{I \in \Lambda} |u|^2_{H^{m+1}(\omega_I \cap \Omega)} \leq \sum_{I \in \Lambda} \sum_{J \in \Lambda_I} |u|^2_{H^{m+1}(\overset{\circ}{\omega}_I \cap \Omega)} \leq N_{max} \sum_{I \in \Lambda} |u|^2_{H^{m+1}(\overset{\circ}{\omega}_I \cap \Omega)}$$

$$= N_{max} |u|^2_{H^{m+1}(\Omega)} \tag{4.178}$$

Estimate (4.171) follows immediately, and consequently, (4.170). ♣

Remark 4.5.1. **1.** When $\beta = 0$, the estimate (4.170) recovers the error estimate for the regular reproducing kernel interpolant .[303] **2.** By taking advantage of the global differential consistency conditions, there is no need to use the notion of "affine equivalence" in the proof, which is a major difference between the current proof and the finite element type proofs. **3.** Because the β-th wavelet kernel satisfies $|\beta| - 1$ order vanishing moment conditions, Theorem (4.5.2) indicates that its sampling range is up to $\rho^{|\beta|}$ scale in the physical space. Apparently, the larger the absolute value $|\beta|$, the finer scale the wavelet kernel can represent, which, in other words, implies that each wavelet kernel has a different bandwidth in the frequency domain [2]. In this sense, the hierarchical partition of unity is a wavelet kernel packet, because

---

[2] Readers may find useful information on vanishing moments condition of a wavelet, or multiplicity zero condition of its Fourier transform, and its effect on bandwidth in Daubechies[129] pp 243-245.

we are basically dealing with a special type of least-square filters. It is note-worthy pointing out the similarity between the wavelet based hierarchical partition of unity and the *wavelet packet* invented by Coifman and Meyer.[122]

5. Applications

In this Chapter, several aspects of applications of meshfree Galerkin methods are discussed. The materials presented here are mainly selected from the authors' research. On the other hand, a comprehesive survey for the important applications of meshfree methods is also presented.

5.1 Explicit Meshfree Computations in Large Deformation

Because of its simplicity, explicit computation is very attractive in practical computations, especially for large scale computations of large deformation problems. However, most inelastic materials are nearly incompressible, which poses some technical difficulties in carrying out displacement based finite element explicite computations. To be more specific, the displacement based Galerkin formulation may induce volumetric locking, which leads to computational failure. In practices, such difficulty is usually handled by using either mixed formulations or enhanced strain methods (e.g. Simo & Rifai[406]), or by some ad-hoc treatments, such as one point (1-pt.) integration/hour-glass control procedure, and selective reduced integration scheme (e.g. the B-bar element proposed by Hughes[209]). Furthermore, to capture strain localizations in inelastic materials, one may have to develop special discontinuous incompatible element, which, to some extent, complicates the implementation since they are usually not suitable for explicit computations. For example, an immediate difficulty is how to adapt the mixed formulations for a quadrilateral (or hexahedral) grid. One of few options available is to use one point (1-pt.) integration with hour-glass control scheme (Nemat-Nasser et al.[351]). However, this leads to other problems as well. For instance, the actual shear-band mode may consist of some hour-glass modes; thus, how to suppress the hour-glass mode while retaining the correct shear-band mode is entirely based on trial and error. Particularly, a commonly used in-elastic constitutive model is the power law governed elasto-viscoplastic solid; it has been found in a study by Watanabe et al.[451] that it may be difficult to suppress hour-glass modes for large power index, m. Besides these drawbacks, there is a major difficulty for explicit finite element algorithm to proceed h-adaptive refinement, while keeping the quadrilateral (or hexahedral) pattern intact. To remedy the inadequacy of finite element methods, a meshfree explicit formulation has been used in simulations of inelastic large deformation of a solid.

Most meshfree simulations of large deformation problems are using the total Lagrangian formulation, which is the hallmark of many particle methods. In inelastic large deformation simulations, two types of consitutive updates are commonly used: (a) hypoelastic-inelastic formulation; and (b) hyperelastic-inelastic formulation. The hypoelastic-inelastic formulations are mainly based on the rate deformations that use additive decomposition (e.g. Peirce et al [1984], Hughes & Winget [1980], Simo & Hughes [1997], and Belytechko, Liu, and Moran [2000]). Whereas the hyperelastic-inelastic formulation is mainly based on multiplicative decomposition (e.g. Simo and Ortiz [1985], Moran et al [1990], and Simo and Hughes [1997]).

Since the focus of this book is meshfree methods, for simplicity, we only demontrate meshfree large deformation in rate formulation.

Consider a body that occupies a region Ω_X with boundary $\Gamma_X = \Gamma_X^u \bigcup \Gamma_X^T$ at time $t = 0$. At the time, t, the deformed body occupies a spatial region, Ω_x. The motion of the continuum is defined as

$$\mathbf{x} = \mathbf{X} + \mathbf{u}(\mathbf{X}, t) \tag{5.1}$$

where $\mathbf{X}$ stands fpr the material coordinates, $\mathbf{x}$ stands for the spatial coordinates, and $\mathbf{u}(\mathbf{X}, t)$ stands for the displacement field.

For the total Lagrangian formulation, the governing equations may be stated as follows:

1. Conservation of mass

$$\rho_0 = \rho J \tag{5.2}$$

   where ρ_0 is the density in the material configuration, whereas ρ is the density in spatial configuration. Note that J is the determinant of the deformation gradient $J = det\{\mathbf{F}\}$, or the determinant of Jacobian between spatial and material coordinates,

$$\mathbf{F} = \frac{\partial x_i}{\partial X_J} \mathbf{e}_i \otimes \mathbf{E}_J \tag{5.3}$$

2. Equation of motion

$$\rho_0 \ddot{\mathbf{u}} = Div\mathbf{P} + \mathbf{B} \tag{5.4}$$

   where $\mathbf{P}$ is the first Piola-Kirchhoff stress tensor, which can be related to Cauchy stress as $\mathbf{P} = J\mathbf{F}^{-1}\boldsymbol{\sigma}$ and the Kirchhoff stress as $\boldsymbol{\tau} = \mathbf{F} \cdot \mathbf{P}$, and $\mathbf{B}$ is the body force per unit volume;

3. Kinematics
   The deformation gradient may decomposed into

$$\mathbf{F} = \mathbf{F}^e \cdot \mathbf{F}^{ine} \tag{5.5}$$

where $\mathbf{F}^e$ describe the elastic deformation and rigid body rotation, and $\mathbf{F}^{ine}$ represents inelastic deformation. The velocity gradient is also decomposed into two parts: the rate of deformation tensor, $\mathbf{D}$, and the spin tensor, $\mathbf{W}$, i.e.

$$\mathbf{L} = \dot{\mathbf{F}}\mathbf{F}^{-1} = \mathbf{D} + \mathbf{W} = \dot{\mathbf{F}}^e \cdot \mathbf{F}^{e-1} + \mathbf{F}^e \cdot \dot{\mathbf{F}}^{ine} \cdot \mathbf{F}^{ine-1} \cdot \mathbf{F}^{e-1} \tag{5.6}$$

where

$$\mathbf{D} = D_{ij}\mathbf{e}_i \otimes \mathbf{e}_j \quad D_{ij} = \frac{1}{2}\left(\frac{\partial v_i}{\partial x_j} + \frac{\partial v_j}{\partial x_i}\right) \tag{5.7}$$

$$\mathbf{W} = W_{ij}\mathbf{e}_i \otimes \mathbf{e}_j \quad D_{ij} = \frac{1}{2}\left(\frac{\partial v_i}{\partial x_j} - \frac{\partial v_j}{\partial x_i}\right) \tag{5.8}$$

4. Constitutive laws:
   For convenience, two commonly used consitutive relations are listed to provide background information of the meshfree simualtions.
   (A) For modeling hyperelastic materials, the second Piola-Kirchhoff stress $\mathbf{S}$ is calculated from the strain energy density function W by

$$S_{ij} = \frac{\partial \Psi}{\partial E_{ij}} = 2\frac{\partial \Psi}{\partial C_{ij}} \tag{5.9}$$

where $\mathbf{E}$ is the Green strain tensor and $\mathbf{C}$ is the right Cauchy-Green tensor. In the meshfree simulations of hyperelastic materials, the following constitutive models have been used: (A) a modified Mooney-Rivlin material (Fried and Johnson [1998]),

$$\Psi = C_1(I_1 - 3I_3^{1/3}) + C_2(I_2 - 3I_3^{2/3}) + \frac{1}{2}\lambda\left(lnI_3\right)^2 \tag{5.10}$$

Neo-Hooken material

$$\Psi = \frac{1}{2}\lambda_0\left(lnJ\right)^2 - \mu_0 lnJ + \frac{1}{2}\mu_0(trace\mathbf{C} - 3) \tag{5.11}$$

(B) The following rate form constitutive equation is often used in large deformation simulation (e.g. Needleman[349])

$$\overset{\triangledown}{\tau} := \mathbf{C}^{elas}\left(\mathbf{D} - \mathbf{D}^{vp}\right), \tag{5.12}$$

where the Jaumann rate of Kirchhoff stress, $\overset{\triangledown}{\tau}$, is defined as

$$\overset{\triangledown}{\tau} = \dot{\tau} - \mathbf{W}\tau + \tau\mathbf{W} \tag{5.13}$$

The yield surface of viscoplastic solid is of von Mises type, which might be changing with time

$$D_{ij}^p := \bar{\eta}(\bar{\sigma}, \bar{\epsilon}) \frac{\partial f}{\partial \tau'_{ij}} \tag{5.14}$$

$$f(\tau', \kappa) = \bar{\sigma} - \kappa = 0 \tag{5.15}$$

$$\bar{\sigma}^2 = \frac{3}{2} \tau' : \tau' , \tag{5.16}$$

$$\tau'_{ij} = \tau_{ij} - \frac{1}{3} tr(\tau) \delta_{ij} \tag{5.17}$$

$$\bar{\epsilon} := \int_0^t \sqrt{\frac{2}{3} \mathbf{D}^p : \mathbf{D}^p} \ dt \tag{5.18}$$

The power law that governs the viscoplastic flow is described as

$$\bar{\eta} = \dot{\epsilon}_0 \left[\frac{\bar{\sigma}}{g(\bar{\epsilon})} \right]^m , \quad g(\bar{\epsilon}) = \sigma_0 \frac{\left[1 + \bar{\epsilon}/\epsilon_0 \right]^N}{1 + \left(\bar{\epsilon}/\epsilon_1 \right)^2} . \tag{5.19}$$

where m is the power index.

5. Boundary conditions:
   The following boundary conditions are specified in the referential configuration,

$$\mathbf{Pn}_0 = \mathbf{T}^0 , \quad \forall \mathbf{X} \in \Gamma_X^T \tag{5.20}$$

$$\mathbf{u} = \mathbf{u}^0 , \quad \forall \mathbf{X} \in \Gamma_X^u \tag{5.21}$$

6. Initial conditions

$$\mathbf{P}(\mathbf{X}, 0) = \mathbf{P}_0(\mathbf{X}), \tag{5.22}$$

$$\mathbf{u}(\mathbf{X}, 0) = \mathbf{u}_0(\mathbf{X}), \tag{5.23}$$

$$\mathbf{v}(\mathbf{X}, 0) = \mathbf{v}(\mathbf{X}) \tag{5.24}$$

Consider a weighted residual form of (5.4)

$$\int_{\Omega_X} \left\{ \rho_0 \ddot{u}_i - P_{iJ,J} - B_i \right\} \delta u_i d\Omega_X = 0 , \tag{5.25}$$

then the following weak form can be derived,

$$\int_{\Omega_X} \rho_0 \ddot{u}_i \delta u_i d\Omega_X + \int_{\Omega_X} P_{Ji} \delta F_{Ji}^T d\Omega_X - \int_{\Omega_X} B_i \delta u_i d\Omega_X$$

$$- \int_{\Gamma_X^T} T_i^0 \delta u_i d\Gamma - \int_{\Gamma_X^u} R_i \delta u_i d\Gamma = 0 . \tag{5.26}$$

Using transform method, one can choose the following the trial, and test
functions,

$$u_i^h(\mathbf{X}, t) = \sum_{I=1}^{NP} N_I(\mathbf{X}) d_{iI}(t) . \tag{5.27}$$

$$\delta u_i^h(\mathbf{X}, t) = \sum_{I=1}^{NP} N_I(\mathbf{X}) \delta d_{iI}(t) . \tag{5.28}$$

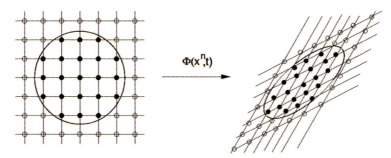

Fig. 5.1. Comparison between the support size of a meshfree interpolant and a FEM interpolant

such that for $I \in \Lambda_b$ $N_I(\boldsymbol{x}_J) = \delta_{IJ}$, $\forall\, J \in \Lambda_b$ and $N_I(\boldsymbol{x}_J) = 0$, $\forall\, J \in \Lambda_{nb}$.

Substituting (5.27)-(5.28) into (5.26), a set of algebraic-differential equations may be formed, which govern the discrete displacement field at each time steps.

The discrete equations of motion can then be put into the standard form,

$$\mathbf{M}\ddot{\mathbf{d}} + \mathbf{f}^{int} = \mathbf{f}^{ext} \tag{5.29}$$

where $\mathbf{M}$ is the mass matrix, and

$$\mathbf{f}_I^{ext} = \int_{\Gamma_X^T} T_i^0(\mathbf{X}, t) N_I(\mathbf{X}) \mathbf{e}_i d\Gamma + \int_{\Omega_X} B_i(\mathbf{X}, t) N_I(\mathbf{X}) \mathbf{e}_i d\Omega \tag{5.30}$$

$$\mathbf{f}_I^{int} = \int_{\Omega_X} P_{Ji} \frac{\partial N_I}{\partial X_J} \mathbf{e}_i d\Omega \tag{5.31}$$

In computation, conventional predictor-corrector scheme is used to update the deformation. The only difference between the meshfree explicit scheme and the conventional FEM explicit scheme is that in each time iteration one has to enforce, or update the essential boundary conditions:

$$\mathbf{d}_i^b(t) = (\mathbf{D}^b)^{-1}\Big(\mathbf{g}_i(t) - \mathbf{D}^{nb}\mathbf{d}_i^{nb}(t)\Big) \tag{5.32}$$

$$\dot{\mathbf{d}}_i^b(t) = (\mathbf{D}^b)^{-1}\Big(\dot{\mathbf{g}}_i(t) - \mathbf{D}^{nb}\dot{\mathbf{d}}_i^{nb}(t)\Big) \tag{5.33}$$

And the essential boundary condition enforcement is accurate, if only there are enough particles distributed along the essential boundary.

Remark 5.1.1. The ability of reproducing kernel shape functions to avoid locking in a displacement based formulation is due to the following reasons:

1. It is a higher order polynomial interpolation. In example 2, the embedded window function is a 2-D cubic spline box function, and the kernel function is constructed by multipling an additional bilinear polynomial basis;

2. The use of sub-reduced integration scheme. All the calculations in this paper have done, unless specified otherwise, are used 2×2 Gauss quadrature integration in 2-D, and $2 \times 2 \times 2$ Gauss quadrature integration in 3-D; They are still reduced integration scheme in principle, however, it appears that no hour-glass mode, nor zero energy mode occurs under such sub-reduced integration scheme.

2. Eqs. (5.32) – (5.33) are a local essential boundary enforcement. In this setting, no global transformation is needed as proposed by Chen et al.[96] One can enforce the essential boundary conditions piece by piece to avoid inverting large algebraic matrix, as illustrated in Fig. 3.12.

5.2 Meshfree Simulation of Large Deformation

An area that meshfree methods have the clear edge over the traditional finite element methods is the simulations of large deformation of solids. The main reasons for this is because meshfree interpolant has a much large support size than FEM interpolant; this means that (a) for the same size of a support area the meshfree interpolant covers more particles than traditional FEM does, and (b) for the same particle density the meshfree interpolant has larger support size than that of tractional FEM interpolant. This property gives meshfree interpolant an advantage to simulate servere local disturtion with sustained computation ability. This is illustrated in Fig. 5.1

In Fig. 5.1, we compare the size of a meshfree interpolant with a FEM interpolant under the same discretization (density of particles) and the same distortion rate. The support of a FEM interpolant does not contain any interior node therefore it is much smaller than than of the meshfree interpolant. Under the same deformation rate, the meshfree interpolant field is much more smooth than FEM interpolant field. Therefore, when deformation rate increases, FEM computation may yield negative Jacobian at certain points inside an element, whereas under the same distortion rate meshfree interpolation still works.

The example is a hyperelastic block under compression in plane strain state. The upper and lower boundaries are assumed to be perfected bonded with a rigid plate and moving twowards each other with a constant velocity $V_0 = 50$ (in/sec), and the two lateral sides are traction-free. Modified Mooney-Rivilin material is used with choice of the following material constants, $\rho = 1.4089 \times 10^{-4}$ (slug), $C_1 = 18.35$ (psi), $C_2 = 1.468$ (psi), and $\lambda = 1.468 \times 10^3$ (psi).

5.2.1 Simulations of Large Deformation of Thin Shell Structures

Another complex issue for nonlinear shell formulation is how to embed inelastic constitutive relation onto manifold. It is usually a nontrivial task to

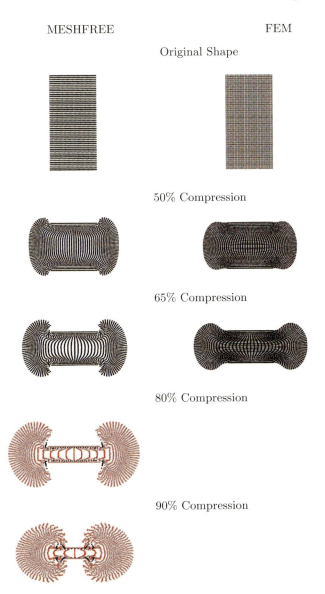

Fig. 5.2. Comparison of the deformations at different time stages for a block of hyperelastic material under compression by using MESHFREE and FEM when $\Delta t = 1 \times 10^{-6}$ (s)

develop an elasto-plastic nonlinear shell theory even for the degenerated approach. Nevertheless, this will not be a problem at all for 3-D direct approach. In this section, the meshfree approach is employed to calculate the thin shell structures that are governed by elasto-plastic constitutive relations. The computational formulas of our computation largely follow from that of Hughes,[208] and Simo & Hughes.[407]

5.2.2 J_2 Hypoelastic-plastic Material at Finite Strain

A rate form hypoelastic J_2 constitutive relation in finite deformation is considered. The J_2 yield criterion is described as

$$f(\boldsymbol{\xi}, \boldsymbol{\alpha}, \epsilon^p) = \|\boldsymbol{\xi}\| - \sqrt{\frac{2}{3}}\kappa(\epsilon^p) = 0 \tag{5.34}$$

$$\mathbf{s} := \boldsymbol{\tau} - \frac{1}{3}tr(\boldsymbol{\tau})\mathbf{1} \tag{5.35}$$

$$\boldsymbol{\xi} := \mathbf{s} - \boldsymbol{\alpha} \tag{5.36}$$

$$\epsilon^p = \int_0^t \sqrt{\frac{2}{3}}\|\mathbf{d}^p(\tau)\|d\tau \tag{5.37}$$

where the Kirchhoff stress $\boldsymbol{\tau} := J\boldsymbol{\sigma}$ and $J = det\mathbf{F}$.

In this simulation, Lie derivative is chosen as the objective rate of stress tensor

$$L_v\boldsymbol{\tau} = \mathbf{c}^{elas} : [\mathbf{d} - \mathbf{d}^p] \tag{5.38}$$

where $\mathbf{c}^{elas}$ is the spatial elasticity tensor; and the Lie derivative is defined as

$$L_v\boldsymbol{\tau} := \dot{\boldsymbol{\tau}} - (\nabla\mathbf{v})\boldsymbol{\tau} - \boldsymbol{\tau}(\nabla\mathbf{v})^t = \mathbf{F}\dot{\mathbf{S}}\mathbf{F}^t \tag{5.39}$$

For isotropic material, the spatial elastic constants remain isotropic under rigid roations, $\mathbf{c}^{elas} = \lambda\mathbf{1} \otimes \mathbf{1} + 2\mu\mathbf{I}$, where $\mathbf{1}$ is the second order identity matrix, and $\mathbf{I}$ is the fourth order identity matrix. The plastic flow is described by the classic J_2 associated flow rule,

$$\mathbf{d}^p = \gamma\hat{\mathbf{n}} \tag{5.40}$$

where

$$\hat{\mathbf{n}} = \frac{\boldsymbol{\xi}}{\|\boldsymbol{\xi}\|} = \frac{\partial f/\partial\boldsymbol{\tau}}{\|\partial f/\partial\boldsymbol{\tau}\|} \tag{5.41}$$

The plastic loading and unloading condition can be expressed in terms of the Kuhn-Tucker condition

$$\gamma \geq 0, \quad f(\boldsymbol{\tau}, \boldsymbol{\alpha}, \epsilon^p) \leq 0, \quad \gamma f(\boldsymbol{\tau}, \boldsymbol{\alpha}, \epsilon^p) = 0 \tag{5.42}$$

The hardening laws are

$$Kinematic\ hardening: \quad L_v\boldsymbol{\alpha} = \frac{2}{3}\gamma\mathbf{n} \tag{5.43}$$

$$Isotropic\ hardening: \quad \kappa(\epsilon^p) = \sigma_Y + K\epsilon^p \tag{5.44}$$

and

$$\dot{\epsilon}^p = \gamma\sqrt{\frac{2}{3}} \tag{5.45}$$

A standard constitutive update for a rate form hypoelastic J_2 theory at finite strain is adopted (See: Simo & Hughes[407] Chapter 8). For the sake of documentation, a brief description of stress update is outlined at following. Define intermediate configuration between time steps, n and $n+1$

$$\mathbf{x}_{n+\theta} := (1-\theta)\mathbf{x}_n + \theta\mathbf{x}_{n+1} . \tag{5.46}$$

where $\theta \in [0,1]$. Consequently,

$$\mathbf{F}_{n+\theta} = (1-\theta)\mathbf{F}_n + \theta\mathbf{F}_{n+1}, \tag{5.47}$$

and the relative deformation gradients, relative incremental displacement gradient, and the relative Eulerian strain tensor are ($\theta \in [0,1]$)

$$\mathbf{f}_{n+\theta} := \mathbf{F}_{n+\theta}\mathbf{F}_n^{-1}; \tag{5.48}$$

$$\mathbf{h}_{n+\theta} := \frac{\partial\bar{\mathbf{u}}(\boldsymbol{x}_{n+\theta})}{\partial\boldsymbol{x}_{n+\theta}}; \tag{5.49}$$

$$\mathbf{e}_{n+\theta} := \frac{1}{2}\left[\mathbf{1} - (\mathbf{f}_{n+\theta}\mathbf{f}_{n+\theta}^T)^{-1}\right] \tag{5.50}$$

and the deformation gradient can be expressed as

$$\mathbf{d}_{n+\theta} = \frac{1}{2\Delta t}\left[\mathbf{h}_{n+\theta} + \mathbf{h}_{n+\theta}^T + (1-2\theta)\mathbf{h}_{n+\theta}^T\mathbf{h}_{n+\theta}\right] \tag{5.51}$$

The corresponding return mapping algorithm is summarized as:

> **Box 3(a) Elastic predictor :**
>
> $\tilde{\mathbf{e}}_{n+\theta} = \Delta t\mathbf{d}_{n+\theta} = \tilde{\mathbf{f}}_{n+\theta}^T\mathbf{e}_{n+1}\tilde{\mathbf{f}}_{n+\theta}$
>
> $\boldsymbol{\tau}_{n+\theta}^{trial} = \mathbf{f}_{n+\theta}\boldsymbol{\tau}_n\mathbf{f}_{n+\theta}^T + \mathbf{c}^{elas} : \tilde{\mathbf{e}}_{n+\theta}$
>
> $\boldsymbol{\alpha}_{n+\theta}^{trial} = \mathbf{f}_{n+\theta}\boldsymbol{\alpha}_n\mathbf{f}_{n+\theta}^T$
>
> $e_{n+\theta}^{ptrial} = e_n^p$
>
> $\boldsymbol{\xi}_{n+\theta}^{trial} = \boldsymbol{\tau}_{n+\theta}^{trial} - \frac{1}{3}tr(\boldsymbol{\tau}_{n+\theta}^{trial})$
>
> $\boldsymbol{\xi}_{n+\theta}^{trial} = \boldsymbol{\tau}_{n+\theta}^{trial} - \boldsymbol{\alpha}_{n+\theta}^{trial}$
>
> $\mathbf{n}_{n+\theta} = \dfrac{\boldsymbol{\xi}_{n+\theta}^{trial}}{\|\boldsymbol{\xi}_{n+\theta}^{trial}\|}$

and

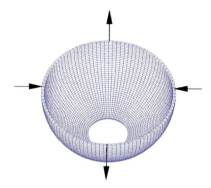

Fig. 5.3. Hemispherical shell under prescribed displacement control

---

Box 3(b) Plastic corrector :

$f_{n+\theta}^{trial} = \|\boldsymbol{\xi}_{n+\theta}^{trial}\| - \sqrt{\frac{2}{3}}(\sigma_Y + Ke_n^p)$

If $(f_{n+\theta}^{trial} > 0)$ *then*

$\qquad \Delta\gamma \quad = \dfrac{|f_{n+\theta}^{trial}|/2\mu}{1 + K/3\mu}$

$\qquad \boldsymbol{\tau}_{n+\theta} = \boldsymbol{\tau}_{n+\theta}^{trial} - 2\mu\Delta\gamma\mathbf{n}_{n+\theta}$

$\qquad \boldsymbol{\alpha}_{n+\theta} = \boldsymbol{\alpha}_{n+\theta}^{trial} + \sqrt{\frac{2}{3}}\Delta\gamma H\mathbf{n}_{n+\theta}$

$\qquad e^p{}_{n+\theta} = e_n^p + \sqrt{\frac{2}{3}}\Delta\gamma$

$\qquad \boldsymbol{\xi}_{n+\theta} \;= \boldsymbol{\tau}_{n+\theta} - \frac{1}{3}tr(\boldsymbol{\tau}_{n+\theta}) - \boldsymbol{\alpha}_{n+\theta}$

Else if $(f_{n+\theta}^{trial} \le 0)$ then

$\qquad \Delta\gamma = 0,$

Endif

$\hspace{10cm}(5.52)$

In all the computations presented in this paper, only isotropic hardening is considered.

5.2.3 Hemispheric Shell under Concentrated Loads

Again this is a problem that belongs to the well-known "standard set of problem" testing finite element accuracy (MacNeal & Harder[316]). The dimensions of the hemispherical shell are listed as follows: its radius is $1.0\ m$, and its thickness is $0.04\ m$. At the bottom part of the spherical shell, there is a hole, which forms a 18° angle from the center of the spheric shell.

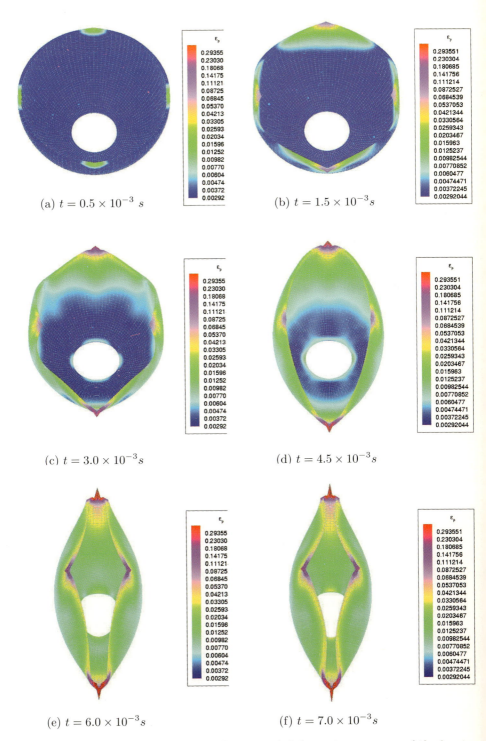

Fig. 5.4. The plastic strain distribution and deformation sequence of the hemispherical shell

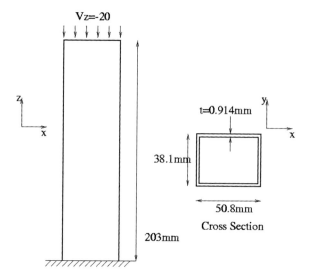

Fig. 5.5. Crash test of a boxbeam

Instead of prescribing concentrated forces on the edge of the spherical shell, we prescribe the displacement at four different locations around the open edge of the hemispherical shell as shown in Fig. 5.3. The prescribed velocity is 100 m/s. A total of 12, 300 particles are used in computation.

In Fig. 5.4, the plastic strain is plotted on the deformed configuration of the hemispheric shell.

5.2.4 Crash Test of a Boxbeam

In this numerical example, we simulate a boxbeam being impacted at one end while the other end being fixed. The rigid impactor is assumed having an infinte mass with a fixed velocity of 10.0 m/s. The Young's modulus of boxbeam is, $E = 2.1 \times 10^{10}$ Pa; Poisson's ratio $\mu = 0.3$, the initial yield stress, $\sigma_0 = 1.06 \times 10^9$ Pa. A linear isotropic hardening law is considered in the numerical simulation, $E_t = 4.09 \times 10^7$.

Neglecting contact and frictions between the impactor and the boxbeam, it is assumed that once the impact occurs, the rigid impactor stay with the boxbeam, i.e. the displacements for both x-direction, and y-direction are constrained at the collision surface.

A total of 7, 952 particles are used in meshfree discretization. A sequence of intermediate results are displayed in Fig. 5.6. Only half of the structure is displayed for a better visulization of the buckling mode at interior region. The accuracy of this particular numerical simulation is typically measured by the locations where the buckling mode appears (e.g. Zeng & Combescure[464]). The experiment results show that the first few buckling modes should appear immediately at the impact location. Our numerical results give the same prediction. It is noted that only 2 layers of particles are used in the thickness direction, which corresponds to one element in finite element simulation.

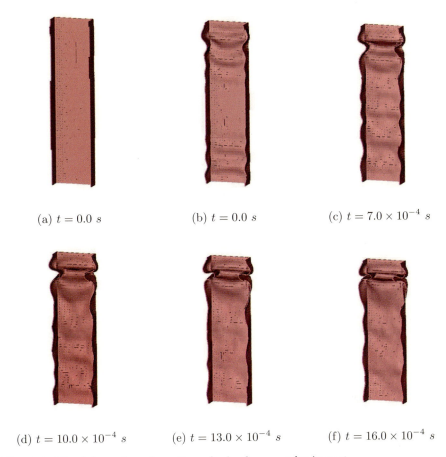

(a) $t = 0.0$ s (b) $t = 0.0$ s (c) $t = 7.0 \times 10^{-4}$ s

(d) $t = 10.0 \times 10^{-4}$ s (e) $t = 13.0 \times 10^{-4}$ s (f) $t = 16.0 \times 10^{-4}$ s

Fig. 5.6. The deformed configuration of a boxbeam under impact

Another simulation with both ends being impacted symmetrically, and simultaneously is conducted and the effective plastic strain is plotted on the deformed configuration, which is shown in Fig. 5.7. One may observe that the maximum plastic deformation occurs at the 90° edge location of the boxbeam, which makes sense because discontinuous curvature of the thin-wall structure could introduce both stress and strain concentrations.

Fig. 5.7. The contour of effective plastic strain on deformed boxbeam

5.3 Simulations of Strain Localization

5.3.1 Model Problems

The first set of model problems of strain localization considered are are tension and compression tests of elasto-viscoplastic specimens under either plane strain, or three-dimensional loading conditions. For plane strain problem, the prescribed displacement/velocity boundary condition is imposed at both ends of a specimen as shown in Fig. 5.8. Numerical results obtained from tension test and compression test under the plane strain condition are displayed in Figs. 5.9–5.10.

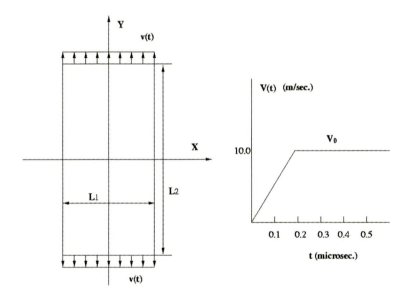

Fig. 5.8. Model Problem: tension test $(v(t) > 0)$; compression test $(v(t) < 0)$.

5.3.2 Mesh-alignment Sensitivity

In fact, mesh-alignment sensitivity would be the first difficulty to encounter, if anyone wishes to use FEM to simulate shear band formations. In the early study,[431] Tvergaard, Needleman, and Lo used the classic quadrilateral element (CST4), which consists of four diagonally crossed constant strain triangle elements, to simulate shear band formations under plane strain condition. CST4 element was originally designed by Nagategal, Parks and Rice [347] to be used in a displacement based formulation to avoid locking for computations

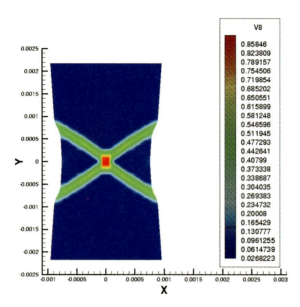

Fig. 5.9. The contours of the effective viscoplastic strain in the tensile bar.

in elasto-plastic materials. In their work,[431] Tvergaard et al. made an optimal arrangement of the aspect ratio of CST4 element, such that the shear-band formations are aligned with the boundary of finite elements, and sharp shear-bands are accurately captured in the computation. Few years later, followed the same philosophy, Tvergaard [432] invented a box-shaped super-element (BST24) consisting 24 tetrahedral element to compute the shear band formation in three-dimensional (3-D) space. Both CST4 and BST24 elements show strong mesh alignment sensitivity, which means that when shear band orientation is oblique to the diagonal line or plane of the quadrilateral/hexahedral element, the computational results deteriorate. Thus, special mesh design is needed to align the finite element boundary properly along the shear band orientation a priori.

To overcome the limitations of CST4/BST24 element, special elements have been considered and designed to relieve locking and offset the undesirable mesh-alignment sensitivity, though sometimes it is difficult to achieve the both ends at the same time. These special elements usually fall into the following two categories:

1. QR4-element: i.e. the four nodes quadrilateral with one point (1-pt.) integration/hour-glass control, which was first used in shear band calculation by Nemat-Nasser, Chung, and Taylor.[351] The 3-D counterpart of QR4-element is the brick element (BR8) with 1-pt./hour-glass control, which was first used by Zbib et al.[463] in 3-D shear band calculation;

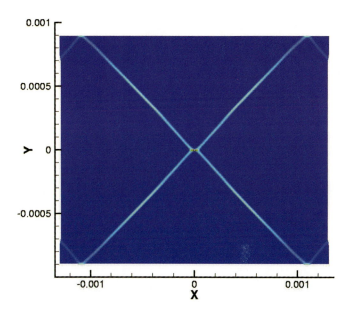

Fig. 5.10. The contours of the effective viscoplastic strain in a slab under the compression test ($r_0 = 0.1$ mm).

2. QLOC-element: (Ortiz, Leroy, and Needleman [363]) and its derivatives, such as QS (Steinmann & Willam.[411]), and the regularized discontinuous element (Simo, Oliver, and Armero.[405]);

In practical computations, all these elements show pros and cons. The QR4/BR8 class element has the best overall performance, but it is difficult to make any h-type refinement, because any valid h-adaptivity will destroy the simple structure of QR4/BR8 elements. Moreover, unlike some other numerical computations, the strain localization modes often contain certain hour-glass modes, in other words, hour-glass modes are not independent from strain localization modes; usually the choice of artificial damping force is completely based on either empirical experiences, or plausible argument, which is at the expense of sacrifice any hope for an accurate prediction on post-bifurcated shear band formation. The QLOC type elements are specially designed to eliminate mesh alignment sensitivity for arbitrary mesh arrangement, and they are theoretically sound and suitable for mixed formulations; but they are complicated to implement, apart from the fact that usually they are required to locate the incipient shear band position, or the strong/weak discontinuous line/surfaces a priori. In general, it is difficult to use them if the

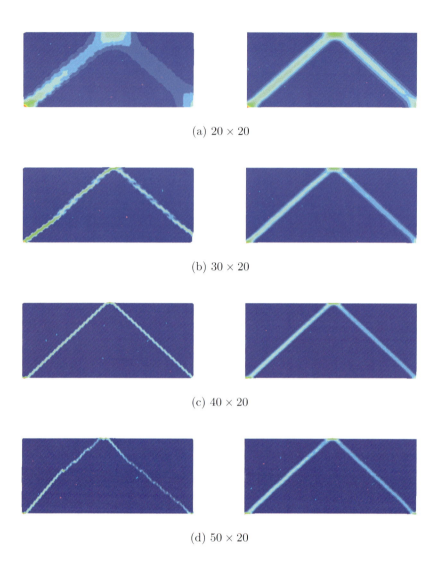

(a) 20×20

(b) 30×20

(c) 40×20

(d) 50×20

Fig. 5.11. Comparison between FEM and RKPM with different aspect ratios in mesh/particle distribution.

singular line/surface has non-zero curvature, or if one deals with complicate shear band patterns, such as the micro-shear band and macro-shear band interaction in crystal plasticity. Therefore, the available remedies for mesh-alignment sensitivity, in our opinion, are either too complex to use, or too ad-hoc and severely limited for a user in general purposes.

In principle, one may think that the mesh structure constraint is an artifice that is coerced subjectively onto the deformed continuum, which might be more than necessary as the physically required compatibility condition of the solid. To the contrary, meshfree methods, which do not have any definite mesh structure, may be free from mesh-alignment constraints. Based on this intuitive notion, numerical experiments have been conducted to test this hypothesis. In a comparison study, a velocity boundary conditions is prescribed on both top/bottom surface of a slab, and an imperfection is planted as reduction of yield stress at the lower left corner the specimen. The computations have been carried out by using both CST4-element, and RKPM shape function at the same specimen with different aspect ratios of element size or size of background cell. The FEM results and RKPM results are juxtaposed in Fig. 5.11. One can find that comparing with the results obtained by using CST4-element, the shear band results obtained by using RKPM shape function always have the same, and correct orientation regardless of the aspect ratio of background cells.

5.3.3 Meshfree Techniques for Simulations of Strain Localization

1. A meshfree hour-glass control strategy

As mentioned above, QR4-element is the only valid option in explicit codes for shear-band computations, and it appears to be the most popular choice used in practices (e.g. Batra et al.,[32] Nemat-Nasser et al.,[351] and Zbib et al.[463]), because of its simplicity. One of the shortcomings of the scheme is how to choose a suitable hour-glass control scheme to suppress the hour-glass modes, while retaining the correct shear-band mode, because shear-band mode may consist of certain of hour-glass modes as well.[351] In fact, it has been pointed out in[351] that constitutive model itself may stimulate spurious deformation and the values of the hour-glass control parameters begin to affect the numerical results, once the deformation becomes unstable. It would be interesting to compare the meshfree wavelet modes presented Chapter 3 and hour-glass modes due to under integration. A set of meshfree wavelet modes can be expressed as,

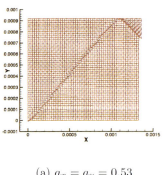

(a) $a_x = a_y = 0.53$

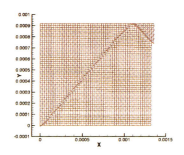

(b) $a_x = a_y = 0.70$

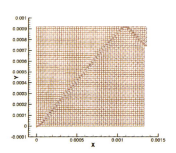

$a_x = a_y = 0.85$

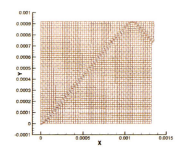

(d) $a_x = a_y = 1.0$

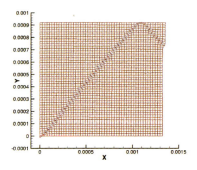

(e) $a_x = a_y = 1.12$

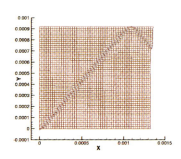

(f) $a_x = a_y = 1.2$

$$\sum_{I \in \Lambda} \mathcal{K}_I^{[10]}(x_I - x) = 0 \tag{5.53}$$

$$\sum_{I \in \Lambda} (x_{1I} - x_1)\mathcal{K}_I^{[10]}(x_I - x) = 1 \tag{5.54}$$

$$\sum_{I \in \Lambda} (x_{2I} - x_2)\mathcal{K}_I^{[10]}(x_I - x) = 0 \tag{5.55}$$

$$\sum_{I \in \Lambda} (x_{1I} - x_1)(x_{2I} - x_2)\mathcal{K}_I^{[10]}(x_I - x) = 0 \tag{5.56}$$

$$\sum_{I \in \Lambda} \mathcal{K}_I^{[01]}(x_I - x) = 0 \tag{5.57}$$

$$\sum_{I \in \Lambda} (x_{1I} - x_1)\mathcal{K}_I^{[01]}(x_I - x) = 0 \tag{5.58}$$

$$\sum_{I \in \Lambda} (x_{2I} - x_2)\mathcal{K}_I^{[01]}(x_I - x) = 1 \tag{5.59}$$

$$\sum_{I \in \Lambda} (x_{1I} - x_1)(x_{2I} - x_2)\mathcal{K}_I^{[01]}(x_I - x) = 0 \tag{5.60}$$

$$\sum_{I \in \Lambda} \mathcal{K}_I^{[11]}(x_I - x) = 0 \tag{5.61}$$

$$\sum_{I \in \Lambda} (x_{1I} - x_1)\mathcal{K}_I^{[11]}(x_I - x) = 0 \tag{5.62}$$

$$\sum_{I \in \Lambda} (x_{2I} - x_2)\mathcal{K}_I^{[11]}(x_I - x) = 0 \tag{5.63}$$

$$\sum_{I \in \Lambda} (x_{1I} - x_1)(x_{2I} - x_2)\mathcal{K}_I^{[01]}(x_I - x) = 1 \tag{5.64}$$

whereas hour-glass modes in a four-nodes quadrilateral element can be represented by the mode shape function, $H_I(x)$ (see Kosloff & Frazier[244]), which satisfies the following conditions:

$$\sum_{I \in \Lambda_e} H_I(x_I - x) = 0 \tag{5.65}$$

$$\sum_{I \in \Lambda_e} (x_{1I} - x_1)H_I(x_I - x) = 0 \tag{5.66}$$

$$\sum_{I \in \Lambda_e} (x_{2I} - x_2)H_I(x_I - x) = 0 \tag{5.67}$$

$$\sum_{I \in \Lambda_e} H_I^2(x_I - x) = 4 \tag{5.68}$$

where Λ_e is the nodal index set in an element. It is clear that hour-glass mode is also a **partition of nullity**, and it can be viewed as a special *wavelet* function as well, provided that the hour-glass modes are also compact supported. It, then, implies that not all hour-glass modes are hazardous, and, as we speculate, energy modes may only furnish a "complete" basis in the discrete functional space for elliptic type partial differential equations (PDEs), but not for hyperbolic, parabolic as well as mixed type PDEs; in other words, the nontrivial zero-energy modes may carry some useful information for non-elliptic PDEs. Thus, the use of viscous force to suppress all the hour-glass modes without discretion can lead to potential errors in numerical simulations. Contrast to FEM, the 1-pt. integration technique can be still used in reproducing kernel particle method without invoking artificial damping,

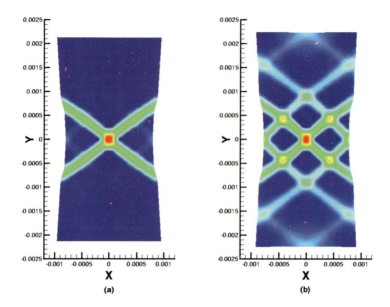

Fig. 5.12. Comparison between of numerical results based on the normal lumped mass and the special lumped mass technique.

or artificial stiffness. Precisely speaking, the undesirable hour-glass modes may be removed, or suppressed by properly adjusting the support size of the shape function, or dilation parameters of the window function, instead of imposing external viscous forces or modifying stiffness matrix. By doing so, one may be able to preserve the accuracy of the post-bifurcated shear-bands without the pollution caused by the artificial hour-glass control. In Fig. 5.3.3, a series of shear band formations are displayed in a quarter specimen of a compressed slab. The computations have been carried out by using 1-pt. integration for regular RKPM shape functions with different support sizes, which is characterized by the dilation parameter, a_x, a_y. One may find that as the normalized dilation parameters, a_x and a_y, increase from 0.53 to 1.2, the undesirable hour-glass modes vanish in the process.

2. Special lumping technique

In explicit calculation, the row summation technique is often adopted to form the lumped mass matrix to avoid inversion of a large size consistent mass matrix, which not only offers computational convenience, but also provides reasonable frequency contents. In this study, we have found that different lumping techniques will produces very different outcomes. In numerical experiments, two types of lumped mass are used in this study: **1.** conventional row-sum technique (see Hughes[209])

$$m_{ij} = \begin{cases} \int_{\Omega} \rho_0 N_i d\Omega & i = j \\ 0 & i \neq j \end{cases} \tag{5.69}$$

2. "special lumping technique" (Hinton et al.[200])

$$m_{ij} = \begin{cases} \alpha \int_{\Omega} \rho_0 N_i^2 d\Omega & i = j \\ 0 & i \neq j \end{cases} \tag{5.70}$$

where

$$\alpha = \frac{\int_{\Omega} \rho_0 d\Omega}{\displaystyle\sum_{I=1}^{NP} \int_{\Omega} \rho_0 N_i^2 d\Omega} \tag{5.71}$$

The justification of Hinton's special lumping technique is that it retains the diagonal part of the consistent mass matrix, and assumes that the diagonal part of the consistent mass matrix covers the correct frequency range of the dynamic response, whereas the non-diagonal part of the consistent mass matrix is not essential for the final results, or at least not in quasi-static cases. This technique ensures the positive definiteness of the mass matrix, and eliminates the singular mode. A possible setback could be that it cuts off the connection, or interaction between the neighboring material particles. However, this setback may be compensated by the nonlocal nature of mesh-free methods, because each material point in meshfree methods is covered by more than one shape function; therefore the interaction between the adjacent particles is always present. As a matter of fact, in our 2-D calculation, as many as sixteen particles to more than one hundred particles share their influences on the movement of a single particle; in 3-D case, as many as more than three hundred particles could be within the domain of influence of a single particle.

In numerical experiments, we simulate the tension test with both row-sum lumping technique and special lumping technique. In the particular test shown in Fig. 5.12, two types of imperfection are planted in the tensile bar: (1) geometric imperfection: a reduction of the width of the tensile bar with the maximum reduction, 5 % at the middle cross section; the tension specimen; (2) yield stress reduction, a distributed reduction of yield stress centered at the middle of the specimen; In this case, two sets of shear-bands will be triggered by different sources of imperfections. The outcome of the numerical computation is dictated by the competition between these two sets of shear-bands. From Fig. 5.12, one may see that the row-sum lumped mass solution predicts the shear-band formation due the reduction of yield stress well, and only leave a hardly-noticed trace of another set of shear-bands, which is due to geometric imperfection, in the background. Whereas for the numerical results obtained from special lumping technique, the two sets of shear bands are equally emphasized, and a great deal of detailed resolution is captured in the numerical solution. Apparently, combining the reproducing kernel interpolation with special lumping technique can provide high quality, detailed resolution shear band solution in numerical simulations. A high resolution shear band in a slab under compression is shown in Fig. 5.13; it is interesting to note that the detailed pattern of the effective plastic strain contour seems to resemble the "patchy slip" pattern in crystals.[373]

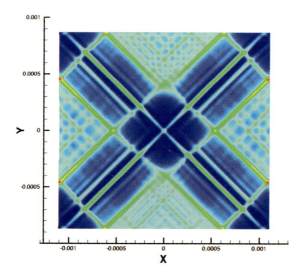

Fig. 5.13. The high resolution shear-band solution obtained by special lumping technique.

5.3.4 Adaptive Procedures

In the following, some adaptive procedures, which are used to seek the refinement of the numerical solutions, are discussed. The attention here is focused on the two different types of adaptive procedures: h-adaptive refinement and a spectral-adaptive (wavelet) refinement.

1. h-adaptive procedure

The h-type refinement procedure has been used to capture shear-band formation for quite a while, notably, Ortiz, et al.,[364] Belytschko et al.[38] and Zienkiewicz et al.[475] . However, technical difficulties have remained in the context of explicit finite element method. The commonly used Delaunay triangulation will certainly destroy the much needed quadrilateral (or hexahedral) pattern, and consequently the refined mesh is not suitable for explicit calculation anymore, though there is a recent attempt to store the triangle (or tetrahedra) element in explicit computation (See Zienkiewicz et al.[474]), nevertheless, triangle element mesh is highly mesh-alignment sensitive to the shear-band formation. On the other hand, meshfree methods enjoy an amazing simplicity in the h-adaptive procedure. For the most part, one can just insert particles into the strain localization zone, and the subsequent numerical solution will be automatically improved.

Since strain localization is a bifurcation problem in nature, elliptic adaptive indicators break down; a primitive or intuitive adaptive criterion is adopted here to determine where the adaptive region should be. We compare the effective plastic strain of every particle with that of its neighboring particles, and choose those regions where particles with higher percentage of relative effective plastic strain to refine the numerical solution. This primitive adaptive index is fairly easy to implement; for the problems that we computed it works efficiently. It should be mentioned that a systematic study of meshfree h-adaptive refinement in shear-band computation has been conducted by Jun & Im.[228] Some convergence issues have been addressed there, and we refer the readers to this recent study. To illustrate the h-adaptive procedure, a two-level, successive h-adaptive solution of a tension test is presented in Figs. 5.14 -5.15. In the tension test shown in Figs. 5.14–5.15, the computation is carried out only in a quartern specimen by enforcing the symmetry conditions. In the zero-level run, 231 particles are used forming a uniformly particle distribution in the specimen, which contains 200 background cells. In each cell, the 2×2 Gauss quadrature integration scheme is used. Based on relative effective plastic strain criterion, an automatic adaptive refinement procedure is implemented: at the first level refinement, all the region that have 25% or higher percentage of relative effective strain are being refined, and the total number of particles increases to 429, with the corresponding 377 integration cells, which bring the quadrature points to 1508; at the second level refinement, all the region that have 12% or higher percentage of relative effective strain are refined, and the total particle number increases to 1264, and total quadrature points increase to 4664 correspondingly.

In order to explain why finite element approximation has difficulties in accommodating h-adaptive refinement in an explicit code, a simple illustration is demonstrated in Fig. 5.16. If the above meshfree discretization has a one-to-one correspondence with a quadrilateral mesh, one can set the fictitious connectivity map for each integration cell, as if they were individual element. After a first level refinement, we plot the deformed mesh in Fig. 5.16; one can find immediately the entanglement and extrusion between the fictitious elements, which hints the break down of FE computation. Of course, in real FE approximation, this can only happen, provided that one can construct higher

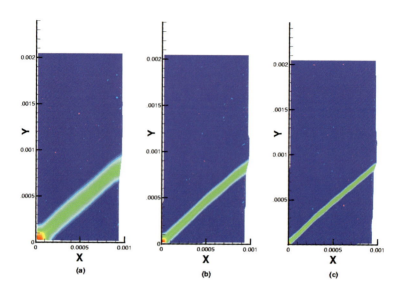

Fig. 5.14. The contours of the effective viscoplastic strain in the tensile bar (only a quartern specimen shown) (a) without any adaptivity; (b) the first level adaptive solution; (c) the second level adaptive solution.

order quadrilateral element along the boundary between coarse mesh and fine mesh. **2.** Spectral (Wavelet) -adaptive procedure Using spectral type of refinement to capture the localization mode can be traced back to the spectral overlay technique proposed by Belytschko et al. [37] The procedure there is to superpose a set of harmonic functions over the original FE shape functions at the place where shear band is supposed to develop.

Contrast to the spectral overlay technique, the adaptive wavelet algorithm proposed here is more general in nature. Instead of using analytical harmonic functions, we use the meshfree hierarchical partition of unity outlined in the previous section, in which the higher order spectral kernel functions are akin to the original RKPM shape functions. Intrinsically, the meshfree hierarchical bases have a distinct distribution of spectral contents of the interpolating object among the different bases; in other words, they consist of a multi-spectral wave packet. As a matter of fact, as shown in,[266] the higher order basis functions do fit into the definition of the pre-wavelet function [1] Furthermore, the orientation of the wavelet basis is isotropic in space, and the enhancement of the numerical solution due to wavelet basis comes out naturally as the outcome of numerical computation, though the adaptive region are selected

---

[1] By "pre-wavelet", we mean that the admissible conditions for the basic wavelet function is satisfied.

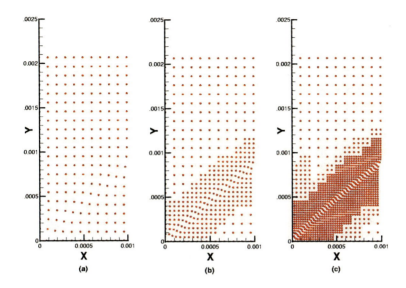

Fig. 5.15. The shear band formation in the tensile bar (only a quarter specimen shown) (a) without any adaptivity; (b) the first level adaptive solution; (c) the second level adaptive solution.

by a given criterion. Since the wavelet basis is genetically connected with the primary interpolation basis, the successive alternating *h-p* refinement process may become possible.

Since the wavelet basis constitute a partition of nullity, it introduces redundant degrees of freedom into the primary shape function basis. Consequently, the resulting stiffness matrix, and mass matrix will become ill-conditioned. In this study, we only use explicit integration scheme, and hence the only problem that we face is a possible singular mass matrix. As a matter of fact, the mass matrix will become singular, if the conventional row summation is used; and the mass matrix will become extremely ill-conditioned when consistent mass matrix is employed. To circumvent this difficulty, again we use the "*special lumping technique*" to form the mass matrix. Denote,

$$\{\Phi_\ell(X)\} = \{\{\Psi_\ell^{[00]}(X)\}, \{\Psi_\ell^{[10]}(X)\}, \{\Psi_\ell^{[01]}(X)\}, \{\Psi_\ell^{[11]}(X)\}\} \qquad (5.72)$$

By using special lumping technique, one is always able to guarantee the positive definiteness of the mass matrix. The formula for mass matrix is given as follows

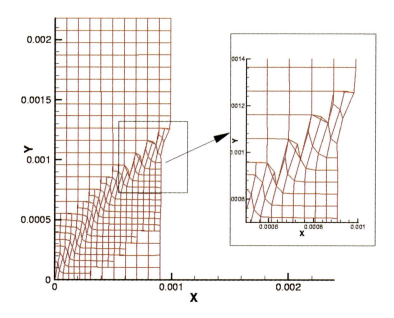

Fig. 5.16. Why does FEM have troubles in h-adaptive refinement ?

$$m_{ij}^s = \begin{cases} \omega\delta_{ij} \displaystyle\int_{\Omega_X} \rho_0 \Phi_j^2 d\Omega & i = j \\[2em] 0 & i \neq j \end{cases} \tag{5.73}$$

where

$$\omega := \frac{\displaystyle\int_{\Omega_X} \rho_0 d\Omega}{\left(\displaystyle\sum_{\ell=1}^{NP} \int_{\Omega_X} \rho_0 \left(\Psi_\ell^{[00]}\right)^2 d\Omega + \sum_{|\alpha|=1}^{\beta} \sum_{\ell}^{NAD} \int_{\Omega_X} \rho_0 \left(\Psi_\ell^{[\alpha]}\right)^2 d\Omega\right)} \tag{5.74}$$

However, there is a setback for this particular proposal. As one may find out, using special lumping technique to avoid singular mass matrix will result the increase of the total mass of a mass conservative system, i.e. an artificial added mass will flow into the system during adaptive procedure, since the added mass is proportional to the added degrees of freedom, or the wavelet shape functions. This could undermine the accuracy of the numerical computations, because of its none-conservative nature of mass. Nevertheless, based on our computational experiences, if the added degrees of freedom is less than 20 %

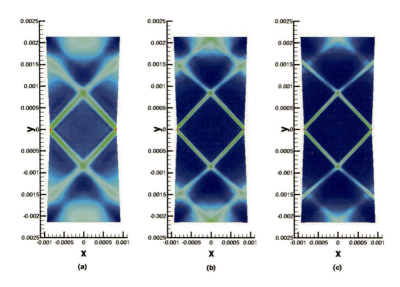

Fig. 5.17. The contours of the effective viscoplastic strain in the tensile bar: (a) without wavelet adaptivity; (b) with wavelet adaptivity ($\alpha = (1,0),(0,1)$); (c) with wavelet adaptivity ($\alpha = (1,0),(0,1),(1,1)$).

of the total degrees of freedom, there is no obvious side-effect on numerical computations. Of course, a further evaluation may be necessary for precise assessment. A numerical tension test is conducted, and the results are plotted in Figs. 5.17–5.18. Only geometric imperfection is planted in the specimen, as the reduction of the width of the tensile bar. The maximum reduction of the width of the tensile bar occurs at the middle cross section, 10%; it, then, linearly varies along the x_2 direction back to the original width. From Figs. 5.17–5.18, one can find that there is significant improvement on the detailed resolution of the numerical solutions due to the wavelet refinement. Note that in Fig. 5.18 (c) the marked particles, i.e. the dark region, are the particles where the higher order wavelet kernels are turned on. A separated account on wavelet-adaptive procedure on shear band formation is presented in.[267]

5.4 Simulations of Dynamics Shearband Propagation

In the meshfree simulations, a thermo-viscoplastic constitutive model is adopted. For simplicity, the heat conduction is neglected; thus, the only regularization agent at the constitutive level is viscosity. This approximation may have certain limitations on some aspects of the simulation, such as an accurate determination of evolving shear band width, ect. . Nevertheless, the problem

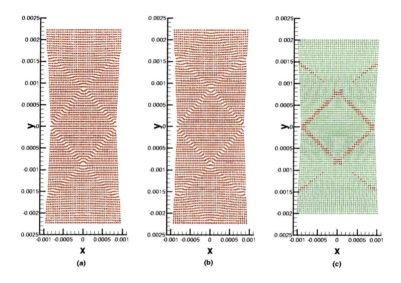

Fig. 5.18. The shear band in the tensile bar represented by particle formation (a) without wavelet adaptivity; (b) with wavelet adaptivity; (c) the adaptive pattern in undeformed configuration .

is well posed in a mathematical sense, and there is a finite, intrinsic length scale associated with the constitutive parameters (see[349, 350, 396] for details). On the other hand, the meshfree interpolant introduces diffusion mechanism as well as numerical length scale into the simulation [2], by the particle distribution density. Some mesh size sensitivity is still present in the meshfree formulation; nevertheless, the mesh alignment sensitivity is significantly suppressed. So far, in most prior simulations of strain localization, the main effort is to simulate the mere appearance of bifurcated deformation modes due to material instability. Nevertheless, how and why such localization zone is able to propagate is still not well understood. A common belief among the researchers is that dynamic shear band propagation is a self-sustained material instability propagation in an auto-catalytic manner.

It should be noted that adiabatic shear band propagation may be viewed as a (weakly) shock wave front *expansion* but different from a (weakly) shock wave front propagation.[188] Based on this philosophy, dynamic shear band propagation may be simulated by a single constitutive model, and it should happen automatically. Unfortunately, most of finite element methods based simulations failed to capture adiabatic shear band propagation.

---

[2] In fact, any discretization will bring numerical length scale into numerical simulation, no matter the discretization is mesh based, or meshfree.

On the other hand, it is most physicists' belief that shear band is a physical entity, within which significant change in material strength and properties take place. For instance, shear band can be identified a possible phase transformation (see[179, 180]). In this spirit, the constitutive behavior in the localized damage zone in the post-bifurcation phase dictates, or controls, shear band propagation. It is believed that the stress collapse in strain localization zone in the post-bifurcation phase is crucial for the subsequent propagation of adiabatic shear band, or expansion of strain localization zone, because it allows the dynamic loading continuously be mounted on the shear band tip, and provides driving force for the shear band propagation. This notion has been implicitly suggested by,[468],[31] and.[385]

The key technical ingredient of simulating dynamic shear band propagation are two: (1) resolving micromechanical length scale that is associated with adiabatic shear band width, and (2) constitutive modeling in post-bifurcation phase, i.e., how to simulate the threshold of stress collapse inside shear band. Note that the threshold of stress collapse is different from the onset strain localization.

5.4.1 Thermal-viscoplastic Model

In inelastic large deformations, the deformation gradient, $\mathbf{F}$, may be decomposed as

$$\mathbf{F} = \mathbf{F}^e \cdot \mathbf{F}^{vp} \tag{5.75}$$

where $\mathbf{F}^e$ describes elastic deformations and rigid body rotations, and $\mathbf{F}^{vp}$ represents viscous inelastic deformations. The rate of deformation tensor, $\mathbf{D}$, and the spin tensor, $\mathbf{W}$, are the symmetric and anti-symmetric parts of the spatial velocity gradient $\mathbf{L} = \dot{\mathbf{F}} \cdot \mathbf{F}^{-1}$, i.e.

$$\mathbf{D} + \mathbf{W} = \dot{\mathbf{F}} \cdot \mathbf{F}^{-1} = \dot{\mathbf{F}}^e \cdot \mathbf{F}^{e-1} + \mathbf{F}^e \cdot \dot{\mathbf{F}}^{vp} \cdot \mathbf{F}^{vp-1} \cdot \mathbf{F}^{e-1} \tag{5.76}$$

and

$$\mathbf{D} := D_{ij}\mathbf{e}_i \otimes \mathbf{e}_j , \qquad D_{ij} := \frac{1}{2}\left(\frac{\partial v_i}{\partial x_j} + \frac{\partial v_j}{\partial x_i}\right) \tag{5.77}$$

$$\mathbf{W} := W_{ij}\mathbf{e}_i \otimes \mathbf{e}_j , \qquad W_{ij} := \frac{1}{2}\left(\frac{\partial v_i}{\partial x_j} - \frac{\partial v_j}{\partial x_i}\right) \tag{5.78}$$

Note that thermal deformation may be considered as eigenstrain, but makes no contribution in geometric decomposition.

We neglect thermo-elastic contribution on internal work, i.e. $\boldsymbol{\tau} : (\mathbf{D}^e + \mathbf{D}^t) \approx 0$, where $\mathbf{D}^t$ is the rate of deformation due to thermal expansion. Assume that the major part of plastic work is converted into heat (Taylor and Quinney[424]). The rate form of balance of energy

$$\int_{\Omega} \rho \dot{e} d\Omega + \frac{d}{dt}\int_{\Omega} \frac{1}{2}\rho \dot{\mathbf{u}} \cdot \dot{\mathbf{u}} d\Omega = \int_{\partial\Omega} \mathbf{t} \cdot \mathbf{v} d\Gamma - \int_{\partial\Omega} \mathbf{n} \cdot \mathbf{q} d\Gamma \tag{5.79}$$

may take the form

$$\int_{\Omega_0} \rho_0 C_p \dot{T} d\Omega_0 = \int_{\Omega_0} \chi \boldsymbol{\tau} : \mathbf{D}^{vp} d\Omega_0 + \int_{\Omega_0} \nabla_{\mathbf{X}} (J \cdot \mathbf{F}^{-1} \cdot \boldsymbol{\kappa} \cdot \mathbf{F}^{-T} \cdot \nabla_{\mathbf{X}} T) d\Omega_0$$

(5.80)

where e is the specific internal energy; $\mathbf{t}$ is the traction; and $\mathbf{q}$ is the heat flux vector through the boundary; C_p is the specific heat at constant pressure used to approximate the specific heat at constant stress.

A local strong form of the energy equation is

$$\rho_0 C_p \frac{\partial T}{\partial t} = \chi \boldsymbol{\tau} : \mathbf{D}^{vp} + \nabla_{\mathbf{X}} \left(J \mathbf{F}^{-1} \cdot \boldsymbol{\kappa} \cdot \mathbf{F}^{-T} \cdot \nabla_{\mathbf{X}} T \right), \quad \forall \, \mathbf{X} \in \Omega_0 \quad (5.81)$$

Because the whole impact process last a few hundred μs, the effect of heat conduction is negligible over the domain of the specimen. By only considering adiabatic heating, we have

$$\rho_0 C_p \frac{\partial T}{\partial t} = \chi \boldsymbol{\tau} : \mathbf{D}^{vp}$$

(5.82)

By doing so, the coupled thermo-elasto-viscoplastic problem is uncoupled. Thus the linear momentum equation (5.26) suffices for constructing Galerkin weak form, and the energy equation (5.82) is only used in the constitutive update.

A rate form constitutive equation is used

$$\overset{\triangledown}{\boldsymbol{\tau}} := \mathbf{C}^{elas} \left(\mathbf{D} - \mathbf{D}^{vp} - \mathbf{D}^t \right),$$

(5.83)

where the Jaumann rate of Kirchhoff stress, $\overset{\triangledown}{\boldsymbol{\tau}} := \dot{\boldsymbol{\tau}} - \mathbf{W} \cdot \boldsymbol{\tau} + \boldsymbol{\tau} \cdot \mathbf{W}$. The thermal rate of deformation, $\mathbf{D}^t$, is given as

$$\mathbf{D}^t = \alpha \dot{T} \mathbf{1}$$

(5.84)

where α is the coefficient of thermal expansion, and $\mathbf{1}$ is the second order unit tensor.

The yield surface of viscoplastic solid is of the von Mises type,

$$f(\bar{\sigma}, \kappa) = \bar{\sigma} - \kappa = 0$$

(5.85)

$$\bar{\sigma}^2 = \frac{3}{2} s_{ij} s_{ij},$$

(5.86)

$$s_{ij} = \tau'_{ij} - \alpha_{ij}$$

(5.87)

$$\tau'_{ij} = \tau_{ij} - \frac{1}{3} tr(\boldsymbol{\tau}) \delta_{ij}$$

(5.88)

$$\bar{\epsilon} := \int_0^t \sqrt{\frac{2}{3} \mathbf{D}^{vp} : \mathbf{D}^{vp}} \; dt$$

(5.89)

where α_{ij} is the back stress for kinematic hardening.

A thermo-elasto-viscoplastic material model is adopted (See[468]), which is described as

$$D_{ij}^{vp} := \bar{\eta}(\bar{\sigma}, \bar{\epsilon}, T) \frac{\partial f}{\partial \sigma_{ij}} \tag{5.90}$$

$$\bar{\eta} = \dot{\epsilon}_0 \left[\frac{\bar{\sigma}}{g(\bar{\epsilon}, T)}\right]^m, \tag{5.91}$$

$$g(\bar{\epsilon}, T) = \sigma_0 \left[1 + \bar{\epsilon}/\epsilon_0\right]^N \left\{1 - \delta\left[\exp\left(\frac{T - T_0}{\kappa}\right) - 1\right]\right\} \tag{5.92}$$

where m is the power index, $\dot{\epsilon}_0$ is the referential strain rate, σ_0 is yield stress, $\epsilon_0 = \sigma_0/E$, and δ is the thermal softening parameter.

The constitutive update largely follows the rate tangent modulus method proposed by,[374] which has been used in the context of thermo-viscoplasticity by.[262, 263] The essence of the rate tangent modulus method is to approximate a function of time in the interval, $t_{n+\theta} \in [t_n, t_{n+1}] \; \theta \in [0, 1]$, as

$$f_\theta := (1 - \theta)f_n + \theta f_{n+1} \tag{5.93}$$

If we choose the predicted velocity field at t_{n+1} as $\mathbf{v}_{n+1}^{trial} = \mathbf{v}_n + \Delta t \mathbf{a}_n$, it follows that

$$\mathbf{v}_\theta = (1 - \theta)\mathbf{v}_n + \theta \mathbf{v}_{n+1}^{trial} = \mathbf{v}_n + \theta \Delta t \mathbf{a}_n \tag{5.94}$$

$$\mathbf{u}_\theta = (1 - \theta)\mathbf{u}_n + \theta \mathbf{u}_{n+1} = \mathbf{u}_n + \Delta t \theta \mathbf{v}_n + \theta^2 \Delta t^2 \mathbf{a}_n \tag{5.95}$$

$$\mathbf{L}_\theta = \mathbf{v}_\theta \, \overline{\nabla} \boldsymbol{x} = \left(\mathbf{v}_\theta \, \overline{\nabla} \mathbf{x}\right) \cdot \mathbf{F}_{n+1}^{-1} \tag{5.96}$$

$$\mathbf{D}_\theta = \frac{1}{2}\left(\mathbf{L}_\theta + \mathbf{L}_\theta^T\right) \tag{5.97}$$

$$\mathbf{W}_\theta = \frac{1}{2}\left(\mathbf{L}_\theta - \mathbf{L}_\theta^T\right) \tag{5.98}$$

For $\theta = 1/2$, the predicted step, or trial step corresponds to the central difference scheme. By following (5.94)–(5.98), the kinematical variables can be calculated at the configuration $t_{n+\theta}$.

The main task here is to update the Kirchhoff stress:

$$\boldsymbol{\tau}_{n+1} = \boldsymbol{\tau}_n + \dot{\boldsymbol{\tau}}_\theta \Delta t \tag{5.99}$$

$$\dot{\boldsymbol{\tau}}_\theta \approx \overset{\triangledown}{\boldsymbol{\tau}}_\theta + \mathbf{W}_\theta \cdot \boldsymbol{\tau}_n + \boldsymbol{\tau}_n \cdot \mathbf{W}_\theta^T \tag{5.100}$$

To accomplish this, we first set

$$\dot{\bar{\epsilon}}_\theta = (1 - \theta)\dot{\bar{\epsilon}}_n + \theta \dot{\bar{\epsilon}}_{n+1} \tag{5.101}$$

where $\dot{\bar{\epsilon}}_{n+1}$ is approximated by a first order Taylor series expansion in $\bar{\sigma}, \bar{\epsilon}$ and T, i.e.

$$\dot{\bar{\epsilon}}_{n+1} = \dot{\bar{\epsilon}}_n + \Delta t \left(\frac{\partial \dot{\bar{\epsilon}}}{\partial \bar{\sigma}}\bigg|_n \dot{\bar{\sigma}}_\theta + \frac{\partial \dot{\bar{\epsilon}}}{\partial \bar{\epsilon}}\bigg|_n \dot{\bar{\epsilon}}_\theta + \frac{\partial \dot{\bar{\epsilon}}}{\partial T}\bigg|_n \dot{T}_\theta\right) \tag{5.102}$$

Evaluate other state variables at $t_{n+\theta} = (n + \theta)\Delta t$,

$$\mathbf{D}_{\theta}^{vp} = \dot{\bar{\epsilon}}_{\theta} \mathbf{p}_n \tag{5.103}$$

$$\overset{\triangledown}{\boldsymbol{\alpha}}_{\theta} = b\mathbf{D}_{\theta}^{vp} \tag{5.104}$$

$$\dot{\boldsymbol{\alpha}}_{\theta} = \overset{\triangledown}{\boldsymbol{\alpha}}_{\theta} + \mathbf{W}_{\theta} \cdot \boldsymbol{\alpha}_n + \boldsymbol{\alpha}_n \cdot \mathbf{W}_{\theta}^T \tag{5.105}$$

$$\boldsymbol{\alpha}_{n+1} = \boldsymbol{\alpha}_n + \dot{\boldsymbol{\alpha}}_{\theta} \Delta t \tag{5.106}$$

where $\mathbf{p} := \dfrac{3}{2} \dfrac{\mathbf{s}'}{\bar{\sigma}}$.

Since $\mathbf{p}_{\theta}$ and $\boldsymbol{\alpha}_{\theta}$ are unknown in the configuration at Ω_n, inconsistent approximations are made $\mathbf{p}_{\theta} \approx \mathbf{p}_n$ $\boldsymbol{\alpha}_{\theta} \approx \boldsymbol{\alpha}_n$ in the explicit calculation.

In isotropic hardening

$$\mathbf{D}^{vp} = \frac{3}{2} \frac{\dot{\bar{\epsilon}}}{\bar{\sigma}} \boldsymbol{\tau}' \tag{5.107}$$

hence

$$\boldsymbol{\tau} : \mathbf{D}^{vp} = \left(\boldsymbol{\tau}' + \frac{1}{3} tr(\boldsymbol{\tau})\mathbf{1}\right) : \left(\frac{3}{2} \frac{\dot{\bar{\epsilon}}}{\bar{\sigma}} \boldsymbol{\tau}'\right) = \frac{3}{2} \frac{\dot{\bar{\epsilon}}}{\bar{\sigma}} \boldsymbol{\tau}' : \boldsymbol{\tau}' = \bar{\sigma}\dot{\bar{\epsilon}} \tag{5.108}$$

Eq. (5.82) becomes

$$\frac{\partial T}{\partial t} = \frac{\chi}{\rho_0 C_p} \bar{\sigma}\dot{\bar{\epsilon}} \tag{5.109}$$

Utilizing (5.109), we propose the following *monolithic* or *simultaneous* rate tangent modulus scheme. Substituting (5.109) at time $t_{n+\theta} = (n + \theta)\Delta t$ into (5.102), we have

$$\dot{\bar{\epsilon}}_{n+1} = \dot{\bar{\epsilon}}_n + \Delta t_n \left\{ \frac{\partial \dot{\bar{\epsilon}}}{\partial \bar{\sigma}} \Big|_n \dot{\bar{\sigma}}_{\theta} + \frac{\partial \dot{\bar{\epsilon}}}{\partial \bar{\epsilon}} \Big|_n \dot{\bar{\epsilon}}_{\theta} + \frac{\partial \dot{\bar{\epsilon}}}{\partial T} \Big|_n \left(\frac{\chi}{\rho_0 C_p} \bar{\sigma}_{\theta} \dot{\bar{\epsilon}}_{\theta}\right) \right\} \tag{5.110}$$

Substituting (5.110) into (5.101) and solving for $\dot{\bar{\epsilon}}_{\theta}$ yield

$$\dot{\bar{\epsilon}}_{\theta} = \frac{\dot{\bar{\epsilon}}_n}{1 + \xi_{\theta}} + \frac{1}{H_{\theta}} \frac{\xi_{\theta}}{1 + \xi_{\theta}} \mathbf{P}_{\theta} : \mathbf{D}_{\theta} \tag{5.111}$$

where

$$H_{\theta} := \frac{3E}{2(1+\nu)} - \frac{\partial \dot{\bar{\epsilon}}/\partial \bar{\epsilon}}{\partial \dot{\bar{\epsilon}}/\partial \bar{\sigma}} \Big|_n - \frac{\partial \dot{\bar{\epsilon}}/\partial T}{\partial \dot{\bar{\epsilon}}/\partial \bar{\sigma}} \Big|_n \frac{\alpha\chi}{\rho_0 C_p} \bar{\sigma}_{\theta}$$

$$\approx \frac{3E}{2(1+\nu)} - \frac{\partial \dot{\bar{\epsilon}}/\partial \bar{\epsilon}}{\partial \dot{\bar{\epsilon}}/\partial \bar{\sigma}} \Big|_n - \frac{\partial \dot{\bar{\epsilon}}/\partial T}{\partial \dot{\bar{\epsilon}}/\partial \bar{\sigma}} \Big|_n \frac{\alpha\chi}{\rho_0 C_p} \bar{\sigma}_n \tag{5.112}$$

$$\mathbf{P}_{\theta} := \mathbf{C}^{elas} : \mathbf{p}_{\theta} \approx \mathbf{C}^{elas} : \mathbf{p}_n \tag{5.113}$$

$$\xi_{\theta} = \theta\Delta t \left(\frac{\partial \dot{\bar{\epsilon}}}{\partial \bar{\sigma}}\right)_n H_{\theta} \tag{5.114}$$

Note that since we do not know the stress state in the configuration $\Omega_{n+\theta}$, in an explicit update we approximate σ_θ by σ_n in the calculations of H_θ, $\mathbf{P}_\theta$, as well as of ξ_θ. This assumption may have been implied in the original derivation of.[374]

Subsequently, the Jaumann rate of the Kirchhoff stress is evaluated as

$$\overset{\triangledown}{\boldsymbol{\tau}}_\theta = \mathbf{C}_\theta^{tan} : \mathbf{D}_\theta - \left\{\frac{\dot{\bar{\epsilon}}_n}{1+\xi_\theta}\right\}\left[\mathbf{P}_\theta + 3K\alpha\frac{\chi\bar{\sigma}_\theta}{\rho C_p}\mathbf{1}\right] \tag{5.115}$$

where

$$\mathbf{C}_\theta^{tan} = \mathbf{C}^{elas} - \frac{\xi_\theta}{H_\theta(1+\xi_\theta)}\left[\mathbf{P}_\theta \otimes \mathbf{P}_\theta + (3\lambda+2\mu)\alpha\frac{\chi\bar{\sigma}_n}{\rho_0 C_p}\mathbf{1}\otimes\mathbf{P}_\theta\right] \tag{5.116}$$

is the adiabatic tangent stiffness, which is not symmetric. Assuming that $\bar{\epsilon}_\theta$ and $\bar{\sigma}_\theta$ are available after the stress update, the temperature can then be updated at each quadrature point as

$$T_{n+1} = T_n + \dot{T}_\theta \Delta t \tag{5.117}$$

5.4.2 Constitutive Modeling in Post-bifurcation Phase

After initial thermal softening, material instability occurs, which leads to strain localization. Based on Marchand and Duffy's well-known experiment on dynamic shear band propagation ([318]), there are three stages in the development of an adiabatic shear band. In stage I, before, or onset of the strain localization, the plastic strain distribution is homogeneous; in stage II, right after initial localization, the plastic strain distribution become inhomogeneous, but the amount and width of the localized deformation remain same; in stage III, there is drastic reduction of flow stress, the nominal strain becomes quite large, from 40 % to more than 1000 %, and changes rapidly from one location to another, indicating spatial oscillation, or fluctuation of strain rate distribution. The most salient characteristics of stage III strain localization is that the material drastically loses its shear stress carrying capacity. This phenomenon is called *stress collapse*, and has been predicted by Wright and his colleagues in their theoretical analysis ([455, 456]). Apparently, there is a critical strain, or threshold for such stress collapse.[328] calculated the critical nominal strain at which the stress collapses. Their result is called the Molinari-Clifton condition. In fact,[178] showed that when the Molinari-Clifton condition is met, steady adiabatic boundary layers collapse into a vortex sheet. Nevertheless, how to implement the Molinari-Clifton condition in numerical simulations is an still open problem. On the other hand, since there is almost a vertical jump in reduction of flow stress carrying capacity (see Figures 8, 12, and 21 in[318]), it would be difficult to represent such drastic change of flow stress by a single constitutive relation that is responsible to describe both thermal-mechanical behaviors of matrix as well as the behaviors of the shear band in the post-bifurcation phase. The underling argument here is that separate constitutive description, or multiple physics model, is more convenient and efficient in constitutive modeling.

In the case of high strain rate loading, it has been speculated that there is an intensified high strain rate zone at the tip of a propagating shear band e.g.[456–458] Moreover, based on the observation of micrograph of shear band surfaces, there is uniform void distribution throughout the shear band ([467]). It is thus suggested that dynamic shear band propagation may be related to damage evolution process, which is controlled by strain rate.

It is highly plausible that the magnitude of a strain rate in front of the shear band tip could be used as a criterion for the onset of stress collapse, or strain localization. It is possible that once the strain rate reaches a certain level, damage occurs within the shear band irreversibly, and the material changes its behavior inside the shear band, subsequently the material loses its stress carrying capability significantly, i.e. the stress collapse. On the other hand, by combining the notion of stress collapse with the notion of wave trapping ([459]), one may be able to explain the autonomous built-up of a high strain rate field in front of the shear band tip.

[468, 469] used a Newtonian flow constitutive relation to model the highly mobile plastic flow inside the shear band.[31] used a compressible ideal fluid to model the shear band; recently,[385] used a shear band softening model, which is based on experimental data,[242, 385] to simulate the post-bifurcation, or stress collapse state of the adiabatic shear band.

Stress collapse inside an adiabatic shear band is triggered by the attainment of a critical strain $\bar{\epsilon}_{cr}$, which is controlled by strain rate. The stress collapses when the critical effective strain

$$\bar{\epsilon}_{cr} = \epsilon_1 + (\epsilon_2 - \epsilon_1) \frac{\dot{\epsilon}_r}{(\dot{\epsilon}_r + \dot{\epsilon})} \tag{5.118}$$

is reached, where ϵ_1, ϵ_2 and $\dot{\epsilon}_r$ are input parameters. After $\bar{\epsilon}$ reaches $\bar{\epsilon}_{cr}$, the damaged material is assumed to behave like a non-Newtonian viscous fluid, so

$$\boldsymbol{\tau} = -\frac{\gamma[1 - J + \alpha(T - T_0)]}{J} \frac{E}{1 - \nu} \mathbf{1} + \mu_d(T)\mathbf{D} \tag{5.119}$$

where γ is the stiffness parameter and μ_d is the viscosity, which may be temperature dependent.

To simulate fracture and crack growth, a simple material damage algorithm is adopted in meshfree simulations. The maximum tensile stress is used as the crack growth criterion. The algorithm works as follows: when the maximum tensile stress at a material point exceeds a certain limit, the crack is assumed to pass through that material point. To model the crack, the stress components are set to zero at that material point, and the value of temperature is set to room temperature.

The main advantages of using the critical stress based material damage algorithm is its simplicity. Because of the nonlocal nature of meshfree approximation, the connectivity relation with respect to the referential configuration needs to be updated to prevent the particles on the other side of the crack from contributing over the crack line. A less accurate, but efficient way to get around a connectivity update is as follows: once a quadrature point within a background cell is damaged, then all the other quadrature points in the same background cell are considered to be damaged.

Since the specimen is 6 mm thick, the crack morphology observed in the experiment tend to be uniform in the thickness direction. Therefore, it may be reasonable to assume that the cleavage fracture toughness in the thermo-elasto-viscoplastic solid is controlled by the maximum circumferential stress, or hoop stress within the plane, which conforms with the conventional theory of brittle fracture (e.g.[158]). In our computation, the following criterion is used

$$\sigma_{max} \geq \sigma_{cr} \tag{5.120}$$

where the critical stress is set as $\sigma_{cr} = 3\sigma_0$ (where σ_0 is the initial yield stress and $3\sigma_0$ is roughly the level of stress triaxiality expected at the tip of a crack in an elasto-plastic solid). When the maximum principal stress reaches $3\sigma_0$ at a material point (Gauss quadrature point), we set

$$\tau_{ij} = 0 \; ; \quad T = T_0 \; , \tag{5.121}$$

5.4.3 Numerical Examples

This example focus on simulating the experiments conducted by Zhou et al.,[467] i.e. the ZRR problem. The experiment involves an asymmetrically impact loading of a pre-notched plate (single notch) by a cylindrical projectile as shown in Fig. 5.19. In this numerical study, two configurations have been used to simulate plate specimens of different sizes, which correspond to two different sets of experiments. The first configuration models the experiment conducted by Zhou, Rosakis, and Ravichandran[467] (see Fig. 5.20 (a)), while the second one models the experiments conducted recently by Guduru, Rosakis, and Ravichandran[187] (see Fig. 5.20 (b)). It may be noted that in the second specimen, there is a *2mm* long fatigue crack in front of the pre-notch, which increases the acuity of the crack. We have conducted both 2D and 3D simulations for the first specimen: a 3D computation with projectile speed at V= 30 m/s, and 3D computation with projectile speed at V = 33 m/s. For the second specimen, we have only carried out a 2D computation with projectile speed at V= 37 m/s.

The primary objectives of this simulation are twofold: (1) to capture failure mode transition; (2) to determine the adiabatic shear band growth criterion and driving force.

Fig. 5.19. An asymmetrically impact loaded plate with a pre-notched crack.

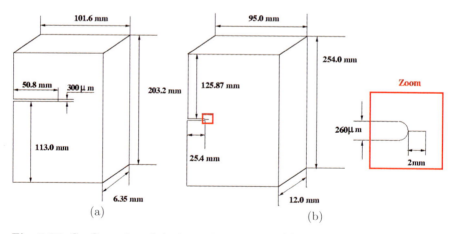

Fig. 5.20. Configuration of single notch specimens: (a) specimen one, (b) specimen two.

5.4.4 Case I: Intermediate Speed Impact ($V = 30\ m/s$)

In the experiment conducted by Zhou et al.,[467] when the impact velocity is in the intermediate range, i.e. $20.0\ m/s < V < 30.0\ m/s$, a shear band initiates from the notch tip, and it is then arrested within the specimen

Parameter	Value	Definition
$\dot{\epsilon}_0$	$1 \times 10^{-3} s^{-1}$	reference strain rate
m	70	rate sensitivity parameter
σ_0	2000 MPa	yield stress
ϵ_0	σ_0/E	
n	0.01	strain hardening exponent
T_0	293 K	reference temperature
δ	0.8	thermal softening parameter
κ	500 K	thermal softening parameter
E	200 GPa	Young's modulus
ν	0.3	Poisson's ratio
ρ	7830 kg m^{-3}	mass density
c_p	448 J (kg·K)$^{-1}$	specific heat
α	11.2×10^{-6} K^{-1}	coefficient of thermal expansion
χ	0.9	the fraction of plastic work converted to heat
ϵ_1	$4.0 \times \epsilon_0$	
ϵ_2	0.3	
$\dot{\epsilon}_r$	4.0×10^4 1/s	in a range $(1.0 \times 10^4$ 1/s $\sim 6.0 \times 10^4$ 1/s$)$

Table 5.1. Material properties of the target plate.

interior. The final failure of the specimen is caused by brittle fracture — a cleavage type (mode I) crack growing from the end of the arrested shear band. This shearband-crack switch under fixed impact speed is intriguing, whose causes are not understood well. It is therefore of considerable challenge to simulate such a failure mode switching phenomenon. The main parameters in our simulations are listed in Table 5.1.

Fig. 5.21(a)-(f) show a sequence of effective stress contours (SS[Pa]) surrounding the failure region following the impact. It can be clearly observed that a strip—a low effective stress zone initiates, and grows starting from the notch-tip, which we identify as the trace of adiabatic shear band. The shear band grows steadily almost in horizontal direction. At a certain point (Fig. 5.21c), it suddenly changes its direction and moves upward. At this time within the strip, the value of effective stress becomes zero, which indicates that a crack is initiated from the tip of the shear band. We identify this turning point as the point at which the shear band transits into an opening crack (see Fig. 5.21 (d), (e) and (f)). To view the shearband-to-crack switch clearly, a 3D view of the plate under impact of the projectile is displayed in Figures 5.22(a),(b). The color conture depicted on the surface of the specimen is effective stress (SS [Pa]), from which one may compare the initial stage of shear band growth with the final stage of crack growth. In Fig. 5.22 (a) (24.0 μs after impact), there is only a shear band in front of pre-notch. As the process continues, from Fig. 5.22 (b), one can observe that a crack is running away from the arrested shear band. In the simulation, the impact between the projectile and the plate is modeled as a real collision between a rigid cylinder projectile and a visco-elasto-plastic solid plate. In the computation, a total 49,086 particles are used to discretize the plate, and 32,080 background cells

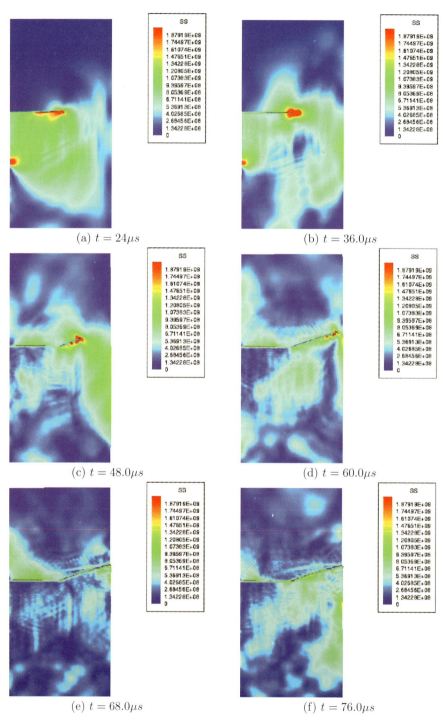

(a) $t = 24\mu s$

(b) $t = 36.0\mu s$

(c) $t = 48.0\mu s$

(d) $t = 60.0\mu s$

(e) $t = 68.0\mu s$

(f) $t = 76.0\mu s$

Fig. 5.21. Brittle-to-Ductile failure mode transition (I) (brittle failure mode at $V = 30.0 m/s$): effective stress contour (SS [Pa]) of a 3D simulation (front plane view).

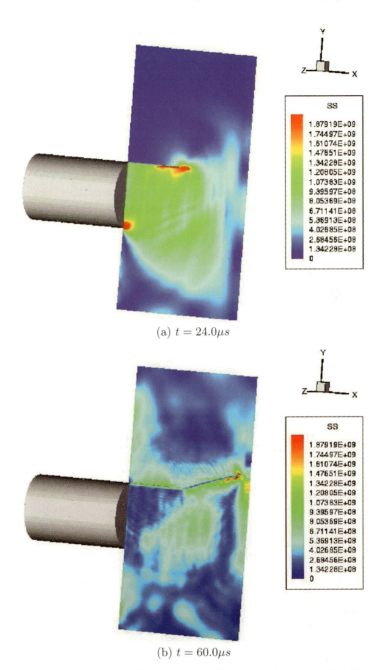

(a) $t = 24.0\mu s$

(b) $t = 60.0\mu s$

Fig. 5.22. Ductile-to-brittle failure mode switch: effective stress (SS [Pa]) of 3D simulation (3D view) with impact velocity $v = 30.0 m/s$.

are allocated for the numerical quadrature. There are eight quadrature points in each background cell, for a total of 256,640 Gauss quadrature points. For the projectile, a total 1299 particles and 792 background cells have been used in discretization. The density of the projectile is $7900kg/m^3$. The projectile is $125mm$ long, and $50mm$ in diameter.

5.4.5 Case II: High Speed Impact ($V = 33 \ m/s$)

When the impact velocity exceeds a certain limit, V_{SB} [3], the cleavage fracture of mode-I crack is suppressed, the shear band initiated from the pre-notch tip never stop, and it propagates through the specimen. Since the impact is due to an unsymmetric collision between the projectile and plate, the shear band propagates slightly towards the lower part of the specimen, rather than propagating straight in the horizontal direction. A sequence of 3D calculations are displayed in Figures 5.23. The color contours represent effective stress (SS[Pa]) value. By examining the effective stress contours, one may notice that a thin strip with lower effective stress value passes through the specimen, which is the trace of the adiabatic shear band.

A sequence of 3D calculations have been displayed in Fig. 5.23, in which the effective stress contours are depicted, one may observe the evolution of the shear band from effective stress contours. By comparing the above results with the intermediate impact speed range, a complete picture of failure mode transition emerges, i.e. a transition from the cleavage fracture of a brittle failure at lower impact speeds to the shear propagation of a ductile failure at higher impact speeds.

As reported by Zhou et al. (1996a), the experimental results indicate that the shear band propagates along a curved surface. Such shear band morphology is very difficult to capture by finite elements, because of the generic mesh alignment sensitivity in such simulations. Using meshfree methods, the curved shearband formation has been accurately captured in the numerical computations presented here.

In Fig. 5.24, the meshfree computation is juxtaposed with experimental observation (Zhou et al.[467]). One may find that the meshfree simulation replicates the experimental observation.

5.5 Simulations of Crack Growth

5.5.1 Visibility Condition

A crucial step to model crack propagation in a numerical simulation is how to represent the evolving crack surface and automatically adjust interpolation

---

[3] In the experiments conducted by Zhou et al. (1996a), this critical velocity is $V_{SB} = 29.6$ m/s for C-300 steel. This value is expected to be sensitive to material properties as well as pre-notch geometry and the size of the specimen used.

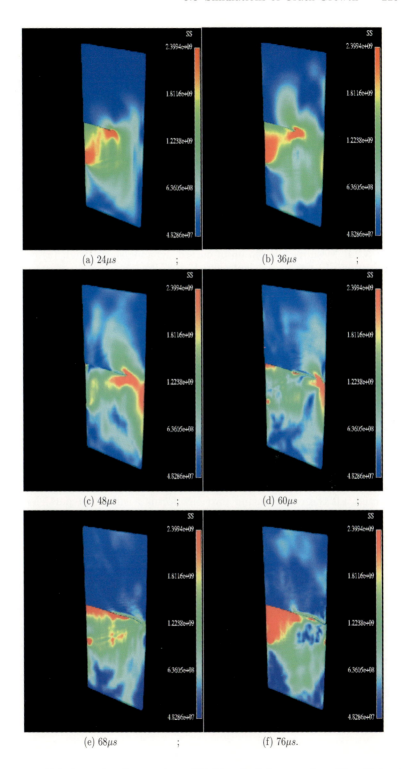

Fig. 5.23. Brittle-to-ductile transition (II) (ductile failure mode at $V = 33.0\ m/s$): effective stress (SS 0.048 GPa $\sim$ 2.399 GPa) contours of a 3D simulation.

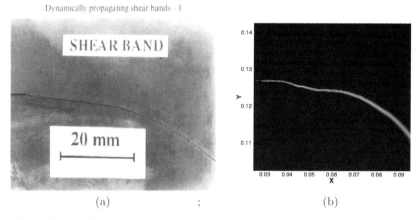

Fig. 5.24. Meshfree simulation of curved dynamic shear band: (a) experimental observation; (b) meshfree calculation ($V = 37.0m/s$).

field around growing crack tip. This process is not only a re-interpolation scheme, but also a process how to model the material re-configuration.

Belytschko and his co-workers (Belytschko et al. [1996]) have developed a so-called visibility condition that can serve as a criterion to automatically adapt topological connectivity map among meshfree particles.

In meshfree discretization, when a crack segment is created in a continuum, the shape and size of the compact support of the original shape functions in the region have to be redefined. The rule for such redefinition of domain of influence for a particle, or for any point in the domain, is the so-called "*visibility condition*" proposed by Belytschko et al. (Belytschko et al. [1996]) An equivalent version of the visibility condition is stated at the following. When the moment matrix in a spatial point X, $X \in \Omega$, is constructed, all the contributing particles forms a subset of particles from the particles in the original domain of influence of the point X; such that one can connect any particle in this subset with the point X in a straight line without intercepting the boundary of the domain, for instance, the crack surfaces. In other words, we reshape the domain of influence of any point X, such that any straight line connecting the point X with a particle in its domain of influence does not penetrate crack surfaces. Fig. 5.25 illustrates how such modified domain of influence is being constructed. In Fig. 5.25, all the particles participating the construction of moment matrix at the point, X, are marked as black circle, and all the particles that are cut from the original domain of influence of the point, X, are marked with the hollow circle.

There are two shortcomings in early meshfree crack surface representation/visibility condition procedure: (1) crack surface re-construction and representation schemes are complicated. The complexity comes from several sources: searching algorithm and re-interpolation algorithm. Because of the

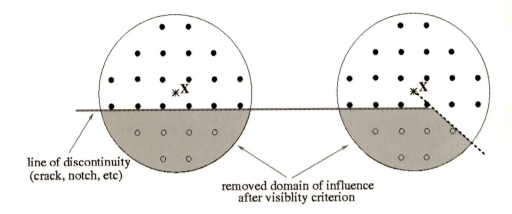

Fig. 5.25. The visibility criterion in determining the domain of influence of a spatial point X.

technical complexity, any generalization of meshfree crack surface modeling of three-dimensional fracture or ductile fracture becomes a formidable task; (2) it has been observed that the meshfree shape functions of re-interpolation field produced by visibility condition may contain strong discontinuities in meshfree shape functions at certain region near a crack tip, although we do not know for certain this is indeed a shortcoming.

To simplify the crack surface modeling procedure, we introduce the following crack surface representation and particle splitting algorithm to model crack surface separation.

5.5.2 Crack Surface Representation and Particle Splitting Algorithm

The two-dimensional crack surface is represented by pairs of piece-wise straight lines as shown in Fig. 5.26. In Fig. 5.26, the particles on the crack surface are marked as square black boxes, except crack tip, whereas other meshfree particles are represented as solid circles. In previous meshfree approaches, when a crack grows, the crack surface is being reconstructed by adding new particles. This is not suitable for ductile crack surface modeling, because one has to re-create state variables and re-distribute mass and volume for any newly added particles.

In our approach, a crack tip is always attached to an existing material/interpolation particle. It only moves from one particle to another as shown in Fig. 5.26.

Assume that the physical criterion to select the new crack tip is available. To find the new crack tip, we first choose a radius R and draw a circle centered at the current crack tip.

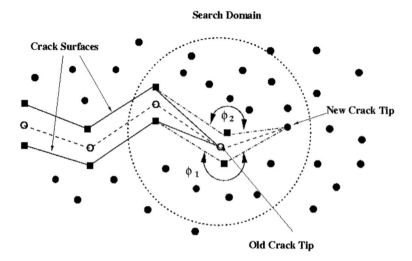

Fig. 5.26. Illustration of Numerical Scheme for Crack Growth.

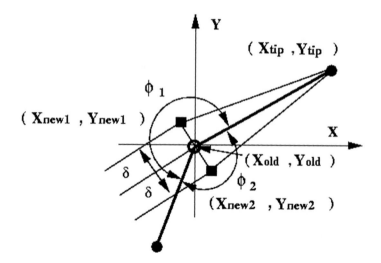

Fig. 5.27. Meshfree particle splitting algorithm.

Then we apply the crack growth criterion to every point inside the circle to decide which point should be the next crack tip, except those points (square boxes) on the crack surfaces, because we do not allow crack surface to become crack tip again (this may happen in some unusual situations).

Once we selected a new crack tip, we split the old crack tip into two points that have the same value of state variables at that particular time. The mass and volume of the two particles are re-assigned according to the following rules, which is called as particle splitting algorithm,

$$Mass_{new1} = \frac{\phi_1}{2\pi} Mass_{old}, \tag{5.122}$$

$$Mass_{new2} = \frac{\phi_2}{2\pi} Mass_{old}; \tag{5.123}$$

and

$$Volume_{new1} = \frac{\phi_1}{2\pi} Volume_{old} \tag{5.124}$$

$$Volume_{new2} = \frac{\phi_2}{2\pi} Volume_{old} \tag{5.125}$$

The kinematic field variables, such as displacements, velocity, and accelerations of the new particles are assigned as

$$\mathbf{Disp}_{new1} = \mathbf{Disp}_{old} + \boldsymbol{\delta} \tag{5.126}$$
$$\mathbf{Disp}_{new2} = \mathbf{Disp}_{old} - \boldsymbol{\delta} \tag{5.127}$$
$$\mathbf{Vel}_{new1} = \mathbf{Vel}_{old} \tag{5.128}$$
$$\mathbf{Vel}_{new2} = \mathbf{Vel}_{old} \tag{5.129}$$
$$\mathbf{Acc}_{new1} = 0.0d0$$
$$\mathbf{Acc}_{new2} = 0.0d0 \tag{5.130}$$

where $\boldsymbol{\delta}$ is vector whose length $|\boldsymbol{\delta}| \ll 1$. It serves the purpose to make a physical distinction of the two new particles once they are separated.

This process is illustrated in Fig. 5.27, in which the point (X_{tip}, Y_{tip}) is the new crack tip, and the old crack tip is split into two particles, (X_{new1}, Y_{new1}) and (X_{new2}, Y_{new2}). A pair of straight lines connect (X_{new1}, Y_{new1}) and (X_{new2}, Y_{new2}) with the new crack tip, (X_{tip}, Y_{tip}).

5.5.3 Parametric Visibility Condition

The meshfree interpolation relies on a local connectivity map to associate one particle with its neighboring particles.

To model crack propagation, one has to develop a numerical algorithm that can automatically modify the local connectivity map and simulate a running crack without user interference.

The following parametric visibility condition is used in the simulation to modify the local meshfree connectivity map to reflect geometric change of domain due to crack growth.

The visibility condition used in this study is illustrated in Fig. 5.28. Figuratively speaking, a crack may be viewed as an opaque wall. A material point at one side of the wall can not "see" the material points in the other side of the wall. This principle is termed as, "visibility condition". To determine whether or not two material points are separated by a crack segment, one can check whether or not the line segment connecting two material points intercept the crack path segment.

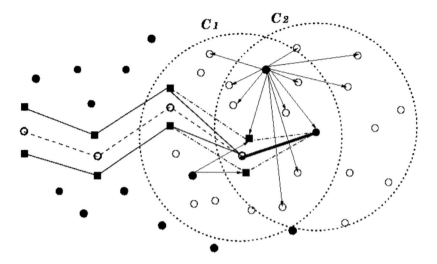

Fig. 5.28. Visibility condition in 2D.

Since crack growth is incremental, one only needs to check and to modify a limited number of particles at current crack tip area, which is defined as the union of two circles centered at the current crack tip and next crack tip (see Fig. 5.28). To modify meshfree connectivity map, one only needs to check the visibility condition inside the union of two circles, denoting as $\mathcal{C} = \mathcal{C}_1 \bigcup \mathcal{C}_2$. It is done by a procedure named "parametric visibility condition".

Suppose that we want to modify connectivity relation between particle $(X_{11}, Y_{11}) \in \mathcal{C}$ and the rest of the particles inside $\mathcal{C}$. We denote an arbitrary point inside $\mathcal{C}$ as (X_{12}, Y_{12}) and two crack tips (old and new) as (X_{21}, Y_{21}) and (X_{22}, Y_{22}).

The parametric equations of the straight line that connects points (X_{11}, Y_{11}) and (X_{12}, Y_{12}) are

$$\begin{cases} X = X_{11} + \lambda_1 \Delta X_1 \\ Y = Y_{11} + \lambda_1 \Delta Y_1 \end{cases} \tag{5.131}$$

where λ_1 is the parametric variable and

$$\Delta X_1 := X_{12} - X_{11} \tag{5.132}$$
$$\Delta Y_1 := Y_{12} - Y_{11} \tag{5.133}$$

On the other hand, the parametric equation for the straight line that connects two crack tips are,

$$\begin{cases} X = X_{21} + \lambda_2 \Delta X_2 \\ Y = Y_{21} + \lambda_2 \Delta Y_2 \end{cases} \tag{5.134}$$

where λ_2 is the parametric variable and

$$\Delta X_2 := X_{22} - X_{21} \tag{5.135}$$
$$\Delta Y_2 := Y_{22} - Y_{21} \tag{5.136}$$

If the two line segments intercept each other, one can equate Eqs. (5.131) and (5.134), and solve for λ_1 and λ_2,

$$\begin{bmatrix} \lambda_1 \\ \lambda_2 \end{bmatrix} = \frac{1}{(\Delta X_1 \Delta Y_2 - \Delta X_2 \Delta Y_1)} \begin{bmatrix} \Delta Y_2 (X_{21} - X_{11}) - \Delta X_2 (Y_{21} - Y_{11}) \\ \Delta Y_1 (X_{21} - X_{11}) - \Delta X_1 (Y_{21} - Y_{11}) \end{bmatrix}$$
$$\tag{5.137}$$

If the two line segments intercept each other, the following **parametric visibility conditions** have to be satisfied,

$$\boxed{0 < \lambda_1 < 1, \quad \text{and} \quad 0 < \lambda_2 < 1 \ .} \tag{5.138}$$

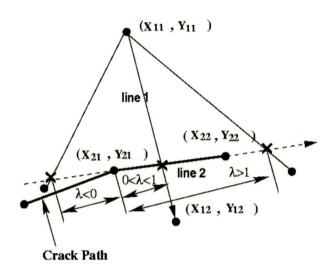

Fig. 5.29. Parametric visibility conditions.

These parametric visibility conditions are illustrated in Fig. 5.29. If both parametric visibility conditions are met, then the line segment between two arbitrary points inside $\mathcal{C}$ will intercept the newly formed crack surfaces and hence one should disconnect the connections between these two points. In other words, either point should be removed from the other point's connectivity map, and it then ensures that there is no non-physical cross-crack interpolation.

In the following, a few artificial examples are shown to display the mesh-free shape functions that are constructed at crack surface via particle splitting algorithm, connectivity modification, and parametric visibility condition.

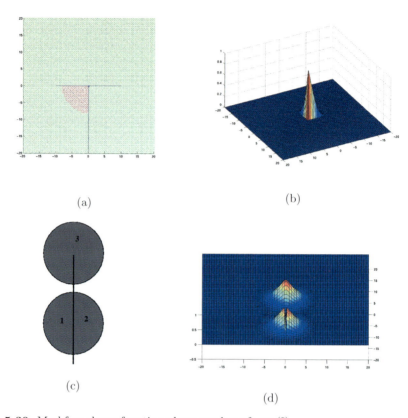

(a)   (b)

(c)

(d)

Fig. 5.30. Meshfree shape function along crack surfaces (I).

In Fig. 5.30 (a) and (b), a meshfree shape function whose support has been cut up to 3/4 by two orthogonally running cracks. In Fig. 5.30 (c) and (d), it is shown that a meshfree function has been cut by a crack into two parts and another meshfree shape is right at the crack tip.

A meshfree shape function whose support size has been cut by a crack up to 1/4 is shown in Fig. 5.31 (a) and (b). In Fig. 5.31 (c) and (d), a meshfree shape function has been severed into three different shape functions.

As reported by Belytschko et al. [1996], there are some abnormalities about these meshfree shape functions whose supports have been modified by visibility conditions. One of them is the apparent strong discontinuity at certain location of the support.

The fracture criteria that we used is a damage based criterion. Initially, we set up a critical damage value as fracture threshold. In this example, that threshold is chosen as $f_{cr} = 0.12$. At each time step, we evaluate damage value of each particle in the neighborhood of the crack tip (the circle in Fig. 5.26)). Once the damage value of a particle exceeds f_{cr}, we declare the particle as

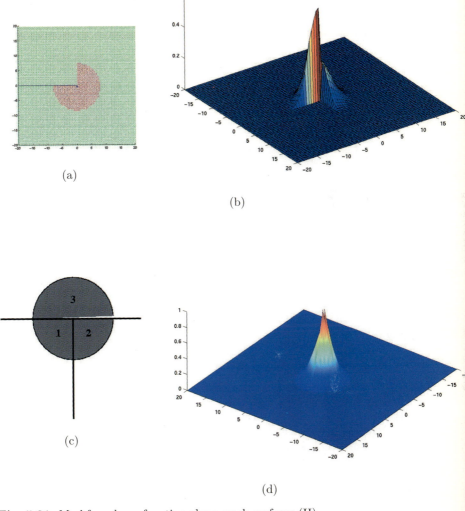

(a)

(b)

(c)

(d)

Fig. 5.31. Meshfree shape function along crack surfaces (II).

the new crack tip. If damage value of two and more particles exceeds f_{cr} at the same time step, there may be the sign to signal crack bifurcation.

In the current simulation, we simply choose the particle that has the largest damage value among all the other particles whose damage value exceeds f_c as the new crack tip.

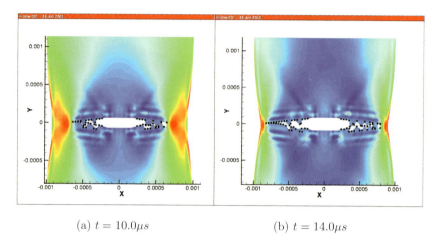

(a) $t = 10.0\mu s$ (b) $t = 14.0\mu s$

Fig. 5.32. Crack surface morphology: σ_{22} contours.

Zooming in the crack region, we can observe crack surface morphology. Fig. 5.32 shows stress distribution, σ_{22}, around a crack region. It is a close snap-shot of crack configurations at two different time instance.

A careful observation of Fig. 5.32 reveals some important features of ductile fracture. First, there is growing blue region that indicates a growing region with small normal stress value, which is an indication of the growth of the traction-free crack surfaces. This fact proves that the crack surfaces constructed by automatically adjusting meshfree interpolation field are indeed traction-free, $\sigma_{22} = 0$, and it provides the right physics around the propagating crack tip. Second, the ductile crack surface shows a zig-zag pattern. This zig-zag pattern of rough crack surface that is the trademark of ductile fracture (see Xia et al. [1995abc]). To the best of the authors' knowledge, such unique feature of ductile fracture has been difficult to capture in previous numerical simulations.

5.5.4 Reproducing Enrichment Technique

Fig. 5.33 displays the profile of a meshfree shape function right in front of a crack tip. One may observe from Fig. 5.33 (b) that there is discontinuity at the back neck of the shape function. One may wonder why this happens,

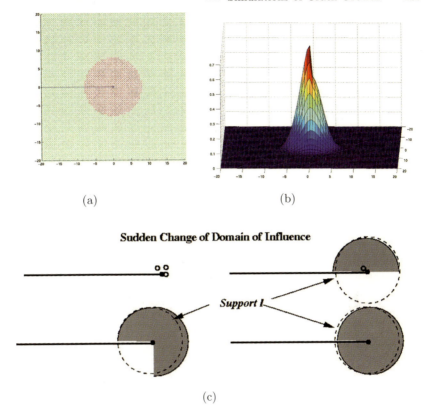

(a)   (b)

(c)

Fig. 5.33. Strong discontinuity of a meshfree shape function at a crack tip.

because based on visibility condition, the support of this meshfree shape function has not changed at all.

As a matter of fact, the visibility condition not only changes the connectivity relations among particles, but also changes the connectivity relations among any material points in the domain (e.g. Gauss quadrature points) and meshfree particles. More precisely speaking, visibility condition is also used to change the domain of influence of any material point in the neighborhood of a crack. The connective domain of influence for different material points inside the support of the meshfree shape function, where the meshfree shape function is evaluated, has changed. Those change may not be continuous as a material point moves toward the crack tip. This situation is illustrated in Fig. 5.33 (c). Shown by Fig. 5.33 (c), as a material point approaches to the crack tip, the domain of influence of a material point suddenly changes from a half circle to 3/4 of a circle, and a slight move towards right or down, the domain of influence of a material point will become almost a full circle except the crack line. This sudden change of domain of influence may be the source of strong discontinuity that appears in the profile of the meshfree shape function.

Some believed that such discontinuity in the meshfree shape function may affect the performance of meshfree shape functions and hence affect the accuracy of the crack tip interpolation field. In fact, Belytschko and his co-workers were developing other methods, for instance the so-called diffraction method, to avoid having discontinuous meshfree shape functions near the crack tip region.

Nonetheless, no definitive evidence has found to link the discontinuity of meshfree shape function with poor interpolation accuracy. It is still an open question to assess the effect of such discontinuity of meshfree shape functions, since the completeness of meshfree interpolation near the crack tip is not affected by such discontinuity. Therefore, in our numerical simulations, close to a crack tip, the meshfree shape function with discontinuity are used without any further modifications.

5.6 Meshfree Contact Algorithm

5.6.1 Contact Detection Algorithm

Before describing the meshfree contact dection algorithm, it would be expedient to recapitulate a moving least square interpolant based meshfree method —-reproducing kernel particle method[303]

$$\mathcal{R}u(x) = \int_\Omega K(y-x)u(y)d\Omega \approx \sum_{I=1}^{NP} K(x_I - x)u_I \Delta x_I \tag{5.139}$$

The kernel function is expressed as

$$\mathcal{K}_I(x) = \mathbf{P}(\frac{x_I - x}{\rho})\boldsymbol{b}(x)\phi_\rho(x_I - x)\Delta V_I \tag{5.140}$$

where $\phi(x)$ is a given window function, which is usually compactly supported, and positive; $\boldsymbol{b}(x)$ is an unknown vector function, and

$$\mathbf{P}(x) := \{P_1(x), P_2(x), P_3(x), \cdots, P_m(x)\} \tag{5.141}$$

is the given polynomial basis, and $P_i(x)$ are monomial functions. To determine the unknown vector function, $\boldsymbol{b}(x)$, one has to solve the following moment equations

$$\mathbf{M}(x)\boldsymbol{b}(x) = \mathbf{1}, \quad \mathbf{1} = \{1, 0, \cdots, 0\}^T \tag{5.142}$$

where the moment matrix is defined as

$$\mathbf{M}(x) := \int_\Omega \mathbf{P}^T(\frac{y-x}{\rho})\phi(\frac{y-x}{\rho})\mathbf{P}(\frac{y-x}{\rho})d\Omega_y \tag{5.143}$$

and

$$\mathbf{M}^h(x) \approx \sum_{I=1}^{NP} \mathbf{P}^T(\frac{x_I - x}{\rho})\phi(\frac{x_I - x}{\rho})\mathbf{P}(\frac{x_I - x}{\rho})\Delta V_I \tag{5.144}$$

Define

$$<f, g> := \int_\Omega f(y-x)g(y-x)\phi_\rho(y-x)d\Omega_y \tag{5.145}$$

$$<f, g>^h := \sum_{I=1}^{NP} f(x_I - x)\phi_\rho(x_I - x)g(x_I - x)\Delta V_I \tag{5.146}$$

Note that $<,>^h$ is not automatically an inner product, unless certain conditions on discretization are met. Eq. (5.143) can be rewritten as

$$\mathbf{M}(x) := \begin{pmatrix} <P_1,P_1>, & <P_1,P_2>, & \cdots & <P_1,P_m> \\ <P_2,P_1>, & <P_2,P_2>, & \cdots & <P_2,P_m> \\ \vdots & & \ddots & \vdots \\ <P_m,P_1>, & <P_m,P_2>, & \cdots & <P_m,P_m> \end{pmatrix}. \tag{5.147}$$

and its discrete counterpart (5.144) as

$$\mathbf{M}^h(x) := \begin{pmatrix} <P_1,P_1>^h, & <P_1,P_2>^h, & \cdots & <P_1,P_m>^h \\ <P_2,P_1>^h, & <P_2,P_2>^h, & \cdots & <P_2,P_m>^h \\ \vdots & & \ddots & \vdots \\ <P_m,P_1>^h, & <P_m,P_2>^h, & \cdots & <P_m,P_m>^h \end{pmatrix}. \tag{5.148}$$

Clearly the moment matrix (5.143) is a *Gram matrix*. Since ϕ is positive and compactly supported, for a given x, if $meas\{supp\{\phi_\rho(y-x)\} \cap \Omega\} \neq 0$. The moment matrix should be positive definite, i.e. $det\{\mathbf{M}\} > 0$. Nevertheless, if $meas\{supp\{\phi_\rho(y-x)\} \cap \Omega\} = 0$, then consequently

$$\int_\Omega P_I(\frac{y-x}{\rho})\phi_\rho(y-x)P_J(\frac{y-x}{\rho})d\Omega = 0 \tag{5.149}$$

the moment matrix will lose its positive definiteness. Mathematically speaking, the determinant of the moment matrix will become zero.

To quantify the concept, it can be stated as $\forall\, x \notin \Omega$

$$meas\{supp[\phi_\rho(y-x)] \cap \Omega\} \to 0, \quad \rho \to 0 \tag{5.150}$$

$$\Rightarrow \quad det\{\mathbf{M}\} \to 0 \tag{5.151}$$

On the other hand, $\forall\, x \in \Omega$

$$meas\{supp[\phi_\rho(y-x)] \cap \Omega\} > 0, \quad \text{if } \rho \neq 0 \tag{5.152}$$

$$\Rightarrow \quad det\{\mathbf{M}\} > 0 \tag{5.153}$$

In the discrete case, things have changed slightly, to recover the above property, we need the following notion of *admissible meshfree discretization*.

Definition 5.6.1. *(Admissible meshfree discretization) (Liu, Li, & Belytschko[303])*
Given a positive window function, $\phi(x)$, and a set of independent functions, $\mathcal{P} = \{1, P_2, P_3, \cdots, P_m\}$. An admissible meshfree discretization satisfies the following conditions:
(1) Every particle of the distribution associates with a compact support

$$S_I := \{|\,X - X_I\,| \leq R\} \tag{5.154}$$

and the union of all the compact support, S_I, generates a covering for the domain $\bar{\Omega}$

$$\bar{\Omega} \subset S := \cup_{I=1}^{NP} S_I \tag{5.155}$$

(2) $\forall\ X \in \bar{\Omega},\ \exists k > 0,$

$$\mathbf{X} \in \cap_{J=1}^{k} supp\{\phi(X - X_J)\}\ , \tag{5.156}$$

where $N_{min} \le k \le N_{max}$ and N_{min}, N_{max} are given;
(3) The particle distribution should be non-degenerated.

We have the following crucial theorem for admissible particle distribution:

Theorem 5.6.1. *A necessary condition for an admissible, non-degenerated particle distribution is that $\forall\ X\ \in\ \bar{\Omega} \subset \mathbb{R}^n$,*

$$\mathbf{X} \in \cap_{J=1}^{k} supp\{\phi(X - X_J)\}\ , \tag{5.157}$$

such that $\max\{m, n+1\} \le N_{min} \le k$, where m is the order of the polynomial basis, and n is the spatial dimension of the domain Ω.

The following proof is partially due to Duarte & Oden[153]
   Proof: Let's show the first part of statement, i.e. $m \le N_{min} \le k$. The discrete moment matrix can be written as

$$\mathbf{M}^h(x) = \begin{pmatrix} M_{11} & M_{12} & \cdots & M_{1m} \\ M_{21} & M_{22} & & M_{2m} \\ \vdots & & \ddots & \vdots \\ M_{m1} & M_{m2} & \cdots & M_{mm} \end{pmatrix} \tag{5.158}$$

For fixed $x \in \Omega$, we assume that it is covered by only k compact supports of particles, numbering in the order $(\alpha_1, \alpha_2, \cdots, \alpha_k)$

$$M_{IJ}(x) = \sum_{\ell=\alpha_1}^{\alpha_k} P_I\left(\frac{x_\ell - x}{\rho}\right)\phi_\rho(x_\ell - x)P_J\left(\frac{x_\ell - x}{\rho}\right)\Delta V_\ell \tag{5.159}$$

Denote $P_I(x_{\alpha_i}) := P_I\left(\frac{x_{\alpha_i} - x}{\rho}\right)$. Then the moment matrix $\mathbf{M}(x)$ can be rewritten as

$$\mathbf{M}^h = \mathbf{P}^h \cdot \Phi \cdot \mathbf{P}^{h^T} = \begin{pmatrix} P_1(x_{\alpha_1}) & P_1(x_{\alpha_2}) & \cdots & P_1(x_{\alpha_k}) \\ P_2(x_{\alpha_1}) & P_2(x_{\alpha_2}) & \cdots & P_2(x_{\alpha_k}) \\ \vdots & \vdots & \ddots & \vdots \\ P_m(x_{\alpha_1}) & P_m(x_{\alpha_2}) & \cdots & P_m(x_{\alpha_k}) \end{pmatrix} \cdot$$

$$\begin{pmatrix} \phi_\rho(x_{\alpha_1} - x) & 0 & \cdots & 0 \\ 0 & \phi_\rho(x_{\alpha_2} - x) & \cdots & 0 \\ \vdots & \vdots & \ddots & \vdots \\ 0 & 0 & \cdots & \phi_\rho(x_{\alpha_k} - x) \end{pmatrix} \cdot \begin{pmatrix} P_1(x_{\alpha_1}) & P_2(x_{\alpha_1}) & \cdots & P_m(x_{\alpha_1}) \\ P_1(x_{\alpha_2}) & P_2(x_{\alpha_2}) & \cdots & P_m(x_{\alpha_2}) \\ \vdots & \vdots & \ddots & \vdots \\ P_1(x_{\alpha_k}) & P_2(x_{\alpha_k}) & \cdots & P_m(x_{\alpha_k}) \end{pmatrix}$$

$$\tag{5.160}$$

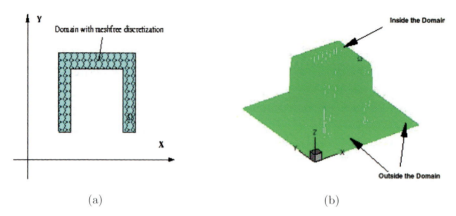

Fig. 5.34. The profiles of determinant of the moment matrix.

From Eq. (5.160), one can find that the rank of $\mathbf{F}$ is most at k, and hence the rank of the moment matrix. Then the positive definiteness of the moment matrix, $\mathbf{M}^h > 0$, requires $k \geq m$.

In the second part of the proof, we show that $k \geq n + 1$ is a necessary condition through an example. Let $\bar{\Omega} \subset \mathbb{R}^2$ and $\mathbf{P} = \{1, xy\}$, and some points, $(x, y) \in \Omega$, are covered by compact supports of only two particles, i.e. $k = 2$. It is obvious that $k \geq m$ but $k < n + 1$. Here $\mathbf{F}$ matrix reads as

$$\mathbf{F} = \begin{pmatrix} 1 & 1 \\ (x_1 - x)(y_1 - y) & (x_2 - x)(y_2 - y) \end{pmatrix} \tag{5.161}$$

Choose $x = \dfrac{1}{2}(x_1 + x_2)$ and $y = \dfrac{1}{2}(y_1 + y_2)$. Then

$$\mathbf{F} = \begin{pmatrix} 1 & 1 \\ \dfrac{1}{4}(x_1 - x_2)(y_1 - y_2) & \dfrac{1}{4}(x_1 - x_2)(y_1 - y_2) \end{pmatrix} \tag{5.162}$$

and the rank of $\mathbf{F}$ reduces to 1. Consequently, the resulting moment matrix will be singular. In fact, in 2-D case, each point in the domain must be covered by compact supports of, at least, three particles, which should form a non-degenerated triangle (i.e. the triangle must have non-zero value of area). Likewise, in 3-D case, each point in the domain should be in the domain of influence of, at least, four distinct particles, which form a non-degenerate tetrahedral volume.   ♣

The above theorem states a fact that any spatial point that is close to the domain of an admissible meshfree discretization the determinant of the moment matrix at that point will have a positive, finite value, because the moment matrix is positive definite in the domain that yields an admissible meshfree discretization. We may call the set of all such spatial point as the *effective domain* of the admissible particle distribution, which could be slightly larger than the exact domain; however, the difference between the effective domain and the exact domain disappear when we refine the particle distribution, or in other words, decrease the support size of the window function[4]

While a point is away from the effective domain, the determinant of the moment matrix ceases to be a finite positive number, i.e. one can not find enough number of particles inside the domain to cover those outside point such that the above necessary condition will soon fail, and the determinant of the moment matrix will approach to zero, which then provide a natural criterion to mark the interior point and the exterior point of any domain under an admissible meshfree discretization. This fact is reflected in the following proposition.

Proposition 5.6.1. *For a given admissible meshfree discretization in the effective domain $\Omega \in \mathbb{R}^n$, if a spatial point, X ($X \notin \Omega$), is sufficiently away from the effective domain Ω, the determinant of the moment matrix at point, X : (X_1, X_2, X_3), will approach to zero; i.e. for given $\delta > 0$ $\exists \epsilon \geq 0$, such that if $dist\{\bar{\Omega}, X\} > \delta$*

$$det\{\mathbf{M}(X)\} < \epsilon, \qquad X = (X_1, X_2, X_3) \tag{5.163}$$

The proposition is evident from Theorem 2.1. As a matter of fact, if one chooses $\delta = \rho$, the average radius of all compact supports, then $dist\{\{\bar{\Omega}\}, X\} > \delta$ leads to $det\{\mathbf{M}\} \rightarrow 0$ immediately. Between Ω and the region that $\{X \mid dist\{\bar{\Omega}, X\} \geq \rho\}$ there is buffer zone, $\{X \mid 0 < dist\{\bar{\Omega}, X\} < \rho\}$, surround the region Ω. Since $det\{\mathbf{M}(x)\}$ is a continuous function in space, its values in the buffet zone could be small positive numbers, nevertheless the closer to Ω the bigger value $det\{\mathbf{M}(x)\}$ attains. Thus, the point in the buffer zone and the point inside the domain can be discriminated by setting a small threshold value. Nevertheless *"the buffer zone"* shrinks as the radius of average compact support decreases.

This intrinsic property of moving least square based meshfree interpolant is illustrated in Fig. 5.34(a) for one dimensional case and Fig. 5.34(b) for two dimensional case. In Fig. 5.34 (b), the value of determinant of the moment matrix computed from meshfree discretization of a concave domain is displayed. One may find that the distribution of the value of determinant of the moment matrix can represent the geometry accurately. Based on this property, we can develop a meshfree contact detection algorithm by a single scalar criterion, which can be used in checking inter-penetration of two different effective domains, as well as two distinct parts of one effective domain.

---

[4] When the density of a particle distribution increases, we always assume that at the same time the support size of the window function decreases, such that the number of the particles inside a compact support, or inside the domain of influence remain approximately the same as in the beginning.

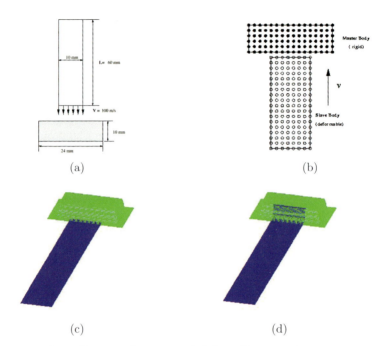

(a)   (b)

(c)   (d)

Fig. 5.35. The collision and contact of deformable solid bar with rigid target: (a) problem statement; (b) admissible particle discretizations in different domain; (c) determinant of the moment matrix in master domain before penetration; (d) determinant of the moment matrix in master domain after penetration.

To illustrate the point, consider a simple Taylor bar impact problem, i.e. a deformable solid bar collides with rigid target as shown in Fig. 5.35(a). Fig. 5.35(b) shows two admissible particle distributions in rigid target (master body) and in deformable Taylor bar (slave body) respectively. By computing the determinant of the moment matrix in master domain for both master body and slave body (slave body is moving in the case), inter-penetration of the two bodies can be easily detected. Fig. 5.35(c) shows that before the Taylor bar collides with the rigid target, the determinant of the slave body in master domain is either zero, or close to zero, whereas the master body itself has an almost constant value of determinant of moment matrix with respect to the master domain particle distribution. After the penetration occurs, one can see from Fig. 5.35(d) that for the contacted slave nodes the determinant of the moment matrix of the master domain is on longer zero, and it has a finite, positive value, which indicates the occurrence of impact. In computation, one can set a proper tolerance to signal such occurrence of inter-penetration of two objects. The detailed computation results for the Taylor bar problem is shown at following.

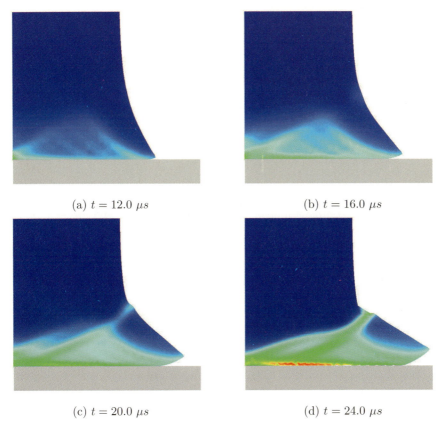

(a) $t = 12.0\ \mu s$ (b) $t = 16.0\ \mu s$

(c) $t = 20.0\ \mu s$ (d) $t = 24.0\ \mu s$

Fig. 5.36. The impact sequence of the Taylor bar via RKPM.

5.6.2 Examples of Contact Simulations

To validate the contact detection algorithm, numerical computations have been carried out testing its viability in computation. We consider the impact problem of the Taylor bar described above. The exact problem statement is described in Fig. 5.35(a). The projectile is a cylindrical rod with radius, $R_s = 5mm$, and height, $L = 60mm$ and the rigid block has radius $R_m = 12mm$ and $10mm$ in height. The material's constitutive model is chosen as the viscoplastic solid described in Section 4. The following material constants are listed in Table (7.5.1) Since the problem is axi-symmetric, only half of the cylinder is modeled in actual computation. Two sets of computations have been carried out: one via meshfree method and the other via finite element method (FEM). In meshfree computation, a total of $11,191$ particles are used in discretization of the Taylor bar, and $1,661$ particles are used to form the meshfree discretization of the rigid target. In time integration, the time step size is chosen as $\Delta t = 2.5 \times 10^{-9} sec$. The dilation parameters

Mass Density	$\rho = 7,800 Kg/m^3$
Young's Modulus	$E = 2,110 MPa$
Poisson's Ratio	$\mu = 0.3$
Yield Stress	$\sigma_0 = 460 MPa$
Hardening Index	$N = 0.1$
Hardening Index	$m = 0.01$
Material Constant	$\epsilon_0 = 2.18 \times 10^{-3}$
Material Constant	$\epsilon_1 = 100\epsilon_0$
Material Constant	$\dot{\epsilon}_0 = 0.0021/sec$

Table 5.2. The List of Material Constants.

in slave body are selected as $\rho_x = 2.24\Delta x$ $\rho_y = 2.24\Delta y$ and $\Delta x = \Delta y = 0.166667mm$; accordingly in master body, they are $\rho_x = 2.24\Delta x$ $\rho_y = 2.24\Delta y$ $\Delta x = 0.08mm$ $\Delta y = 1.0mm$. Coulomb friction coefficient is chosen as $\mu = 0.1$, and the threshold value for contact detection is set at $\epsilon_{cr} = 5 \times 10^{-10}$. A sequence of snap shots are taken from the numerical results at a different time instance, and shown in Figs 5.36 (a)-(d).

The same meshfree contact detection algorithm can be also used in finite element computation as well by assuming the existing finite element grid is a valid particle distribution. In this particular example, we only need to check the intrusion of projectile into the rigid target; therefore, the projectile, i.e. the Taylor bar is discretized via finite element mesh, and there is a meshfree particle distribution on rigid target. In general, one can construct moment matrix upon the finite element mesh by assign appropriate dilation parameter for the nodal distribution.

In the finite element computation, the classical CST4 element are used in simulation, i.e. a quadrilateral box consisting of four diagonally crossed triangle elements. A total of $21,991$ nodes and $43,200$ elements are used in computation (there are $10,800$ quadrilateral boxes). The time step increment is chosen to be $\Delta t = 6.25 \times 10^{-10}s$ due to the small size of the element. Similar deformation patterns to that of meshfree computation (Fig. 5.36) are observed (See Fig. 5.37). One may observe from both Figs 5.36 and 5.37 that there is a mushroom region at the bottom of the Taylor bar, and there is a visible, cup shape shear band formation across the radius direction of the Taylor bar, which is due to plastic deformation during the high speed impact. Furthermore, numerical simulations in both meshfree and finite element computations predict the separation of the projectile and the rigid target at the edge of the Taylor bar in the contact region, which is an indication of excellent performance of the contact detection algorithm. The results reported here are consistent with the numerical results reported early by Batra & Stevens,[32] or by Benson[54]) in finite element computations, in which the conventional contact detection algorithm, such as Benson-Hallquist algorithm is used.

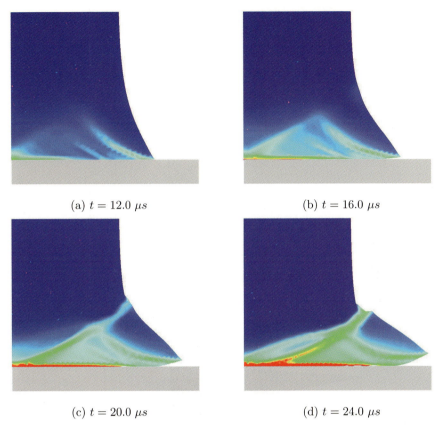

(a) $t = 12.0\ \mu s$ (b) $t = 16.0\ \mu s$

(c) $t = 20.0\ \mu s$ (d) $t = 24.0\ \mu s$

Fig. 5.37. The impact sequence of the Taylor bar via FEM.

5.7 Meshfree Simulations of Fluid Dynamics

5.7.1 Meshfree Stabilization Method

In this section, the synchronized reproducing kernel interpolant constructed from the combination of wavelet kernels is used as a weighting function in a Petrov-Galerkin formulation to compute some typical pathological problems in numerical computations.

Advection-diffusion Equations. We consider the following multi-dimensional advection-diffusion equation described as follows,

$$\mathcal{L}\varphi = -(\kappa_{ij}\varphi_{,j})_{,i} + u_i\varphi_{,i} = f\ ,\qquad \forall\ \boldsymbol{x} \in \Omega \tag{5.164}$$

$$\varphi = g\ ,\qquad \forall\ \boldsymbol{x} \in \Gamma_g \tag{5.165}$$

$$n_i\kappa_{ij}\varphi_{,j} = h\ ,\qquad \forall\ \boldsymbol{x} \in \Gamma_h\ , \tag{5.166}$$

where $\{\kappa_{ij}\}$ is the diffusivity, and $\{u_i\}$ is the given velocity of the flow field.

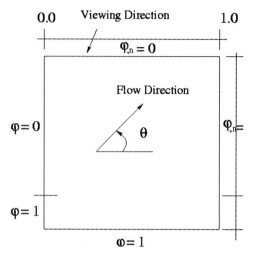

Fig. 5.38. Advection skew to the "mesh": Problem statement.

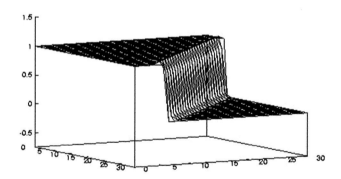

Fig. 5.39. Numerical results for advection skew to the "mesh" via wavelet Petrov-Galerkin.

Define

$$(f, g) := \int_{\Omega} fg d\Omega \qquad (5.167)$$

$$\mathbf{B}(w, \varphi) := \int_{\Omega} \left(wu_{,i}\varphi_{,i} + w_{,i}\kappa_{ij}\varphi_{,j} \right) d\Omega \qquad (5.168)$$

$$\mathbf{L}(w) := \int_{\Omega} wf d\Omega + \int_{\Gamma_h} whd\Gamma_h \ , \qquad (5.169)$$

Let

$$\varphi^h(\boldsymbol{x}) := \sum_{I \in \Lambda} \varphi_I^h \Psi_I^{[00]}(\boldsymbol{x}) \qquad (5.170)$$

$$w^h(\boldsymbol{x}) := \sum_{I \in A} c_I \Psi_I^{[00]}(\boldsymbol{x}) \qquad (5.171)$$

$$\bar{w}^h(\boldsymbol{x}) := w^h(\boldsymbol{x}) + \tau \tilde{w}^h(\boldsymbol{x}) \qquad (5.172)$$

where τ is a stability control parameter and is $\mathcal{O}(1)$.
For $|\beta| = 1$ $(\beta = (1, 0), (0, 1))$,

$$\tilde{w}^h(\boldsymbol{x}) := \hat{u}_j \tilde{w}_j^h(\boldsymbol{x}) \ , \qquad (5.173)$$

$$\tilde{w}_j^h(\boldsymbol{x}) := \sum_{I \in \Lambda} c_I \Psi_I^{[\beta_j]}(\boldsymbol{x}) \ , \qquad (5.174)$$

$$\hat{u}_j = \frac{u_j}{\|u\|} \ , \quad \text{and} \quad \|u\|^2 := u_i u_i \qquad (5.175)$$

where $0 \le j \le n$ and $|\beta_j| = 1$.
Then, a consistent weighted residual form is

$$\left(\bar{w}^h, \mathcal{L}\varphi^h - f \right) = 0 \ . \qquad (5.176)$$

Integration by parts yields the weak formulation

$$\mathbf{B}(\bar{w}^h, \varphi^h) - \int_{\Gamma_g} \bar{w}^h n_i \kappa_{ij} \varphi_{,j}^h d\Gamma_g = \mathbf{L}(\bar{w}^h) \ . \qquad (5.177)$$

where $\bar{w}^h$ is a synchronized interpolant. The first numerical test is the so-called advection skew to the "*mesh*". The problem statement is described in Fig. 5.38, and the result is displayed in Fig. 5.39. In Fig. 5.39, a uniform particle distribution, 21×21, is used; no shock capturing term is involved. The results show that the numerical solutions are stable.

The second test is the so-called cosine hill problem — advection in a rotating flow field. The numerical results are shown in Fig. 5.40. Part (a) of Fig. 5.40 displays the profile of the function φ, and the part (b) of the Figure shows the contour of advection-diffusion field φ. The computation is performed on a 30 by 30 particle distribution. In Fig. 5.40, one can observe that there is no phase error caused by numerical instability. In both cases, the diffusive coefficient κ is taken as 10^{-6}. A proof of stability and convergence for using the above wavelet Petrov-Galerkin method in numerical computation of advection-diffusion problem is presented in the next section.

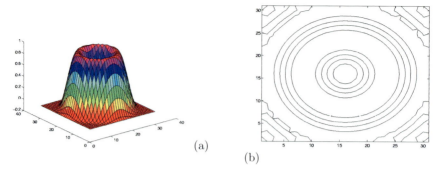

Fig. 5.40. Numerical results for advection in a rotating field; (a) The profile of φ; (b) The contours of φ.

Stokes Flow Problem. The mathematical formulation of the problem is as follows,

$$\nabla \cdot \boldsymbol{\sigma} + \mathbf{f} = \mathbf{0} \tag{5.178}$$
$$\nabla \cdot \mathbf{u} = \mathbf{0} \tag{5.179}$$

where

$$\boldsymbol{\sigma} = -p\mathbf{I} + 2\mu\boldsymbol{\epsilon}, \qquad \boldsymbol{\epsilon} = \frac{1}{2}\left(\nabla\mathbf{u} + (\nabla\mathbf{u})^{t}\right).$$

with the boundary conditions

$$\mathbf{u}(\boldsymbol{x}) = \mathbf{g}(\boldsymbol{x}) \qquad \forall\, \boldsymbol{x} \in \Gamma_{g} \tag{5.180}$$
$$(\boldsymbol{\sigma} \cdot \mathbf{n})(\boldsymbol{x}) = \mathbf{h}(\boldsymbol{x}) \qquad \forall\, \boldsymbol{x} \in \Gamma_{h} \tag{5.181}$$

It is well-known that the velocity based computational formulation of the Stokes flow problem suffers from instability in numerical simulations, if the conventional Bubnov-Galerkin procedure is adopted. An efficient remedy in computational strategy is to use the so-called mixed method. However, in a mixed formulation, there are certain restrictions that have to be met for displacement interpolation and pressure interpolation. The celebrated Babuška-Brezzi condition is such a requirement imposed by stability criterion. For example, it excludes the convenient equal-order interpolation in computations. Hughes, Franca, & Balestra (1986) proposed a Petrov-Galerkin formulation to circumvent the Babuška-Brezzi condition (CBB), such that any consistent interpolation schemes can be employed in computations. Here, following almost the same Petrov-Galerkin formulation, we use the synchronized reproducing kernel interpolant as weighting function in the computation, instead of using the gradient of the pressure trial function as used in Hughes' stabilized FEM formulation.[211] Assume that the set Λ can be decomposed as[5]

---

[5] This is not always possible in multi-dimensional problems; see discussion in [265].

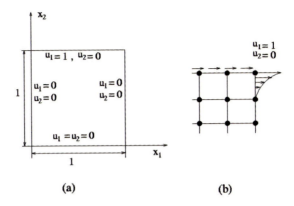

Fig. 5.41. The cavity problem; (a) Problem statement; (b) Detailed boundary conditions at the corner.

$$\Lambda = \overset{\circ}{\Lambda} \oplus \partial\Lambda \tag{5.182}$$

where

$$\overset{\circ}{\Lambda} := \{I \in \Lambda, \ \Psi_I^{[00]}(\boldsymbol{x}) = 0, \ \forall \boldsymbol{x} \in \partial\Omega\} \tag{5.183}$$

$$\partial\Lambda := \{I \in \Lambda, \ \Psi_I^{[00]}(\boldsymbol{x}) \neq 0, \ \forall \ \boldsymbol{x}_I \in \partial\Omega\} \tag{5.184}$$

We choose the following equal-order interpolation for both displacements and pressure:

$$\boldsymbol{u}^h(\boldsymbol{x}) := \sum_{I \in \overset{\circ}{\Lambda}} \boldsymbol{u}_I \Psi_I^{[00]}(\boldsymbol{x}) + \sum_{I \in \partial\Lambda} \boldsymbol{g}_I \Psi_I^{[00]}(\boldsymbol{x}) = \boldsymbol{v}^h(\boldsymbol{x}) + \boldsymbol{g}^h(\boldsymbol{x}) \tag{5.185}$$

$$\boldsymbol{w}^h(\boldsymbol{x}) := \sum_{I \in \overset{\circ}{\Lambda}} \boldsymbol{w}_I \Psi_I^{[00]}(\boldsymbol{x}) \ . \tag{5.186}$$

and

$$p^h(\boldsymbol{x}) := \sum_{I \in \Lambda} p_I \Psi_I^{[00]}(\boldsymbol{x}) \tag{5.187}$$

$$q^h(\boldsymbol{x}) := \sum_{I \in \Lambda} q_I \Psi_I^{[00]}(\boldsymbol{x}) \ . \tag{5.188}$$

For $\beta = 1$, let

$$\tilde{\boldsymbol{q}}^h(\boldsymbol{x}) := \{\tilde{q}_1^h(\boldsymbol{x}), \tilde{q}_2^h(\boldsymbol{x})\} \tag{5.189}$$

$$\tilde{q}_j^h(\boldsymbol{x}) := \frac{\varrho_j}{2\mu} \sum_{I \in \Lambda} \tilde{q}_{Ij} \Psi_I^{[\beta_j]}(\boldsymbol{x}) \ , \quad j = 1, 2; \tag{5.190}$$

where there is no summation on j, and in 2-D

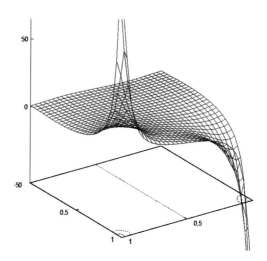

Fig. 5.42. Pressure elevation: Petrov-Galerkin solution.

$$\beta_j = (1,0), (0,1) \ . \tag{5.191}$$

Then, the following CBB type of Petrov-Galerkin weak form is used in the computation,

$$\mathbf{B}_\tau(\mathbf{w}^h, q^h, \tilde{\mathbf{q}}^h; \mathbf{v}^h, p^h) = \mathbf{L}_\tau(\mathbf{w}^h, q^h, \tilde{\mathbf{q}}^h) \tag{5.192}$$

where

$$\mathbf{B}_\tau(\mathbf{w}^h, q^h, \tilde{\mathbf{q}}^h; \mathbf{v}^h, p^h) := \big(\boldsymbol{\epsilon}(\mathbf{w}^h), 2\mu\boldsymbol{\epsilon}(\mathbf{v}^h)\big) - (\nabla \cdot \mathbf{w}^h, p^h) + (q^h, \nabla \cdot \mathbf{v}^h)$$
$$+ \big(\tau\tilde{\mathbf{q}}^h, \nabla p^h - 2\mu\nabla \cdot \boldsymbol{\epsilon}(\mathbf{v}^h)\big) \tag{5.193}$$

$$\mathbf{L}_\tau(\mathbf{w}^h, q^h, \tilde{\mathbf{q}}^h) := \big(\mathbf{w}^h + \tau\tilde{\mathbf{q}}^h, \mathbf{f}\big) + \big(\mathbf{w}^h, h\big)_{\Gamma_h} - \big(\boldsymbol{\epsilon}(\mathbf{w}^h), 2\mu\boldsymbol{\epsilon}(\mathbf{g}^h)\big)$$
$$- \big(q^h, \nabla \cdot \mathbf{g}^h\big) + \big(\tau\tilde{\mathbf{q}}^h, 2\mu\nabla \cdot \boldsymbol{\epsilon}(\mathbf{g}^h)\big) \tag{5.194}$$

where τ is the stability control parameter.

In numerical experiments, the well-known cavity problem has been tested. It is a driven cavity flow problem with "leaky lid" boundary condition. The problem statement is shown in Fig. 5.41. The wavelet reproducing kernel

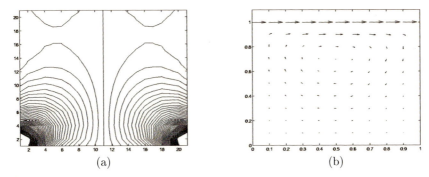

Fig. 5.43. Numerical results of cavity problem: (a) Pressure contours; (b) Velocity field.

interpolants developed in Chapter 3 are employed in the computation. The results presented in Fig. 5.42 and Fig. 5.43 are based on an 11×11 particle distribution on a unit square. In all the computations, 2-D cubic spline is used as the window function, and the dilation vector is chosen as $\varrho_1 = 1 \cdot h_{x_1}$, $\varrho_2 = 1 \cdot h_{x_2}$. Fig. 5.42 shows the pressure profile of the cavity problem. Part (a) is the numerical result obtained from Petrov-Galerkin formulation based on Eq. (5.192). Part (b) is the numerical result obtained from regular Galerkin method, from which, the pressure distribution exhibits apparent spurious pressure mode. In Fig. 5.43, the pressure contour and the velocity field, or the field of streamline, are displayed; both of them are obtained from the Petrov-Galerkin formulation (5.192).

5.7.2 Multiscale Simulation of Fluid Flows

As its name suggested, the RKPM approximation can be interpreted as filering or sampling in a numerical signal process, and the RKPM shape functions built on different particle densities may be regarded as the filers with different spatial resolutions.

Two computational strategies have been developed for meshfree multi-scale computations of fluid flows. The first strategy is the so-called meshfree wavelet method introduced in previous sections. In addition to use "meshfree wavelet" developed in the previous sections, another meshfree multi-scale procedure is often used in computation uitilizing the flexibility of meshfree kernel functions to adjust its dilation parameters.

In principle, a numerical solution with mixed scale information can be decomposed into a low scale solution and the corresponding high scale solution,

$$u^h(\boldsymbol{x}) = u_0(\boldsymbol{x}) \qquad\qquad\qquad \text{the finest level}$$
$$= u_1(\boldsymbol{x}) + w_1(\boldsymbol{x}) \qquad\qquad \text{two} - \text{level decomposition}$$
$$= u_1(\boldsymbol{x}) + w_1(\boldsymbol{x}) + w_2(\boldsymbol{x}) \quad \text{three} - \text{level decomposition}$$

$$\vdots \qquad\qquad \vdots$$

$$= u_m(\boldsymbol{x}) + \sum_{i=1}^{m} w_i(\boldsymbol{x}), \qquad (m+1) - level\ decomposition \quad (5.195)$$

where

$$u^m(\boldsymbol{x}) = \int_{\Omega} \mathcal{K}_{2^m h}(\boldsymbol{x} - \bar{\boldsymbol{x}}) u(\bar{\boldsymbol{x}}) d\Omega_{\bar{\boldsymbol{x}}} \qquad\qquad (5.196)$$

and the high scale solution is

$$w_i(\boldsymbol{x}) = \int_{\Omega} \Psi_i(\boldsymbol{x} - \bar{\boldsymbol{x}}) u(\bar{\boldsymbol{x}}) d\Omega_{\bar{\boldsymbol{x}}} \qquad\qquad (5.197)$$

with

$$\Psi_i(\boldsymbol{x} - \bar{\boldsymbol{x}}) = \mathcal{K}_{2^{i-1} h}(\boldsymbol{x} - \bar{\boldsymbol{x}}) - \mathcal{K}_{2^i h}(\boldsymbol{x} - \bar{\boldsymbol{x}}) \qquad\qquad (5.198)$$

which is often referred to as "the wavelet function" in the engineering sense.

Assume that the meshfree interpolants at both scales can reproduce vector basis function, $\mathbf{P}(\boldsymbol{x})$, i.e.

$$\mathcal{P}_{i-1}\mathbf{P}(\boldsymbol{x}) = \mathbf{P}(\boldsymbol{x}) \qquad\qquad (5.199)$$
$$\mathcal{P}_i\mathbf{P}(\boldsymbol{x}) = \mathbf{P}(\boldsymbol{x}) \qquad\qquad (5.200)$$

and the high-scale operator (wavelet operator) is defined as

$$\mathcal{H}_i = \mathcal{P}_{i-1} - \mathcal{P}_i \qquad\qquad (5.201)$$

Obviously,

$$\mathcal{H}_i\mathbf{P}(\boldsymbol{x}) = \int_{\Omega} \Psi_i(\boldsymbol{x} - \bar{\boldsymbol{x}})\mathbf{P}(\bar{\boldsymbol{x}}) d\Omega_{\bar{\boldsymbol{x}}}$$
$$= \left(\mathcal{P}_{i-1} - \mathcal{P}_i\right)\mathbf{P}(\boldsymbol{x}) = 0 , \qquad\qquad (5.202)$$

which means that operator $\mathcal{H}_i$ is orthogonal to reproduced basic function space.

In Fig. 5.44, a 2D meshfree shape function is decomposed into low scale part and high scale part. The corresponding Fourier transforms of these shape functions are depicted right below their spatial distributions. Note that this method does not need two different particle distributions, or two meshes. The lower scale components are obtained by simply adjusting the dilation parameters of the shape functions.

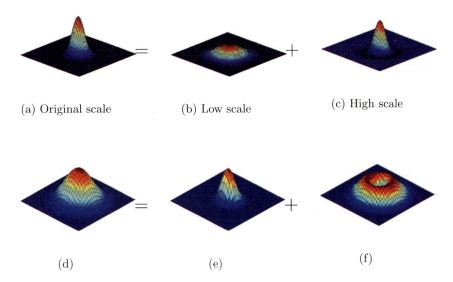

(a) Original scale (b) Low scale (c) High scale

(d) (e) (f)

Fig. 5.44. Decomposition of a 2D meshfree shape function into high scale and lower scale parts, (a), (b), (c); and their Fourier transforms (d), (e), and (f).

This meshfree multi-scale decomposition technique is used in two types of applications. The first type of applications is to use the following multi-scale interpolant in numerical computations. Suppose that there are total NP particles in a coarse scale meshfree discretization. We can construct the following multi-scale interpolation field only based on coarse scale particle distribution,

$$
u^h(\boldsymbol{x}) = \sum_{I=1}^{NP} N_I(\boldsymbol{x}, \rho) u_I + \sum_{I=1}^{NP} \Big(N_I(\boldsymbol{x}, \rho/2) - N_I(\boldsymbol{x}, \rho) \Big) w_I
$$

$$
= \sum_{I=1}^{NP} N_I(\boldsymbol{x}) u_I + \sum_{I=1}^{NP} W_I(\boldsymbol{x}) w_I \tag{5.203}
$$

Comparing this multi-scale method to previous meshfree wavelet method, one may find that this method has its simplicity. However, the dilation parameter can not be too small, otherwise, the fine scale of meshfree shape functions, $\{N_I(\boldsymbol{x}, \rho/2)\}_{I=1}^{NP}$, will not be able to reproduce the polynomial basis $\mathbf{P}(\boldsymbol{x})$.

The second application is the multi-scale post-process treatment. Suppose that we have obtained a computation result with mixed scales. We can use meshfree multi-scale method to decompose or separate the numerical computation result into the coarse scale and the fine scale, i.e.

$$u^h(\boldsymbol{x}) = \sum_{I=1}^{NP} N_I(\boldsymbol{x}, \rho)u_I = \sum_{I=1}^{NP} N_I(\boldsymbol{x}, 2\rho)u_I + \sum_{I=1}^{NP} \Big[N_I(\boldsymbol{x}, \rho) - N_I(\boldsymbol{x}, 2\rho) \Big] u_I$$

$$= \sum_{I=1}^{NP} N_I^{low}(\boldsymbol{x})u_I + \sum_{I=1}^{NP} N_I^{high}(\boldsymbol{x})u_I = u^{low}(\boldsymbol{x}) + u^{high}(\boldsymbol{x}) \quad (5.204)$$

Fig. 5.45 shows a meshfree decomposition of a numerical solution into differ-

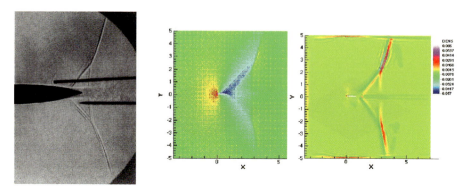

Fig. 5.45. Physical and computed phenomena for transonic flow. The total solution and high-scale solution are shown for comparison to an experimental result.

ent scales, which is a simulation result of a supersonic flow passing through an airfoil.[186] One may observe that the spatial separation of a low scale solution and a high scale solution identifies the shock location and thus aids the understanding of the physical phenomena involved. As can be seen from high-scale solution (far right), the transonic pocket is identified and compares well with the experimental observation (far left).

5.8 Implicit RKPM Formulation

5.8.1 The Governing Equations

We consider here incompressible, Newtonian fluids with constant density ρ and dynamic viscosity μ, flowing in a domain Ω with boundary Γ. The Navier-Stokes equations are written in index form in terms of the velocity u_i and pressure p as follows:

1. Conservation of Mass

$$u_{i,i} = 0 \text{ in } \Omega, \quad (5.205)$$

2. Conservation of Momentum

$$\rho u_{i,t} + \rho u_j u_{i,j} = \sigma_{ij,j} + \rho b_i. \qquad (5.206)$$

where b_i is the ith body force component. The Cauchy stress tensor σ_{ij} is defined as:

$$\sigma_{ij} = -p\delta_{ij} + 2\mu e_{ij},$$
$$e_{ij} = \frac{1}{2}(u_{i,j} + u_{j,i}) \text{ in } \Omega \qquad (5.207)$$

3. Boundary Conditions
   In addition to these equations of motion, we stipulate the following boundary conditions:

$$u_i = g_i, \ x \in \Gamma_{g_i} \qquad (5.208a)$$
$$\sigma_{ij} n_j = h_i, \ x \in \Gamma_{h_i} \qquad (5.208b)$$

where g_i and h_i are given functions. The fluid boundary Γ is partitioned into Γ_{g_i} and Γ_{h_i} such that for each degree of freedom i, $\Gamma = \Gamma_{g_i} \bigcup \Gamma_{h_i}$ while $\Gamma_{g_i} \bigcap \Gamma_{h_i} = 0$ (i.e. the Dirichlet and Neumann condition boundaries span the entire fluid surface but do not intersect).

The transformation of the strong form of the conservation equations (5.205) and (5.206) into weak forms suitable for solution with RKPM requires two test functions: the velocity test function δu_i and the pressure test function δp. The velocity test function δu_i satisfies the homogeneous boundary condition:

$$\delta u_i = 0, \ x \in \Gamma_{g_i} \qquad (5.209)$$

The standard Galerkin method for advection-dominated flows of incompressible fluids leads to undesirable oscillations in the velocity and pressure solutions. To reduce or eliminate these oscillations, the test functions are augmented with stabilization terms ([205][211]):

$$\delta \tilde{v}_i = \delta u_i + \tau^m u_k \delta v_{i,k} + \tau^c \delta p_{,i} \qquad (5.210a)$$
$$\delta \tilde{p} = \delta p + \tau^c \delta u_{i,i} \qquad (5.210b)$$

where τ^m and τ^c are scalar stabilization parameters that are in general functions of the computational grid, the time step size, and the flow variables.[210]

The weak form of the continuity equation (5.205) is obtained by multiplying by the pressure test function $\delta \tilde{p}$ and integrating over Ω:

$$\int_\Omega (\delta p + \tau^c \delta u_{j,j}) u_{i,i} d\Omega = 0 \qquad (5.211)$$

Similarly, the weak form of the momentum equation (5.206) is derived by multiplying by the velocity test function $\delta \tilde{v}_i$:

$$\int_{\Omega} (\delta u_i + \tau^m u_k \delta u_{i,k} + \tau^c \delta p_{,i})(\rho u_{i,t} + \rho u_j u_{i,j} - \sigma_{ij,j} - \rho b_i)d\Omega = 0 \quad (5.212)$$

By rearranging (5.212) and using integration by parts, the final weak form of the momentum equation becomes:

$$\int_{\Omega} \rho(\delta u_i + \tau^m u_k \delta u_{i,k} + \tau^c \delta p_{,i})(\dot{u}_i + u_j u_{i,j})d\Omega - \int_{\Gamma_{h_i}} \delta v_i h_i d\Gamma_{h_i}$$

$$- \int_{\Gamma_g} \delta v_i \sigma_{ij} n_j d\Gamma_g + \int_{\Omega} \delta u_{i,j} \sigma_{ij} d\Omega + \int_{\Omega} (\tau^m u_k \delta v_{i,k} + \tau^c \delta p_{,i}) p_{,i}$$

$$- \int_{\Omega} (\tau^m u_k \delta v_{i,k} + \tau^c \delta p_{,i})\mu(u_{i,jj} + u_{j,ij})d\Omega = 0 \quad (5.213)$$

Note that the last integral on the left hand side requires the calculation of the second derivative of the velocity, $u_{i,jj}$. For linear finite elements, the term is either equal to zero or approximately zero on element interiors, depending on the grid regularity; $u_{i,jj}$ is a delta function on element boundaries. For RKPM using cubic splines, $u_{i,jj}$ is a smooth function, and so this final term can in theory be computed. In practice, however, we neglect this term due to the cost of the computations and memory space required to calculate and store the second derivatives of the RKPM shape functions.[445] The neglect of these second-derivative stabilization terms is common practice in finite element computations and has only a small effect on accuracy for most problems (see for example[429]). One reason for this is that the stabilization parameters τ^m and τ^c are designed to go to zero with decreasing nodal spacing; in many regions of the flow where second derivatives are large, such as the boundary layer, the nodal spacing and therefore the stabilization term are small.

The final weak form that will be solved is the combination of the momentum equation (5.211) and the continuity equation (5.213):

$$\int_{\Omega} \rho(\delta u_i + \tau^m u_k \delta u_{i,k} + \tau^c \delta p_{,i})(\dot{u}_i + u_j u_{i,j})d\Omega - \int_{\Gamma_{h_i}} \delta v_i h_i d\Gamma_{h_i}$$

$$- \int_{\Gamma_g} \delta v_i \sigma_{ij} n_j d\Gamma_g + \int_{\Omega} \delta u_{i,j} \sigma_{ij} d\Omega + \int_{\Omega} (\tau^m u_k \delta v_{i,k} + \tau^c \delta p_{,i}) p_{,i} d\Omega$$

$$+ \int_{\Omega} (\delta p + \tau^c \delta u_{j,j}) u_{i,i} d\Omega = 0 \quad (5.214)$$

5.8.2 Essential Boundary Conditions

In order to efficiently enforce essential boundary conditions despite this inconvenience, we begin by decomposing the total solution u^h into two functions:

$$u^h(x) = g^h_{bound}(x) + v^h(x). \quad (5.215)$$

where $g^h_{bound}(x)$ is a known function that is related to (to be defined) the prescribed essential boundary condition $u(x) = g(x)$ on Γ_g, while $v^h(x)$ should depend on the nodal degrees of freedom u_I but should go to zero at nodes on the essential boundary.

The simplest way to construct $g^h_{bound}(x)$ is to interpolate using a set of shape functions that satisfy the Kronecker delta property on the essential boundary, e.g. the linear finite element shape functions. Because only the shape functions associated with the essential boundary nodes are needed, we define B as the set of nodes on Γ_g and write:

$$g^h_{bound}(x) = \sum_{I \in B} N_I(x) g_I \tag{5.216}$$

where $N_I(x)$ is the finite element shape function at node I and $g_I = g(x_I)$.

In order to construct $v^h(x)$, we first note that equation (5.216) is a projection of $g(x)$ onto the set of finite element shape functions at the boundary. We will therefore subtract from the RKPM approximation,

$$u^h(\mathbf{x}) = \sum_I \Phi_I(\mathbf{x}) u_I, \tag{5.217}$$

its own projection onto the same set of shape functions, giving as desired a $v^h(x)$ that is zero on the boundary nodes. Temporarily defining $w^h(x) \equiv \sum_{I \in A} \Phi_I(x) u(x_I)$, where A is the set of all nodes (including those on the boundary):

$$
\begin{aligned}
v^h(x) &= w^h(x) - \sum_{I \in B} N_I(x) w^h(x_I) \\
&= \sum_{I \in A} \Phi_I(x) u_I - \sum_{I \in B} N_I(x) \left[\sum_{J \in A} \Phi_J(x_I) u_J \right] \\
&= \sum_{I \in A} \left[\Phi_I(x) - \sum_{J \in B} N_J(x) \Phi_I(x_J) \right] u_I
\end{aligned}
\tag{5.218}
$$

The projection of the RKPM approximation on the set of boundary finite element shape functions is the "bridging scale" referred to in the name of the method.[443]

Substituting (5.216) and (5.218) into (5.215):

$$u^h(x) = \sum_{I \in B} N_I(x) g_I + \sum_{I \in A} \tilde{\Phi}_I(x) u_I \tag{5.219}$$

where

$$\tilde{\Phi}_I(x) \equiv \Phi_I(x) - \sum_{J \in B} N_J(x) \Phi_I(x_J) \tag{5.220}$$

It can be easily shown that for a node K on the essential boundary, $u^h(x_K) = g_K$. Thus the essential boundary conditions can be applied directly through the coefficients g_I. The summation in the definition of the modified shape functions (5.220) is non-zero only within elements that contain one or more nodes lying on the essential boundary. Therefore, throughout most of the domain the usual meshfree shape functions can be used unchanged. In addition, there is no need to create an entire finite element mesh, as only one layer of elements on the essential boundary is necessary.

An example of both unmodified and modified 1-D RKPM shape functions whose supports span 7 nodes are shown in Fig. 5.46, along with the FEM boundary shape functions. Note that after modification, only the FEM functions are non-zero at the boundary nodes.

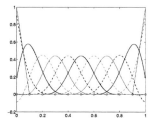 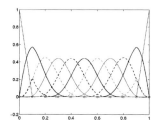

Fig. 5.46. Unmodified (above) and modified (below) 1-D RKPM shape functions and FEM boundary shape functions. FEM shape functions are the piecewise linear functions at either end. Different line styles are included only to distinguish neighboring shape functions. Note that only the finite element shape functions are non-zero at the boundary in the modified plot.

Remarks:

- This method of applying boundary conditions preserves the Nth order accuracy of the RKPM shape functions.[442]
- When used in a Galerkin solution, the order of convergence of the RKPM method is preserved as long as it is noted that boundary integrals such as $\int_{\Gamma_g} \delta u_i \sigma_{ij} n_j d\Gamma_g$ in the weak form (5.214) are non-zero, since the boundary conditions are enforced only on the boundary nodes and not on the entire boundary ([442,192]).
- There is no matrix inversion or transformation necessary for the application of boundary conditions, as is needed for many methods ([442,185]). For this reason, this method provides better performance in parallel algorithms than other methods such as the corrected collocation method,[442] in which the most memory-efficient means of solving (e.g. matrix LU decomposition and back-substitution) result in inherently serial procedures.
- The function $g^h_{bound}(x)$ is a known function, independent of the nodal degrees of freedom u_I, and will therefore contribute to the right hand side in a Galerkin solution. Furthermore, variations of $u^h(x)$ to be used as test functions reduce to:

$$\delta u^h(x) = \delta u^h(x) = \sum_{I \in A} \tilde{\Phi}_I(x) \delta u_I \tag{5.221}$$

5.8.3 Discretization of the Weak Form

The discretizations of the conservation equations are written using the modified shape functions $\tilde{\Phi}(\mathbf{x})$ to enforce essential boundary conditions. The velocity and pressure functions, $u_i(\mathbf{x})$ and $p(\mathbf{x})$, along with the test functions $\delta u_i(\mathbf{x})$ and $\delta p(\mathbf{x})$, are interpolated as:

$$u_i^h(\mathbf{x}) = \sum_{I \in B_{u_i}} N_I(\mathbf{x}) g_{iI} + \sum_{I \in A} \tilde{\Phi}_I(\mathbf{x}) u_{iI} \tag{5.222a}$$

$$\delta u_i^h(\mathbf{x}) = \sum_{I \in A} \tilde{\Phi}_I(\mathbf{x}) \delta u_{iI} \tag{5.222b}$$

$$p^h(\mathbf{x}) = \sum_{I \in B_p} N_I(\mathbf{x}) s_I + \sum_{I \in A} \tilde{\Phi}_I(\mathbf{x}) p_I \tag{5.222c}$$

$$\delta p^h(\mathbf{x}) = \sum_{I \in A} \tilde{\Phi}_I(\mathbf{x}) \delta p_I \tag{5.222d}$$

Subscripts on the boundary node sets B are included as a reminder that the essential boundary conditions may be applied at different nodes for different variables. The coefficients s_I are included in equation (5.222c) for those cases in which an essential boundary condition for the pressure is enforced. For many simulations there is no such condition on the pressure; in these cases, the set B_p is empty, and $\tilde{\Phi}(\mathbf{x}) = \Phi(\mathbf{x})$ for the pressure interpolation. Substituting Eq. (5.222b) and Eq. (5.222d) into Eq. (5.214) gives:

$$\sum_{I \in A} \int_{\Omega} \rho(\delta u_{iI} \tilde{\Phi}_I + \tau^m u_k^h \delta u_{iI} \tilde{\Phi}_{I,k} + \tau^c \delta p_I \tilde{\Phi}_{I,i})(\dot{u}_i^h + u_j^h u_{i,j}^h) d\Omega$$

$$+ \sum_{I \in A} \int_{\Omega} \delta u_{iI} \tilde{\Phi}_{I,j} \sigma_{ij}^h d\Omega + \sum_{I \in A} \int_{\Omega} (\tau^m u_k^h \delta u_{iI} \tilde{\Phi}_{I,k} + \tau^c \delta p_I \tilde{\Phi}_{I,i}) p_{,i}^h d\Omega$$

$$- \sum_{I \in A} \int_{\Gamma_{h_i}} \delta u_{iI} \tilde{\Phi}_I h_i d\Gamma_{h_i} - \sum_{I \in A} \int_{\Gamma_{g_i}} \delta u_{iI} \tilde{\Phi}_I \sigma_{ij} n_j d\Gamma_{g_i}$$

$$+ \sum_{I \in A} \int_{\Omega} (\delta p_I \tilde{\Phi}_I + \tau^c \delta u_{iI} \tilde{\Phi}_{I,i}) u_{i,i}^h d\Omega = 0, \tag{5.223}$$

where u_i^h and p^h are calculated from (5.222a) and (5.222c) respectively. We avoid making this substitution explicitly in order to emphasize that no large system matrices are computed or stored in our formulation. By the arbitrariness of the test function degrees of freedom δu_{iI} and δp_I, we have four equations at each node I:

$$r_{iI}^u = 0, \tag{5.224a}$$

$$r_I^p = 0 \tag{5.224b}$$

where the residual vectors r_{iI}^u and r_I^p are:

$$r_{iI}^u = \int_\Omega \rho(\tilde{\Phi}_I + \tau^m u_k^h \tilde{\Phi}_{I,k})(\dot{u}_i^h + u_j^h u_{i,j}^h)d\Omega - \int_{\Gamma_{h_i}} \tilde{\Phi}_I h_i d\Gamma_{h_i} - \int_{\Gamma_{g_i}} \tilde{\Phi}_I \sigma_{ij} n_j d\Gamma_{g_i}$$
$$+ \int_\Omega \tilde{\Phi}_{I,j}\sigma_{ij}^h d\Omega + \int_\Omega \tau^m u_k^h \tilde{\Phi}_{I,k} p_{,i}^h d\Omega + \int_\Omega \tau^c \tilde{\Phi}_{I,i} u_{j,j}^h d\Omega$$

$$(5.225a)$$

$$r_I^p = \int_\Omega \rho \tau^c \tilde{\Phi}_{I,i}(\dot{u}_i^h + u_j^h u_{i,j}^h)d\Omega + \int_\Omega \tau^c \tilde{\Phi}_{I,i} p_{,i}^h d\Omega + \int_\Omega \tilde{\Phi}_I u_{i,i}^h d\Omega$$

$$(5.225b)$$

The residuals in Eq. (5.225) are evaluated at each iteration in our solution algorithm (see section 5.8.7) by first computing u_i^h and p^h and their derivatives at every integration point according to (5.222a) and (5.222c). The integration points can be chosen based either on the elements of a corresponding finite element grid, or in a regular array unrelated to the nodal distribution. In our work, we use the former method; generally, we find that we require one integration point at the center of each tetrahedral element.

5.8.4 Time Integration Scheme

The velocity and pressure are to be solved at every time step using the residual equations (5.225) derived in the previous section. Specifically, we calculate Δu_{iI}^m and p_I^{m+1}, where the superscript m denotes the current time step and

$$\Delta u_{iI}^m = u_{iI}^{m+1} - u_{iI}^m \tag{5.226}$$

The increment of the velocity at time m, $\Delta\mathbf{u}^m$, is used to calculate the time derivative of the velocity approximation, $\dot{u}_i^h$:

$$\dot{u}_i^h = \frac{u_i^{h,m+1} - u_i^{h,m}}{\Delta t} = \frac{1}{\Delta t}\sum_{I \in A} \tilde{\Phi}_I \Delta u_{iI}^m \tag{5.227}$$

The value of the velocity approximation u_i^h in equation (5.225) is evaluated as:

$$u_i^h = \alpha u_i^{h,m+1} + (1-\alpha)u_i^{h,m}$$
$$= \sum_{I \in B} N_I g_{iI} + \sum_{I \in A} \tilde{\Phi}_I(u_{iI}^m + \alpha \Delta u_{iI}^m) \tag{5.228}$$

where $0 \le \alpha \le 1$. In this work we use $\alpha = \frac{1}{2}$ for a central difference scheme. Note that p is always computed at $m+1$; there is no central difference on p.

Newton's method is used at each time step for this implicit algorithm with initial guesses $\Delta\mathbf{u}^m = \mathbf{0}$ and $\mathbf{p}^{m+1} = \mathbf{p}^m$. The vectors of increments in velocity and pressure, $\Delta\mathbf{u}^{incr}$ and $\mathbf{p}^{incr}$, are solved iteratively to satisfy the nonlinear residual equations (5.225):

$$\mathbf{r}_i^w(\mathbf{u}^m, \Delta\mathbf{u}^m + \mathbf{u}^{incr}, \mathbf{p}^{m+1} + \mathbf{p}^{incr}) = 0 \tag{5.229a}$$

$$\mathbf{r}^q(\mathbf{u}^m, \Delta\mathbf{u}^m + \mathbf{u}^{incr}, \mathbf{p}^{m+1} + \mathbf{p}^{incr}) = 0 \tag{5.229b}$$

Taylor expanding (5.229) gives:

$$\mathbf{r}_i^w(\mathbf{u}^m, \Delta\mathbf{u}^m, \mathbf{p}^{m+1}) + \mathbf{r}_{i,\mathbf{u}}^w(\mathbf{u}^m, \Delta\mathbf{u}^m, \mathbf{p}^{m+1})\Delta\mathbf{u}^{incr}$$
$$+ \mathbf{r}_{i,p}^w(\mathbf{u}^m, \Delta\mathbf{u}^m, \mathbf{p}^{m+1})\mathbf{p}^{incr} \approx 0 \tag{5.230a}$$

$$\mathbf{r}^q(\mathbf{u}^m, \Delta\mathbf{u}^m, \mathbf{p}^{m+1}) + \mathbf{r}_{,\mathbf{u}}^q(\mathbf{u}^m, \Delta\mathbf{u}^m, \mathbf{p}^{m+1})\Delta\mathbf{u}^{incr}$$
$$+ \mathbf{r}_{,p}^q(\mathbf{u}^m, \Delta\mathbf{u}^m, \mathbf{p}^{m+1})\mathbf{p}^{incr} \approx 0 \tag{5.230b}$$

The resulting matrix equation, linear in $\Delta\mathbf{u}^{incr}$ and $\mathbf{p}^{incr}$, is:

$$\begin{bmatrix} \mathbf{r}_{i,\mathbf{u}}^w & \mathbf{r}_{i,p}^w \\ \mathbf{r}_{,\mathbf{u}}^q & \mathbf{r}_{,p}^q \end{bmatrix} \left\{ \begin{array}{c} \Delta\mathbf{u}^{incr} \\ \mathbf{p}^{incr} \end{array} \right\} = \left\{ \begin{array}{c} -\mathbf{r}_i^w \\ -\mathbf{r}^q \end{array} \right\} \tag{5.231}$$

The unknowns $\Delta\mathbf{u}^m$ and $\mathbf{p}^m$ can be updated as:

$$\Delta\mathbf{u}^m \rightarrow \Delta\mathbf{u}^m + \Delta\mathbf{u}^{incr} \tag{5.232a}$$

$$\mathbf{p}^{m+1} \rightarrow \mathbf{p}^{m+1} + \mathbf{p}^{incr} \tag{5.232b}$$

This Newton step is repeated for a fixed number of iterations or until the norms of the residual vector $\mathbf{r}_i^w$ and $\mathbf{r}^q$ are below a given tolerance before proceeding to the next time step.

We solve the linear system of Eq. (5.231) using the Generalized Minimum Residual (GMRES) algorithm with diagonal preconditioning. The minimization property of this method ensures that even an incomplete GMRES procedure decreases the residual. In combination with Newton's method, a very small Krylov subspace is frequently enough to obtain fast convergence to the solution of a nonlinear problem. In the solution of a matrix equation $\mathbf{A}\mathbf{x} = \mathbf{b}$, the GMRES algorithm requires the repeated multiplication of a given vector by the matrix $\mathbf{A}$. In our implementation, we compute each of these matrix-vector products without evaluating the individual elements of the matrix. Instead, the analytical expression for the residual derivatives is used to directly compute the product of the matrix on the left hand side of Eq. (5.231) with a given vector. This strategy greatly reduces the amount of memory necessary for storage, at the cost of a slight increase in the total number of computations.

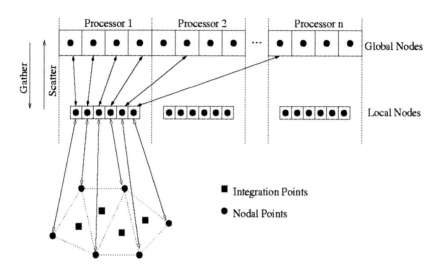

Fig. 5.47. Gather and scatter operation for meshfree method.

5.8.5 Communication Structure

Because the physical domain is partitioned between processors, communication is required. There are two sets of geometrical entities that must be distributed. First, the nodes themselves are assigned to different processors. Each processor stores the "official" data for its own nodes, and updates at each time step. In addition, numerical computation of the integrals in the conservation equations must be done by evaluating at a discrete set of integration points. To compute efficiently, each processor should therefore also receive a subset of the integration points, and be responsible for calculating the contribution to the residual vectors at those points. In order to evaluate a function at a given integration point, a processor must have information about all nodes under whose domain of influence that integration point falls. The integration point in turn contributes to the residual computed at each of those same influence nodes. These nodes are not necessarily among the ones "owned" by that processor, but are distributed among different processors, necessitating communication at each iteration. For every evaluation of the residual vectors, each processor seeks from the other processors the data at the nodes that are needed by its own integration points. This is known as a *gather* operation. Once this data has been used to compute the residual vectors, each processor has a piece of the residual vector at all nodes whose

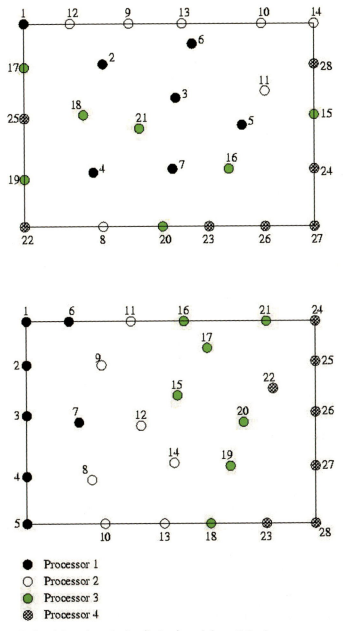

Fig. 5.48. Before(above) and after(below) nodal partitioning.

domains include any one of that processor's integration points. The goal, however, is for each processor to have the entire residual at each of its own nodes. So the reverse of the previous communication step must take place; each processor sends the residual it has computed at a given node to the processor that is responsible for that node, where the total residual is stored. This is called a *scatter* operation. Both gather and scatter operations, which are sometimes referred to as a swap and add procedure, must be performed at every iteration. This process is shown in Fig. 5.47.

5.8.6 Partitioning Schemes

In the code, the nodes and integration points are assigned to the processors based on a domain decomposition algorithm, in which each processor receives, and is responsible for, computing on a partition of the domain Ω. Due to the large domain of influence of each node, meshfree methods require more communication time than similarly sized parallel finite element computations. In order to minimize the required amount of communication, the domain is partitioned before the analysis in such a way that each processor "owns" a large number of integration points and nodes that neighbor each other. In actual computations, the nodes and integration points are sorted according to their x-coordinates in ascending order. Using this algorithm, the nodes are numbered from left to right in the geometry domain that is to be solved as in Fig. 5.48. This saves communication time between processors because in most cases, a node does not need to communicate with neighbors that may belong to another processor; most often, neighbors belong to the same processor after partitioning.

This partitioning method is used in our code for simplicity. There are a variety of different partitioning codes or softwares such as Metis that can be easily used as a "black box".[142]

5.8.7 Outline of Procedures

This section includes the algorithm for implementing RKPM implicit analysis.

1. Serial Pre-Processing
    a) Read geometry data
    b) Partition the nodes and integration points onto each processor
2. Parallel Analysis on Each Processor
    a) Read the input partitioned data
    b) Set up data structure for local analysis
    c) Calculate shape functions $\Phi_I(\mathbf{x})$ and their derivatives;
    d) Calculate the modified shape functions $\tilde{\Phi}(\mathbf{x})$ and their derivatives, Eq. (5.220)
3. Time Increment Loop

    a) Initialize variables: $\Delta \mathbf{d}_i^m = \mathbf{0}$, $\mathbf{p}^{m+1} = \mathbf{p}^m$

    b) Newton Iteration Loop

        i. Calculate the residuals, Eqs. (5.225)

        ii. Check residual, if converges, then go to (iii)

        iii. Solve for $\Delta \mathbf{d}^{incr}$ and $\mathbf{p}^{incr}$ using GMRES, Eq. (5.231)

        iv. Update the variables, $\mathbf{d}^m$ and $\mathbf{p}^{m+1}$, Eq. (5.232)

    c) Output the results to files

4. Serial Post-Processing

    a) Create a single output file for processing .

5.9 Numerical Examples of Meshfree Simulations

5.9.1 Simple 3-D Flow Past a Circular Cylinder

The first example is the meshfree simulation of a uniform flow past a 3-D cylinder. This flow has been the subject of many theoretical, experimental and computational investigations; two notable examples to which we will compare our results are those of Collins and Dennis ([123][124]) and Bar-Lev and Yang,[27] who studied analytically the early time history of flow past a cylinder initially at rest. Our simulations employ a parallel version of RKPM with the enrichment boundary condition implementation. For comparision, we have solved the same problem using both finite elements and RKPM with the "corrected collocation" method for boundary conditions proposed in.[442]

    A cylinder with a diameter of 1.5cm with its axis in the z-direction is placed in a uniform x-directional flow. The dimensions of the computational domain are $21.5cm \times 14cm \times 4cm$, and the cylinder is located $4.5cm$ downstream of the inflow; see Fig. 5.49. Two different discretizations are used: a coarse discretization with 2,236 nodes and 11,628 integration points, and a fine discretization with 15,447 nodes and 87,646 integration points. For both nodal distributions, the discretizaton is finest near the cylinder surface in order to resolve the boundary layer.

    Initially, the velocity is uniform with speed U_0 everywhere. For $t > 0$, $\mathbf{u} = \mathbf{0}$ is enforced on the surface of the cylinder. The inflow boundary condition at $x = -4.5cm$ is uniform flow of speed U_0, while the outflow at $x = 17cm$ is a zero-stress boundary. The top, bottom, and sides of the computational domain have zero penetration conditions, but allow slip parallel to the walls.

    Fig. 5.50 shows computed drag coefficients for both nodal distribution plotted against Reynolds number ranging from 10 to 1,000 compared with the values obtained from experiments (data from many sources collected by,[452] p.266). As shown in the Figure, using the enrichment method of Section 5.8.7 to implement the essential boundary condition yields results closer to the experimental values than the one without, but both methods are more accurate than FEM. The values of drag coefficient for different Reynolds numbers are tabulated in Table 5.9.1. Our results seem to indicate that even

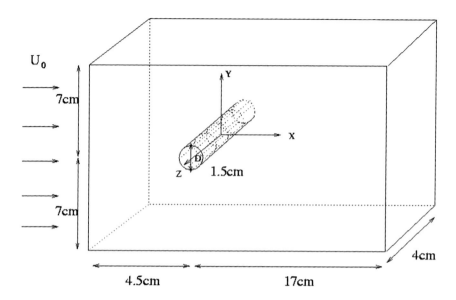

Fig. 5.49. Flow past a cylinder.

the coarse discretization is sufficiently fine to obtain accurate results using RKPM, while the FEM solution is not fully resolved even with the fine mesh.

The calculated drag coefficients for the early time history are plotted for Reynolds numbers of 40, 200 and 1,000 respectively, in Fig. 5.51. The time $T = U_0 t/a$ is non-dimensionalized based on the uniform velocity U_0 and the radius of the cylinder, a. The computational results using RKPM and FEM are compared with the analytical results of Bar-Lev and Yang[27] and Collins and Dennis.[124] It can be seen that RKPM more closely matches the theory than FEM, most likely due to the ability of meshfree shape functions to represent the sharp boundary layer.

The 3-D streamline contours of the flow field for Reynolds numbers of 200 and 1,000 at different times are presented in figures 5.52 and 5.53 for both FEM and RKPM solutions. The vortices and their developments through the early times can be clearly seen in the figures. This shows that the method used in this paper has the ability to capture the characteristics of the flow.

5.9.2 3-D Flow past a Building

A parallel RKPM code is used to calculate the uniform flow past a building to demonstrate the parallel or scalable ability of meshfree code and its application to large scale computations.

The shape of the building is shown in Fig. 5.54. The structure is 250 feet in height and has a square base with sides of 50 feet.

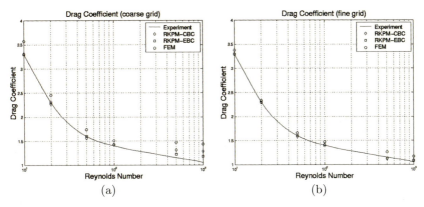

Fig. 5.50. Drag coefficient of the cylinder vs. Reynolds number. CBC: Collocation Boundary Condition implementation; EBC: Enrichment Boundary Condition. (a) Coarse Grid (2,236 nodes, 11,628 integration points) and (b) Fine Grid (15,447 nodes, 87,646 qudrature points).

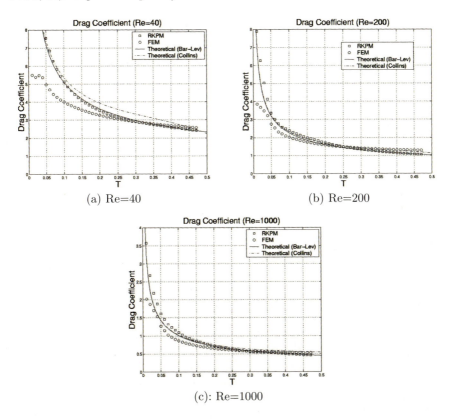

Fig. 5.51. Early time history of the drag coefficient for different Reynolds numbers. Bar-Lev:;[27] Collins:[124] $T = U_0 t/a$.

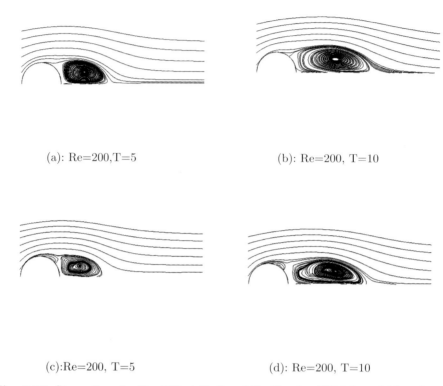

(a): Re=200,T=5   (b): Re=200, T=10

(c):Re=200, T=5   (d): Re=200, T=10

Fig. 5.52. Streamlines for Re=200 at T=5 and T=10 using FEM (a and b) and Meshfree method (c and d).

The computational domain extends 350 feet downstream of the building, and 150 feet upstream and to each side. It extends vertically from the base of the building to a point 100 feet above the top of the building. The domain is discretized using 38,359 nodes and 845,456 integration points. The discretization is much finer at and near the surface of the building than it is far away. A photo of the real building is shown in Fig. 5.55 (a). The boundary conditions are the same as the previous example, and the computed pressure profile may be traced in Fig. 5.55 (b).

The parallel performance on the Cray T3E for this problem is shown in Fig. 5.56. Communication takes a large percentage of the total time when using 128 processors; therefore, speedup cannot be seen past 64 processors. Comparing the performance of the RKPM method with both boundary condition implementations shows that the enrichment boundary condition implementation is more parallelizeable.

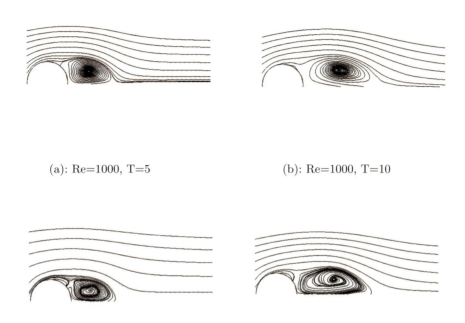

(a): Re=1000, T=5   (b): Re=1000, T=10

(c): Re=1000, T=5   (d): Re=1000, T=10

Fig. 5.53. Streamlines for Re=1000 at T=5 and T=10 using FEM (a and b) and Meshfree method (c and d).

Table 5.3. Drag coefficients for different Reynolds numbers. CBC: Collocation Boundary Condition implementation; EBC: Enrichment Boundary Condition

Method	Re=10	Re=20	Re=50	Re=100	Re=500	Re=1000
Experimental values	3.3	2.3	1.6	1.4	1.15	1.05
2256 nodes						
FEM	8.2%	6.6%	8.7%	7.7%	28%	36%
RKPM-CBC	0.38%	1.4%	2.5%	1.5%	14%	22%
RKPM-EBC	0.0%	0.0%	0.31%	1.1%	8.1%	12%
15,447 nodes						
FEM	2.5%	1.4%	3.7%	4.9%	14%	20%
RKPM-CBC	0.041%	0.45%	1.3%	0.79%	2.6%	3.8%
RKPM-EBC	0.0%	0.0%	0.25%	0.36%	1.7%	2.9%

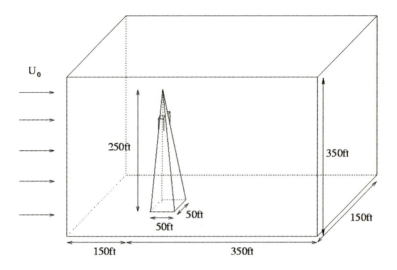

Fig. 5.54. Flow past a building.

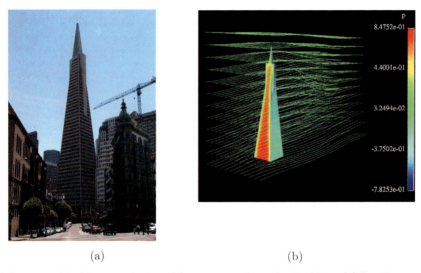

(a)   (b)

Fig. 5.55. Meshfree simulation of flow passing through a building; (a) San Francisco TransAmerica Tower, (b) building pressure tracer.

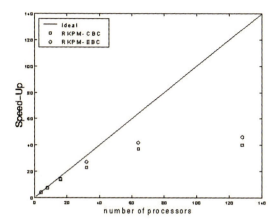

Fig. 5.56. Parallel performance on CrayT3E of RKPM for flow past a building problem, 38,359 nodes and 845,456 integration points (speedup is calculated based on the calculation time of using 4 processors). CBC: Collocation Boundary Condition implementation; EBC: Enrichment Boundary Condition implementation.

6. Reproducing Kernel Element Method (RKEM)

6.1 Introduction

Interests in constructing versatile finite element interpolants, meshfree interpolants, or general partition of unity shape functions is the current trend in improving the-state-of-the-art finite element technology (see Belytschko, Liu and Moran[49]) and meshfree technology (see Li and Liu[274] and Babuška, Banerjee, and Osborn[25]). In this chapter, we introduce and analyze a new method called the reproducing kernel element method (RKEM), which is constructed by combining the virtues of finite element approximations and reproducing kernel particle approximations (RKPM,[96, 297, 298, 303]).

Since the invention of finite element methods (FEM) in the 1950s, there have been demands in construction of smooth finite element (FE) shape functions over discretizations of an arbitrary domain of multiple dimensions. This is because in many engineering applications the FEM Galerkin weak formulations are involved with higher order derivatives of unknown functions. For instances to simulate static and dynamic behaviors of beams, plates, and shells structures, e.g. simulation of a DNA string, simulation of a thin film on a silicon substrate, or geometric modeling of a human torso. A Galerkin weak formulation with second order derivatives of the unknown displacement is required in these simulations. This has a few consequences: (1) it may be required to interpolate the first or the second derivatives of the unknown function on the boundary (which is denoted as I^1 or I^2 type interpolation); (2) it requires a globally $C^1(\Omega)$ conforming interpolation field.

Similar situations occur in computations of gradient elasticity and gradient plasticity problems as well, which are primary theoretical and computational models in nanoscale and microscale constitutive modelings in order to represent cohesive force or dislocation interactions. Furthermore, the classical force method of structural engineering is deeply rooted in complementary variational principle. To use stress functions based complementary variational weak formulations in FEM computations requires a globally conforming $C^1(\Omega)$ interpolation field as well. It is because this very reason that force method based finite element methods have never thrived.

In the current finite element method technology, the smoothness of FEM shape functions is limited by the inter-element boundary continuities. For example, to solve a fourth order differential equation, one needs C^1 elements in a standard conforming method. However, as it is well-known, it is not practical to use C^1 elements for problems over two or higher dimensional domains. The proposed RKEM eliminates this difficulty, and it provides a systematic procedure to construct higher order conforming finite element shape functions in multiple dimensions. On the other hand, for most of meshfree methods, the treatment of Dirichlet boundary conditions is problematic due to the loss of the Kronecker delta property of meshfree shape functions. In the literature, a variety of techniques were proposed and analyzed for enforcing Dirichlet boundary conditions, e.g., Lagrangian multiplier technique ([40]), transformation technique ([104]), hierarchical enrichment technique ([192,441]), reproducing kernel interpolation technique ([106]), singular kernel function technique ([104,235]), collocation technique ([442]), window or correction function ([182]), and use of D'Alembert's principle ([185]). Nevertheless, most of these techniques do not have good scalability in parallel computations. For the proposed RKEM, the Kronecker delta property is kept as long as certain conditions on the support size of the kernel function are satisfied. Thus the treatment of Dirichlet boundary conditions in RKEM is straightforward. The proposed RKEM enjoys some distinguished features:

1. The smoothness of the global basis functions is solely determined by that of the kernel function, which can be made arbitrary according to the requirements of the applications ($C^n, n \geq 1$);
2. The global basis functions of RKEM have the Kronecker delta property at the associated nodes, provided that some conditions on the support size of the kernel function are met. In fact, one can easily construct a RKEM interpolant in multi-dimension that has arbitrarily higher order Kronecker delta property, i.e. $I^m, m \geq 1$;
3. In principle, the reproducing properties of RKEM interpolant can be also made arbitrary, $P^k, k \geq 1$.

Note that in this book the term, I^m *interpolation field*, refers to the interpolant that can interpolate the derivatives of an unknown function up to m-th order, whereas the term, C^n *interpolation field*, refers to the interpolant having globally continuous derivatives up to n-th order, and the term, P^k *interpolation field*, refers to the interpolant that can reproduce complete k-th order polynomials.

This method is aimed at achieving the following objectives: (1) no numerically induced discontinuity between elements; (2) no special treatment required for enforcing essential boundary conditions; and (3) high order smooth interpolation in arbitrary domains of multiple dimensions.

Note that objective (3) has also been an outstanding problem in computational geometry.

This chapter consists of two parts: Sec. 6.2 and Sec. 6.3-6.5. Sec. 6.2 focuses on the discussion of the theoretical foundation of RKEM, along with some numerical examples; Sec. 6.3-6.5 focus on how to construct multi-dimensional $I^m/C^n/P^k$ RKEM interpolants.

6.2 Reproducing Kernel Element Interpolant

Let $\Omega \subset \mathbb{R}^d$ be an open, bounded domain with a Lipschitz continuous boundary $\Gamma = \partial \Omega$. Let there be a subdivision $\{\Omega_n\}_{n=1}^N$ of the domain $\overline{\Omega} = \Omega \cup \Gamma$, i.e.:

1. Each Ω_n is a closed set with a non-empty interior.
2. $\overline{\Omega} = \cup_{n=1}^N \Omega_n$.
3. For $m \neq n$, $\overset{\circ}{\Omega}_m \cap \overset{\circ}{\Omega}_n = \emptyset$, where $\overset{\circ}{\Omega}_m$ denotes the interior of Ω_m.

6.2.1 Global Partition Polynomials

To construct a RKEM shape function, we first introduce a notion of the so-called *global partition polynomial*.

Consider a FEM discretization, $\Omega_e, e \in \Lambda_E := \{1, 2, 3, \cdots, n_{el}\}$ where n_{el} is the total number of elements. We assume that each element, Ω_e, has n_{de} number of vertices, or nodes, i.e. $\{\mathbf{x}_{e,i}\}_{i \in \Lambda_e} \subset \Omega_e$, here $\Lambda_e := \{1, 2, 3, \cdots, n_{de}\}$ is the element nodal index set. We further assume that there are linearly independent functions $\{\psi_{e,i}\}_{i \in \Lambda_e}$ and such that the following *reproducing property of order k* holds:

$$\sum_{i \in \Lambda_e} \psi_{e,i}(\boldsymbol{x})\, \boldsymbol{x}_{e,i}^{\gamma} = \boldsymbol{x}^{\gamma} \quad \forall \gamma : |\gamma| \leq k, \ \forall \boldsymbol{x} \in \overline{\Omega}. \tag{6.1}$$

In most cases, $\psi_{e,i}$ are globally defined polynomial functions, we call them *the global partition polynomials*. One important property of the global polynomials is that they are C^∞ functions.

Before we proceed further, it may be useful to review how the FEM shape function is constructed.

Finite element shape functions, $\phi_{e,i}$, $e \in \Lambda_E := \{1, 2, \cdots, n_{el}\}$, and $i \in \Lambda_e = \{1, \cdots, n_{ne}\}$, are compactly supported, i.e. $supp\{\phi_{e,i}\} = \Omega_e$. When $\mathbf{x} \notin \Omega_e$, $\phi_{e,i}(\mathbf{x}) = 0$. For a large class of FEM shape functions, this condition is enforced by multiplying Heaviside function with certain global polynomial functions, or more precisely,

$$\phi_{e,i}(x) = \psi_{e,i}(x)\chi_e(x) \tag{6.2}$$

where function, $\chi_e(x)$, is the characteristic function of element e, i.e.

$$\chi_e(x) := \begin{cases} 1 \ x \in \Omega_e \\ 0 \ x \notin \Omega_e \end{cases} \tag{6.3}$$

The characteristic function in (6.3) truncates the analytical polynomial functions such that FEM shape functions are localized in compact supports. Because of the presence of discontinous characteristic function, $\chi_e(x)$, most of

FEM shape functions, $\phi_{e,i}(x)$, are $C^0(\Omega)$ functions, instead of $C^\infty(\Omega)$ functions, in multiple dimensions.

In other words, the FEM shape function is constructed by combining global partition polynomials with element characteristic function. Therefore, the so-called *global partition polynomial* may be viewed as the continuous extension of regular FEM polynomial shape function[1], and it is defined in $\mathbb{R}^d$. By construction, the FEM shape function has one-to-one correspondence with the global partition polynomials, i.e.

$$\phi_{e,i} \in C^k(\Omega_e) \leftrightarrow \psi_{e,i} \in C^\infty(\mathbb{R}^d) \tag{6.4}$$

In addition to reproducing condition (6.1), the global partition polynomial has two special properties. First, if a FEM shape function has the following Kronecker delta properties

$$D^\alpha \phi_{e,i}^{(\beta)}\Big|_{\boldsymbol{x}=\boldsymbol{x}_j} = \delta_{ij}\delta_{\alpha\beta}, \quad \boldsymbol{x}_i,\ \boldsymbol{x}_j \in \bar{\Omega}, \ \ |\alpha|, |\beta| \le m, \tag{6.5}$$

then the corresponding global polynomial function has the same properties,

$$D^\alpha \psi_{e,i}^{(\beta)}\Big|_{\boldsymbol{x}=\boldsymbol{x}_j} = \delta_{ij}\delta_{\alpha\beta}, \quad \boldsymbol{x}_i,\ \boldsymbol{x}_j \in \bar{\Omega}_e, \ \ |\alpha|, |\beta| \le m \ . \tag{6.6}$$

In fact, in most cases, (6.5) is a consequence or built in property of (6.6).

Second, unlike FEM shape functions, global partition polynomials are not a global partition of unity under global index, i.e.

$$\sum_{e\in\Lambda_E}\sum_{i\in\Lambda_e} \phi_{e,i}(\boldsymbol{x}) = 1 \ , \quad \leftrightarrow \quad \sum_{e\in\Lambda_E}\sum_{i\in\Lambda_e} \psi_{e,i}(\boldsymbol{x}) \ne 1 \ . \tag{6.7}$$

However, they are partition of unity under local index, whereas the FEM shape function isn't,

$$\sum_{i\in\Lambda_e} \phi_{e,i}(\boldsymbol{x}) \ne 1 \ , \ \ \forall \boldsymbol{x} \in \Omega \quad \leftrightarrow \quad \sum_{i\in\Lambda_e} \psi_{e,i}(\boldsymbol{x}) = 1 \ , \ \ \forall \boldsymbol{x} \in \Omega \ . \tag{6.8}$$

In general, if

$$\sum_{i\in\Lambda_e} \phi_{e,i}(\mathbf{x})\mathbf{x}_i^\beta = \mathbf{x}^\beta, \ \ \forall\, \mathbf{x} \in \Omega_e, \ \ e \in \Lambda_E \tag{6.9}$$

then one also has

$$\sum_{i\in\Lambda_e} \psi_{e,i}(\mathbf{x})\mathbf{x}_i^\beta = \mathbf{x}^\beta, \ \ \forall\, \mathbf{x} \in \mathbb{R}^d, \ \ e \in \Lambda_E \tag{6.10}$$

For instance, for a one-dimensional setting, with $\Omega_e = [x_{e-1}, x_e]$, we let

---

[1] In this book, we only consider polynomial type of FEM shape functions

$$\psi_{e,1}(x) = \frac{x_e - x}{x_e - x_{e-1}},$$

$$\psi_{e,2}(x) = \frac{x - x_{e-1}}{x_e - x_{e-1}}$$

defined for any $x \in \mathbb{R}$.

A two dimensional example of global partition shape function set is:

$$\psi_{e,1}(\xi(\mathbf{x}), \eta(\mathbf{x})) = \frac{1}{4}(1 - \xi)(1 - \eta) \qquad (6.11)$$

$$\psi_{e,2}(\xi(\mathbf{x}), \eta(\mathbf{x})) = \frac{1}{4}(1 + \xi)(1 - \eta) \qquad (6.12)$$

$$\psi_{e,3}(\xi(\mathbf{x}), \eta(\mathbf{x})) = \frac{1}{4}(1 + \xi)(1 + \eta) \qquad (6.13)$$

$$\psi_{e,4}(\xi(\mathbf{x}), \eta(\mathbf{x})) = \frac{1}{4}(1 - \xi)(1 + \eta) \qquad (6.14)$$

with $-\infty < \xi, \eta < \infty$ and functions $\xi(\mathbf{x})$ and $\eta(\mathbf{x})$ can be found by its inverse relationship,

$$\mathbf{x} = \sum_{i=1}^{4} \psi_{e,i}(\boldsymbol{\xi})\mathbf{x}_{e,i} \qquad (6.15)$$

Linear global partition polynomials in 1-D and bilinear global partition polynomials in 2-D are plotted in Figure 6.1 (a) and (b) respectively.

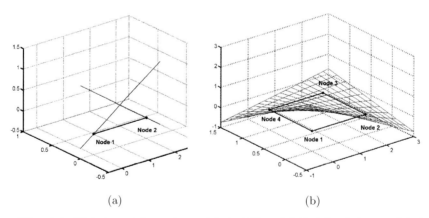

(a)   (b)

Fig. 6.1. Global partition polynomials in 1-D and 2-D: (a) 1-D linear global partition polynomials; (b) 2-D bilinear global partition polynomials.

Unlike FEM shape function, which the global partition polynomials are patched together with element characteristic functions, to construct RKEM shape function, we patch the global polynomial functions together by associating them with compactly supported meshfree shape functions, e.g. RKPM interpolant. By doing so, we achieve the following three goals: (1) the continuity of RKEM shape functions is solely controlled by compactly supported meshfree interpolant, and they are no longer piece-wise continuous; (2) the RKEM shape function is still localized or compactly supported, and (3) the reproducing properties of the RKEM is controlled by the global partition polynomials.

To construct a meshfree interpolation basis on a finite element mesh, let us introduce a kernel function $K_\rho(z; x)$ such that it is nonzero only when $\|z\| < \rho$. The positive number ρ represents the support size of the kernel function with respect to its first argument. Later on, we will be more specific on the form of the function $K_\rho(z; x)$. Then we define the following quasi-interpolation operator on a continuous function $v \in C(\overline{\Omega})$:

$$\mathcal{I}v(x) = \sum_{e \in \Lambda_E} \left[\int_{\Omega_e} K_\rho(y - x; x) \, dy \sum_{i \in \Lambda_e} \psi_{n,i}(x) \, v(x_{n,i}) \right]. \tag{6.16}$$

As shown in Figure 6.2, the nodes involved in the evaluation of $\mathcal{I}v(x)$ at a point x depend on the support size ρ.

For the interpolation operator $\mathcal{I}$, we have the following result on its polynomial reproducing property.

Proposition 6.2.1. *Assume the reproducing property* (6.1) *for each subdomain. Then the interpolation operator $\mathcal{I}$ defined in* (6.16) *has the reproducing property of order k:*

$$\mathcal{I}x^\gamma = x^\gamma \quad \forall \gamma : |\gamma| \le k, \ x \in \overline{\Omega} \tag{6.17}$$

if and only if it has the reproducing property of order 0:

$$\mathcal{I}1 = 1. \tag{6.18}$$

Proof. We only need to show that (6.18) implies (6.17). Let γ be such that $|\gamma| \le k$. Then using the assumption (6.1), we have

$$\mathcal{I}x^\gamma = \sum_{n=1}^{N} \left[\int_{\Omega_n} K_\rho(y - x; x) \, dy \sum_{i=1}^{I_n} \psi_{n,i}(x) \, x_{n,i}^\gamma \right]$$

$$= \sum_{n=1}^{N} \left[\int_{\Omega_n} K_\rho(y - x; x) \, dy \, x^\gamma \right]$$

$$= x^\gamma \mathcal{I}1.$$

Since $\mathcal{I}1 = 1$, we conclude $\mathcal{I}x^\gamma = x^\gamma$.

The condition (6.18) can be more conveniently restated as

$$\int_{\Omega} K_\rho(\boldsymbol{y} - \boldsymbol{x}; \boldsymbol{x}) \, d\boldsymbol{y} = 1 \quad \forall \, \boldsymbol{x} \in \overline{\Omega}. \tag{6.19}$$

For some integer $m \geq k$, we require the interpolation operator $\mathcal{I}$ defined in (6.16) to have a reproducing property of order m. By Proposition 6.2.1, this is equivalent to the conditions

$$\mathcal{I} \boldsymbol{x}^\alpha = \boldsymbol{x}^\alpha \quad \forall \alpha : |\alpha| = 0, k+1, \cdots, m, \ \forall \, \boldsymbol{x} \in \Omega. \tag{6.20}$$

From now on, we will focus on the following particular choice for the kernel function

$$K_\rho(\boldsymbol{z}; \boldsymbol{x}) = \frac{1}{\rho^d} \, \phi\left(\frac{\boldsymbol{z}}{\rho}\right) \boldsymbol{p}\left(\frac{\boldsymbol{z}}{\rho}\right)^T \boldsymbol{b}(\boldsymbol{x}), \tag{6.21}$$

where,

$$\boldsymbol{p}(\boldsymbol{z}) = (1, z_1^{k+1}, z_1^k z_2, \ldots, z_d^{k+1}, z_1^{k+2}, \ldots, z_d^m)^T$$

is the vector of constant 1 and the monomials of degree between $(k+1)$ and m: $\boldsymbol{z}^\alpha$ for $|\alpha| = k+1, \ldots, m$. The variable vector $\boldsymbol{b}$ has the same dimension as the vector $\boldsymbol{p}$. The function ϕ is the *window function*, and has a support size ρ. The interpolant of a continuous function $v \in C(\overline{\Omega})$ can then be written as

$$\mathcal{I}v(\boldsymbol{x}) = \sum_{n=1}^N \left[\int_{\Omega_n} \frac{1}{\rho^d} \, \phi\left(\frac{\boldsymbol{y} - \boldsymbol{x}}{\rho}\right) \boldsymbol{p}\left(\frac{\boldsymbol{y} - \boldsymbol{x}}{\rho}\right)^T \boldsymbol{b}(\boldsymbol{x}) \, d\boldsymbol{y} \sum_{i=1}^{I_n} \psi_{n,i}(\boldsymbol{x}) \, v(\boldsymbol{x}_{n,i}) \right]. \tag{6.22}$$

Use this form of the kernel function, the conditions (6.20) can be rewritten as a linear system for the unknown variable coefficient vector $\boldsymbol{b}(\boldsymbol{x})$:

$$\sum_{n=1}^N \left[\int_{\Omega_n} \frac{1}{\rho^d} \, \phi\left(\frac{\boldsymbol{y} - \boldsymbol{x}}{\rho}\right) \boldsymbol{p}\left(\frac{\boldsymbol{y} - \boldsymbol{x}}{\rho}\right)^T d\boldsymbol{y} \sum_{i=1}^{I_n} \psi_{n,i}(\boldsymbol{x}) \, \boldsymbol{x}_{n,i}^\alpha \right] \boldsymbol{b}(\boldsymbol{x}) = \boldsymbol{x}^\alpha \tag{6.23}$$
$$\forall \alpha : |\alpha| = 0, k+1, \ldots, m.$$

Equivalently, the system can be rewritten as

$$\sum_{n=1}^N \left[\int_{\Omega_n} \frac{1}{\rho^d} \, \phi\left(\frac{\boldsymbol{y} - \boldsymbol{x}}{\rho}\right) \boldsymbol{p}\left(\frac{\boldsymbol{y} - \boldsymbol{x}}{\rho}\right)^T d\boldsymbol{y} \sum_{i=1}^{I_n} \psi_{n,i}(\boldsymbol{x}) \left(\frac{\boldsymbol{x} - \boldsymbol{x}_{n,i}}{\rho}\right)^\alpha \right]$$
$$\boldsymbol{b}(\boldsymbol{x}) = \delta_{|\alpha|,0} \forall \alpha : |\alpha| = 0, k+1, \ldots, m. \tag{6.24}$$

Here, we first illustrate, in detail, the procedure to construct the global RKEM shape function with the first order polynomial reproducing capacity. Consider an 1-D example of RKEM interpolant with the first order reproducing condition and linear global partition polynomials. Since the basis function satisfies linear consistency, by Proposition 6.2.1, we need only to reproduce a constant to satisfy the first order reproducing conditions. In this case, $\boldsymbol{p}(z) = 1$ and $\boldsymbol{b}(x) = b_0(x)$. By (6.21), the kernel function is

$$K_\rho(y - x; x) = b_0(x)\frac{1}{\rho}\phi\left(\frac{y - x}{\rho}\right), \tag{6.25}$$

while the interpolation function (6.16) becomes

$$\mathcal{I}v(x) = \sum_{e \in \Lambda_E}\left[\int_{\Omega_e} b_0(x)\frac{1}{\rho}\phi(\frac{y - x}{\rho})dy \sum_{i \in \Lambda_e}\psi_{e,i}(x)v(x_{e,i})\right]. \tag{6.26}$$

The coefficient function $b_0(x)$ is determined by the zero-th consistency condition

$$1 = \sum_{e \in \Lambda_E}\int_{\Omega_e} b_0(x)\frac{1}{\rho}\phi\left(\frac{y - x}{\rho}\right)dy,$$

and we have

$$b_0(x) = \left[\int_\Omega \frac{1}{\rho}\phi\left(\frac{y - x}{\rho}\right)dy\right]^{-1}. \tag{6.27}$$

The proposed approximation can be explicitly expressed as

$$\mathcal{I}v(x) = \sum_{e \in \Lambda_E} b_0(x)\left[\int_{\Omega_n} \frac{1}{\rho}\phi\left(\frac{y - x}{\rho}\right)dy \sum_{i \in \Lambda_e}\varphi_{e,i}(x)v(x_{e,i})\right]$$

$$= \sum_{I=1}^{NP}\Psi_I(x)v_I. \tag{6.28}$$

Reproducing kernel element approximation for evaluation point x in a 1-D uniform partition with different support sizes and a linear basis function are constructed as shown in Figure 6.2, and a reproducing kernel element interpolation for evaluation point x in a 1-D uniform partition with quadratic basis function is constructed as shown in Figure 6.3.

6.2.2 Some Properties

In this section, we assume the system (6.23) or (6.24) is uniquely solvable at any $\boldsymbol{x} \in \overline{\Omega}$. We now examine two properties of the reproducing kernel element method introduced in Section 6.2. The first is the Kronecker delta property of the global basis functions associated with the interpolation operator $\mathcal{I}$. We will denote the collection of the nodes of the method by $\{\boldsymbol{x}_j\}_{j=1}^{NP}$. For each $j = 1, \cdots, NP$, the node $\boldsymbol{x}_j$ equals $\boldsymbol{x}_{e,i}$ for some n and some i, $1 \leq i \leq n_{de}$, $1 \leq e \leq n_{e\ell}$. Some of the nodes $\boldsymbol{x}_j$ belong to more than one subdomain (element). For the node $\boldsymbol{x}_j$, we denote $B(\boldsymbol{x}_j; \rho)$ the ball of radius ρ centered at $\boldsymbol{x}_j$, and denote $\Omega(\boldsymbol{x}_j)$ the union of the subdomains that contain $\boldsymbol{x}_j$ as a node.

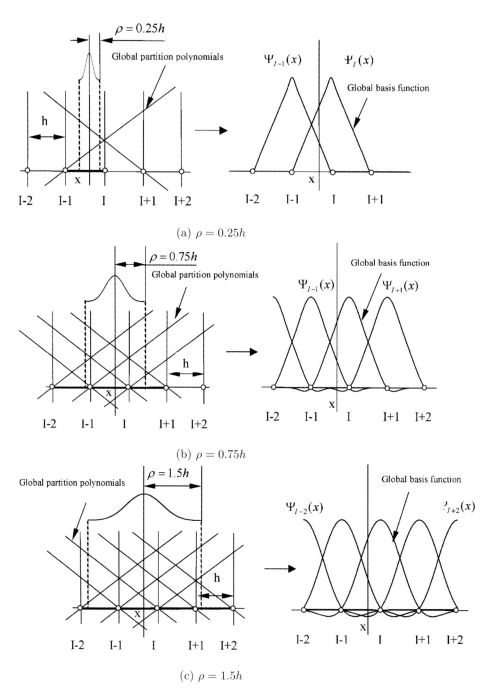

(a) $\rho = 0.25h$

(b) $\rho = 0.75h$

(c) $\rho = 1.5h$

Fig. 6.2. A reproducing kernel element approximation with a different support size for evaluation point x: (a): $\mathcal{I}v(x) = \Psi_{I-1}(x)v_{I-1} + \Psi_I(x)v_I$, (b): $\mathcal{I}v(x) = \Psi_{I-2}(x)v_{I-2} + \Psi_{I-1}(x)v_{I-1} + \Psi_I(x)v_I + \Psi_{I+1}(x)v_{I+1}$, (c): $\mathcal{I}v(x) = \Psi_{I-2}(x)v_{I-2} + \Psi_{I-1}(x)v_{I-1} + \Psi_I(x)v_I + \Psi_{I+1}(x)v_{I+1} + \Psi_{I+2}(x)v_{I+2}$.

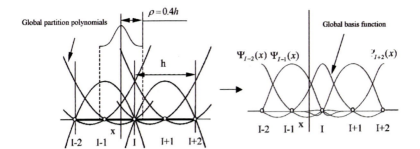

Fig. 6.3. A reproducing kernel element interpolation with a quadratic basis function for evaluation point x: $\mathcal{I}v(x) = \Psi_{I-2}(x)v_{I-2} + \Psi_{I-1}(x)v_{I-1} + \Psi_I(x)v_I + \Psi_{I+1}(x)v_{I+1} + \Psi_{I+2}(x)v_{I+2}$ with quadratic basis function.

Proposition 6.2.2. *Assume that the basis function* $\{\psi_{e,i}\}_{i \in \Lambda_e}$ *are nodal basis functions, i.e., they satisfy*

$$\psi_{e,i}(\boldsymbol{x}_{e,j}) = \delta_{i,j}, \quad \forall i, j \in \Lambda_e \tag{6.29}$$

Also assume

$$B(\boldsymbol{x}_j; \rho) \subset \Omega(\boldsymbol{x}_j), \quad 1 \leq j \leq NP. \tag{6.30}$$

where NP *is the total number of nodes in a mesh.*

Then the interpolation operator $\mathcal{I}$ *defined in (6.16) enjoys the interpolation condition: for any continuous function* $v \in C(\overline{\Omega})$,

$$\mathcal{I}v(\boldsymbol{x}_j) = v(\boldsymbol{x}_j), \quad 1 \leq j \leq NP. \tag{6.31}$$

Proof. We have

$$\mathcal{I}v(\boldsymbol{x}_j) = \sum_{e \in \Lambda_E} \left[\int_{\Omega_e} K_\rho(\boldsymbol{y} - \boldsymbol{x}_j; \boldsymbol{x}_j) \, d\boldsymbol{y} \sum_{i \in \Lambda_e} \psi_{e,i}(\boldsymbol{x}_j) \, v(\boldsymbol{x}_{e,i}) \right].$$

Note that $K_\rho(\boldsymbol{y} - \boldsymbol{x}_j; \boldsymbol{x}_j)$ is non-zero only if $\|\boldsymbol{y} - \boldsymbol{x}_j\| \leq \rho$. Using the assumptions (6.30) and (6.29), we find

$$\mathcal{I}v(\boldsymbol{x}_j) = \int_{\Omega(\boldsymbol{x}_j)} K_\rho(\boldsymbol{y} - \boldsymbol{x}_j; \boldsymbol{x}_j) \, v(\boldsymbol{x}_j) \, d\boldsymbol{y}$$

$$= v(\boldsymbol{x}_j) \int_\Omega K_\rho(\boldsymbol{y} - \boldsymbol{x}_j; \boldsymbol{x}_j) \, d\boldsymbol{y}.$$

By the zero-th order reproducing property (6.19), we then obtain the interpolation property (6.31).

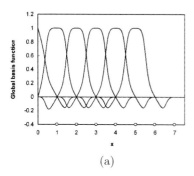 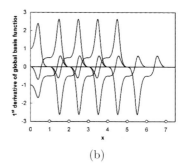

(a)   (b)

Fig. 6.4. A set of global basis functions and their derivative in 1-D domain: (a) RKEM global basis function, and (b) the 1st derivative of RKEM global basis function.

The condition (6.30) imposes a restriction on the size of ρ: it has to be small enough. In the case of linear elements and only vertices of the elements are used as the nodes, then ρ must be smaller than a constant times $\min_{1\leq e\leq n_{e\ell}} h_e$. The constant depends on the shape regularity of the finite element partition. If side mid-points are used as the nodes (quadratic elements), then ρ should be smaller than the constant times $0.5 \min_{1\leq e\leq n_{e\ell}} h_e$. Now if we like to have the Kronecker delta property only for the global basis functions associated with the nodes on the Dirichlet boundary, then $\min_{1\leq e\leq n_{e\ell}} h_e$ can be replaced by $\min_{1\leq e\leq n_0} h_e$, where $\Omega_1, \cdots, \Omega_{n_0}$ denote the elements on the boundary that contain Dirichlet boundary nodes. To be specific, in the 1-D case with a uniform partition and linear reproducing, the condition is $\rho \leq h$.

The second property is on the differentiability of the interpolation function $\mathcal{I}v$, i.e., the differentiability of the global basis function associated with the interpolation operator $\mathcal{I}$.

Proposition 6.2.3. *Assume the basis functions $\psi_{n,i} \in C^{k_1}(\overline{\Omega})$, $1 \leq i \leq I_n$, $1 \leq n \leq N$. Assume the window function $\phi \in C^{k_2}(\mathbb{R}^d)$. Here, k_1 and k_2 are two non-negative integers. Then for any continuous function $v \in C(\overline{\Omega})$, its interpolant $\mathcal{I}v \in C^{\min(k_1,k_2)}(\overline{\Omega})$.*

Proof. From (6.24), we see that each component of $\boldsymbol{b}$ is a $C^{\min(k_1,k_2)}(\overline{\Omega})$ function. The result of the proposition follows immediately from the representation formula (6.22).

We notice that in the context of triangular finite elements, the basis functions $\psi_{n,i}$ are polynomials, and are thus infinitely smooth. Then the regularity of the global basis functions of the new method is the same as that of the window function ϕ.

A set of global basis functions and its derivative for 1-D case with support size $\rho = 0.95\,h$, and the linear basis function and cubic B-spline kernel are plotted in Figure 6.4 (a),(b) respectively. The Kronecker delta property and continuity of the global basis functions are clearly illustrated.

Note that based on the definition, the RKE interpolation formula (6.16) is involved with an integral. There are several ways to evaluate this integral to obtain the explicit expression for a RKE shape function, $\Psi_I(x)$, see Eq. (6.28). The RKE shape functions and their derivatives shown in Fig. 6.4 (a) and (b) are constructed based on a nodal integration scheme proposed in Part II of this series (see:[275]). The advantage of using nodal integration to evaluate the integral is that it can provide an explicit expression of RKE interpolant. Nonetheless, the RKE shape functions used in the numerical examples of this paper are all constructed by evaluating the integral via Gauss quadrature.

6.2.3 Error Analysis of the Method with Linear Reproducing Property

In this section, we give a detailed analysis of the method with $m = k+1$ and when linear finite elements are used for the local approximations. To simplify the exposition, we let $d \leq 3$. The method for more general cases is being analyzed. Throughout the section, c denotes a generic constant that does not depend on the discretization parameters and functions under consideration. We first list various assumptions.

We assume that $\{\{\Omega_e\}_{e \in \Lambda_E}$ is a family of quasi-uniform finite element partition of the domain $\overline{\Omega}$ into triangular or tetrahedral elements. For a mesh $\{\Omega_e\}_{e \in \Lambda_E}$ in the family, we let h be the mesh size, and let h_e be the diameter of Ω_e. Since the mesh family is quasiuniform, we have a constant $c > 0$ such that

$$c\,h \leq h_e \leq h, \quad 1 \leq e \leq n_{el}. \tag{6.32}$$

Associated with each mesh $\{\Omega_e\}_{e \in \Lambda_E}$, we define a parameter ρ, intended for the support size of the reproducing kernel function. We assume that there exists a constant $c \geq 1$ such that

$$c^{-1}h \leq \rho \leq c\,h. \tag{6.33}$$

For the window function ϕ, we assume

$$\begin{cases} \mathrm{supp}(\phi)B_1, \\ \phi(\boldsymbol{x}) > 0 \text{ for } \|\boldsymbol{x}\| < 1, \\ \phi \in C^l(\mathbb{R}^d) \text{ for some } l \geq 1. \end{cases} \tag{6.34}$$

Here, $B_1 = \{\,\boldsymbol{x} \in \mathbb{R}^d \mid \|\boldsymbol{x}\| \leq 1\,\}$ is the unit ball.

For a continuous function $v \in C(\overline{\Omega})$, the interpolant is (cf. (6.22))

$$\mathcal{I}_1 v(\boldsymbol{x}) = \sum_{e \in \Lambda_E} \left[\int_{\Omega_e} \frac{1}{\rho^d} \phi\left(\frac{\boldsymbol{y} - \boldsymbol{x}}{\rho}\right) d\boldsymbol{y} \sum_{i \in \Lambda_e} \psi_{e,i}(\boldsymbol{x})\, v(\boldsymbol{x}_{e,i}) \right] b_0(\boldsymbol{x}), \tag{6.35}$$

where the coefficient function $b_0(\boldsymbol{x})$ is determined by the zero-th reproducing condition

$$b_0(\boldsymbol{x}) = \left[\int_{\Omega} \frac{1}{\rho^d} \phi\left(\frac{\boldsymbol{y} - \boldsymbol{x}}{\rho}\right) d\boldsymbol{y} \right]^{-1}. \tag{6.36}$$

Note that by Proposition 6.2.1, we have

$$\mathcal{I}_1 v(\boldsymbol{x}) = v(\boldsymbol{x})$$

for any linear function v.

From the formula (6.36) and assumptions (6.34), (6.33), it is readily verified that

$$\max_{\boldsymbol{x} \in \overline{\Omega}} |b_0(\boldsymbol{x})| + h \max_{\boldsymbol{x} \in \overline{\Omega}} \max_{1 \leq i \leq d} |\partial_i b_0(\boldsymbol{x})| \leq c. \tag{6.37}$$

Now let $u \in H^2(\Omega)$ and let us bound the interpolation error $u - \mathcal{I}_1 u$. By the Sobolev embedding theorem, we have $u \in C(\overline{\Omega})$ and so $\mathcal{I}_1 u$ is well defined. We first bound the interpolation error on a typical element Ω_e. Define

$$\tilde{\Omega}_e = \{ \boldsymbol{x} \in \overline{\Omega} \mid \text{dist}(\boldsymbol{x}, \Omega_e) \leq \rho \}.$$

We denote $\psi_{e,i}(\boldsymbol{x})$, $1 \leq i \leq n_{de}$, the natural extensions of the linear element shape functions corresponding to the n_{de} nodes of the element Ω_e. Note that $\{\psi_{e,i}(\boldsymbol{x})\}_{i \in \Lambda_e}$ are linear functions defined on the whole space $\mathbb{R}^d$. Using (6.32), we find that

$$\max_{i \in \Lambda_e} \|\psi_{e,i}\|_{C(\tilde{\Omega}_e)} + h \max_{i \in \Lambda_e} \|\psi_{e,i}\|_{C^1(\tilde{\Omega}_e)} \leq c. \tag{6.38}$$

Introduce the linear interpolation function

$$\mathcal{I}_{1,e} u(\boldsymbol{x}) = \sum_{i \in \Lambda_e} \psi_{e,i}(\boldsymbol{x}) \, u(\boldsymbol{x}_{e,i}).$$

By a standard scaling argument (cf.[84, 110]), we have the error estimate

$$\|u - \mathcal{I}_{1,e} u\|_{L^2(\tilde{\Omega}_e)} + h \, \|u - \mathcal{I}_{1,e} u\|_{H^1(\tilde{\Omega}_e)} + h^{d/2} \|u - \mathcal{I}_{1,e} u\|_{L^\infty(\tilde{\Omega}_e)} \leq c \, h^2 \, |u|_{H^2(\tilde{\Omega}_e)}. \tag{6.39}$$

Since $\mathcal{I}_1$ reproduces linear functions, we have

$$\mathcal{I}_1(\mathcal{I}_{1,e} u) = \mathcal{I}_{1,e} u.$$

Write

$$u - \mathcal{I}_1 u = u - \mathcal{I}_{1,e} u - \mathcal{I}_1(u - \mathcal{I}_{1,e} u). \tag{6.40}$$

By the definition (6.35),

$$\mathcal{I}_1(u - \mathcal{I}_{1,e} u)(\boldsymbol{x}) = \sum_{e \in \Lambda_E} \left[\int_{\Omega_e} \frac{1}{\rho^d} \phi\left(\frac{\boldsymbol{y} - \boldsymbol{x}}{\rho}\right) d\boldsymbol{y} \right.$$

$$\left. \sum_{i \in \Lambda_e} \psi_{e,i}(\boldsymbol{x}) \, (u - \mathcal{I}_{1,e} u)(\boldsymbol{x}_{e,i}) \right] b_0(\boldsymbol{x}). \tag{6.41}$$

Note that for $\boldsymbol{x} \in \Omega_e$, the term with the index n in the summation is possibly nonzero only if $\Omega_n \cap \tilde{\Omega}_j \neq \emptyset$. Using the bounds (6.38) and (6.37), we then have

$$\|\mathcal{I}_1(u - \mathcal{I}_{1,e}u)\|_{H^1(\Omega_e)} \leq c\, h^{-1} \|u - \mathcal{I}_{1,e}u\|_{L^\infty(\tilde{\Omega}_e)} |\tilde{\Omega}_e|^{1/2}.$$

Here, $|\tilde{\Omega}_e|$ denotes the volume of the region $\tilde{\Omega}_e$, $|\tilde{\Omega}_e| \leq c\, h^d$. Applying the estimate (6.39), we obtain

$$\|\mathcal{I}_1(u - \mathcal{I}_{1,e}u)\|_{H^1(\Omega_e)} \leq c\, h\, |u|_{H^2(\tilde{\Omega}_e)}. \tag{6.42}$$

A similar argument shows that

$$\|\mathcal{I}_1(u - \mathcal{I}_{1,e}u)\|_{L^2(\Omega_e)} \leq c\, h^2 |u|_{H^2(\tilde{\Omega}_e)}. \tag{6.43}$$

From the decomposition (6.40), the error estimates (6.39), (6.42) and (6.43), we obtain

$$\|u - \mathcal{I}_1 u\|_{L^2(\Omega_e)} + h\, \|u - \mathcal{I}_1 u\|_{H^1(\Omega_e)} \leq c\, h^2 |u|_{H^2(\tilde{\Omega}_e)}, \quad e \in \Lambda_E$$

Then we have the following global interpolation error estimates:

$$\|u - \mathcal{I}_1 u\|_{L^2(\Omega)} + h\, \|u - \mathcal{I}_1 u\|_{H^1(\Omega)} \leq c\, h^2 |u|_{H^2(\Omega)}. \tag{6.44}$$

Theorem 6.2.1. *Consider a spatial discretization satisfying the assumptions (6.32) and (6.33). Construct the RKEM interpolant with a window function satisfying condition (6.34). Then for any $u \in C^0(\Omega) \cap H^2(\Omega)$, the interpolation error estimate (6.44) holds.*

Consider solving a linear second-order elliptic boundary value problem. The weak formulation is:

$$\text{Find } u \in V, \quad \text{such that } a(u, v) = \ell(v) \quad \forall\, v \in V, \tag{6.45}$$

where $V \subset H^1(\Omega)$. The bilinear form $a(\cdot, \cdot)$ is continuous and V-elliptic, and the linear form ℓ is continuous on V. In the setting described at the beginning of the section, we define X_h to be the space of functions of the form

$$\sum_{e \in \Lambda_E} \left[\int_{\Omega_e} \frac{1}{\rho^d} \phi\left(\frac{y - x}{\rho} \right) dy \sum_{i \in \Lambda_e} \psi_{e,i}(x)\, \xi_{e,i} \right] b_0(x),$$

where the coefficient function $b_0(x)$ is given in (6.36), $\xi_{e,i} \in \mathbb{R}$, and if $x_{e_1,i_1} = x_{e_2,i_2}$ is a node common to two elements, then $\xi_{e_1,i_1} = \xi_{e_2,i_2}$. Then we let $V_h = V \cap X_h$ and approximate the continuous problem (6.45) by

$$\text{Find } u_h \in V_h, \quad \text{such that } a(u_h, v_h) = \ell(v_h), \quad \forall\, v_h \in V_h. \tag{6.46}$$

By the Lax-Milgram theorem (see[84]), both (6.45) and (6.46) have unique solutions. To estimate error, we can use Céa's inequality,

$$\|u - u_h\|_{H^1(\Omega)} \leq c \inf_{v_h \in V_h} \|u - v_h\|_{H^1(\Omega)}.$$

Suppose the boundary condition is of Neumann or Robin type, or of Dirichlet type in the one-dimensional case. Then we can replace the term

$$\inf_{v_h \in V_h} \|u - v_h\|_{H^1(\Omega)}$$

by $\|u - \mathcal{I}_1 u\|_{H^1(\Omega)}$ in Céa's inequality, and conclude that

$$\|u - u_h\|_{H^1(\Omega)} \le c \|u - \mathcal{I}_1 u\|_{H^1(\Omega)} \le c\, h\, |u|_{H^2(\Omega)}$$

if the exact solution $u \in H^2(\Omega)$. Furthermore, the standard duality argument can be employed to show that

$$\|u - u_h\|_{L^2(\Omega)} \le c\, h^2 |u|_{H^2(\Omega)}.$$

In a sequel paper, we will extend the error analysis above to the more general cases. Loosely speaking, under similar assumptions on the finite element partitions and kernel functions, if the reproducing degree is m and the regularity index l in (6.34) is not smaller than m, then for the reproducing interpolant defined in (6.22), we have the error estimates

$$\|v - \mathcal{I}v\|_{H^j(\Omega)} \le c\, h^{m+1-j} \|v\|_{H^{m+1}(\Omega)}, \quad j = 0, 1, \ldots, m, \ \forall v \in H^{m+1}(\Omega).$$

6.2.4 Numerical Examples

In this section, we report numerical results for the performance of the proposed RKEM in solving various boundary value problems of the differential equations with special features.

A Problem with Rough Solution. To validate the method, a special 1-D benchmark problem is solved first by using the proposed method. This problem was originally proposed by Rachford and Wheeler[384] to test the convergence property of the H^1-Galerkin method, and was used again by Babuška et al.[22] to test the mixed-hybrid finite element method, and by Liu et al.[303] to test the meshfree reproducing kernel particle method. The boundary value problem is

$$-u'' + u = f(x), \quad x \in (0, 1), \tag{6.47}$$
$$u'(0) = \alpha/(1 + \alpha^2 \bar{x}^2), \tag{6.48}$$
$$u'(1) = -\left[\arctan(\alpha(1 - \bar{x})) + \arctan(\alpha\bar{x})\right]. \tag{6.49}$$

The right side function in (6.47) is chosen as

$$f(x) = \frac{2\alpha \left[1 + \alpha^2(1 - \bar{x})(x - \bar{x})\right]}{[1 + \alpha^2(x - \bar{x})^2]^2} + (1 - x)\left[\arctan(\alpha(x - \bar{x})) + \arctan(\alpha\bar{x})\right]$$

so that the exact solution of this problem is

$$u(x) = (1 - x)\left[\arctan(\alpha(x - \bar{x})) + \arctan(\alpha\bar{x})\right].$$

The solution changes its roughness as the parameter α varies. It becomes smoother as the parameter α gets smaller, and the graph of the solution has a sharp knee at location $x = \bar{x}$ when α is very large. For the numerical example here, α and $\bar{x}$ are chosen as 50.0 and 0.40, respectively. The comparison of an exact solution and numerical solution with 80 nodes is plotted in Fig. 6.5.

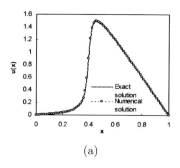
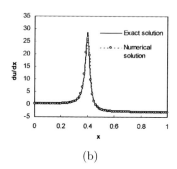

(a)   (b)

Fig. 6.5. Comparison between exact and numerical solutions of the benchmark problem: (a) the exact and numerical solution; (b) the derivative of exact and numerical solution.

Convergence rates of the numerical solutions are first examined for global RKEM interpolants satisfying a linear consistency condition. A cubic B-spline kernel function is used to construct the reproducing kernel function. Different spatial discretizations, in which the number of nodes uniformly varies from 11 to 2,561, are analyzed. Convergence rates in terms of L_2 and H^1 interpolation error norm for different support sizes are plotted in Figs. 6.6 (a) and (b) respectively. Although the interpolation solutions are more accurate for normalized support size from 0.8 to 1.5 than that of FEM, the convergence rates are 2 and 1 in L_2 and H^1 interpolation error norm respectively. When the optimal support size $1.99h$ is chosen, convergence rate 2 in both L_2 and H^1 interpolation error norm is observed for this problem.

The same spatial discretizations are used to test the convergence rates of the proposed method. As shown in Fig. 6.7 (a), for this example the numerical solution in the L_2 error norm is improved compared with the FEM solution even though they have roughly the same convergence rate index 2. The improvement in the H^1 error norm is more dramatic due to the high order continuity of global basis functions as illustrated in Fig. 6.7 (b).

An Example of a Fourth Order Differential Equation. We consider a boundary value problem of a one-dimensional fourth-order partial differential equation in this example. For conforming approximations, we need C^1 shape functions. It is difficult to construct a C^1 shape function for FEM in two

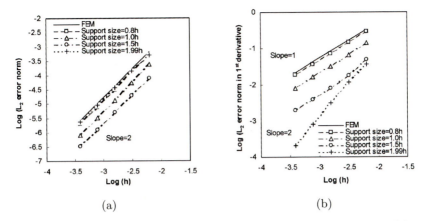

Fig. 6.6. Convergence rates of interpolation with different support sizes: (a) convergence rate in L_2 error norm; (b) convergence rate in L_2 error norm with first derivative.

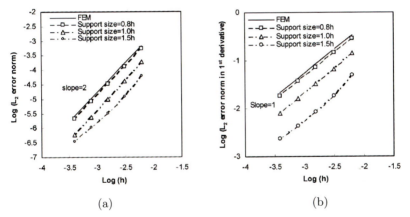

Fig. 6.7. Convergence rates of Galerkin solutions with different support sizes: (a) Convergence rate measured in L_2 error norm; (b) Convergence rate measured in L_2 error norm with the first derivative.

or higher dimension domain. However, shape functions with any smoothness degree can be easily constructed in any dimension for RKEM. To show the performance of RKEM on solving fourth-order problems, we apply it to the following problem:

$$u^{(4)} + u = f \quad \text{in } (0,1), \tag{6.50}$$

$$u^{(2)}(0) = u^{(3)}(0) = 1, \tag{6.51}$$

$$u^{(2)}(1) = u^{(3)}(1) = e, \tag{6.52}$$

where $f(x)2\,e^x$. The exact solution of this problem is $u(x) = e^x$.

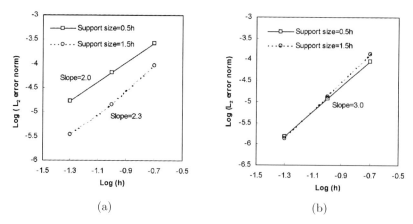

Fig. 6.8. L_2 norm errors for Galerkin solution and interpolation: (a) L_2 norm errors for Galerkin solution; (b) L_2 norm errors for interpolation.

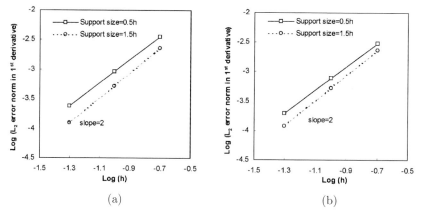

Fig. 6.9. L_2 norm errors for the 1st derivative of Galerkin solutions and interpolations: (a) L_2 norm errors for the 1st derivative of Galerkin solutions; (b) L_2 norm errors for the 1st derivative of interpolations.

Quadratic element is used to construct the basis function. Several uniform discretizations with quadratic elements are chosen in convergence study. The L_2 norms of the error in the Galerkin solution, its first and second derivatives are plotted in Figs. 6.8(a), 6.9(a) and 6.10(a), respectively. The corresponding interpolation L_2 error norms in primary variable and its first and second derivatives are given in Figs. 6.8(b), 6.9(b) and 6.10(b). The theoretical convergence orders for the RKEM interpolation errors in L_2, H^1 and H^2 are 3, 2, and 1. For RKEM solutions, the convergence rate in H^2 norm is 1. The convergence rate of numerical solutions match the theoretical results.

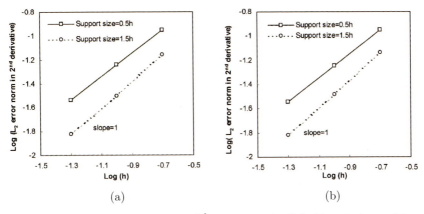

Fig. 6.10. L_2 norm errors for the 2nd derivative of a Galerkin solution and interpolation: (a) L_2 norm errors for the 2nd derivative of Galerkin solution; (b) L_2 norm errors for the 2nd derivative of interpolation.

A Two-dimensional Dirichlet Boundary Value Problem. The purpose of this example is to show that for Dirichlet boundary value problems, the proposed RKEM (i) can be applied directly; and (ii) maintains the optimal convergence order for any problem dimension and reproducing degree. We solve the following two-dimensional example by RKEM:

$$-\Delta u + u = f \quad \text{in } \Omega, \tag{6.53}$$
$$u = g \quad \text{on } \partial\Omega, \tag{6.54}$$

where $\Omega = (0,1)^2$, $f(x,y) = (1 - x^2 - y^2)e^{xy}$, and $g(x,y) = e^{xy}$.

The exact solution of this problem is $u(x,y) = e^{xy}$. A set of three spatial discretizations consisting 4×4, 8×8 and 16×16 quadratic rectangular/square elements is used in convergence study. Denote h the side of the corresponding square element. We consider two cases depending on the support size of the kernel function.

Case 1. When the support size is less than $0.5h$, the RKEM interpolant enjoys the Kronecker delta property. The numerical solution of the new method is compared with that of FEM. As shown in Figs. 6.11(a) and 6.12(a), almost the same convergence rate is observed for both methods. Unlike meshfree interpolant with quadratic basis, the convergence rate of RKEM solution in H^1 error norm is around 2.

Case 2. When the support size is larger than $0.5h$, RKEM interpolant losses the Kronecker delta property. To enforce the boundary conditions, a similar technique used in meshfree methods is adopted. As shown in Figs. 6.11(a) and 6.12(a), the accuracy of the numerical solution via RKEM in this case is actually improved compared with that of FEM.

The comparison of interpolation results in L_2 and H^1 error norm for RKEM and FEM is given in Figs. 6.11(b) and 6.12(b).

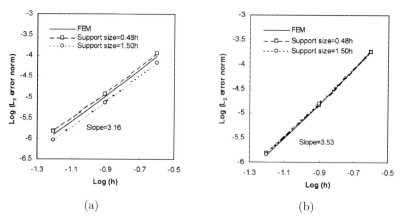

(a)   (b)

Fig. 6.11. L_2 norm errors for Galerkin solutions and interpolations: (a) L_2 norm errors for Galerkin solutions; (b) L_2 norm errors for interpolations.

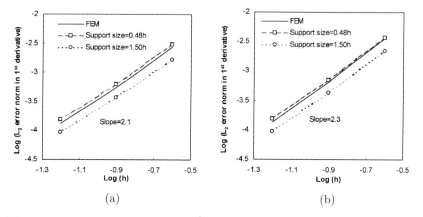

(a)   (b)

Fig. 6.12. L_2 norm errors for the 1st derivative of Galerkin solutions and interpolations: (a) L_2 norm errors for the 1st derivative of Galerkin solutions; (b) L_2 norm errors for the 1st derivative of interpolations.

Cantilever Beam Problem. Consider a linear elastic cantilever beam with external load P acting on its right end. The cantilever beam has a depth $D1$

and length $L4$ as shown in Fig. 6.13. Using RKEM, we solve it numerically as a plane strain problem with material properties: Young's modulus
$E = 3.0 \times 10^7$ and Poisson's ratio $\nu = 0.3$. The traction boundary conditions
at $x0$ and $x = L$ are prescribed according to an exact solution. Three uniformly spatial discretizations, with 85, 297 and 1,105 nodes respectively as
shown in Fig. 6.14 (a), (b) and (c) are used for convergence study. A bilinear basis was adopted for FEM and for RKEM as well to generate RKEM
interpolants with a cubic B-spline kernel function. For comparison, the numerical results of FEM and RKEM, numerical errors are measured by both
L2 norm in displacement and energy norm, which are displayed in Figs. 6.15
(a), (b) respectively. Based on numerical results, the RKEM solution is more
accurate than the FEM solution, especially in derivatives, with comparable
computational cost.

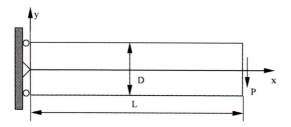

Fig. 6.13. Problem statement of a cantilever beam.

In the analysis of almost incompressible materials, locking behavior will be
observed in the numerical methods with the pure displacement formulation.
In RKPM, however locking behavior is reduced or even avoided by choosing
a proper dilation parameter. The beam problem with $\nu = 0.4999$ is analyzed
by FEM, RKEM and RKPM. The comparison among the numerical results
by using these different methods for solving the cantilever beam with almost
incompressible material is made in a table (see Table 6.1).

Methods	5×17 nodes	9×33 nodes	17×65 nodes
FEM	23.6%	27.1%	27.6%
RKEM(with ρ=2.0h)	36.2%	66.0%	87.7%
RKEM(with ρ=2.9h)	72.7%	87.3%	95.0%
RKPM(with ρ=2.0h)	40.8%	78.7 %	89.5%
RKPM(with ρ=2.9h)	89.0%	95.7%	96.6%

Table 6.1. Tip deflection accuracy (%) for FEM, RKEM and RKPM in solving
beam problem with incompressible material.

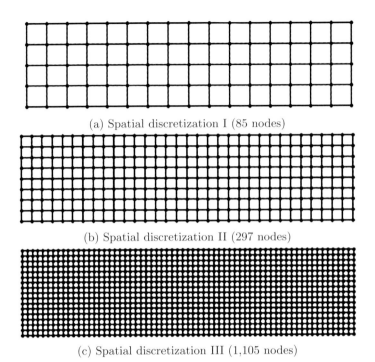

(a) Spatial discretization I (85 nodes)

(b) Spatial discretization II (297 nodes)

(c) Spatial discretization III (1,105 nodes)

Fig. 6.14. Model discretizations.

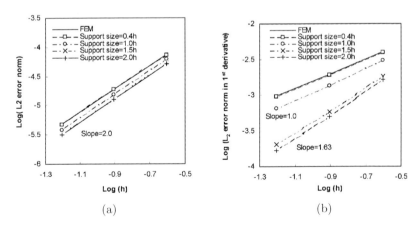

Fig. 6.15. Error norms for Galerkin solutions of FEM and RKEM (a) L_2 error norm in displacement for Galerkin solutions; (b) error norm in energy of Galerkin solutions.

6.3 Globally Conforming I^m/C^n Hierarchies

Constructing a globally conforming $I^m/C^n, (n, m > 1)$ interpolation field in multiple dimension was *the challenge* in the early development of finite element methods. It attracted a group of very creative engineers and mathematicians working on the subject. Some of them were intellectual heavy weights of the time, e.g. Clough and Tocher [1965], Bazeley et al. [1965], Fraeijs de Veubeke [1965], Argyris et al. [1968], Irons [1969], Felippa and Clough [1970], Bramble and Zlámal [1970], Birkhoff [1971], Birkhoff and Mansfield [1974], among others.

However, the problem has never been solved in a satisfactory manner, as Thomas J. R. Hughes[209] commented in his critically acclaimed finite element textbook,

> "*Continuous (i.e., C^0) finite element interpolations are easily constructed. The same cannot be said for multi-dimensional C^1-interpolants. It has taken considerable ingenuity to develop compatible C^1-interpolation schemes for two-dimensional plate elements based on classical theory, and the resulting scheme have always been extremely complicated in one way or another.*"

During the past half century, it has been an outstanding problem to construct compatible higher order continuous finite element interpolants in multiple dimensions. Although there were a few compatible C^1 elements constructed in two-dimensional case, e.g. Argyris element[14] or Bell's element,[35] they usually require adding extra degrees of freedom on either nodal points or along the boundary or in the interior of the element to make it work (or eventually do not work well because of their complexity).

In fact, the problem is still one of open problems in computational geometry, needless to say the construction of an arbitrary continuous $I^m/C^n(\Omega)$ interpolation field. In order to circumvent the difficulties in constructing higher order continuous interpolants, various mixed formulations have been developed over the years to relax the continuity requirement on finite element interpolation spaces. However, the problem does not entirely go away. To a large extent, it is translated into another problem — stability of mixed formulations. Most mixed weak formulations may not be coercive unless finite element interpolants used satisfy certain pre-requisites, e.g. Babuška-Brezzi condition. As it is well known, most lower order interpolation fields do not satisfy such condition.

As a matter of fact, in many engineering applications, the incompatible element is the only viable choice in numerical computations. During the past fifty years, a main theme of finite element method research is to develop suitable finite element shape functions that can be used in various mixed formulations, or can be used with incompatible modes. Nevertheless, no general solution has been found. Even though engineers invented the so-called patch test to examine applicability of various incompatible elements,[219] serious-minded mathematicians view it as a *"variational crime"*. It is fair to say that this predicament in FEM has more or less hinged the advancement of finite element technology.

In this section, we present systematic procedures to construct I^m/C^n interpolants in the framework of reproducing kernel element method.

6.4 Globally Conforming I^m/C^n Hierarchy I

In this section, we construct the first globally conforming I^m/C^n hierarchy. Consider the nodal point i of the element $e \in \Lambda_E$. Assume that one can find a global partition polynomial basis, $\psi_{e,i}^{(0)}(\boldsymbol{x})$, such that

$$D^\alpha \psi_{e,i}^{(0)} \Big|_{\boldsymbol{x}=\boldsymbol{x}_j} = \delta_{\alpha 0}\delta_{ij}, \qquad |\alpha| \leq q . \tag{6.55}$$

where the superscript (0) indicating that this is a global partition polynomial at zero-th level, which implies that it only interpolates the value of an unknown function, not for its derivatives. In our notation, the integer number in the superscript of an interpolant denotes the order of the derivatives that it interpolates.

We can then construct an RKEM interpolant basis as

$$\Psi_I^{(0)}(\boldsymbol{x}) = \mathop{\mathbf{A}}_{e\in\Lambda_E} \left(\sum_{j\in\Lambda_e} \mathcal{K}_{e,j}^\rho(\boldsymbol{x})\Delta x_{e,j} \right) \psi_{e.i}^{(0)}(\boldsymbol{x}) \tag{6.56}$$

where the subscript I is the global nodal index number, and the FEM connectivity $(e, i) \rightarrow I$ is implied. The meshfree kernel functions satisfy the following condition,

$$\mathop{\mathbf{A}}_{e\in\Lambda_E} \left(\sum_{j\in\Lambda_e} \mathcal{K}_{e,j}^\rho(\boldsymbol{x})\Delta x_{e,j} \right) \sum_{i\in\Lambda_e} \psi_{e.i}^{(0)}(\boldsymbol{x}) = 1 . \tag{6.57}$$

Note that it may be not necessary to require the global partition polynomials, $\{\psi_{e,i}^{(0)}\}_{i\in\Lambda_e}$, to form a partition of unity.

We construct higher order RKEM bases by multiplying the zero-th order basis with some polynomials, i.e. $I \in \Lambda_P$,

$$\Psi_I^{(1)}(\boldsymbol{x}) = (\boldsymbol{x} - \boldsymbol{x}_I)\Psi_I^{(0)}(\boldsymbol{x}) \tag{6.58}$$

$$\Psi_I^{(2)}(\boldsymbol{x}) = \frac{1}{2}(\boldsymbol{x} - \boldsymbol{x}_I)^2\Psi_I^{(0)}(\boldsymbol{x}) \tag{6.59}$$

$$\cdots\cdots$$

$$\Psi_I^{(\ell)}(\boldsymbol{x}) = \frac{1}{\ell!}(\boldsymbol{x} - \boldsymbol{x}_I)^\ell\Psi_I^{(0)}(\boldsymbol{x}) . \tag{6.60}$$

The proposed RKEM I^m/C^n interpolation hierarchy can be written as

$$\mathcal{I}^m f = \sum_{I \in \Lambda_P} \left(\Psi_I^{(0)}(\boldsymbol{x}) f_I + \Psi_I^{(1)}(\boldsymbol{x}) Df \Big|_I + \cdots + \Psi_I^{(m)}(\boldsymbol{x}) D^m f \Big|_I \right) \quad (6.61)$$

In this construction, $m = k$, which means that the globally conforming interpolants are only capable of reproducing a complete mth order polynomials.

The main result of this interpolation is summarized in the following proposition:

Proposition 6.4.1. *Assume that* $\exists \psi_{e,i}^{(0)}(\boldsymbol{x})$, $\forall e \in \Lambda_E$, $i \in \Lambda_e$, *such that*

$$D^\alpha \psi_{e,i}^{(0)} \Big|_{\boldsymbol{x}=\boldsymbol{x}_j} = \delta_{\alpha 0}\delta_{ij}, \quad |\alpha| \le \ell, i, j \in \Lambda_e \quad (6.62)$$

and define,

$$\Psi_I^{(0)}(\boldsymbol{x}) = \mathbf{A}_{e \in \Lambda_E} \left(\sum_{j \in \Lambda_e} \mathcal{K}_{e,j}^\rho(\boldsymbol{x}) \Delta x_{e,j} \right) \psi_{e,i}^{(0)}(\boldsymbol{x}) \quad (6.63)$$

$$\Psi_I^{(1)} = (\boldsymbol{x} - \boldsymbol{x}_I) \Psi_I^{(0)}(\boldsymbol{x}) \quad (6.64)$$

$$\Psi_I^{(2)} = \frac{1}{2!}(\boldsymbol{x} - \boldsymbol{x}_I)^2 \Psi_I^{(0)}(\boldsymbol{x}) \quad (6.65)$$

$$\cdots\cdots$$

$$\Psi_I^{(\ell)} = \frac{1}{\ell!}(\boldsymbol{x} - \boldsymbol{x}_I)^\ell \Psi_I^{(0)}(\boldsymbol{x}) \quad (6.66)$$

where $(e, i) \to I$ *and*

$$\sum_{I \in \Lambda_P} \Psi_I^{(0)}(\boldsymbol{x}) := \mathbf{A}_{e \in \Lambda_E} \left(\sum_{j \in \Lambda_e} \mathcal{K}_{e,j}^\rho(\boldsymbol{x}) \Delta x_{e,j} \right) \left(\sum_{i \in \Lambda_e} \psi_{e,i}^{(0)}(\boldsymbol{x}) \right) = 1 . \quad (6.67)$$

The interpolation scheme (6.61) has the following properties:

1. $D^\alpha \Psi_I^{(\beta)} \Big|_{\boldsymbol{x}=\boldsymbol{x}_J} = \delta_{\alpha\beta}\delta_{IJ}$, $I, J \in \Lambda_P$, $|\alpha|, |\beta| \le m$;
2. $\mathcal{I}^m \boldsymbol{x}^\lambda = \boldsymbol{x}^\lambda$, $\forall \boldsymbol{x} \in \Omega$, $|\lambda| \le m$.

Proof:
We first show that $D^\alpha \Psi_I^{(0)} \Big|_{\boldsymbol{x}=\boldsymbol{x}_J} = \delta_{\alpha 0}\delta_{IJ}$, $|\alpha| \le m$.
By the product rule,

$$D^\alpha \Psi_I^{(0)} \Big|_{\boldsymbol{x}=\boldsymbol{x}_J} = D^\alpha \left\{ \mathbf{A}_{e \in \Lambda_E} \left[\left(\sum_{j \in \Lambda_e} \mathcal{K}_{e,j}^\rho(\boldsymbol{x}) \Delta x_{e,j} \right) \psi_{e,i}^{(0)}(\boldsymbol{x}) \right] \right\} \Big|_{\boldsymbol{x}=\boldsymbol{x}_J}$$

$$= \mathbf{A}_{e \in \Lambda_E} \left\{ \sum_{|\gamma| \le |\alpha|} \binom{\alpha}{\gamma} \left(\sum_{j \in \Lambda_e} D^{\alpha-\gamma} \mathcal{K}_{e,j}^\rho(\boldsymbol{x}) \Delta x_{e,j} \right) \left(D^\gamma \psi_{e,i}^{(0)}(\boldsymbol{x}) \right) \right\} \Big|_{\boldsymbol{x}=\boldsymbol{x}_J}$$

$$= \mathbf{A}_{e \in \Lambda_E} \left\{ \sum_{|\gamma| \le |\alpha|} \binom{\alpha}{\gamma} \delta_{\gamma 0}\delta_{IJ} \left(\sum_{j \in \Lambda_e} D^{\alpha-\gamma} \mathcal{K}_{e,j}^\rho(\boldsymbol{x}) \Delta x_{e,j} \right) \right\} \quad (6.68)$$

Note that in line 3, Eq.(6.62) is used if $\boldsymbol{x}_J$ is a nodal point of Ω_e. If $\boldsymbol{x}_J$ is not a nodal point of Ω_e, by locality of meshfree basis functions,

$$\boldsymbol{x}_J \notin \bar{\Omega}_e \Rightarrow \boldsymbol{x}_J \notin \bigcup_{j \in \Lambda_e} supp\{\mathcal{K}^\rho_{e,j}\} \tag{6.69}$$

and therefore, we have the term δ_{IJ} instead of δ_{Ij}. A pictorial illustration of this point is shown in Fig. 6.16.

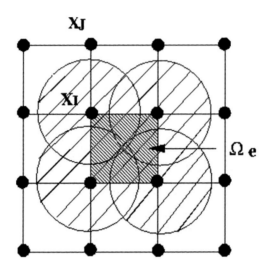

Fig. 6.16. A pictorial proof of $\boldsymbol{x}_J \notin \bar{\Omega}_e \Rightarrow \boldsymbol{x}_J \notin \bigcup_{j \in \Lambda_e} supp\{\mathcal{K}^\rho_{e,j}\}$.

Since the terms associated with $\gamma \neq 0$ are all zero, the only non-zero term left in (6.68) is the term corresponding to $\gamma = 0$. Consider $\begin{pmatrix} \alpha \\ 0 \end{pmatrix} = 1$. We have

$$D^\alpha \Psi_I^{(0)} = \delta_{IJ} \underset{e \in \Lambda_E}{\boldsymbol{A}} \left\{ \sum_{j \in \Lambda_e} D^\alpha \mathcal{K}^\rho_{e,j}(\boldsymbol{x}) \Delta \boldsymbol{x}_{e,j} \right\} \Big|_{\boldsymbol{x} = \boldsymbol{x}_J} \tag{6.70}$$

and we now show that $\underset{e \in \Lambda_E}{\boldsymbol{A}} \left\{ \sum_{j \in \Lambda_e} D^\alpha \mathcal{K}^\rho_{e,j}(\boldsymbol{x}) \Delta \boldsymbol{x}_{e,j} \right\} \Big|_{\boldsymbol{x} = \boldsymbol{x}_J} = \delta_{\alpha 0}.$

Even though in general,

$$\sum_{i \in \Lambda_e} \psi^{(0)}_{e,i}(\boldsymbol{x}) \neq 1,$$

but at the nodal point, Kronecker delta property ensures that

$$\sum_{i \in \Lambda_e} \psi_{e,i}^{(0)}(\boldsymbol{x}) \Big|_{\boldsymbol{x}=\boldsymbol{x}_J} = 1, \tag{6.71}$$

Consider the partition of unity condition (6.67).

$$\mathop{\mathbf{A}}_{e \in \Lambda_E} \left[\sum_{j \in \Lambda_e} \mathcal{K}_{e,j}^{\rho}(\boldsymbol{x}) \Delta \boldsymbol{x}_{e,j} \right] \left[\sum_{i \in \Lambda_e} \psi_{e,i}^{(0)}(\boldsymbol{x}) \right] = 1 \,. \tag{6.72}$$

It implies that

$$\mathop{\mathbf{A}}_{e \in \Lambda_E} \left\{ \sum_{j \in \Lambda_e} D^{\alpha} \mathcal{K}_{e,j}^{\rho}(\boldsymbol{x}) \Delta \boldsymbol{x}_{e,j} \right\} \Big|_{\boldsymbol{x}=\boldsymbol{x}_J} = 1, \quad \text{when} \quad \alpha = 0 \,. \tag{6.73}$$

When $\alpha \neq 0$, differentiation of (6.72) yields,

$$D^{\alpha} \mathop{\mathbf{A}}_{e \in \Lambda_E} \left\{ \sum_{j \in \Lambda_e} \mathcal{K}_{e,j}^{\rho}(\boldsymbol{x}) \Delta \boldsymbol{x}_{e,j} \right\} \Big|_{\boldsymbol{x}=\boldsymbol{x}_J}$$

$$= \mathop{\mathbf{A}}_{e \in \Lambda_E} \left\{ \sum_{|\gamma| \leq |\alpha|} \binom{\alpha}{\gamma} \left[\sum_{j \in \Lambda_e} D^{\alpha-\gamma} \mathcal{K}_{e,j}^{\rho}(\boldsymbol{x}) \Delta \boldsymbol{x}_{e,j} \right] \left[\sum_{i \in \Lambda_e} D^{\gamma} \psi_{e,i}^{(0)}(\boldsymbol{x}) \right] \right\} \Big|_{\boldsymbol{x}=\boldsymbol{x}_J}$$

$$= \mathop{\mathbf{A}}_{e \in \Lambda_E} \left\{ \sum_{|\gamma| \leq |\alpha|} \binom{\alpha}{\gamma} \left[\sum_{j \in \Lambda_e} D^{\alpha-\gamma} \mathcal{K}_{e,j}^{\rho}(\boldsymbol{x}) \Delta \boldsymbol{x}_{e,j} \right] \Big|_{\boldsymbol{x}=\boldsymbol{x}_J} \left[\sum_{i \in \Lambda_e} \delta_{\gamma 0} \delta_{IJ} \right] \right\}$$

$$= \delta_{IJ} \mathop{\mathbf{A}}_{e \in \Lambda_E} \left\{ \binom{\alpha}{0} \left[\sum_{j \in \Lambda_e} D^{\alpha} \mathcal{K}_{e,j}^{\rho}(\boldsymbol{x}) \Delta \boldsymbol{x}_{e,j} \right] \right\} \Big|_{\boldsymbol{x}=\boldsymbol{x}_J} = 0 \,,$$

$$\Rightarrow \mathop{\mathbf{A}}_{e \in \Lambda_E} \left\{ \sum_{j \in \Lambda_e} D^{\alpha} \mathcal{K}_{e,j}^{\rho}(\boldsymbol{x}) \Delta \boldsymbol{x}_{e,j} \right\} \Big|_{\boldsymbol{x}=\boldsymbol{x}_J} = 0, \quad \text{when} \quad \alpha \neq 0 \,. \tag{6.74}$$

We now show that

$$D^{\alpha} \Psi_I^{(\beta)}(\boldsymbol{x}) \Big|_{\boldsymbol{x}=\boldsymbol{x}_J} = \delta_{\alpha\beta} \delta_{IJ}, \quad \text{when} \quad \beta \neq 0 \,. \tag{6.75}$$

where $\Psi_I^{(\beta)}(\boldsymbol{x}) = \dfrac{(\boldsymbol{x} - \boldsymbol{x}_I)^{\beta}}{\beta!} \Psi_I^{(0)}(\boldsymbol{x})$.

It is readily to show that

$$D^{\alpha} \Psi_I^{(\beta)}(\boldsymbol{x}) = \sum_{|\gamma| < |\alpha|} \binom{\alpha}{\gamma} D^{\alpha-\gamma} \Psi_I^{(0)}(\boldsymbol{x}) D^{\gamma} \left((\boldsymbol{x} - \boldsymbol{x}_I)^{\beta} \right)$$

$$= \sum_{|\gamma| \leq |\alpha|} \left(\binom{\alpha}{\gamma} D^{\alpha-\gamma} \Psi_I^{(0)}(\boldsymbol{x}) \frac{\beta!}{(\beta - \gamma - 1)! \beta!} \right.$$

$$\left. (\boldsymbol{x} - \boldsymbol{x}_I)^{<\beta-\gamma>} \right) \Big|_{\boldsymbol{x}=\boldsymbol{x}_J} \tag{6.76}$$

where

$$< \beta - \gamma >= \begin{cases} \beta - \gamma, & |\beta - \gamma| \geq 0 \\ 0, & |\beta - \gamma| < 0 \end{cases} \tag{6.77}$$

Consequently,

$$D^\alpha \Psi_I^{(\beta)}(\boldsymbol{x}) = \sum_{|\gamma| \leq |\alpha|} \binom{\alpha}{\gamma} D^{\alpha-\gamma} \Psi_I^{(0)}(\boldsymbol{x}) \frac{1}{(\beta - \gamma)!} (\boldsymbol{x} - \boldsymbol{x}_I)^{<\beta-\gamma>} \Big|_{\boldsymbol{x}=\boldsymbol{x}_J}$$

$$= \sum_{|\gamma| \leq |\alpha|} \binom{\alpha}{\gamma} \delta_{\alpha-\beta 0} \delta_{IJ} \frac{1}{(\beta - \gamma)!} (\boldsymbol{x}_J - \boldsymbol{x}_I)^{<\beta-\gamma>}$$

$$= \binom{\alpha}{\beta} \delta_{\alpha\beta} \delta_{IJ} = \delta_{\alpha\beta} \delta_{IJ} \ . \tag{6.78}$$

We now show that $\mathcal{I}\boldsymbol{x}^\lambda = \boldsymbol{x}^\lambda$. By construction $|\lambda| < m$,

$$\mathcal{I}\boldsymbol{x}^\lambda = \sum_{I \in \Lambda_P} \left(\Psi_I^{(0)}(\boldsymbol{x}) \boldsymbol{x}^\lambda + \Psi_I^{(1)}(\boldsymbol{x}) \lambda \boldsymbol{x}^{<\lambda-1>} + \cdots \right.$$

$$+ \Psi_I^{(\gamma)}(\boldsymbol{x}) \lambda(\lambda - 1) \cdots (\lambda - \gamma + 1) \boldsymbol{x}_I^{<\lambda-\gamma>} + \cdots$$

$$\left. + \Psi_I^{(m)}(\boldsymbol{x}) \lambda(\lambda - 1) \cdots (\lambda - m + 1) \boldsymbol{x}_I^{<\lambda-m>} \right) \tag{6.79}$$

Consider $\Psi_I^{(\gamma)}(\boldsymbol{x}) := \frac{1}{\gamma!} (\boldsymbol{x} - \boldsymbol{x}_I)^\gamma \Psi_I^{(0)}(\boldsymbol{x})$ and the last term $< \lambda - \gamma >\geq 0$ is the term $\gamma = \lambda$. We then have

$$\mathcal{I}\boldsymbol{x}^\lambda = \sum_{I \in \Lambda_P} \Psi_I^{(0)}(\boldsymbol{x}) \left(\binom{\lambda}{0} \boldsymbol{x}_I^\lambda + \binom{\lambda}{1} \boldsymbol{x}_I^{\lambda-1} (\boldsymbol{x} - \boldsymbol{x}_I) + \cdots + \right.$$

$$\left. + \cdots + \binom{\lambda}{\gamma} \boldsymbol{x}_I^{\lambda-\gamma} (\boldsymbol{x} - \boldsymbol{x}_I)^\gamma + \cdots + \binom{\lambda}{\lambda} (\boldsymbol{x} - \boldsymbol{x}_I)^\lambda \right)$$

$$= \sum_{I \in \Lambda_P} \Psi_I^{(0)}(\boldsymbol{x}) \left(\sum_{|\gamma| \leq |\lambda|} \binom{\lambda}{\gamma} \boldsymbol{x}_I^{\lambda-\gamma} (\boldsymbol{x} - \boldsymbol{x}_I)^\gamma \right)$$

$$= \sum_{I \in \Lambda_P} \Psi_I^{(0)}(\boldsymbol{x}) \boldsymbol{x}^\lambda = \boldsymbol{x}^\lambda \tag{6.80}$$

where the partition of unity condition $\sum_{I \in \Lambda_P} \Psi_I^{(0)}(\boldsymbol{x}) = 1$ is used.

♣

6.4.1 1D I^2/C^n Interpolation

Let $\xi = (x - x_e)/L_e$ and $L_e = x_{e+1} - x_e$. Choose the continuous extension of the first two 1D fifth order Hermite interpolants as the global partition polynomials

$$\begin{cases} \psi_{e,1}(x) = H_1^{(0)}(\xi) = 1 - 10\xi^3 + 15\xi^4 - 6\xi^5 \\ \psi_{e,2}(x) = H_2^{(0)}(\xi) = 10\xi^3 - 15\xi^4 + 6\xi^5 \end{cases} \qquad (6.81)$$

The fifth order Hermite interpolants satisfy the following interpolation conditions,

$$\text{partition of unity} : \sum_{i=1}^{2} H_{e,i}^{(0)}(x) = 1; \qquad (6.82)$$

$$\frac{d^\alpha H_{e,i}^{(0)}}{dx^\alpha}\bigg|_{x=x_j} = \delta_{\alpha 0}\delta_{ij}, \quad i,j = 1,2; \quad 0 \le \alpha \le 2 \qquad (6.83)$$

Suppose that the finite element connectivity map has the form

$$\Lambda_E \times \Lambda_e \to \Lambda_P : (e_1, i_1), (e_2, i_\ell) \to I \qquad (6.84)$$

Note that $\ell = 1, 2$. In the interior $\ell = 2$ and on the boundary $\ell = 1$.

The 1D I^2/C^n RKEM interpolants are constructed as follows:

$$\Psi_I^{(0)}(x) = \sum_{k=1}^{\ell} \left(\sum_{j \in \Lambda_{e_k}} \mathcal{K}_{e_k,j}^\varrho(x) \Delta x_{e_k,j} \right) H_{e_k,i_k}^{(0)}(x) \qquad (6.85)$$

$$\Psi_I^{(1)}(x) = \sum_{k=1}^{\ell} \left(\sum_{j \in \Lambda_{e_k}} \mathcal{K}_{e_k,j}^\varrho(x) \Delta x_{e_k,j} \right) (x - x_I) H_{e_k,i_k}^{(0)}(x) \qquad (6.86)$$

$$\Psi_I^{(2)}(x) = \sum_{k=1}^{\ell} \left(\sum_{j \in \Lambda_{e_k}} \mathcal{K}_{e_k,j}^\varrho(x) \Delta x_{e_k,j} \right) \frac{(x - x_I)^2}{2!} H_{e_k,i_k}^{(0)}(x) \qquad (6.87)$$

The hybrid RKEM interpolation can be expressed as

$$\mathcal{I}f(x) = \sum_{I \in \Lambda_P} \left(\Psi_I^{(0)}(x) f_I + \Psi_I^{(1)}(x) f_I' + \Psi_I^{(2)}(x) f_I'' \right) \qquad (6.88)$$

where $f_I' := \dfrac{df}{dx}\bigg|_{x=x_I}$ and $f_I'' := \dfrac{d^2 f}{dx^2}\bigg|_{x=x_I}$

We choose fifth order spline as meshfree window function. All three I^2/C^4 RKEM shape functions and their derivatives are plotted in Fig. 6.17.

The proposed I^2/C^n RKEM interpolants satisfy Hermite interpolation conditions (6.82)-(6.83), and they can reproduce polynomial, $x^\lambda, \lambda = 0, 1, 2$.

6.4.2 2D I^0/C^n Quadrilateral Element

We now consider a two-dimensional I^0 quadrilateral element (see Fig. 6.18 (a)). By nodal integration (trapezoidal rule),

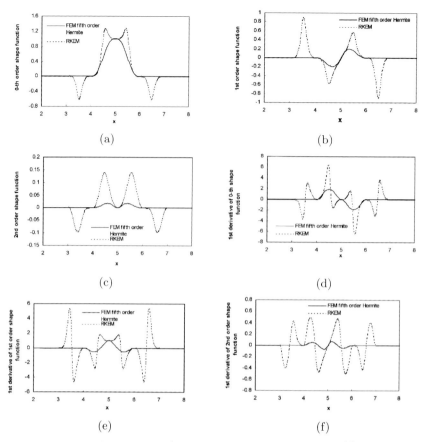

Fig. 6.17. An 1D I^2/C^4 RKEM interpolant: (a) shape function $\Psi_I^{(0)}(x)$; (b) shape function $\Psi_I^{(1)}(x)$; (c) shape function $\Psi_I^{(2)}(x)$; (d) the 1st derivative of $\Psi_I^{(0)}(x)$; (e) the 1st derivative of $\Psi_I^{(1)}(x)$; and (f) the 1st derivative of $\Psi_I^{(2)}(x)$.

$$\int_{\Omega_e} \mathcal{K}_\varrho(\mathbf{y} - \mathbf{x}; \mathbf{x}) d\mathbf{y} \approx \frac{\Delta V_e}{4} \left(\mathcal{K}_{e,1}^\varrho(\mathbf{x}) + \mathcal{K}_{e,2}^\varrho(\mathbf{x}) + \mathcal{K}_{e,3}^\varrho(\mathbf{x}) + \mathcal{K}_{e,4}^\varrho(\mathbf{x}) \right)$$

$$= \sum_{j=1}^{4} \mathcal{K}_{e,j}^\varrho(\mathbf{x}) \Delta V_{e,j} \tag{6.89}$$

where $\Delta V_{e,j} = \frac{1}{4}\Delta V_e, \;\; j = 1, 2, 3, 4$, and

$$\mathcal{K}_{e,j}^\rho(\mathbf{x}) := \frac{1}{\varrho^2} w\left(\frac{\mathbf{x} - \mathbf{x}_{e,j}}{\varrho} \right) b(\mathbf{x}) \tag{6.90}$$

The partition of unity condition yields the equality

$$\mathop{\mathbf{A}}_{e \in \Lambda_E} \left(\sum_{j=1}^{4} \frac{\Delta V_{e,j}}{\varrho^2} w\left(\frac{\mathbf{x} - \mathbf{x}_{e,j}}{\varrho} \right) \right) b(\mathbf{x}) \left(\sum_{i=1}^{4} \psi_{e,i}(\mathbf{x}) \right) = 1 \tag{6.91}$$

Solving $b(\mathbf{x})$, one may find that

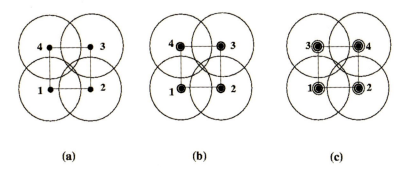

Fig. 6.18. Two dimensional quadrilateral RKEM hierarchical interpolation field: (a) I^0 element, (b) I^1 element, and (c) I^2 element.

$$K_{e,j}^{(0)}(\mathbf{x}) := \frac{1}{\varrho^2} w\left(\frac{\mathbf{x}_{e,j} - \mathbf{x}}{\varrho}\right) \left\{ \mathop{\mathbf{A}}_{e \in \Lambda_E} \left[\sum_{j=1}^{4} \frac{\Delta V_{e,j}}{\varrho^2} w\left(\frac{\mathbf{x}_{e,j} - \mathbf{x}}{\varrho}\right) \left(\sum_{i=1}^{4} \psi_{e,i}(\mathbf{x})\right)\right] \right\}^{-1}$$

(6.92)

Consider the connectivity map

$$(e_1, i_1), \cdots, (e_k, i_k), \cdots, (e_\ell, i_\ell) \to I \tag{6.93}$$

The two-dimensional quadrilateral reproducing kernel element shape function can be expressed as

$$\Psi_I^{(0)}(\mathbf{x}) = \sum_{k=1}^{\ell} \left(\sum_{j=1}^{4} \frac{\Delta V_{e_k,j}}{\varrho} w\left(\frac{\mathbf{x} - \mathbf{x}_{e_k,j}}{\varrho}\right) \right.$$

$$\left\{ \mathop{\mathbf{A}}_{e \in \Lambda_E} \left(\sum_{j=1}^{4} \frac{\Delta V_{e,j}}{\varrho^2} w\left(\frac{\mathbf{x} - \mathbf{x}_{e,j}}{\varrho}\right)\right) \left(\sum_{i=1}^{4} \psi_{e,i}(x)\right) \right\}^{-1}$$

$$\left. \psi_{e_k,i_k}(\mathbf{x}) \right). \tag{6.94}$$

For element e with four nodes $(x_1, y_1), (x_2, y_2), (x_3, y_3), (x_4, y_4)$, and the global partition shape functions are:

$$\psi_{e,1}(\xi(\mathbf{x}), \eta(\mathbf{x})) = \frac{1}{4}(1 - \xi)(1 - \eta) \tag{6.95}$$

$$\psi_{e,2}(\xi(\mathbf{x}), \eta(\mathbf{x})) = \frac{1}{4}(1 + \xi)(1 - \eta) \tag{6.96}$$

$$\psi_{e,3}(\xi(\mathbf{x}), \eta(\mathbf{x})) = \frac{1}{4}(1 + \xi)(1 + \eta) \tag{6.97}$$

$$\psi_{e,4}(\xi(\mathbf{x}), \eta(\mathbf{x})) = \frac{1}{4}(1 - \xi)(1 + \eta) \tag{6.98}$$

with $-\infty < \xi, \eta < \infty$ and functions $\xi(\mathbf{x})$ and $\eta(\mathbf{x})$ can be found by its inverse relationship,

$$\mathbf{x} = \sum_{i=1}^{4} \psi_{e,i}(\boldsymbol{\xi})\mathbf{x}_{e,i} \tag{6.99}$$

6.4.3 Globally Compatible Q12P1I1 Quadrilateral Element

This is a multiple dimensional compatible interpolant. It only has a minimal twelve degrees of freedom over four nodes. It is an I^1 interpolation, i.e. the primary variable and its first order derivatives are sampled. If the fifth order spline is chosen as the meshfree window function, its smoothness is C^4.

Consider a four-nodes quadrilateral element. Choosing the same meshfree kernel function as in Eq. (6.92), except that

$$\psi_{e,1}^{(0)}(\mathbf{x}) = H_{c1}^{(0)}(\xi)H_{c1}^{(0)}(\eta) \tag{6.100}$$

$$\psi_{e,2}^{(0)}(\mathbf{x}) = H_{c2}^{(0)}(\xi)H_{c1}^{(0)}(\eta) \tag{6.101}$$

$$\psi_{e,3}^{(0)}(\mathbf{x}) = H_{c2}^{(0)}(\xi)H_{c2}^{(0)}(\eta) \tag{6.102}$$

$$\psi_{e,4}^{(0)}(\mathbf{x}) = H_{c1}^{(0)}(\xi)H_{c2}^{(0)}(\eta) \tag{6.103}$$

where $H_{c1}^{(0)}(\zeta) = 1 - 3\zeta^2 + 2\zeta^3$, $H_{c2}^{(0)}(\zeta) = 3\zeta^2 - 2\zeta^3$, and $\zeta = \xi$ and η.

The coordinate transformation between (x, y) and (ξ, η) is bilinear,

$$x(\xi, \eta) = \alpha_0 + \alpha_1\xi + \alpha_2\eta + \alpha_3\xi\eta \tag{6.104}$$

$$y(\xi, \eta) = \beta_0 + \beta_1\xi + \beta_2\eta + \beta_3\xi\eta \tag{6.105}$$

and

$$\begin{cases} \alpha_0^e = x_{e,1}, \\ \alpha_1^e = x_{e,2} - x_{e,1}, \\ \alpha_2^e = x_{e,4} - x_{e,1}, \\ \alpha_3^e = x_{e,1} - x_{e,2} + x_{e,3} - x_{e,4}; \end{cases} \quad \text{and} \quad \begin{cases} \beta_0^e = y_{e,1}, \\ \beta_1^e = y_{e,2} - y_{e,1}, \\ \beta_2^e = y_{e,4} - y_{e,1}, \\ \beta_2^e = y_{e,1} - y_{e,2} + y_{e,3} - y_{e,4}. \end{cases} \tag{6.106}$$

Subsequently, one may find the related global partition polynomials,

$$x = \sum_{j=1}^{4} N_{e,j}(\xi, \eta)x_{e,j}, \quad y = \sum_{j=1}^{4} N_{e,j}(\xi, \eta)y_{e,j}, \tag{6.107}$$

where $N_{e,1}(\xi, \eta) = 1 - \xi - \eta + \xi\eta$, $N_{e,2}(\xi, \eta) = \xi(1 - \eta)$, $N_{e,3}(\xi, \eta) = \xi\eta$, and $N_{e,4}(\xi, \eta) = \eta(1 - \xi)$. Furthermore, it can be readily found that

$$\frac{\partial\xi}{\partial x} = \frac{1}{\Delta}(\beta_2 + \beta_3\xi), \quad \frac{\partial\xi}{\partial y} = \frac{-1}{\Delta}(\alpha_2 + \alpha_3\xi), \tag{6.108}$$

$$\frac{\partial\eta}{\partial x} = \frac{-1}{\Delta}(\beta_1 + \beta_3\eta), \quad \frac{\partial\eta}{\partial y} = \frac{1}{\Delta}(\alpha_1 + \alpha_3\eta), \tag{6.109}$$

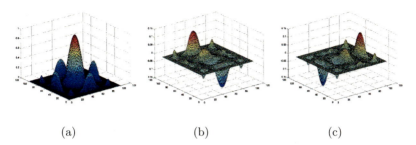

Fig. 6.19. Smooth quadrilateral I^1/C^4 RKEM interpolant: (a) the 1st shape function, $\Psi_I^{(00)}(\boldsymbol{x})$; (b) the 2nd shape function $\Psi_I^{(10)}(\boldsymbol{x})$; and (c) the 3rd shape function $\Psi_I^{(01)}(\boldsymbol{x})$.

where $\Delta = (\alpha_1 + \alpha_3\eta)(\beta_2 + \beta_3\xi) - (\alpha_2 + \alpha_3\xi)(\beta_1 + \beta_3\eta)$.

Three I^1/C^4 quadrilateral RKEM shape functions are plotted in Fig. 6.19.

The zero-th order shape functions in an element are

$$\Psi_{e,i}^{(00)}(\mathbf{x}) = \left(\sum_{j=1}^{4} \mathcal{K}_{e,j}^{\varrho}(\mathbf{x})\Delta V_{e,j}\right)\psi_{e,i}^{(0)}(\xi,\eta), \quad i = 1,2,3,4 \tag{6.110}$$

The higher order basis functions are

$$\Psi_{e,i}^{(10)}(\mathbf{x}) = (x - x_{e,i})\Psi_{e,i}^{(00)}(\mathbf{x}), \quad i = 1,2,3,4 \tag{6.111}$$

$$\Psi_{e,i}^{(01)}(\mathbf{x}) = (y - y_{e,i})\Psi_{e,i}^{(00)}(\mathbf{x}), \quad i = 1,2,3,4 \tag{6.112}$$

Suppose that the finite element connectivity map has the form,

$$\Lambda_E \times \Lambda_e \to \Lambda_P : (e_1, i_1), \cdots, (e_\ell, i_\ell) \to I \tag{6.113}$$

The zero-th order global shape function has the form

$$\Psi_I^{(00)}(\mathbf{x}) = \sum_{k\in\Lambda_I}\left(\sum_{j\in\Lambda_{e_k}} \mathcal{K}_{e_k,j}^{\varrho}(\mathbf{x})\Delta V_{e_k,j}\right)\psi_{e_k,i_k}^{(0)}(\mathbf{x})$$

$$= \sum_{k\in\Lambda_I} \Psi_{e_k,i_k}^{(00)}(\boldsymbol{x}) \tag{6.114}$$

where $\Lambda_I = \{1, 2, \cdots, \ell\}$.

The global higher order basis functions have simple form

$$\Psi_I^{(10)}(\mathbf{x}) = (x - x_I)\Psi_I^{(00)}(\mathbf{x}) \tag{6.115}$$

$$\Psi_I^{(01)}(\mathbf{x}) = (y - y_I)\Psi_I^{(00)}(\mathbf{x}) \tag{6.116}$$

One can easily write down 16 degrees of freedom bilinear RKEM quadrilateral interpolant function by adding four additional higher order partition polynomials,

$$\Psi_{e,i}^{(11)}(\mathbf{x}) = (x - x_{e,i})(y - y_{e,i})\Psi_{e,i}^{(00)}(\mathbf{x}), \quad i = 1, 2, 3, 4 \qquad (6.117)$$

or by adding one global shape function,

$$\Psi_I^{(11)}(\mathbf{x}) = (x - x_I)(y - y_I)\Psi_I^{(00)}(\mathbf{x}). \qquad (6.118)$$

It can be shown that the above compatible RKEM Interpolants can reproduce polynomials, $1, x, y,$ and xy.

6.4.4 Globally Compatible Q16P2I2 Quadrilateral Element

Choosing the same meshfree kernel function as in Eq. (6.92), except that

$$\psi_{e,1}^{(0)}(\mathbf{x}) = H_{f1}^{(0)}(\xi)H_{f1}^{(0)}(\eta) \qquad (6.119)$$

$$\psi_{e,2}^{(0)}(\mathbf{x}) = H_{f2}^{(0)}(\xi)H_{f1}^{(0)}(\eta) \qquad (6.120)$$

$$\psi_{e,3}^{(0)}(\mathbf{x}) = H_{f2}^{(0)}(\xi)H_{f2}^{(0)}(\eta) \qquad (6.121)$$

$$\psi_{e,4}^{(0)}(\mathbf{x}) = H_{f2}^{(0)}(\xi)H_{f1}^{(0)}(\eta) \qquad (6.122)$$

where $H_{f1}^{(0)}(\zeta) = 1 - 10\zeta^3 + 15\zeta^4 - 6\zeta^5$, and $H_{f2}^{(0)}(\zeta) = 10\zeta^3 - 15\zeta^4 + 6\zeta^5$. The coordinate transformation are bilinear, which is exactly the same as Eq. (6.104)-(6.107).

We can construct a compatible I^2/C^n RKEM interpolant with a minimal 24 degrees of freedom. The first four zero-order basis functions are

$$\Psi_{e,i}^{(00)}(\mathbf{x}) = \left(\sum_{j=1}^{4} \mathcal{K}^{\varrho}{}_{e,j}(\mathbf{x})\right)\psi_{e,i}^{(0)}(\mathbf{x}), \quad i = 1, 2, 3, 4 \qquad (6.123)$$

and higher order basis functions

$$\Psi_{e,i}^{(10)}(\mathbf{x}) = (x - x_{e,i})\Psi_{e,i}^{(00)}(\mathbf{x}), i = 1, 2, 3, 4 \qquad (6.124)$$

$$\Psi_{e,i}^{(01)}(\mathbf{x}) = (y - y_{e,i})\Psi_{e,i}^{(00)}(\mathbf{x}), \quad i = 1, 2, 3, 4 \qquad (6.125)$$

and

$$\Psi_{e,i}^{(20)}(\mathbf{x}) = \frac{1}{2}(x - x_{e,i})^2\Psi_{e,i}^{(00)}(\mathbf{x}), \quad i = 1, 2, 3, 4 \qquad (6.126)$$

$$\Psi_{e,i}^{(11)}(\mathbf{x}) = (x - x_{e,i})(y - y_{e,i})\Psi_{e,i}^{(00)}(\mathbf{x}), \quad i = 1, 2, 3, 4 \qquad (6.127)$$

$$\Psi_{e,i}^{(02)}(\mathbf{x}) = \frac{1}{2}(y - y_{e,i})^2\Psi_{e,i}^{(00)}(\mathbf{x}), \quad i = 1, 2, 3, 4 \qquad (6.128)$$

The six I^2/C^4 quadrilateral RKEM shape functions are plotted in Fig. 6.20.

The zero-th order global shape function has the form,

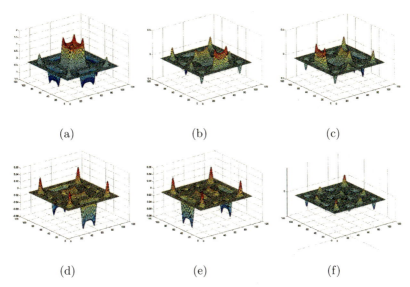

(a)   (b)   (c)

(d)   (e)   (f)

Fig. 6.20. Compatible quadrilateral $I^2/C^4/P^2$ RKPM interpolants: (a) the 1st shape function, $\Psi_I^{(00)}(\boldsymbol{x})$; (b) the 2nd shape function $\Psi_I^{(10)}(\boldsymbol{x})$; (c) the 3rd shape function $\Psi_I^{(01)}(\boldsymbol{x})$; (d) the 4th shape function, $\Psi_I^{(20)}(\boldsymbol{x})$; (e) the 5th shape function $\Psi_I^{(02)}(\boldsymbol{x})$; and (f) the 6th shape function $\Psi_I^{(11)}(\boldsymbol{x})$.

$$\Psi_I^{(00)}(\mathbf{x}) = \sum_{k=1}^{\ell} \Psi_{e_k,i_k}^{(00)}(\boldsymbol{x}) \tag{6.129}$$

$$\Psi_I^{(10)}(\mathbf{x}) = (x - x_I)\Psi_I^{(00)}(\mathbf{x}) \tag{6.130}$$

$$\Psi_I^{(01)}(\mathbf{x}) = (y - y_I)\Psi_I^{(00)}(\mathbf{x}) \tag{6.131}$$

$$\Psi_I^{(20)}(\mathbf{x}) = \frac{1}{2!}(x - x_I)^2\Psi_I^{(00)}(\mathbf{x}) \tag{6.132}$$

$$\Psi_I^{(11)}(\mathbf{x}) = (x - x_I)(y - y_I)\Psi_I^{(00)}(\mathbf{x}) \tag{6.133}$$

$$\Psi_I^{(02)}(\mathbf{y}) = \frac{1}{2!}(y - y_I)^2\Psi_I^{(00)}(\mathbf{x}) \tag{6.134}$$

6.4.5 Smooth I^0/C^n Triangle Element

We now construct RKEM triangle elements. For I^0 triangle (see: Fig. 6.21 (a)), the minimal degrees of freedom is nine. By nodal integration (Lobatto quadrature rule),

$$\int_{\Omega_e} \mathcal{K}_\varrho(\mathbf{y} - \mathbf{x}; \mathbf{x}) d\mathbf{y} \approx \frac{\Delta V_e}{3} \left(\mathcal{K}^\varrho_{e,1}(\mathbf{x}) + \mathcal{K}^\varrho_{e,2}(\mathbf{x}) + \mathcal{K}^\varrho_{e,3}(\mathbf{x}) \right)$$

$$= \sum_{j=1}^{3} \mathcal{K}^\varrho_{e,j}(\mathbf{x}) \Delta V_{e,j} \tag{6.135}$$

where $\Delta V_{e,j} = \frac{1}{3} \Delta V_e$, $j = 1, 2, 3$, and

$$\mathcal{K}^\varrho_{e,j}(\mathbf{x}) := \frac{1}{\varrho^2} w \left(\frac{\mathbf{x}_{e,j} - \mathbf{x}}{\varrho} \right) b(\mathbf{x}) \tag{6.136}$$

The partition of unity condition requires that

$$\mathop{\mathbf{A}}_{e \in \Lambda_E} \left(\sum_{j=1}^{3} \frac{\Delta V_e}{3\varrho^2} w \left(\frac{\mathbf{x}_{e,j} - \mathbf{x}}{\varrho} \right) \right) b(\mathbf{x}) \left(\sum_{i=1}^{4} \psi^{(0)}_{e,i}(\mathbf{x}) \right) = 1 \tag{6.137}$$

Subsequently, one may find that

$$\mathcal{K}^\varrho_{e,j}(\mathbf{x}) := \frac{1}{\varrho^2} w \left(\frac{\mathbf{x}_{e,j} - \mathbf{x}}{\varrho} \right) \left\{ \mathop{\mathbf{A}}_{e \in \Lambda_E} \left[\sum_{j=1}^{3} \frac{\Delta V_e}{3\varrho^2} w \left(\frac{\mathbf{x}_{e,j} - \mathbf{x}}{\varrho} \right) \left(\sum_{i=1}^{3} \psi^{(0)}_{e,i}(\mathbf{x}) \right) \right] \right\}^{-1} \tag{6.138}$$

For each element, there are three nodes $(x_1, y_1), (x_2, y_2), (x_3, y_3)$, the global partition polynomials are expressed in terms of area coordinates,

$$\psi^{(0)}_{e,i}(\mathbf{x}) = \xi_{e,i}(\mathbf{x}), \quad i = 1, 2, 3 \quad \text{and} \quad \sum_{i=1}^{3} \xi_{e,i} = 1, \tag{6.139}$$

$$\begin{bmatrix} 1 \\ x \\ y \end{bmatrix} = [\mathbf{T}] \begin{bmatrix} \xi_{e,1} \\ \xi_{e,2} \\ \xi_{e,3} \end{bmatrix} \quad \text{and} \quad \begin{bmatrix} \xi_{e,1} \\ \xi_{e,2} \\ \xi_{e,3} \end{bmatrix} = [\mathbf{T}]^{-1} \begin{bmatrix} 1 \\ x \\ y \end{bmatrix} \tag{6.140}$$

$$[\mathbf{T}] = \begin{bmatrix} 1 & 1 & 1 \\ x_1 & x_2 & x_3 \\ y_1 & y_2 & y_3 \end{bmatrix} \quad \text{and} \quad [\mathbf{T}]^{-1} = \frac{1}{\det \mathbf{T}} \begin{bmatrix} x_2 y_3 - x_3 y_2 & y_{23} & x_{32} \\ x_3 y_1 - x_1 y_3 & y_{31} & x_{13} \\ x_1 y_2 - x_2 y_1 & y_{12} & x_{21} \end{bmatrix} \tag{6.141}$$

where $\det \mathbf{T} = x_{21} y_{31} - x_{31} y_{21}$ and $x_{ij} := x_i - x_j$, $y_{ij} := y_i - y_j$.

Consider the connectivity map and by the Chain rule

$$\frac{\partial}{\partial x} = \frac{\partial}{\partial \xi_{e,i}} \frac{\partial \xi_{e,i}}{\partial x} = \frac{\epsilon_{ijk} y_{jk}}{2 \det \mathbf{T}} \frac{\partial}{\partial \xi_{e,i}} \tag{6.142}$$

$$\frac{\partial}{\partial y} = \frac{\partial}{\partial \xi_{e,i}} \frac{\partial \xi_{e,i}}{\partial y} = \frac{\epsilon_{ijk} x_{jk}}{2 \det \mathbf{T}} \frac{\partial}{\partial \xi_{e,i}} \tag{6.143}$$

where ϵ_{ijk} is the permutation symbol, and Einstein summation rule is implied here.

Consider the connectivity map

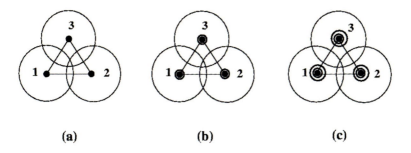

Fig. 6.21. Two dimensional triangle RKEM hierarchical interpolation field: (a) I^0 element; (b) I^1 element; and (c) I^2 element.

$$(e_1, i_1), \cdots, (e_k, i_k), \cdots, (e_\ell, i_\ell) \to I \tag{6.144}$$

The two-dimensional triangle RKEM shape function can be expressed as

$$\Psi_I^{(0)}(\mathbf{x}) = \sum_{k=1}^{\ell} \left(\sum_{j=1}^{3} \frac{\Delta V_{e_k}}{3\varrho} w\left(\frac{\mathbf{x} - \mathbf{x}_{e_k,j}}{\varrho} \right) \right.$$

$$\left\{ \underset{e \in \Lambda_E}{\mathbf{A}} \left(\sum_{j=1}^{3} \frac{\Delta V_e}{3\varrho^2} w\left(\frac{\mathbf{x} - \mathbf{x}_{e,j}}{\varrho} \right) \right) \left(\sum_{i=1}^{4} \xi_{e,i}(x) \right) \right\}^{-1}$$

$$\left. \xi_{e_k, i_k}(\mathbf{x}) \right). \tag{6.145}$$

Note that again, the smoothness of the interpolant is C^n and n is the order of continuity of the window function.

6.4.6 Globally Compatible T9P1I1 Triangle Element

To construct a 2D compatible, I^1/C^n triangle element, the meshfree kernel function can be chosen the same form as in Eq. (6.138), except that

$$\psi_{e,i}^{(0)}(\mathbf{x}) = 3\xi_{e,i}^2 - 2\xi_{e,i}^3, \quad i = 1, 2, 3 \tag{6.146}$$

The relationship between $\xi_{e,i} \sim x, y$ are described in Eqs. (6.140) and (6.141).

The global partition polynomials in an element satisfy the interpolation conditions:

$$\psi_{e,i}^{(0)}(\mathbf{x}_j) = \delta_{ij}; \quad \frac{\partial \psi_{e,i}^{(0)}}{\partial x}(\mathbf{x}_j) = 0, \quad \frac{\partial \psi_{e,i}^{(0)}}{\partial y}(\mathbf{x}_j) = 0. \tag{6.147}$$

Note that $\sum_{i=1}^{3} \psi_{e,i}^{(0)}(\mathbf{x}) \neq 1$. Therefore, the global partition polynomials in this case do not form a partition of unity.

The zero-th order basis functions in an element are

$$\Psi_{e,i}^{(00)}(\mathbf{x}) = \left(\sum_{j=1}^{3} \mathcal{K}^{\varrho}{}_{e,j}(\mathbf{x})\Delta V_{e,j}\right)\left(3\xi_{e,i}^{2} - 2\xi_{e,i}^{3}\right), \quad i = 1, 2, 3 \qquad (6.148)$$

The higher order bases functions in an element are

$$\Psi_{e,i}^{(10)}(\mathbf{x}) = (x - x_{e,i})\Psi_{e,i}^{(00)}(\mathbf{x}), i = 1, 2, 3 \qquad (6.149)$$

$$\Psi_{e,i}^{(01)}(\mathbf{x}) = (y - y_{e,i})\Psi_{e,i}^{(00)}(\mathbf{x}), i = 1, 2, 3 \qquad (6.150)$$

The element has a minimal 9 degrees of freedom. One can show that the higher order basis functions satisfy the following interpolation properties,

$$\Psi_{e,i}^{(10)}(\mathbf{x}_i) = 0, \quad \text{and} \quad \Psi_{e,i}^{(01)}(\mathbf{x}_i) = 0, \quad \forall \, i = 1, 2, 3, \qquad (6.151)$$

$$\frac{\partial \Psi_{e,i}^{(10)}}{\partial x}\bigg|_{\mathbf{x}=\mathbf{x}_j} = \delta_{ij}, \quad \frac{\partial \Psi_{e,i}^{(01)}}{\partial y}\bigg|_{\mathbf{x}=\mathbf{x}_j} = 0, \quad i, j = 1, 2, 3 \qquad (6.152)$$

$$\frac{\partial \Psi_{e,i}^{(10)}}{\partial x}\bigg|_{\mathbf{x}=\mathbf{x}_j} = 0, \quad \frac{\partial \Psi_{e,i}^{(01)}}{\partial y}\bigg|_{\mathbf{x}=\mathbf{x}_j} = \delta_{ij}, \quad i, j = 1, 2, 3 \qquad (6.153)$$

Suppose that the finite element connectivity map has the form,

$$\Lambda_E \times \Lambda_e \to \Lambda_P : (e_1, i_1), \cdots, (e_\ell, i_\ell) \to I \qquad (6.154)$$

The zero-th order global shape function has the form

$$\Psi_I^{(00)}(\mathbf{x}) = \sum_{k \in \Lambda_I}\left(\sum_{j \in \Lambda_{e_k}} \mathcal{K}^{\varrho}{}_{e_k,j}(\mathbf{x})\Delta V_{e_k,j}\right)\psi_{e_k,i_k}^{(0)}(\mathbf{x}) \qquad (6.155)$$

The global higher order basis functions have very simple form

$$\Psi_I^{(10)}(\mathbf{x}) = (x - x_I)\Psi_I^{(00)}(\mathbf{x}) \qquad (6.156)$$

$$\Psi_I^{(01)}(\mathbf{x}) = (y - y_I)\Psi_I^{(00)}(\mathbf{x}) \qquad (6.157)$$

The interpolation formula is

$$\mathcal{I}^1 f(\mathbf{x}) = \sum_{I \in \Lambda}\left(\Psi_I^{(00)}(\mathbf{x})f_I + \Psi_I^{(10)}(\mathbf{x})f_{Ix} + \Psi_I^{(01)}(\mathbf{x})f_{Iy}\right) \qquad (6.158)$$

The interpolation scheme (6.158) can reproduce 2D linear polynomial exactly, i.e. $\mathcal{I}^1 f(\mathbf{x}) = f(\mathbf{x})$ if $f(\mathbf{x}) = a + bx + cy$.

Remark 6.4.1. There is an alternative approach that can restore the partition of unity property for the global partition polynomials. Let,

$$\psi_{e,1}^{(0)} = 3\xi_{e,1}^2 - 2\xi_{e,1}^3$$

$$\psi_{e,2}^{(0)} = 3\xi_{e,2}^2 - 2\xi_{e,2}^3$$

$$\psi_{e,3}^{(0)} = 1 - \psi_{e,1}^{(0)} - \psi_{e,2}^{(0)} . \qquad (6.159)$$

It is easy to verify that $\sum_{i=1}^{3} \psi_{e,i}^{(0)}(\boldsymbol{x}) = 1$.

6.4.7 Globally Compatible T18P2I2 Triangle Element

Construct meshfree kernel function as formulated in Eq. (6.138), in which we choose

$$\psi_{e,i}^{(0)}(\mathbf{x}) = 10\xi_{e,i}^3 - 15\xi_{e,i}^4 + 6\xi_{e,i}^5, \quad e \in \Lambda_E, \quad i = 1,2,3 \tag{6.160}$$

Again, area coordinates are chosen to represent FEM shape functions, where $0 \le \xi_{e,i} \le 1$ are the relationship between $\xi_{e,i} \sim x, y$ are given in Eqs. (6.140) and (6.141).

The global partition polynomials in an element satisfy the following interpolation conditions $\forall\ i,j = 1,2,3$:

$$\psi_{e,i}^{(0)}(\mathbf{x}_j) = \delta_{ij}; \tag{6.161}$$

$$\frac{\partial \psi_{e,i}^{(0)}}{\partial x}(\mathbf{x}_j) = 0, \quad \frac{\partial \psi_{e,i}^{(0)}}{\partial y}(\mathbf{x}_j) = 0. \tag{6.162}$$

$$\frac{\partial^2 \psi_{e,i}^{(0)}}{\partial x^2}(\mathbf{x}_j) = 0, \quad \frac{\partial^2 \psi_{e,i}^{(0)}}{\partial x \partial y}(\mathbf{x}_j) = 0, \quad \frac{\partial^2 \psi_{e,i}^{(0)}}{\partial y^2}(\mathbf{x}_j) = 0. \tag{6.163}$$

Note that if (6.160) is used, the global partition polynomial may not be a partition of unity, i.e. $\sum\limits_{i=1}^{3} \psi_{e,i}^{(0)}(\mathbf{x}) \ne 1$. Nevertheless, the scheme should work theoretically. If one choses $\psi_{e,i}^{(0)}(\mathbf{x}) = 10\xi_{e,i}^3 - 15\xi_{e,i}^4 + 6\xi_{e,i}^5, \quad i = 1,2, e \in \Lambda_E$ and $\psi_{e,3}^{(0)} = 1 - \psi_{e,1}^{(0)} - \psi_{e,2}^{(0)}$, one can then recover the property of partition of unity.

The local zero-th order basis functions are

$$\Psi_{e,i}^{(00)}(\mathbf{x}) = \left(\sum_{j=1}^{3} \mathcal{K}^\varrho_{e,j}(\mathbf{x}) \Delta V_{e,j}\right) \psi_{e,i}^{(0)}(\mathbf{x}), \quad i = 1,2,3 \tag{6.164}$$

The first order partition polynomial basis functions are

$$\Psi_{e,i}^{(10)}(\mathbf{x}) = (x - x_{e,i}) \Psi_{e,i}^{(0)}(\mathbf{x}), i = 1,2,3 \tag{6.165}$$

$$\Psi_{e,i}^{(01)}(\mathbf{x}) = (y - y_{e,i}) \Psi_{e,i}^{(0)}(\mathbf{x}), i = 1,2,3 \tag{6.166}$$

The second order partition polynomial basis functions are

$$\Psi_{e,i}^{(20)}(\mathbf{x}) = \frac{(x - x_{e,i})^2}{2!} \Psi_{e,i}^{(00)}(\mathbf{x}), \quad i = 1,2,3 \tag{6.167}$$

$$\Psi_{e,i}^{(11)}(\mathbf{x}) = (x - x_{e,i})(y - y_{e,i}) \Psi_{e,i}^{(00)}(\mathbf{x}), \quad i = 1,2,3 \tag{6.168}$$

$$\Psi_{e,i}^{(02)}(\mathbf{x}) = \frac{(y - y_{e,i})^2}{2!} \Psi_{e,i}^{(00)}(\mathbf{x}), \quad i = 1,2,3 \tag{6.169}$$

There are a total of 18 degrees of freedom.

Expressing in global form, one may write the zero-th order global shape function as

$$\Psi_I^{(00)}(\mathbf{x}) = \sum_{k \in \Lambda_I} \left(\sum_{j \in \Lambda_{e_k}} \mathcal{K}^{\varrho}_{e_k,j}(\mathbf{x}) \Delta V_{e_k,j} \right) \psi^{(0)}_{e_k,i_k}(\mathbf{x}) \tag{6.170}$$

The 1st order basis functions are

$$\Psi_I^{(10)}(\mathbf{x}) = (x - x_I)\Psi_I^{(00)}(\mathbf{x}) \tag{6.171}$$

$$\Psi_I^{(01)}(\mathbf{x}) = (y - y_I)\Psi_I^{(00)}(\mathbf{x}) \tag{6.172}$$

The second order basis functions are

$$\Psi_I^{(20)}(\mathbf{x}) = \frac{1}{2!}(x - x_I)^2\Psi_I^{(00)}(\mathbf{x}) \tag{6.173}$$

$$\Psi_I^{(11)}(\mathbf{x}) = (x - x_I)(y - y_I)\Psi_I^{(00)}(\mathbf{x}) \tag{6.174}$$

$$\Psi_I^{(02)}(\mathbf{y}) = \frac{1}{2!}(y - y_I)^2\Psi_I^{(00)}(\mathbf{x}) \tag{6.175}$$

The compatible I^2 interpolation scheme is

$$\mathcal{I}^2 f(\mathbf{x}) = \sum_{I \in \Lambda_P} \left\{ \Psi_I^{(0)}(\mathbf{x})f_I + \Psi_I^{(10)}(\mathbf{x})f_{Ix} + \Psi_I^{(01)}(\mathbf{x})f_{Iy} \right.$$

$$\left. + \Psi_I^{(20)}(\mathbf{x})f_{Ixx} + \Psi_I^{(11)}(\mathbf{x})f_{Ixy} + \Psi_I^{(02)}(\mathbf{x})f_{Iyy} \right\} \tag{6.176}$$

One can show that the above interpolation formula has quadratic consistency.

6.5 Globally Conforming I^m/C^n Hierarchy II

The globally conforming I^m/C^n hierarchy constructed in the previous section has its limitations. The globally conforming hierarchy is in fact $I^m/C^n/P^m$, i.e. $k = m$. This means that the reproducing property of the interpolant is limited by the interpolation index order m, or the enrichment order. It can only reproduce complete m-th order polynomials. In order to achieve higher order accuracy, another (the second) globally conforming $I^m/C^n/P^k(k \geq m)$ hierarchy is proposed. This interpolation hierarchy can reproduce a complete kth order polynomials with $k \geq m$.

6.5.1 Construction

Assume that there exists a set of Hermite type global polynomials, $\{\psi_{e,i}^{(0)}, \psi_{e,i}^{(1)}, \cdots, \psi_{e,i}^{(m)}\}$,

such that within the element e, they can reproduce λ-th order polynomials, i.e.

$$\sum_{i\in\Lambda_e}\left\{\psi_{e,i}^{(0)}(\boldsymbol{x})\boldsymbol{x}_i^\lambda + \lambda\psi_{e,i}^{(1)}(\boldsymbol{x})\boldsymbol{x}_i^{\lambda-1} + \cdots + \frac{\lambda!}{(\lambda-m)!}\psi_{e,i}^{(m)}(\boldsymbol{x})\boldsymbol{x}_i^{\lambda-m}\right\} = \boldsymbol{x}^\lambda ,$$
$$|\lambda| \leq k \tag{6.177}$$

Assume that the mesh topology has a subjective connectivity map such that

$$\Lambda_E \times \Lambda_e \to \Lambda_P : (e_1, i_1), \cdots, (e_\ell, i_\ell) \to I . \tag{6.178}$$

The global RKEM basis functions are constructed as follows,

$$\Psi_I^{(0)}(\boldsymbol{x}) = \sum_{k=1}^{\ell}\left(\sum_{j\in\Lambda_{e_k}} \mathcal{K}_{e_k,j}^\rho(\boldsymbol{x})\Delta V_{e_k,j}\right)\psi_{e_k,i_k}^{(0)}(\boldsymbol{x})$$

$$\Psi_I^{(1)}(\boldsymbol{x}) = \sum_{k=1}^{\ell}\left(\sum_{j\in\Lambda_{e_k}} \mathcal{K}_{e_k,j}^\rho(\boldsymbol{x})\Delta V_{e_k,j}\right)\psi_{e_k,i_k}^{(1)}(\boldsymbol{x})$$

$$\cdots\cdots \tag{6.179}$$

$$\Psi_I^{(m)}(\boldsymbol{x}) = \sum_{k=1}^{\ell}\left(\sum_{j\in\Lambda_{e_k}} \mathcal{K}_{e_k,j}^\rho(\boldsymbol{x})\Delta V_{e_k,j}\right)\psi_{e_k,i_k}^{(m)}(\boldsymbol{x}) \tag{6.180}$$

where $I \in \Lambda_P$ and

$$\mathop{\mathbf{A}}_{e\in\Lambda_E} \sum_{j\in\Lambda_e}\left(\mathcal{K}_{e,j}^\rho(\boldsymbol{x})\Delta V_{e,j}\right) = 1. \tag{6.181}$$

Again, the proposed RKEM I^m/C^n interpolation scheme can be written as

$$I^m f = \sum_{I \in \Lambda_P} \left(\Psi_I^{(0)}(\boldsymbol{x}) f_I + \Psi_I^{(0)}(\boldsymbol{x}) D f \Big|_I + \cdots + \Psi_I^{(m)}(\boldsymbol{x}) D^m f \Big|_I \right) \quad (6.182)$$

The main properties of this globally conforming I^m/C^n hierarchy are summarized in the following proposition.

Proposition 6.5.1. *Assume that* $\exists \{\psi_{e,i}^{(0)}, \psi_{e,i}^{(1)}, \cdots \psi_{e,i}^{(m)}\}$, $\forall e \in \Lambda_E, i \in \Lambda_e$, *such that*

1.

$$D^\alpha \psi_{e,i}^{(\beta)} \Big|_{\boldsymbol{x}=\boldsymbol{x}_j} = \delta_{\alpha\beta} \delta_{ij}, \quad |\alpha|, |\beta| \leq m, \text{ and } i, j \in \Lambda_e; \quad (6.183)$$

2.

$$\sum_{i \in \Lambda_e} \left\{ \psi_{e,i}^{(0)}(\boldsymbol{x}) \boldsymbol{x}_i^\lambda + \lambda \psi_{e,i}^{(1)}(\boldsymbol{x}) \boldsymbol{x}_i^{\lambda-1} + \cdots + \frac{\lambda!}{(\lambda-m)!} \psi_{e,i}^{(m)}(\boldsymbol{x}) \boldsymbol{x}_i^{\lambda-m} \right\} = x^\lambda .$$

$$(6.184)$$

Consider the local-global connectivity map for a given mesh,

$$\Lambda_E \times \Lambda_e \to \Lambda_P : (e_1, i_1) \cdots (e_k, i_k) \cdots (e_\ell, i_\ell) \to I . \quad (6.185)$$

We construct the following hybrid meshfree/FEM shape functions,

$$\Psi_I^{(0)}(\boldsymbol{x}) = \sum_{k=1}^\ell \left(\sum_{j \in \Lambda_{e_k}} \mathcal{K}_{e_k,j}^\rho(\boldsymbol{x}) \Delta V_{e_k,j} \right) \psi_{e_k,i_k}^{(0)}(\boldsymbol{x})$$

$$\Psi_I^{(1)}(\boldsymbol{x}) = \sum_{k=1}^\ell \left(\sum_{j \in \Lambda_{e_k}} \mathcal{K}_{e_k,j}^\rho(\boldsymbol{x}) \Delta V_{e_k,j} \right) \psi_{e_k,i_k}^{(1)}(\boldsymbol{x})$$

$$\cdots\cdots \quad (6.186)$$

$$\Psi_I^{(m)}(\boldsymbol{x}) = \sum_{k=1}^\ell \left(\sum_{j \in \Lambda_{e_k}} \mathcal{K}_{e_k,j}^\rho(\boldsymbol{x}) \Delta V_{e_k,j} \right) \psi_{e_k,i_k}^{(m)}(\boldsymbol{x}) \quad (6.187)$$

where $I \in \Lambda_P$.

Then the interpolation scheme (6.182) has the following properties,

1. $D^\alpha \Psi_I^{(\beta)} \Big|_{\boldsymbol{x}=\boldsymbol{x}_J} = \delta_{\alpha\beta} \delta_{IJ}, \quad I, J \in \Lambda_P, |\alpha|, |\beta| \leq m;$

2. $I^m \boldsymbol{x}^\lambda = \boldsymbol{x}^\lambda, \forall \boldsymbol{x} \in \Omega, |\lambda| \leq k.$

Proof:
1. We first show $D^\alpha \Psi_I^{(\beta)} \Big|_{\boldsymbol{x}=\boldsymbol{x}_J} = \delta_{\alpha\beta} \delta_{IJ}, \quad \forall I, J \in \Lambda_P$. By definition,

$$\Psi_I^{(\beta)}(\boldsymbol{x}) = \sum_{k=1}^{\ell} \bigg(\sum_{j \in \Lambda_{e_k}} \mathcal{K}_{e_k,j}^\rho(\boldsymbol{x}) \Delta V_{e_k,j} \bigg) \psi_{e_k,i_k}^{(\beta)}$$

$$\Rightarrow D^\alpha \Psi_I^{(\beta)}(\boldsymbol{x}) = \sum_{|\gamma| \le |\alpha|} \left\{ \binom{\alpha}{\gamma} \sum_{k=1}^{\ell} D^{\alpha-\gamma} \bigg(\sum_{j \in \Lambda_{e_k}} \mathcal{K}_{e_k,j}^\rho(\boldsymbol{x}) \Delta V_{e_k,j} \bigg) D^\gamma \psi_{e_k,i_k}^{(\beta)} \right\} \bigg|_{\boldsymbol{x}=\boldsymbol{x}_J}$$

If $\boldsymbol{x}_J \in \Omega_{e_k}$, based on (6.183) $D^\gamma \psi_{e_k,i_k}^{(\beta)}(\boldsymbol{x}_J) = \delta_{\alpha\beta}\delta_{IJ}$. If $\boldsymbol{x}_J \notin \Omega_{e_k}$, the restriction on the support size of meshfree kernels requires that

$$\mathcal{K}_{e_k,i_k}^\rho(\boldsymbol{x}_J) \equiv 0 \tag{6.188}$$

Therefore,

$$D^\alpha \Psi_I^{(\beta)}(\boldsymbol{x}) = \sum_{|\gamma| \le |\alpha|} \left\{ \binom{\alpha}{\gamma} \sum_{k=1}^{\ell} D^{\alpha-\gamma} \bigg(\sum_{j \in \Lambda_{e_k}} \mathcal{K}_{e_k,j}^\rho(\boldsymbol{x}) \Delta V_{e_k,j} \bigg) \delta_{IJ}\delta_{\gamma\beta} \right\} \bigg|_{\boldsymbol{x}=\boldsymbol{x}_J}$$

Since

$$\sum_{k=1}^{\ell} \bigg(\sum_{j \in \Lambda_{e_k}} \mathcal{K}_{e_k,j}^\rho(\boldsymbol{x}) \Delta V_{e_k,j} \bigg) = 1, \tag{6.190}$$

if $|\alpha - \gamma| > 0$,

$$\sum_{k=1}^{\ell} D^{\alpha-\gamma} \bigg(\sum_{j \in \Lambda_{e_k}} \mathcal{K}_{e_k,j}^\rho(\boldsymbol{x}) \Delta V_{e_k,j} \bigg) = 0. \tag{6.191}$$

That is

$$\binom{\alpha}{\gamma} \sum_{k=1}^{\ell} D^{\alpha-\gamma} \bigg(\sum_{j \in \Lambda_{e_k}} \mathcal{K}_{e_k,j}^\rho(\boldsymbol{x}) \Delta V_{e_k,j} \bigg) \bigg|_{\boldsymbol{x}=\boldsymbol{x}_J} = \delta_{\alpha\gamma} \tag{6.192}$$

Hence

$$D^\alpha \Psi_I^{(\beta)}(\boldsymbol{x}) \bigg|_{\boldsymbol{x}=\boldsymbol{x}_J} = \delta_{\alpha\beta}\delta_{IJ} \ . \tag{6.193}$$

Next, we show $\mathcal{I}\boldsymbol{x}^\lambda = \boldsymbol{x}^\lambda$, $|\lambda| \le k$. Note that here $|\lambda| < k$. $|\lambda|$ can be greater than m and in most cases it is. This is the reason why the second I^m/C^n hierarchy is more accurate.

Based on the construction,

$$\mathcal{I}^m \boldsymbol{x}^\lambda = \sum_{I \in \Lambda_P} \left(\Psi_I^{(0)}(\boldsymbol{x})\boldsymbol{x}_I^\lambda + \Psi_I^{(1)}(\boldsymbol{x})\lambda \boldsymbol{x}_I^{<\lambda-1>} + \cdots + \frac{\lambda!}{(\lambda-\gamma)!}\Psi_I^{(\gamma)}(\boldsymbol{x})\boldsymbol{x}_I^{<\lambda-\gamma>} \right.$$

$$\left. + \cdots + \frac{\lambda!}{(\lambda-m)!}\Psi_I^{(m)}(\boldsymbol{x})\boldsymbol{x}_I^{<\lambda-m>} \right)$$

$$= \mathbf{A}_{e \in \Lambda_E} \left(\sum_{j \in \Lambda_e} \left(\mathcal{K}_{e,j}^\rho \varDelta V_{e,j} \right) \sum_{i \in \Lambda_e} \left(\psi_{e,i}^{(0)}\boldsymbol{x}_{e,i}^\lambda + \psi_{e,i}^{(1)}\boldsymbol{x}_{e,i}^{\lambda-1} + \cdots \right.\right.$$

$$\left.\left. + \frac{\lambda!}{(\lambda-\gamma)!}\psi_{e,i}^{(\gamma)}\boldsymbol{x}_{e,i}^{<\lambda-\gamma>} + \cdots + \frac{\lambda!}{(\lambda-m)!}\psi_{e,i}^{(m)}\boldsymbol{x}_{e,i}^{<\lambda-m>} \right)\right) \qquad (6.194)$$

Based on the assumptions that

1.

$$\sum_{i \in \Lambda_e} \left(\psi_{e,i}^{(0)}\boldsymbol{x}_{e,i}^\lambda + \cdots + \frac{\lambda!}{(\lambda-\gamma)!}\psi_{e,i}^{(\gamma)}\boldsymbol{x}_{e,i}^{<\lambda-\gamma>} + \cdots + \frac{\lambda!}{(\lambda-m)!}\psi_{e,i}^{(m)}\boldsymbol{x}_{e,i}^{<\lambda-m>} \right) = \boldsymbol{x}^\lambda;$$

2.

$$\mathbf{A}_{e \in \Lambda_E} \sum_{j \in \Lambda_e} \left(\mathcal{K}_{e,j}^\rho \varDelta V_{e,j} \right) = 1.$$

We conclude that

$$\mathcal{I}^m \boldsymbol{x}^\lambda = \mathbf{A}_{e \in \Lambda_E} \sum_{j \in \Lambda_e} \left(\mathcal{K}_{e,j}^\rho \varDelta V_{e,j} \right) \boldsymbol{x}^\lambda = \boldsymbol{x}^\lambda . \qquad (6.197)$$

♣

6.5.2 1D Example: An $I^1/C^4/P^3$ Interpolant

For tutorial purpose, we first construct an 1D $I^1/C^4/P^3$ hybrid RKEM interpolant.

Consider a two nodes element, $e \to [x_e, x_{e+1}]$. Let $\xi = (x - x_e)/L_e$ where $L_e = x_{e+1} - x_e$. We choose 1D cubic Hermite polynomials as the global partition polynomials, which are

$$\psi_{e,1}^{(0)}(\xi) = 1 - 3\xi^2 + 2\xi^3, \quad \psi_{e,2}^{(0)}(\xi) = 3\xi^2 - 2\xi^3,$$

$$\psi_{e,1}^{(1)}(\xi) = L_e(\xi - 2\xi^2 + \xi^3), \quad \psi_{e,2}^{(1)}(\xi) = L_e(-\xi^2 + \xi^3) .$$

For a sufficient smooth function $f(x)$, the hybrid RKEM interpolant is expressed as

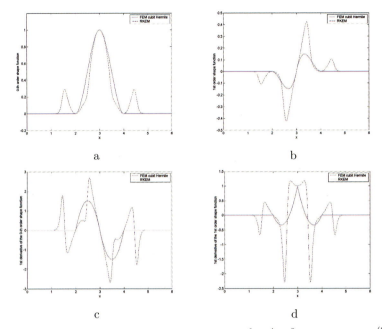

Fig. 6.22. The global shape functions of 1D $I^1/C^4/P^3$ element: (a) $\Psi_I^{(0)}(x)$; (b) $\Psi_I^{(1)}(x)$; (c) $d\Psi_I^{(0)}/dx$; (d) $d\Psi_I^{(1)}/dx$.

$$
\begin{aligned}
\mathcal{I}^1 f(x) = \mathop{\mathbf{A}}_{e \in \Lambda_E} & \left\{ \left(\sum_{j=1}^{2} \mathcal{K}_{e,j}^{\rho}(x) \Delta x_{e,j} \right) \left(\sum_{i=1}^{2} \psi_{e,i}^{(0)}(x) f_{e,i} \right) \right. \\
& \left. + \left(\sum_{j=1}^{2} \mathcal{K}_{e,j}^{\rho}(x) \Delta x_{e,j} \right) \left(\sum_{i=1}^{2} \psi_{e,i}^{(1)}(x) f_{e,i}' \right) \right\} \\
= & \sum_{I \in \Lambda_P} \left(\Psi_I^{(0)}(x) f_I + \Psi_I^{(1)}(x) f_I' \right)
\end{aligned}
\tag{6.198}
$$

Suppose that the mesh connectivity map has the form,

$$
\Lambda_E \times \Lambda_e \to \Lambda_P : (e_1, i_1), (e_2, i_2) \to I
$$

The global RKEM shape functions are

$$
\Psi_I^{(0)}(x) = \sum_{k=1}^{2} \left(\sum_{j \in \Lambda_{e_k}} \mathcal{K}_{e_k,j}^{\rho}(x) \Delta x_{e_k,j} \right) \psi_{e_k,i_k}^{(0)}
\tag{6.199}
$$

$$
\Psi_I^{(1)}(x) = \sum_{k=1}^{2} \left(\sum_{j \in \Lambda_{e_k}} \mathcal{K}_{e_k,j}^{\rho}(x) \Delta x_{e_k,j} \right) \psi_{e_k,i_k}^{(1)}
\tag{6.200}
$$

In order that $\Psi_I^{(i)} \in C^4(\Omega), i = 1, 2$ and $I \in \Lambda_E$, we choose the fifth order spline function as the window function of the meshfree kernel. The global shape functions and their first order derivatives are plotted in Fig. 6.22.

6.5.3 2D Example I: Compatible Gallagher Element

In this example, we choose the global partition polynomials derived from the previous 2D incompatible rectangular element, or Gallagher element.[174]

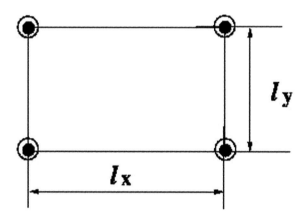

Fig. 6.23. A globally conforming Gallagher element.

To derive the global partition polynomials, we consider the following element interpolation field in a four nodes rectangular element,

$$\mathcal{I}_{loc}f = \sum_{i=1}^{4} \psi_{e,i}^{(00)}(\boldsymbol{x})f_{e,i} + \psi_{e,i}^{(10)}(\boldsymbol{x})\frac{\partial f}{\partial x}\bigg|_{e,i} + \psi_{e,i}^{(01)}(\boldsymbol{x})\frac{\partial f}{\partial y}\bigg|_{e,i}$$
$$= \boldsymbol{\psi}^T \mathbf{f} \tag{6.201}$$

where $\boldsymbol{\psi}$ is denoted as the local shape function array and vector $\mathbf{f}$ is the nodal data array, i.e.

$$\boldsymbol{\psi}^T = \{\psi_{e,1}^{(00)}, \psi_{e,1}^{(10)}, \psi_{e,1}^{(01)}, \psi_{e,2}^{(00)}, \psi_{e,2}^{(10)}, \psi_{e,2}^{(01)}, \psi_{e,3}^{(00)}, \psi_{e,3}^{(10)}, \psi_{e,3}^{(01)} \psi_{e,4}^{(00)},$$
$$\psi_{e,4}^{(10)}, \psi_{e,4}^{(01)}\} \tag{6.202}$$

$$\mathbf{f}^T = \{f_{e1}, f_{e1,x}, f_{e1,y}, f_{e2}, f_{e2,x}, f_{e2,y}, f_{e3}, f_{e3,x}, f_{e3,y},$$
$$f_{e4}, f_{e4,x}, f_{e4,y}\} \tag{6.203}$$

Based on Przemieniecki,[377] we have the explicit expressions for the chosen global partition polynomials,

$$\psi^T = \begin{bmatrix} 1 - \xi\eta - (3 - 2\xi)\xi^2(1 - \eta) - (1 - \xi)(3 - 2\eta)\eta^2 \\ (1 - \xi)\eta(1 - \eta)^2\ell_{ey} \\ -\xi(1 - \xi)^2(1 - \eta)\ell_{ex} \\ (1 - \xi)(3 - 2\eta)\eta^2 + \xi(1 - \xi)(1 - 2\xi)\eta \\ -(1 - \xi)(1 - \eta)\eta^2\ell_{ey} \\ -\xi(1 - \xi)^2\eta\ell_{ex} \\ (3 - 2\xi)\xi^2\eta - \xi\eta(1 - \eta)(1 - 2\eta) \\ -\xi(1 - \eta)\eta^2\ell_{ey} \\ (1 - \xi)\xi^2\eta\ell_{ex} \\ (3 - 2\xi)\xi^2(1 - \eta) + \xi\eta(1 - \eta)(1 - 2\eta) \\ \xi\eta(1 - \eta)^2\ell_{ey} \\ (1 - \xi)\xi^2(1 - \eta)\ell_{ex} \end{bmatrix} \qquad (6.204)$$

where $-\infty < \xi, \eta < \infty$.

Assume that the mesh connectivity map is,

$$\Lambda_E \times \Lambda_e \rightarrow \Lambda_P : (e_1, i_1) \cdots (e_\ell, i_\ell) \rightarrow I .$$

The RKEM global shape function for the rectangular can be expressed as follows:

$$\Psi_I^{(\alpha)}(\boldsymbol{x}) = \sum_{k=1}^{\ell} \sum_{j \in \Lambda_{e_k}} \left(\mathcal{K}_{e_k,j}^{\rho}(\boldsymbol{x}) \Delta V_{e_k,j} \right) \psi_{e_k,i_k}^{(\alpha)}(\boldsymbol{x}) . \qquad (6.205)$$

6.5.4 2D Example II: $T12P3I(4/3)$ Triangle Element

The above rectangular element can not discretize an arbitrary domain. Here, we propose a two-dimensional globally conforming, bilinear, 12 degrees of freedom triangle element. It is illustrated in Fig. 6.24. The notation $I^{4/3}$ means that at each nodal point we interpolate the unknown function, say $f(x, y)$, its two first order derivatives, $\dfrac{\partial f}{\partial x}$ and $\dfrac{\partial f}{\partial x}$ and its mixed derivative, $\dfrac{\partial^2 f}{\partial x \partial y}$, which is one-third of the second derivatives. We denote the one-third of second derivative as a cross in Fig. 6.24. Since we interpolant one third of the second derivatives, the interpolation scheme is neither I^1 nor I^2. We denote it as $I^{(1+1/3)} = I^{(4/3)}$.

The global partition polynomials in an element can form a local interpolation,

$$\mathcal{I}_{ele}f = \sum_{i=1}^{3} \left(\psi_{e,i}^{(00)} f_{e,i} + \psi_{e,i}^{(10)} \frac{\partial f}{\partial x} \Big|_{(e,i)} + \psi_{e,i}^{(01)} \frac{\partial f}{\partial y} \Big|_{(e,i)} + \psi_{e,i}^{(11)} \frac{\partial^2 f}{\partial x \partial y} \Big|_{(e,i)} \right)$$

$$= c_1 + c_2 x + c_3 y + c_4 x^2 + c_5 xy + c_6 y^2 + c_7 x^3 + c_8 x^2 y + c_9 xy^2$$

$$+ c_{10} y^3 + c_{11} x^2 (x^2 + xy + y^2) + c_{12} y^2 (x^2 + xy + y^2) \qquad (6.206)$$

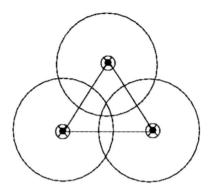

Fig. 6.24. An bilinear triangle (T12P3I4/3) element.

To link the nodal data with constants,$c's$, we define vectors, $\boldsymbol{\psi}$, $\mathbf{f}$, $\mathbf{p}$, and $\mathbf{c}$:

$$\boldsymbol{\psi}_e^T(\boldsymbol{x}) := \{\psi_{e,1}^{(00)}, \psi_{e,1}^{(10)}, \psi_{e,1}^{(01)}, \psi_{e,1}^{(11)}, \psi_{e,2}^{(00)}, \psi_{e,2}^{(10)}, \psi_{e,2}^{(01)}, \psi_{e,2}^{(11)},$$

$$\psi_{e,3}^{(00)}, \psi_{e,3}^{(10)}, \psi_{e,3}^{(01)}, \psi_{e,3}^{(11)}\} \tag{6.207}$$

$$\mathbf{f}_e^T := \{f_1, f_{1x}, f_{1y}, f_{1xy}, f_2, f_{2x}, f_{2y}, f_{2xy}, f_3, f_{3x}, f_{3y}, f_{3xy}\}_e \tag{6.208}$$

$$\mathbf{p}^T(\boldsymbol{x}) := \{1, x, y, x^2, xy, y^2, x^3, x^2y, xy^2, y^3, x^2(x^2+xy+y^2),$$

$$y^2(x^2+xy+y^2)\} \tag{6.209}$$

$$\mathbf{c}_e^T := \{c_1, c_2, c_3, c_4, c_5, c_6, c_7, c_8, c_9, c_{10}, c_{11}, c_{12}\}_e \tag{6.210}$$

We can write that

$$\mathcal{I}_{loc}f = \boldsymbol{\psi}_e^T \mathbf{f}_e = \mathbf{p}^T \mathbf{c}_e. \tag{6.211}$$

The nodal values are related with coefficients c's by a set of 12 simultaneous linear algebraic equations

$$\mathbf{f}_e = \mathbf{C}_e \mathbf{c}_e \tag{6.212}$$

where matrix $\mathbf{C}_e$ is defined by Eq. (6.213).

One can then find vector $\mathbf{c}_e$ by solving the linear algebraic equation,

$$\mathbf{c}_e = \mathbf{C}_e^{-1}\mathbf{f} \tag{6.214}$$

Then, the global partition polynomials can be found as

$$\boldsymbol{\psi}_e(\boldsymbol{x}) = \mathbf{C}_e^{-T}\mathbf{p}(\boldsymbol{x}) \tag{6.215}$$

Assume that the mesh connectivity map is as usual

$$\Lambda_E \times \Lambda_e \to \Lambda_P : (e_1, i_1) \cdots (e_\ell, i_\ell) \to I . \tag{6.216}$$

$$
\mathbf{C}_e =
\left[
\begin{array}{cccccccccccc}
1 & x_1 & y_1 & x_1^2 & x_1 y_1 & y_1^2 & x_1^3 & x_1^2 y_1 & x_1 y_1^2 & y_1^3 & x_1^2(x_1^2+x_1 y_1+y_1^2) & y_1^2(x_1^2+x_1 y_1+y_1^2) \\
0 & 1 & 0 & 2x_1 & y_1 & 0 & 3x_1^2 & 2x_1 y_1 & y_1^2 & 0 & x_1(4x_1^2+3x_1 y_1+2y_1^2) & y_1^2(2x_1+y_1) \\
0 & 0 & 1 & 0 & x_1 & 2y_1 & 0 & x_1^2 & 2x_1 y_1 & 3y_1^2 & x_1^2(x_1+2y_1) & y_1(4y_1^2+3x_1 y_1+2x_1^2) \\
0 & 0 & 0 & 0 & 1 & 0 & 0 & 2x_1 & 2y_1 & 0 & 3x_1^2+4x_1 y_1 & 3y_1^2+4x_1 y_1 \\[4pt]
1 & x_2 & y_2 & x_2^2 & x_2 y_2 & y_2^2 & x_2^3 & x_2^2 y_2 & x_2 y_2^2 & y_2^3 & x_2^2(x_2^2+x_2 y_2+y_2^2) & y_2^2(x_2^2+x_2 y_2+y_2^2) \\
0 & 1 & 0 & 2x_2 & y_2 & 0 & 3x_2^2 & 2x_2 y_2 & y_2^2 & 0 & x_2(4x_2^2+3x_2 y_2+2y_2^2) & y_2^2(2x_2+y_2) \\
0 & 0 & 1 & 0 & x_2 & 2y_2 & 0 & x_2^2 & 2x_2 y_2 & 3y_2^2 & x_2^2(x_2+2y_2) & y_2(4y_2^2+3x_2 y_2+2x_2^2) \\
0 & 0 & 0 & 0 & 1 & 0 & 0 & 2x_2 & 2y_2 & 0 & 3x_2^2+4x_2 y_2 & 3y_2^2+4x_2 y_2 \\[4pt]
1 & x_3 & y_3 & x_3^2 & x_3 y_3 & y_3^2 & x_3^3 & x_3^2 y_3 & x_3 y_3^2 & y_3^3 & x_3^2(x_3^2+x_3 y_3+y_3^2) & y_3^2(x_3^2+x_3 y_3+y_3^2) \\
0 & 1 & 0 & 2x_3 & y_3 & 0 & 3x_3^2 & 2x_3 y_3 & y_3^2 & 0 & x_3(4x_3^2+3x_3 y_3+2y_3^2) & y_3^2(2x_3+y_3) \\
0 & 0 & 1 & 0 & x_3 & 2y_3 & 0 & x_3^2 & 2x_3 y_3 & 3y_3^2 & x_3^2(x_3+2y_3) & y_3(4y_3^2+3x_3 y_3+2x_3^2) \\
0 & 0 & 0 & 0 & 1 & 0 & 0 & 2x_3 & 2y_3 & 0 & 3x_3^2+4x_3 y_3 & 3y_3^2+4x_3 y_3
\end{array}
\right]
\tag{6.213}
$$

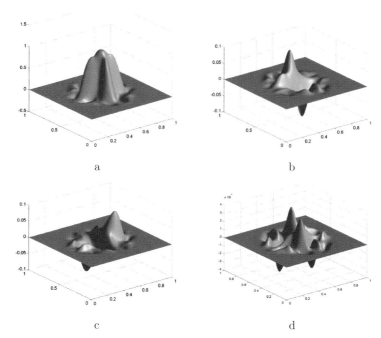

Fig. 6.25. The four global shape functions of T12P3I(4/3) element: (a) $\Psi_I^{(00)}(\boldsymbol{x})$; (b) $\Psi_I^{(10)}(\boldsymbol{x})$; (c) $\Psi_I^{(01)}(\boldsymbol{x})$; (d) $\Psi_I^{(11)}(\boldsymbol{x})$.

The RKEM global shape functions of the triangle resume the same formula,

$$\Psi_I^{(\alpha)}(\boldsymbol{x}) = \sum_{k=1}^{\ell} \sum_{j \in \Lambda_{e_k}} \left(\mathcal{K}_{e_k,j}^{\rho}(\boldsymbol{x}) \Delta V_{e_k,j} \right) \psi_{e_k,ik}^{(\alpha)}(\boldsymbol{x}) \ . \tag{6.217}$$

6.5.5 2D Example III: Q12P3I1 Quadrilateral Element

One can consider constructing quadrilateral element by using the same technique. To illustrate an example, we outline the procedure to construct a twelve degrees of freedom quadrilateral element, $Q12P3I1$. The element interpolation provided by the global partition polynomials in an element is

$$\begin{aligned}
\mathcal{I}_{ele} f &= \sum_{i=1}^{4} \left(\psi_{e,i}^{(0)} f_{e,i} + \psi_{e,i}^{(10)} \frac{\partial f}{\partial x} \Big|_{(e,i)} + \psi_{e,i}^{(01)} \frac{\partial f}{\partial y} \Big|_{(e,i)} \right) \\
&= c_1 + c_2 x + c_3 y + c_4 x^2 + c_5 xy + c_6 y^2 + c_7 x^3 + c_8 x^2 y + c_9 xy^2 \\
&\quad + c_{10} y^3 + c_{11} x^3 (x+y) + c_{12} y^3 (x+y)
\end{aligned} \tag{6.218}$$

To link the nodal data with constants, $c's$, vectors, $\boldsymbol{\psi}$, $\mathbf{f}$, $\mathbf{p}$, and $\mathbf{c}$ are defined

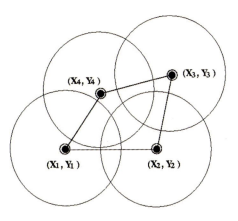

Fig. 6.26. Illustration of Q12P3I1 quadrilateral element.

$$\boldsymbol{\psi}_e^T(\boldsymbol{x}) := \{\psi_{e,1}^{(00)}, \psi_{e,1}^{(10)}, \psi_{e,1}^{(01)}, \psi_{e,2}^{(00)}, \psi_{e,2}^{(10)}, \psi_{e,2}^{(01)}, \psi_{e,3}^{(00)}, \psi_{e,3}^{(10)}, \psi_{e,3}^{(01)},$$
$$\psi_{e,4}^{(00)}, \psi_{e,4}^{(10)}, \psi_{e,4}^{(01)}\} \tag{6.219}$$

$$\mathbf{f}_e^T := \{f_1, f_{1x}, f_{1y}, f_2, f_{2x}, f_{2y}, f_3, f_{3x}, f_{3y}, f_4, f_{4x}, f_{4y}\}_e \tag{6.220}$$

$$\mathbf{p}^T(\boldsymbol{x}) := \{1, x, y, x^2, xy, y^2, x^3, x^2y, xy^2, y^3, x^3(x+y), y^3(x+y)\} \tag{6.221}$$

$$\mathbf{c}_e^T := \{c_1, c_2, c_3, c_4, c_5, c_6, c_7, c_8, c_9, c_{10}, c_{11}, c_{12}\}_e \tag{6.222}$$

We can write that

$$\mathcal{I}_{loc}f = \boldsymbol{\psi}_e^T \mathbf{f}_e = \mathbf{p}^T \mathbf{c}_e. \tag{6.223}$$

Again, the nodal values are related with coefficients c's by a system of 12

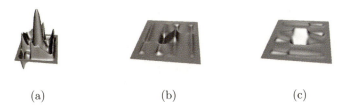

| (a) | (b) | (c) |

Fig. 6.27. The shape functions of Q12P3I1 element.

linear algebraic equations

$$\mathbf{f}_e = \mathbf{C}_e \mathbf{c}_e \tag{6.224}$$

where matrix $\mathbf{C}_e$ is defined by Eq. (6.225).

$$
\mathbf{C}_e =
\begin{vmatrix}
1 & x_1 & y_1 & x_1^2 & x_1y_1 & y_1^2 & x_1^3 & x_1^2y_1 & x_1y_1^2 & y_1^3 & x_1^3(x_1+y_1) & y_1^3(x_1+y_1) \\
0 & 1 & 0 & 2x_1 & y_1 & 0 & 3x_1^2 & 2x_1y_1 & y_1^2 & 0 & 4x_1^3+3x_1^2y_1 & y_1^3 \\
0 & 0 & 1 & 0 & x_1 & 2y_1 & 0 & x_1^2 & 2x_1y_1 & 3y_1^2 & x_1^3 & 4y_1^3+3y_1^2x_1 \\
1 & x_2 & y_2 & x_2^2 & x_2y_2 & y_2^2 & x_2^3 & x_2^2y_2 & x_2y_2^2 & y_2^3 & x_2^3(x_2+y_2) & y_2^3(x_2+y_2) \\
0 & 1 & 0 & 2x_2 & y_2 & 0 & 3x_2^2 & 2x_2y_2 & y_2^2 & 0 & 4x_2^3+3x_2^2y_2 & y_2^3 \\
0 & 0 & 1 & 0 & x_2 & 2y_2 & 0 & x_2^2 & 2x_2y_2 & 3y_2^2 & x_2^3 & 4y_2^3+3y_2^2x_2 \\
1 & x_3 & y_3 & x_3^2 & x_3y_3 & y_3^2 & x_3^3 & x_3^2y_3 & x_3y_3^2 & y_3^3 & x_3^3(x_3+y_3) & y_3^3(x_3+y_3) \\
0 & 1 & 0 & 2x_3 & y_3 & 0 & 3x_3^2 & 2x_3y_3 & y_3^2 & 0 & 4x_3^3+3x_3^2y_3 & y_3^3 \\
0 & 0 & 1 & 0 & x_3 & 2y_3 & 0 & x_3^2 & 2x_3y_3 & 3y_3^2 & x_3^3 & 4y_3^3+3y_3^2x_3 \\
1 & x_4 & y_4 & x_4^2 & x_4y_4 & y_4^2 & x_4^3 & x_4^2y_4 & x_4y_4^2 & y_4^3 & x_4^3(x_4+y_4) & y_4^3(x_4+y_4) \\
0 & 1 & 0 & 2x_4 & y_4 & 0 & 3x_4^2 & 2x_4y_4 & y_4^2 & 0 & 4x_4^3+3x_4^2y_4 & y_4^3 \\
0 & 0 & 1 & 0 & x_4 & 2y_4 & 0 & x_4^2 & 2x_4y_4 & 3y_4^2 & x_4^3 & 4y_4^3+3y_4^2x_4
\end{vmatrix}
$$

$$(6.225)$$

By solving the linear algebraic equation, one can then find vector $\mathbf{c}_e$

$$
\mathbf{c}_e = \mathbf{C}_e^{-1}\mathbf{f} \tag{6.226}
$$

and then the global partition polynomials

$$
\boldsymbol{\psi}_e(\boldsymbol{x}) = \mathbf{C}_e^{-T}\mathbf{p}(\boldsymbol{x}) . \tag{6.227}
$$

6.6 Numerical Examples

To validate the method, the proposed RKEM interpolants are tested in Galerkin procedures to solve various Kirchhoff plate problems, because the Galerkin weak formulation of a Kirchhoff plate involves second derivatives and a global $C^1(\Omega)$ interpolation field is the minimum requirement. Moreover, the boundary conditions of Kirchhoff plate problems involve interpolating boundary data of both the first order derivative (slopes) and the second order derivative (curvatures), it provides a severe test to the newly proposed RKEM triangle interpolants. For more information on how to impose boundary conditions for finite element computation of thin plates, readers are referred to Hughes' finite element book[209] pages 324 - 327.

We consider three problems: (1) a simply supported triangle plate subjected uniform load; (2) a simply supported square plate subjected uniform load; and (3) a clamped circular plate under uniform load.

6.6.1 Equilateral Triangular Plate

To validate the method, an equilateral triangular thin plate under uniform load is solved first by using the proposed method. The coordinate axes are taken as shown in Fig. 6.28(a). In the case of a uniformly loaded plate with simply supported ends, the deflection of surface is given as:[435]

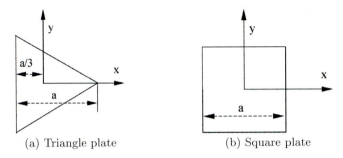

(a) Triangle plate   (b) Square plate

Fig. 6.28. Problem description for plate.

$$w = \frac{p}{64aD}\left[x^3 - 3y^2x - a(x^2 + y^2) + \frac{4}{27}a^3\right]\left(\frac{4}{9}a^2 - x^2 - y^2\right) \quad (6.228)$$

where a=1 is the height of the triangular plate.

Three discretizations with 9, 36, and 144 T12P3I3 elements shown in Fig. 6.29 are used for the convergency study. Due to the generalized Kronecker delta property of globally conforming interpolation, it is easy to exactly impose the simply supported boundary conditions at boundary nodes. The maximum deflection at the center for the three cases are, $1.08428070E - 03\frac{pa^4}{D}$, $1.03376896E - 03\frac{pa^4}{D}$, and $1.02974603E - 03\frac{pa^4}{D}$, respectively; and the exactsolution is $1.02880658E - 03\frac{pa^4}{D}$. The deflection surfaces corresponding to the three models are shown in Fig. 6.30. The L_2 error norms in the primary variable, and its first and second derivatives are shown in Fig. 6.31 for interpolation and Galerkin solutions, respectively. The convergency rates in terms of error norms L_2, H^1 and H^2 are 3.2, 2.8, and 1.7 for the Galerkin solution, respectively. They match well with the convergency rates for the interpolation solution, which are 3.4, 2.4, and 1.4 in terms of error norms L_2, H^1 and H^2, respectively.

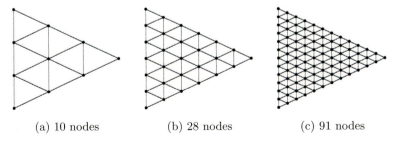

(a) 10 nodes   (b) 28 nodes   (c) 91 nodes

Fig. 6.29. Model discretizations for triangle thin plate.

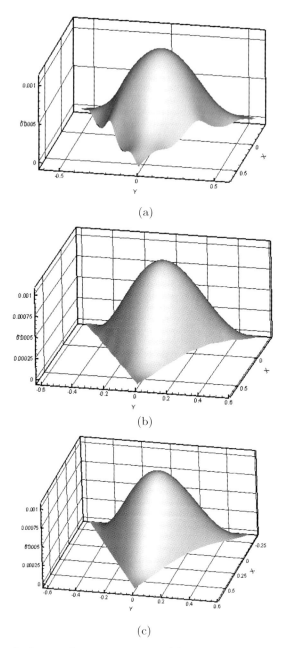

Fig. 6.30. Deflection for triangle plate: (a) 9 elements; (b) 36 elements; and (c) 144 elements.

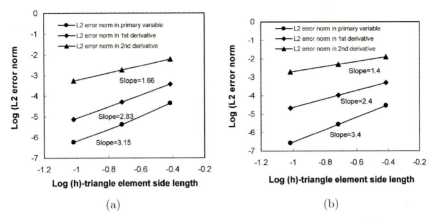

Fig. 6.31. Convergency rates of interpolation and Galerkin solutions: (a) L_2 error norms for a Galerkin solution; (b) L_2 error norms for interpolation solution.

6.6.2 Clamped Circular Plate

We now solve a clamped unit diameter circular Kirchhoff plate, as depicted in Fig. 6.32.

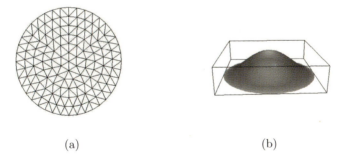

Fig. 6.32. Problem domain and triangle mesh.

The exact solution is given in[437] for a plate of radius a as

$$w(x,y) = \frac{p}{64D} \left(a^2 - x^2 - y^2\right)^2$$

The deformed shape of the circular plate is juxtaposed with the finite element mesh depicted in Fig. 6.32.

The convergence rates of L_2, H_1 and H_2 error norm are computed and are tabulated in Table 6.2.

L_2	H_1	H_2
2.6812	2.6495	1.4606

Table 6.2. Rates of convergence for clamped circular plate.

7. Molecular Dynamics and Multi-scale Methods

Due to the advance of nano-science and nano-technology, computational nano-mechanics has emerged as a major research area in the field of computational mechanics. The main research activities include: developing algorithms for contemporary molecular dynamics, multi-scale methods, and bridging scale methods. In this chapter, some basic concepts and algorithmics of molecular dynamics are introduced first, and then the coupling between molecular dynamics and finite element methods or meshfree methods are discussed.

Molecular dynamics (MD) is probably the most important and most widely used particle method in scientific and engineering fields.[9, 91, 117, 358, 390] There are two types of molecular dynamics: the-first-principle-based molecular dynamics, or ab initio molecular dynamics; and semi-empirical molecular dynamics. Both molecular dynamics have been applied to traditional engineering areas such as mechanical engineering, aerospace engineering, electrical engineering, and environmental engineering, among others. In this Chapter, we briefly outline the main technical ingredients of several molecular dynamics, and introduce multi-scale and bridging scale methods.[312, 366, 379, 446, 447]

7.1 Classical Molecular Dynamics

Molecular dynamics (MD) was first used in thermodynamics and physical chemistry to calculate the collective or average thermochemical properties of various physical systems including gases, liquids, and solids. It has been recently applied to simulate the instantaneous atomic behavior of a material system. There are two basic assumptions made in standard molecular dynamics simulations.[9, 390]

1. Molecules or atoms are described as a system of interacting material points, whose motion is described dynamically with a vector of instantaneous positions and velocities. The atomic interaction has a strong dependence on the spatial orientation and distances between separate atoms. This model is often referred to as the soft sphere model, where the softness is analogous to the electron clouds of atoms.

2. No mass changes in the system. Equivalently, the number of atoms in
   the system remains the same. The simulated system is usually treated
   as an isolated domain system with conserved energy. However, non-
   conservative techniques [53-58] are available which model the dissipation
   of the kinetic energy into the surrounding media. These techniques have
   been proven to be useful in the multiple scale simulation methods, and
   they will be presented in later sections.

7.1.1 Lagrangian Equations of Motion

The equation of motion of a system of interacting material points (particles,
atoms), having in total s degrees of freedom, can be most generally written
in terms of a Lagrangian function L, e.g.,[257]

$$\frac{d}{dt}\frac{\partial L}{\partial \dot{q}_\alpha} - \frac{\partial L}{\partial q_\alpha} = 0, \quad \alpha = 1, 2, \cdots, s. \tag{7.1}$$

Here q are the generalized coordinates, the arbitrary observables that uniquely
define spatial positions of the atoms, and the superposed dot denotes time
derivatives. As will be discussed later, Eq. (7.1) can be rewritten in terms
of the generalized coordinates and momenta, and successively utilized within
the statistical mechanics formulation. The molecular dynamics simulation
is most typically run with the reference to a Cartesian coordinate systems,
where equations (7.1) can be simplified to give

$$\frac{d}{dt}\frac{\partial L}{\partial \dot{\mathbf{r}}_i} - \frac{\partial L}{\partial \mathbf{r}_i} = 0, \quad i = 1, 2, \cdots, N. \tag{7.2}$$

Here, $\mathbf{r}_i = (x_i, y_i, z_i)$ is the radius vector of atom i, Fig. 7.1, and N is the
total number of simu-lated atoms, $N = s/3$ within the three-dimensional
settings. The spatial volume occupied by these N atoms is usually referred
to as the MD domain. Due to the homogeneity of time and space, and also
isotropy of space in inertial coordinate systems, the equations of motion (7.2)
must not depend on the choice of initial time of observation, the origin of
the coordinate system, and directions of its axes. These basic principles are
equivalent to the requirements that the Lagrangian function cannot explicitly
depend on time, directions of the radius and velocity vectors $\mathbf{r}_i$ and $\dot{\mathbf{r}}_i$, and
it can only depend on the absolute value of the velocity vector $\dot{\mathbf{r}}_i$. In order to
provide identical equations of motions in all inertial coordinate systems, the
Lagrangian function must also comply with the Galilean relativity principle.
One function satisfying all these requirements reads[257]

$$L = \sum_{i=1}^{N}(\dot{x}_i^2 + \dot{y}_i^2 + \dot{z}_i^2) \equiv \sum_{i=1}^{N}\frac{m_i\dot{\mathbf{r}}_i^2}{2} \tag{7.3}$$

for a system of free, non-interacting, particles; m_i is the mass of particle
i. Interaction between the particles can be described by adding to (7.3) a

particular function of atomic coordinates U, depending on properties of this interaction. Such a function is defined with a negative sign, so that the system's Lagrangian acquires the form:

$$L = \sum_{i=1}^{N} \frac{m_i \dot{\mathbf{r}}_i^2}{2} - U(\mathbf{r}_1, \mathbf{r}_2, \cdots, \mathbf{r}_N), \tag{7.4}$$

where the two terms represent the system's kinetic and potential energy, respectively (note the additivity of the kinetic energy term). This gives the general structure of Lagrangian for a con-servative system of interacting material points in Cartesian coordinates. It is important to note two features of this Lagrangian: the additivity of the kinetic energy term and the absence of ex-plicit dependence on time. The fact that the potential energy term only depends on spatial con-figuration of the particles implies that any change in this configuration results in an immediate effect on the motion of all particles within the simulated domain. The inevitability of this as-sumption is related to the relativity principle. Indeed, if such an effect propagated with a finite speed, the former would depend on the choice of an inertial system. In this case the laws of motion (in particular, the MD solutions) would be dissimilar in various systems; that would contradict the relativity principle. By substituting the Lagrangian (7.4) to equations (7.2), the equations of motion can finally be written in the Newtonian form,

$$m_i \ddot{\mathbf{r}}_i = -\frac{\partial U(\mathbf{r}_1, \mathbf{r}_2, \cdots, \mathbf{r}_N)}{\partial \mathbf{r}_i} \equiv \mathbf{F}_i, \quad i = 1, 2, \cdots, N \tag{7.5}$$

The force $\mathbf{F}_i$ is usually referred to as the internal force, i.e. the force exerted on atom i due to specifics of the environment it is exposed to. Eqs. (7.5) are further solved for a given set of initial conditions to get trajectories of the atomic motion in the simulated system. An important issue arising in MD simulations is accounting for a mechanism of the conduc-tion of heat away from a localized area of interest. The MD domain is usually far too small to properly describe this process within a conservative system. Modern computer power allows modeling domains with a maximum of only several hundred million atoms; that corresponds to a material specimen of size only about $0.1 \times 0.1 \times 0.1 \mu m$. MD simulations are most often per-formed using periodic boundary conditions, implying that the total energy in the system remains constant. One common solution to this problem is to apply damping forces to a group of atoms along the boundaries of the MD domain. That is known as the heat bath technique, see reference[61] for details. However, this approach cannot capture the true mechanism of dissipation in real systems. Furthermore, the potential energy term shown in Eq. (7.4), having no explicit dependence in time, implies the use of conservative models. According to some recent studies,[89, 238, 447] non-conservative models can also be constructed, using this basic form of the Lagrangian and implementing the so-called "wave-transmitting" boundary conditions to describe energy dis-sipation from the molecular dynamics domain into the surround-ing media. The basic idea of such an approach is to calculate response of the outer region to excitations originating from the MD domain at each time step of the sim-ulation. The outward heat flow is then cancelled due to negative work done by the corresponding response forces applied to boundary atoms within the localized area of interest. The classical Lagrangian formulation, discussed in this section, is typically used for those molecular dynamics simulations aimed on the analysis of detailed atomic motion, rather than on obtaining averaged (statistical) characteristics.[189] In the latter case, the Hamiltonian formulation can be alternatively used, as will be discussed in the next section.

7.1.2 Hamiltonian Equations of Motion

The Lagrangian formulation for the MD equations of motion discussed in previous section assumes description of the mechanical state of simulated system by means of generalized coordinates and velocities. This description, however, is not the only one possible. An alternative de-scription, in terms of the generalized coordinates and momentum, is utilized within the Hamil-tonian formulation, e.g.[257] The former provides a series of advantages, par-ticularly in studying general or averaged features of the simulated systems, such as the specifics of energy distribution and thermal flow, as well as in computing the physical observables (thermodynamic quantities), such as tem-perature, volume and pressure. In the latter case, the methods of statistical mechanics methods are employed, and those typically utilize the Hamiltonian formulation in describing the state and evolution of many-particle systems. Transition to the new set of independent variables can be accomplished as the following. First employ the complete differential of the Lagrangian function of Eq. (7.1),

$$dL = \sum_\alpha \frac{\partial L}{\partial q_\alpha} dq_\alpha + \sum_\alpha \frac{\partial L}{\partial \dot{q}_\alpha} d\dot{q}_\alpha, \quad \alpha = 1, 2, \cdots, s, \tag{7.6}$$

and rewrite this as

$$dL = \sum_\alpha \dot{q}_\alpha dq_\alpha + \sum_\alpha p_\alpha d\dot{q}_\alpha, \tag{7.7}$$

where the generalized momenta are defined to be

$$p_\alpha = \frac{\partial L}{\partial \dot{q}_\alpha}. \tag{7.8}$$

The right-hand side of equation (7) can be rearranged as

$$dL = \sum_\alpha \dot{p}_\alpha dq_\alpha + d\left(\sum_\alpha p_\alpha \dot{q}_\alpha\right) - \sum_\alpha \dot{q}_\alpha dp_\alpha, \tag{7.9}$$

and

$$d\left(\sum_\alpha p_\alpha \dot{q}_\alpha - L\right) = \sum_\alpha \dot{q}_\alpha dp_\alpha - \sum_\alpha \dot{p}_\alpha dq_\alpha, \tag{7.10}$$

where the function

$$H(p, q, t) = \sum_\alpha p_\alpha \dot{q}_\alpha - L \tag{7.11}$$

is referred to as the (classical) Hamiltonian of the system. The value of the Hamiltonian function is an integral of motion for conservative systems, and it is defined to be the total energy of the system in terms of the generalized coordinates and momenta. Thus, we have obtained

$$dH = \sum_\alpha \dot{q}_\alpha dp_\alpha - \sum_\alpha \dot{p}_\alpha dq_\alpha \tag{7.12}$$

and therefore

$$\dot{q}_\alpha = \frac{\partial H}{\partial p_\alpha}, \quad \dot{p}_\alpha = -\frac{\partial H}{\partial q_\alpha} . \tag{7.13}$$

These are the Hamiltonian equations of motion in terms of new variable p and q. They comprise a system of 2s first order ODEs on 2s unknown functions $p(t)$ and $q(t)$. A set of values of these functions at a given time represents the state of system at this time. This set can also be viewed as a vector in a 2s-dimensional vector space known as the phase space. A complete set of these vectors, observed in the course of temporal evolution of the system, defines a hyper-surface in the phase space, known as the phase space trajectory. The phase space trajectory provides a complete description of the system's dynamics. Although both the kinetic and potential energies do usually vary or fluctuate in time, the phase space trajectory determined from Eqs. (7.13) conserves the total energy of the system. Indeed, the time rate of change of the Hamiltonian is equal to zero,

$$\frac{dH}{dt} = \frac{\partial H}{\partial t} + \sum_\alpha \frac{\partial H}{\partial q_\alpha}\dot{q}_\alpha + \sum_\alpha \frac{\partial H}{\partial p_\alpha}\dot{p}_\alpha = \frac{\partial H}{\partial t} = 0, \tag{7.14}$$

since it has no explicit dependence on time in the case of a conservative system, as follows from (7.11) and (7.13). For a conservative system of N interacting atoms in a Cartesian coordinate system, the Ham-iltonian description acquires the following form:

$$H(\mathbf{r}_1, \mathbf{r}_2, \cdots, \mathbf{r}_N; \mathbf{p}_1, \mathbf{p}_2, \cdots, \mathbf{p}_N) = \sum_i \frac{\mathbf{p}_i^2}{2m_i} + U(\mathbf{r}_1, \mathbf{r}_2, \cdots, \mathbf{r}_N) \tag{7.15}$$

and

$$\dot{\mathbf{r}}_i = \frac{\partial H}{\partial \mathbf{p}_i}, \quad \dot{\mathbf{p}}_i = \frac{\partial H}{\partial \mathbf{r}_i}, \tag{7.16}$$

where the momenta are related to the radius vectors as $\mathbf{p}_i = m_i \dot{\mathbf{r}}_i$. If the Hamiltonian function and an initial state of the atoms in the system are known, one can compute the instantaneous positions and momentums of the atoms at all successive times, solv-ing Eqs. (7.16). That gives the phase space trajectory of the atomic motion, which can be of particular importance in studying the dynamic evolution of atomic structure and bonds, as well as the thermodynamic states of system. We note, however, that the Newtonian equations (7.5), following from the Lagrangian formulation (7.1), can be more appropriate in studying particular de-tails of the atomic processes, especially in solids. Newtonian formulation is usually more con-venient in terms of imposing external forces and constraints (for instance, periodic boundary conditions), as well as the post-processing and visualization of the results.

7.1.3 Interatomic Potentials

According to Eq. (7.5), the general structure of the governing equations for molecular dy-namics simulations is given by a straightforward second order ODE. However, the potential function for (7.5) can be an extremely com-plicated object, when accurately representing the atomic interaction within the simulated system. The nature of this interaction is due to complicated quantum effects taking place at the subatomic level that are responsible for chemical properties such as valence and bond energy; quantum effects also are responsible for the spatial arrange-ment (topology) of the interatomic bonds, their formation and breakage. In order to obtain reli-able results in molecular dynamic simulations, the classical interatomic potential should accurately ac-count for these quantum mechanical processes, though in an averaged sense. The issues related to the form of the potential function for various classes of atomic systems have been extensively discussed in literature. General struc-ture of this function is presented by the following:

$$U(\mathbf{r}_1, \mathbf{r}_2, \cdots, \mathbf{r}_N) = \sum_i V_1(\mathbf{r}_i) + \sum_{i,j>i} V_2(\mathbf{r}_i, \mathbf{r}_j) + \sum_{i,j>i,k>j} V_3(\mathbf{r}_i, \mathbf{r}_j, \mathbf{r}_k) + \cdots, \tag{7.17}$$

where $\mathbf{r}_n$ is the radius vector of the nth particle, and function V_m is called the m-body potential. The first term of (7.17) represents the energy due an external force field, such as gravitational or electrostatic, which the system is immersed into; the second shows pair-wise interaction of the particles, the third gives the three-body components, etc. In practice, the external field term is usually ignored, while all the multi-body effects are incorporated into V_2 in order to reduce the computational expense of the simulations.

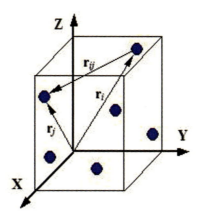

Fig. 7.1. Coordination in atomic systems

7.1.4 Two-body (pair) Potentials

At the subatomic level, the electrostatic field due to the positively charged atomic nucleus is neutralized by the negatively charged electron clouds surrounding the nucleus. Within the quan-tum mechanical description of electron motion, a probabilistic approach is employed to evaluate the probability densities at which the electrons can occupy particular spatial locations. The term "electron cloud" is typically used in relation to spatial distributions of these densities. The negatively charged electron clouds, however, experience cross-atomic attraction, which grows as the distance between the nuclei decreases. On reaching some particular distance, which is referred to as the bond length, this attraction is equilibrated by the repulsive force due to the positively charged nuclei. A further decrease in the inter-nuclei distance results in a quick growth of the resultant repulsive force. There exist a variety of mathematical models to describe the above physical phenomena. In 1924, Jones[229, 230] proposed the following potential function to describe pair-wise atomic inter-actions:

$$V(\mathbf{r}_i, \mathbf{r}_j) = V(r) = 4\epsilon \left[\left(\frac{\sigma}{r} \right)^{12} - \left(\frac{\sigma}{r} \right)^6 \right], \quad r = |\mathbf{r}_{ij}| = |\mathbf{r}_i - \mathbf{r}_j|. \qquad (7.18)$$

This model is currently known as the Lennard-Jones (LJ) potential, and it is used in simulations of a great variety of atomic systems and processes. Here, $\mathbf{r}_{ij}$ is the interatomic radius-vector, see Fig. 7.1, σ is the collision diameter, the distance at which $V(r) = 0$, and ϵ shows the bond-ing/dislocation energy — the minimum of function 7.18 to occur for an atomic pair in equilibrium.

The first term of this potential represents atomic repulsion, dominating at small separation dis-tances while the second term shows attraction (bonding) between two atoms. Since the square bracket quantity is dimensionless, the choice of units for V depends on the definition of ϵ. Typically, joule (J), therefore it is more convenient to use a smaller energy unit, such as electron volt (eV), rather than joules. $1eV = 1.602 \times 10^{-19} J$, which represents the work done if an elementary charge is accelerated by an electrostatic field of a unit voltage. The energy ϵ represents the amount of work that needs to be done in order to remove one of two coupled atoms from its equilibrium position ρ to infinity. The value ρ is also known as the equilibrium bond length, and it is related to the collision diameter as $\rho = \sqrt[6]{2}\sigma$. In a typical atomic system, the collision diameter as is equal to several angstrom ($\circ A$), $1\circ A = 10^{-10}m$. The corresponding force between the two atoms can be

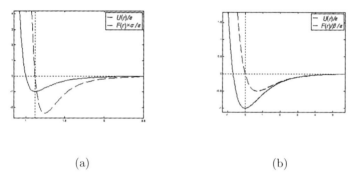

(a)   (b)

Fig. 7.2. Pair-wise potentials and the interatomic forces: (a) Lennard-Jones, (b) Morse.

expressed as a function of the inter-atomic distance,

$$F(r) = -\frac{\partial V(r)}{\partial r} = 24\frac{\epsilon}{\sigma}\left[2\left(\frac{\sigma}{r}\right)^{13} - \left(\frac{\sigma}{r}\right)^{7}\right]. \tag{7.19}$$

The potential and force functions (7.18) and (7.19) are plotted versus the interatomic distance in Fig. 7.2 a, using dimensionless quantities. Another popular model for pair-wise interaction is known as the Morse potential, Fig. 7.2 b:

$$V(r) = \epsilon\left[e^{2\beta(\rho-r)} - 2e^{\beta(\rho-r)}\right], \quad F(r) = 2\epsilon\beta\left[e^{2\beta(\rho-r)} - e^{\beta(\rho-r)}\right], \tag{7.20}$$

where ρ and ϵ are the equilibrium bond length and dislocation energy respectively; β is an inverse length scaling factor. Similar to the Lennard-Jones model, the first term of this potential is repulsive and the second is attractive, which is interpreted as representing bonding. The Morse potential (7.20) has been adapted for modeling atomic interaction in various types of materials and interfaces; examples can be found elsewhere e.g.[439]

The Lennard-Jones and Morse potentials are most commonly used in molecular dynamics simulations, based on the pair-wise approximation, in chemistry, physics and engineering.

Cut-off Radius of the Potential Function. One important issue arising from molecular dynamics simulations relates to the truncation of the potential functions, such as (7.18) and (7.20). Note that computing the internal force for the equations of motion (7.5) due only to pair-wise interaction requires $(N^2 - N)/2$ terms, where N is the total number of atoms. This value corresponds to the case when one takes into account the interaction of each current atom i with all other simulated atoms $j \neq i$; this can be expensive even for considerably small systems. Assuming that the current atom interacts with just its nearest neighbors can reduce the computational effort significantly. Therefore, a cut-off radius, R, is typically introduced and defined as some maximum value of the modulus of the radius vector. The truncated pair-wise potential can then be written as the following:

$$V^{(tr)}(r) = \begin{cases} V(r), & r \leq R, \\ 0, & r > R \end{cases} \qquad (7.21)$$

If each atom interacts with only n atoms in its R-vicinity, the evaluation of the internal pair-wise forces will involve $nN/2$ terms. To assure continuity (differentiability) of V, according to (7.5), a "skin" factor can be alternatively introduced for the truncated potential by means of a smooth step-like function f_c, which is referred to as the cut-off function,

$$V^{(tr)}(r) = f_c(r)V(r) \qquad (7.22)$$

f_c assures a smooth and quick transition from 1 to 0 when the value of r approaches R, and is usually chosen to take the form of a simple analytical function of the interatomic distance r. One example of a trigonometric cut-off function is shown in the next section.

Multi-body Interaction. The higher order terms of the potential function (7.17) are typically employed in simulations of solids and complex molecular structures to account for chemical bond formation, their topology and spatial arrangement, as well as the chemical valence of atoms. However, practical implemen-tation of the multi-body interaction can be extremely involved. As a result, all the multi-body terms of the order higher than three are usually ignored. Essentially, the three-body potential V_3 is intended to provide contributions to the potential energy due to the change of angle between radius vectors $\mathbf{r}_{ij} = \mathbf{r}_i - \mathbf{r}_j$, in addition to the change of absolute values. This accounts for changes in molecular shapes and bonding geometries in atomic structure, e.g.[412] However, the general three-body potentials, such as V_3 in

Eq. (7.17), have been criticized for their inability to describe the energetics of all possible bonding geometries,[68, 426] while a general form for a four- and five-body potential appears intractable, and would contain too many free parameters. As a result, a variety of advanced two-body potentials have been proposed to efficiently account for the specifics of a local atomic environment by incorporating some par-ticular multi-body dependence inside the function V_2, known as bond order functions, rather than introducing the multi-body potential functions $V_{m>2}$. These terms implicitly include the angular dependence of interatomic forces by introducing the so-called bond-order function, while the overall pair-wise formulation is preserved. Also, these potentials are usually treated as short-range ones, i.e. accounting for interaction between a current atom and several neighbors only. Some of the most common models of this type are the following: the Tersoff potential[426, 427] for a class of covalent systems, such as carbon, silicon and germanium, the Brenner[82] and REBO[83] potentials for carbon and hydrocarbon molecules, and the Finnis-Sinclair potential for BCC metals.[164]

In spite of the variety of existing local environment potentials, all of them feature a common overall structure, given by the following expression:

$$V_2(\mathbf{r}_i, \mathbf{r}_j) \equiv V_{ij} = (V_R(r_{ij}) - B_{ij}V_A(r_{ij})), \quad r_{ij} = |\mathbf{r}_{ij}|, \tag{7.23}$$

where V_R and V_A are pair-wise repulsive and attractive interactions, respectively, and the bond order function B is intended to represent the multi-body effects by accounting for spatial ar-rangements of the bonds in a current atom's vicinity. The silicon potential model by Tersoff[426] gives an example of the local environment approach:

$$V_{ij} = f_c(r_{ij})(Ae^{-\lambda_1 r_{ij}} - B_{ij}e^{-\lambda_2 r_{ij}}), \tag{7.24}$$

$$B_{ij} = (1 + \beta^n \zeta_{ij}^n)^{-1/2n} \tag{7.25}$$

$$\zeta_{ij} = \sum_{k \neq i,j} f_c(r_{ik})g(\theta_{ijk})e^{\lambda_3^3(r_{ij}-r_{ik})^3}, \tag{7.26}$$

$$g(\theta) = 1 + c^2/d^2 - c^2/[d^2 + (h - \cos\theta)^2]. \tag{7.27}$$

Here, the cutoff function is chosen as

$$f_c(r) = \frac{1}{2} \begin{cases} 2 & r < R - D, \\ 1 - \sin(\pi(r-R)/2D) & R - D < r < R + D, \\ 0 & r > R + D, \end{cases} \tag{7.28}$$

where the middle interval function is known as the "skin" of the potential. Note that if the local bond order is ignored, so that $B = 2A = const$, and $\lambda_1 = 2\lambda_2$, potential (7.24) reduces to the Morse model (7.20) at $r < R - D$. In other words, all deviations from a simple pair potential are ascribed to the dependence of the function B on the local atomic environment. The value of this function is determined by the number of competing bonds, the strength λ of the bonds and the angles θ between them (θ_{ijk} shows the angle between bonds ij and ik). The function ζ of (7.24) is a weighted measure of the number of bonds competing with the bond i-j, and the parameter n shows how much the closer neighbors are favored over more distant ones in the competition to form bonds. The potentials proposed by Brenner and co-workers[82, 83] are usually viewed as more accu-rate, though more involved, extensions of the Tersoff models.[426, 427] The Brenner potentials include more detailed terms V_A, V_R and B_{ij} to account for different types of chemical bonds that occur in the diamond and graphite phases of the carbon, as well in hydrocarbon molecules. Another special form of the multi-body potential is provided by the embedded atom method (EAM) for metallic systems.[143, 144] One appealing aspect of the EAM is its physical picture of metallic bonding, where each atom is embedded in a host electron gas created by all neighboring atoms. The atom-host interaction is inherently more complicated than the simple pair-wise model. This interaction is described in a cumulative way, in terms of an empirical embedding energy function. The embedding function incorporates some important many-atom effects by providing the amount of energy (work) required to insert one atom into the electron gas of a given density. The total potential energy U includes the embedding energies G of all of atoms in the system, and the electrostatic pair-wise interaction energies V:

$$U = \sum_i G_i\left(\sum_{j \neq i} \rho_j^a(r_{ij})\right) + \sum_{i,j>i} V_{ij}(r_{ij}) . \tag{7.29}$$

Here, ρ_j^a is the averaged electron density for a host atom j, viewed as a function of the distance between this atom and the embedded atom i. Thus, the host electron density is employed as a linear superposition of contributions from individual atoms, which in turn are assumed to be spherically symmet-ric. The embedded atom method has been applied successfully to studying defects and fracture, grain boundaries, interdiffusion in alloys, liquid metals, and other metallic systems and processes.[144]

7.1.5 Energetic Link between MD and Quantum Mechanics

Within the molecular dynamics method, the interacting particles are viewed either as material points exerting potential forces into their vicinity, or as solid spheres with no internal structure. In other words, the internal state of the atoms and molecules does not vary in the course of the simulation, and there is no energy exchange between the MD system and the separate sub-atomic objects, the electrons and nuclei. However, each of the atoms within the MD system represents a complicated physical domain that can evolve in time and switch its internal state by exchanging energy with the surrounding media. Most importantly, the nature of averaged interatomic forces that are employed in the MD simulations is in fact determined by characteristics of the subatomic processes and states. The dependence of the potential function

U on the separation between atoms and molecules and their mutual orientation can in principle be obtained from quantum mechanical (QM) calculations. The further use of this function within a classical MD simulation provides an "energetic link" between the atomic and subatomic scales. These arguments are employed in any multis-cale approach designed to accurately relate the MD and QM simulations. Indeed, in the absence of information about the trajectories of particles within a quantum mechanical model, the energy arguments are solely appropriate for establishing the exchange of the information between the MD and QM subsystems. To illustrate the general idea of MD/QM coupling, consider a simple example with two inter-acting hydrogen atoms. Those comprise one hydrogen molecule H2, which consists of two proton nuclei (+) and two electrons (-). The positions of the electrons with respect to each other and the nuclei are defined by the lengths r_{12} and $r_{\alpha i}$, $\alpha = a, b, i = 1, 2$, and the separation distance be-tween two atoms is given by r, i as depicted in Fig. 7.3

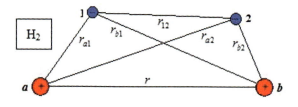

Fig. 7.3. Coordination in the Hydrogen Molecule

Obviously, the total energy E of this system consists of the energies of two unbound hydrogen atoms, and E_b, plus an atomic binding energy term U,

$$E = E_a + E_b + U \tag{7.30}$$

Since the classical MD models assume no energy absorption by the simulated atoms, the values E, E_a, and E_b should relate to the atomic states with minimum possible energies, i.e. the so-called ground states. Provided that these energies are known from quantum mechanical calculations, and the full energy of the coupled system is available for various values r, one obtains a dependence $E(r)$, and therefore the energy of pair-wise atomic interactions as a function of r:

$$U(r) \equiv V_2(r) = E(r) - E_a - E_b . \tag{7.31}$$

When necessary, this function can be interpolated with a smooth curve, and next utilized in the classical molecular dynamics equations of motion (7.5) or (7.15) and (7.16); that is the general idea of estab-lishing the link between the quantum mechanical and MD simulations. The energies E_a and E_b can be found by solving the stationary Schrödinger wave equa-tion [81-84] for each non-interacting hydrogen atom, i.e. when they are formally put at the infi-nite separation distance, $r \to \infty$. This equation gives the functional eigenvalue problem

$$\hat{h}\psi_\alpha = E_\alpha \psi_\alpha, \tag{7.32}$$

$$\hat{h} = -\frac{\hbar^2}{2m}\Delta_\alpha - \frac{e^2}{r_\alpha}, \quad \Delta_\alpha = \frac{\partial^2}{\partial r^2}, \quad \alpha = a, b . \tag{7.33}$$

Here $\hbar, m$ and e are Planck's constant, the electron mass and charge, respec-tively; Δ is the Laplace operator, and $r_a \equiv r_{a1}, r_b \equiv r_{b2}$. $\hat{h}$ is the one-electron Hamiltonian operator. This operator resembles the Hamiltonian function of classical dynamics (7.15) which represents the total en-ergy of a system in terms of the coordinates and momenta of the particles. Similarly, the first term in (7.33) is the kinetic energy operator, and the second term gives the Coulomb potential of the electrostatic electron-proton interaction. Obviously, $E_a = E_b$ and $\psi_a = \psi_b$ for two identical atoms in the ground states. We will nevertheless preserve the above notation for generality, be-cause similar ar-guments hold also for a pair of distinct atoms. The wave function solution, such as ψ_α, provides the complete description of a quantum mechanical sys-tem in the corresponding energy state. At the same time, the wave function itself gives no immediate physical insight. It serves as a mathematical tool only and cannot be determined experimentally. It is used in further calcu-lations in order to obtain observable quantities. For instance, the product $\psi_\alpha^* \psi_\alpha$, where the star notation means complex conjugate, provides a real-valued probability-density function of electron. Its integration over a spatial domain in the vicinity of the nucleus of the unbound atom α results in the probability of finding the electron in this spatial domain. The one-electron wave function ψ_α is often referred to as the hydrogen atomic orbital. All hy-drogen orbitals and the corresponding energy levels are known in closed form that can be found elsewhere in quantum mechanics textbooks, e.g.[256]

In principle, the full energy E of the bound diatomic system H_2 in the ground state can be obtained for a given separation distance r between the nuclei by solving the molecular two-electron Schrödinger equation,

$$\hat{H}\Psi = E\Psi \tag{7.34}$$

$$\hat{H} = -\frac{\hbar^2}{2m}\Delta - \sum_{\alpha,i}\frac{e^2}{r_{\alpha i}} + \frac{e^2}{r_{12}} + \frac{e^2}{r}, \tag{7.35}$$

where the Laplacian Δ involves all the electronic degrees of freedom, and $\hat{H}$ and Ψ are the molecular Hamiltonian and the two-electron wave function, describing the ground state of the coupled system.

One may in fact use molecular dynamics to solve the resultant Schrödinger equations, for instance the hydrogen molecules. The molecular dynamics methods used to solve the resultant Schödinger equations are called ab initio molecular dynamics, which shall be discussed next.

7.2 Ab initio Methods

Based on our view of the hierarchical structure of the universe, it is believed that if one can understand the mechanics of a small length scale, then one can understand the mechanics at all scales. Though this fool-proof philosophy may be debatable, its simplicity is attractive, especially as we have entered into a new era of super-computing. According to our current knowledge, there are four forces in the universe,

1. strong interaction (nuclear force);
2. Coulomb force (electrostatic force);
3. weak interaction (the force related to β decay); and
4. gravitational force.

Forces (1) and (3) are short-ranged. They can be neglected in conventional engineering applications. The so-called first-principle calculations, or ab initio calculations only take into account of forces (2) and (4) in the framework of non-relativistic quantum mechanics. Technically speaking, ab initio methods are used to determine the electron density distribution, and the atomic structures of various materials. By doing so, one may be able to predict the various properties of a material at the atomic level.

Comparing to continuum mechanics, atomic scale simulation is indeed ab initio. However, non-relativistic quantum mechanics may not be the ultimate theory; besides, there are often many approximations involved in simulations of the quantum state of many-electron systems. The connotation of first-principle is used within a specific context. Ultimately, as Ohno et al.,[358] put, *"only God can use the true methodology represented by the term, 'first principle methods'; humans have to use a methodology which is fairly reliable but not exact. "*.

Quantum Mechanics of Many-electron Systems. In quantum mechanics, the state of an N-electron particle system can be described by its wave functions (e.g.[147, 162, 256]). Denoting the Hamiltonian of the system as H, and its ξ-th eigenfunction (wavefunction) as $\Psi_\xi(1, 2, \cdots, N)$, if we write the Hamiltonian for the i-th electron as H_i, the total Hamiltonian reads

$$H = H_1 + H_2 + \cdots + H_N \tag{7.36}$$

which may be explicitly written as

$$H = -\frac{1}{2}\sum_{i=1}^{N}\nabla_i^2 + \sum_{i>j}^{N}\frac{1}{|\mathbf{r}_i - \mathbf{r}_j|} + \sum_{i=1}^{N}v(\mathbf{r}_i) \tag{7.37}$$

Note that the atomic units of ($e = \hbar = m = 1$) is used in (7.37). The first term in (7.37) represents the electron kinetic energy, the second term is due to the electron-electron Coulomb interaction, and the third term $v(\mathbf{r}_i)$ denotes the Coulomb potential caused by the nuclei. The electron distribution can be determined by solving the following steady state Schrödinger equation

$$H\Psi_{\lambda_1,\lambda_2,\cdots,\lambda_N}(1, 2, \cdots, N) = E_{\lambda_1,\lambda_2,\cdots,\lambda_N}\Psi_{\lambda_1,\lambda_2,\cdots,\lambda_N}(1, 2, \cdots, N) \tag{7.38}$$

where $E_{\lambda_1,\lambda_2,\cdots,\lambda_N} = \epsilon_{\lambda_1} + \epsilon_{\lambda_2} + \cdots + \epsilon_{\lambda_N}$ and ϵ_{λ_i} is the eigenvalue of the one electron Schrödinger equation $H_i\psi_{\lambda_i}(i) = \epsilon_{\lambda_i}\psi_{\lambda_i}(i)$.

Though the complete Hamiltonian H for any complex molecule is easily determined, solving the resultant Schrödinger equation is usually difficult even for simple cases, such as the hydrogen molecule discussed. A variety of numerical methods have been developed to obtain approximate multiatom/multi-electron wave functions.

The tight binding method utilizes the exact hydrogen orbitals to give the so-called molecular orbital $\tilde{\psi}$, an approximate wave function solution for a single electron interacting with several arbitrarily arranged nuclei. The Hartree-Fock and related methods employ these molecular orbitals to provide an approximate wave function $\tilde{\Psi}$ for the entire molecule, i.e. for several electrons interacting with the same group of nuclei. In principle, the tight binding method "adds nuclei/atoms", while the Hartree-Fock method adds the electrons to the hydrogen-like system. The molecular shape can also be investigated by finding the configuration with a minimum of the total energy. In all cases, when an approximate N-electron wavefunction $\tilde{\Psi}$ is available, the total energy of the system can be computed as an integral over the electronic variables, instead of directly solving the Schrödinger equation 7.34. For the H2 molecule, it gives

$$E \approx \frac{\int \cdots \int \tilde{\Psi}^* \hat{H}\tilde{H}dr_{a1}\cdots dr_{b2}dr_{12}}{\int \cdots \int \tilde{\Psi}^*\tilde{\Psi}dr_{a1}\cdots dr_{b2}dr_{12}} \tag{7.39}$$

where the star notation implies the complex conjugate. The formula (7.39) is obtained by premul-tiplying the wave equation (7.34) with the complex conjugate solution $\tilde{\Psi}^*$ and integrating it over all electronic degrees of freedom.

The configuration integrals, such as those in (7.39), are usually written in quantum mechanics concisely as

$$E \approx \frac{< \tilde{\Psi}|\hat{H}|\tilde{\Psi} >}{< \tilde{\Psi}|\tilde{\Psi} >} \tag{7.40}$$

which implies integration over all electronic coordinates. According to (7.31) and (7.40), the MD/QM linkage then gives

$$U(r) = \frac{< \tilde{\Psi}(r)|\hat{H}|\tilde{\Psi}(r) >}{< \tilde{\Psi}(r)|\tilde{\Psi}(r) >} - E_a - E_b \tag{7.41}$$

where the polyatomic multielectron wave function $\tilde{\Psi}$ depends on the inter-atomic distance r pa-rametrically. Finally, the system of coupled MD/QM equations may be expressed as

$$\hat{H}\Psi = \left(U + \sum_{\alpha} E_\alpha\right)\Psi, \quad m_i\ddot{\mathbf{r}}_i = \frac{\partial U}{\partial \mathbf{r}_i} \tag{7.42}$$

that represent the concurrent coupling between the subatomic and atomic simulations of nanostructured systems.

The density functional methods are based on alternative arguments. Instead of evaluating the multielectron wave function, an approximate electron density function $\rho(\mathbf{r})$ is derived to give the probability density of finding electrons in the vicinity of a group of nuclei. In contrast to the molecular wave function, the function $\rho(\mathbf{r}$ of any system depends only on three spatial variables, the components of a radius vector x, y and z. Deriving a proper form of $\rho(\mathbf{r}$ is an important intermediate task in this method. The ground state energy E of a molecular system is then found as a functional operator over $\rho(\mathbf{r})$ without involving the multielectron wave function formulation. Collectively, the density functional and Hartree-Fock methods are often re-ferred to as the ab initio methods.

Two approximations are commonly used in ab initio calculations: the Hartree-Fock approximation and the density functional theory.

The Hartree-Fock Approximation. The Hartree-Fock approximation,[167, 194, 195] is a Ritz variational approximation. Since the exact solution of (7.38) is obtained by setting the following quadratic functional to minimum:

$$< \Psi|H|\Psi > = \sum_{s_1}\sum_{s_2}\cdots\sum_{s_N} \int \Psi^*(1, 2, \cdots, N)H\Psi(1, 2, \cdots, N)d\mathbf{r}_1 d\mathbf{r}_2 \cdots d\mathbf{r}_N$$
$$= min\{E\} = E_0 \tag{7.43}$$

The Hartree-Fock approximation is to solve the following one electron form of the Hartree-Fock equation instead of Eq. (7.43),

$$H_0\psi_\lambda(i) + \left[\sum_{\nu=1}^{N}\sum_{s_i}\int \psi_\nu^*(j)U(i,j)\psi_\nu(j)d\mathbf{r}_j\right]\psi_\lambda(i)$$

$$-\left[\sum_{\nu=1}^{N}\sum_{s_i}\int \psi_\nu^*(j)U(i,j)\psi_\lambda(j)d\mathbf{r}_j\right]\psi_\mu(i) = \epsilon_\lambda\psi_\lambda(i) \tag{7.44}$$

Here $\psi_{\lambda_i}(i)$ is a one-electron solution of one-electron Schrödinger equation, $H_0(i) = -\frac{1}{2}\nabla_i^2 + v(\mathbf{r}_i)$, and

$$U(i,j) = \frac{1}{|\mathbf{r}_i - \mathbf{r}_j|}; \tag{7.45}$$

$$v(\mathbf{r}_i) = -\sum_i \frac{Z_j}{|\mathbf{r}_i - \mathbf{R}_j|} \tag{7.46}$$

where Z_j is the nucleus charge of the j-th atom, and $\mathbf{R}_j$ is the spatial coordinate of the j-th atom.

In,[422] the accuracy of large-scale (10,000 basis size) ab initio Hartree-Fock calculation is assessed. There is a large body of literature on Hartree-Fork quantum molecular dynamics simulations,[276, 414, 436] A good survey on research work done at the IBM Research Laboratory is presented by Clementi,[120] who has done pioneering work in this field.

7.2.1 Density Functional Theory

An alternative method to solving an N-particle electron system is the Density Functional Theory.[196, 202, 241] The idea is similar to SPH: instead of studying a discrete N-body particle system, one assumes that there is a continuous electron density cloud, $\rho(\mathbf{r})$, such that the system's thermodynamic potential can be expressed as

$$\Omega = \int v(\mathbf{r})\rho(\mathbf{r})d\Omega + T[\rho(\mathbf{r})] + U[\rho(\mathbf{r})] + E_{xc}[\rho(\mathbf{r})] - \mu\int\rho(\mathbf{r})d\mathbf{r} \tag{7.47}$$

where $v(\mathbf{r})$ is the external potential, $T[\rho(\mathbf{r})]$ is the electron kinetic energy, $U[\rho(\mathbf{r})]$ the Coulomb potential, $E_{xc}[\rho(\mathbf{r})]$ is the exchange-correlation energy functional, and μ is the chemical potential. Based on this continuous representation, one may be able to solve the N-electron system by determining the solution of the following effective one-electron Schrödinger equation—Kohn-Sham equation

$$\left\{-\frac{1}{2}\nabla^2 + v(\mathbf{r}) + \int\frac{\rho(\mathbf{r}')}{|\mathbf{r}-\mathbf{r}'|}d\mathbf{r}' + \mu_{xc}[\rho](\mathbf{r})\right\}\psi_\lambda(\mathbf{r}) = \epsilon_\lambda\psi_\lambda(\mathbf{r}) \tag{7.48}$$

where $\mu_{xc}[\rho](\mathbf{r}) = \delta E_{xc}/\delta\rho(\mathbf{r})$.

There are other ab initio methods such as pseudopotential approach, APW approach, Green's function method, etc. One may consult the monograph by Ohno et al.[358] for detailed discussions.

7.2.2 Ab initio Molecular Dynamics

As a particle method, ab initio molecular dynamics is used to study material's properties at the atomic coordinate level. In ab initio molecular dynamics, one needs to compute the wavefunctions of electrons as well as the movement of the nuclei. The velocity and the position of an atom is primarily determined by the position of the nucleus, which is not only influenced by the nuclei of other atoms surrounding it, but also by the electrons surrounding it. On the other hand, the wavefunction of an electron is also influenced by the presence of the nuclei nearby.

In most ab initio molecular dynamics, the so-called Born-Oppenheimer (BO) adiabatic approximation[75] is used. The approximation assumes that the temperature is very low, and hence only the ground state of electrons is considered, and in addition, the interaction between nuclei and electrons is neglected. In fact, up to today, ab initio molecular dynamics can only deal with the systems that obey the Born-Oppenheimer condition. In electron-nuclear system, nuclei behave like an Newtonian particles, but the wavefunction of an electron is governed by the Schrödinger equation. A popular algorithm is the Car-Parrinello method.[90] Imagine that a small fictitious mass is attached to each electron; the steady state Schrödinger equation will become a hyperbolic equation. Then one can find both the electron wave function, ψ_λ, as well as the atomic coordinates, $\mathbf{R}_i$, by integrating the Newtonian equation of motion. When the fictitious mass attached to each electron approaches zero, the solution should converge to the solution of the coupled electron-nucleus many-body system. The computational task is to integrate the following equations

$$
\begin{cases}
\mu \dfrac{d^2}{dt^2}\psi_\lambda = -H\psi_\lambda + \sum_\nu \Lambda_{\lambda\nu}\psi_\nu, & (a) \\[2mm]
M_i \dfrac{d^2}{dt^2}\mathbf{R}_i = -\nabla_i E, & (b)
\end{cases}
\tag{7.49}
$$

where $\nabla_i E$ is the force acting the nucleus, which is determined by density functional theory as

$$
-\nabla_i E = -\nabla_i \sum_{j \neq i} \frac{Z_i Z_j}{|\mathbf{R}_i - \mathbf{R}_j|} - \int \rho(\mathbf{r})\nabla_i v_i(|\mathbf{r} - \mathbf{R}|)d\mathbf{r}
$$

$$
- \int \frac{\delta E\{\rho\}}{\delta \rho}\nabla_i \rho(\mathbf{r})d\mathbf{r}
\tag{7.50}
$$

The time integration of the electron wave function is carried out by the following predictor-corrector algorithm:

$$
\psi_\lambda^{n+1} = \phi^{n+1} + \frac{(\Delta t)^2}{\mu}\sum_\nu \Lambda_{\lambda\nu}\psi_\nu^n
\tag{7.51}
$$

$$
\psi^{n+1} = 2\psi^n - \psi^{n-1} + \frac{(\Delta t)^2}{\mu}H\psi_\lambda^n
\tag{7.52}
$$

where n is the time step number. The unknown Lagrangian multiplier $\Lambda_{\lambda\nu}$ can be obtained from the orthogonality condition by solving nonlinear algebraic equations. This method is called the Ryckaert method.[395] Eq. (7.49b) can be integrated using either leapfrog or Verlet method.[438]

A brief review of quantum molecular dynamics on the simulation of nucleic acids can be found in.[414] A parallelization of general quantum mechanical molecular dynamics (QMMD) is presented in[197] Simulations on liquid chemicals are reported in.[51, 220]

7.2.3 Tight Binding Method

Between ab initio methods and classical molecular dynamics, there are other semi-empirical methods, such as the Tight-Binding Method.[12, 13, 408] The Tight-Binding method is a quantum mechanics method, because the forces acting on each atom are based on quantum mechanics, but it uses empirical parameters in the construction the Hamiltonian. Those parameters can be obtained from either experiments or ab initio simulations.

The tight binding method, or the method of linear combination of atomic orbitals (LCAO) was originally proposed by Bloch[71] and later revised by Slater and Koster[408] in the context of periodic potential problems. The objective of this method is to construct an approximate wave function of a single electron in a non-central field due to two or more point sources (nuclei). Such a wave function is referred to as the molecular orbital (MO) and is further used in obtain-ing approximate trial functions for the corresponding multielectron systems within the Hartree-Fock and related methods. The tight binding method is based on the assumption that the molecular orbital can be ap-proximated as a linear combination of the corresponding atomic orbitals, i.e. from the readily available hydrogen type orbitals for each of the nuclei comprising the given molecular configura-tion. For the hydrogen ion H_2^+, consisting of two proton nuclei and one electron, the tight binding method provides the following approximate molecular orbital, e.g.:[254, 375]

$$\tilde{\psi} = c_a \psi_a + c_b \psi_b . \tag{7.53}$$

The physical interpretation of this approximation is that in the vicinities of nuclei a and b, the sought molecular orbital should resemble the atomic orbitals ψ_a and ψ_b respectively.

According to the variational principle of quantum mechanics, the energy for a given (chosen) approximate wavefunction is always greater than for the true, or a more accurate, wave function. Therefore, the coefficients c_a and c_b for (7.53) can be found by minimizing the integral (7.39). Due to (7.39) and (7.53), the approximate ground state energy $\tilde{E} \geq E$ gives

$$\tilde{E} = \frac{< \tilde{\psi}|\hat{h}|\tilde{\psi} >}{< \tilde{\psi}|\tilde{\psi} >} = \frac{c_a^2 H_{aa} + c_b^2 H_{bb} + 2c_a c_b H_{ab}}{c_a^2 + c_b^2 + 2c_a c_b S_{ab}} \tag{7.54}$$

where

$$S_{\alpha\beta} = <\psi_\alpha|\psi_\beta>, \quad H_{\alpha\beta} =<\psi_\alpha|\hat{h}|\psi_\beta>, \quad \alpha = a, b \text{ and } \beta = a, b \quad (7.55)$$

$$\hat{h} = -\frac{\hbar^2}{2m}\Delta - \frac{e^2}{r_{a1}} - \frac{e^2}{r_{b1}} + \frac{e^2}{r}, \quad (7.56)$$

where $\hat{h}$ is the Hamiltonian operator for a single electron in the field of two protons and r is the separa-tion between two protons. In deriving (7.54), it was also assumed that the atomic orbitals obey the normalization condition

$$S_{\alpha\alpha} =<\psi_\alpha|\psi_\alpha>= 1 \quad (7.57)$$

as well as the symmetries $S_{\alpha\beta} = S_{\beta\alpha}$ and $H_{\alpha\beta} = H_{\beta\alpha}$ that hold for the hydrogen ion. At the variational minimum, we have the conditions $\partial\tilde{E}/\partial c_a = 0$, $\partial\tilde{E}/\partial c_b = 0$. Employing these conditions and the symmetry $H_{ab} = H_{ba}$, we obtain

$$c_a^2 - c_b^2 = 0, \quad (7.58)$$

which is only possible when

$$c_a = c_b, \quad \text{or} \quad c_a = -c_b \quad (7.59)$$

Thus, there exist two molecular orbitals for the H_2^+ ion, one symmetric and one antisymmet-ric:

$$\tilde{\psi}^+ = N^+(\psi_a + \psi_b), \quad \tilde{\psi}^- = N^-(\psi_a - \psi_b), \quad (7.60)$$

where $N^\pm$ are normalization factors, which can be found from the condition similar to (7.57). According to (40) and (35) these molecular states are characterized by the energies

$$\tilde{E}^+ = \frac{H_{aa} + H_{ab}}{1 + S_{ab}}, \quad \tilde{E}^- = \frac{H_{aa} - H_{ab}}{1 - S_{ab}}. \quad (7.61)$$

7.2.4 Numerical Examples

Nanotubes Filled with Fullerenes. The recent resurgence of molecu-lar dynamics, both quantum and classical, is largely due to the emergence of nano-technology. Materials at the nanoscale have demonstrated impres-sive physical and chemical properties, thus suggesting a wide range of areas for applications. For instance, carbon nanotubes are remarkably strong, and have better electrical conductance, as well as heat conductivity than copper at room temperature. Moreover, nanotubes are such light weight and high-strength (TPa) materials that they eventually will play an important role in reinforced fiber composites, and as both devices and nanowires. In particu-lar, nanotubes having fullerenes inside could have different physical properties compared to empty nanotubes. Such structure also hold promise for use in potential functional devices at nanometer scale: nano-pistons, nano-bearings, nano-writing devices, and nano-capsule storage system.

Modeling of nanotubes filled with fullerenes has two aspects: (1) the bonded interaction between fullerenes and nanotubes; (2) the bonded interactions among the carbon atoms of the nanotubes. Recently, Qian et al.[379] used combined molecular dynamics and meshfree Galerkin approach to simulate interaction between fullerenes and a nanotube. In the non-bonded interaction, the nanotube is modeled as a continuum governed by the Cauchy-Born rule (e.g. Tadmor et al. 1996[421] and Milstein 1982[326]). For the bonded interaction, a modified potential is used to simulate interactions among carbon atoms. Specifically, Tersoff-Brenner model (Tersoff 1988,[427] Brenner 1990[82]) is used in simulation,

$$\Phi_{ij}(R_{ij}) = \Phi_R(R_{ij}) - \bar{B}_{ij}\Phi_A(R_{ij}) \tag{7.62}$$

where Φ_R and Φ_A represent the repulsive and attractive potential respectively,

$$\Phi_R(R_{ij}) = f(R_{ij})\frac{D_{ij}^{(e)}}{(S_{ij}-1)}exp\left\{-\sqrt{2S_{ij}}\beta_{ij}(R_{ij}-R_{ij}^{(e)})\right\} \tag{7.63}$$

$$\Phi_A(R_{ij}) = f(R_{ij})\frac{D_{ij}^{(e)}S_{ij}}{(S_{ij}-1)}exp\left\{-\sqrt{2/S_{ij}}\beta_{ij}(R_{ij}-R_{ij}^{(e)})\right\} \tag{7.64}$$

For carbon-carbon bonding, $D_{ij}^{(e)} = 6.0eV$, $S_{ij} = 1.22$, $\beta_{ij} = 2.1A^{-1}$, $R_{ij}^e = 1.39A$, and

$$f(r) = \begin{cases} 1 & r < R_{ij}^{(1)} \\ \frac{1}{2}\left(1+\cos\left[\frac{\pi(r-R_{ij}^{(1)})}{R_{ij}^{(2)}-R_{ij}^{(1)}}\right]\right) & R_{ij}^{(1)} \leq r \leq R_{ij}^{(2)} \\ 0 & r > R_{ij}^{(2)} \end{cases} \tag{7.65}$$

The effect of bonding angle is taking into account in term $\bar{B}_{ij}$ (See Brenner[82] and Qian et al[379]). The length of the nanotubes are $L = 129\mathring{A}$, and the diameter of the nanotube is $6.78\mathring{A}$ (5,5), which is close to the diameter of C_{60}.

Nano-indentation. The second example is the simulation of nano-indentation processes as shown in Fig. 7.5. In a nano-indentation experiment, the size of a typical indenter is of the order of tens of nanometers. To minimize the boundary effects, the substrate for the MD simulations must be at least an order of magnitude larger than the indenter. A model for this system would easily fall beyond the affordable range of the modern computer power. To reduce the computational requirements, a virtual potential is often introduced to mimic the indenter. The effective domain of this potential is much smaller than that of a real indenter. Rigid boundary conditions are typically applied

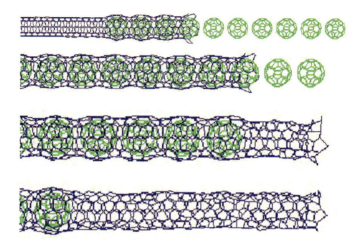

Fig. 7.4. Molecular dynamics simulations (EAM) of C_{60} passing through nanotube.[379]

on the bottom of the substrate with periodic boundary conditions in the indentation plane. These boundary conditions artificially stiffen the material, which suppresses the nucleation of dislocations. Furthermore, the evolution of any emitted dislocations may also be affected by these boundary conditions. The validity of the corresponding force versus indentation depth curve becomes questionable, especially when a small domain is simulated.

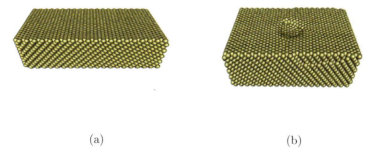

(a)   (b)

Fig. 7.5. Indentation pattern in a gold substrate: MD simulation. Actual imprint size can be tens-to-hundreds of nanometers.

7.3 Coupling between MD and FEM

As an essential part of nano-science and nano-technology, the multi-scale method provides an unique means to simulate, to observe, to synthesize, and to analyze many important physical processes at different length and time scales in a unified framework. Multiscale methods and simulations have become an important research field within the past decade. Much of this is due to the fact that multi-scale phenomena have become increasingly important in our daily life. Well-known examples of multi-scale phenonmena include: mechanics of DNA and RNA, mechanics of an epitaxial thin film, initiation of a crack or a shear band, mechanics of nano-devices and structures, the motion of a dislocation, etc. In these examples, the fine scale information at a particular local region is crucial to the overall macro-scale responses of the system, which is manifested by the interaction between the fine scale motion in a local region and the global macro-scale behaviors. This class of multi-scale phenomena can not be described by a phenomenological theory, nor is a first-principle based theory capable of providing sensible solution, because a complete fine scale micro-mechanics simulation is usually prohibitively expensive.

Therefore, the logical step to find solution of this type of problems is to couple the simulations with different scale concurrently, e.g. coupling a continuum simulation where there the overall responses are the ultimate objective with an atomic simulation at a specific region where it is important to model the fine scale physics accurately.

We define concurrent multi-scale methods as the methods that combine information available from distinct length and time scales into a single coherent, coupled, and simultaneous simulations.

7.3.1 MAAD

One pioneering multi-scale approach was the work by Abraham *et al.*[3] The idea was to concurrently link tight binding (TB), molecular dynamics (MD) and finite elements (FE) together in a unified approach called MAAD (macroscopic, atomic, *ab initio* dynamics). Concurrent linking here means that all three simulations run at the same time, and dynamically transmit necessary information to and receive information from the other simulations. In this approach, the FE mesh is graded down until the mesh size is on the order of the atomic spacing, at which point the atomic dynamics are governed via MD. Finally, at the physically most interesting point, i.e. at a crack tip, TB is used to simulate the atomic bond breaking processes. The interactions between the three distinct simulation tools are governed by conserving energy in the system[86]

$$H_{TOT} = H_{FE} + H_{FE/MD} + H_{MD} + H_{MD/TB} + H_{TB} \qquad (7.66)$$

More specifically, the Hamiltonian, or total energy of the MD system can be written as

$$H_{MD} = \sum_{i<j} V^{(2)}(r_{ij}) + \sum_{i,(j<k)} V^{(3)}(r_{ij}, r_{ik}, \Theta_{ijk}) + K \tag{7.67}$$

where the summation is over all atoms in the system, K is the kinetic energy of the system, r_{ij} and r_{ik} indicate the distance between two atoms and Θ_{ijk} is the bonding angle between three atoms. The summation convention $i < j$ is performed so that each atom ignores itself in finding its nearest neighbors. Here the potential energy is comprised of two parts. The first $(V^{(2)})$ is two-body interactions, for example nearest neighbor spring interactions in 1D. The second part are the three-body interactions $(V^{(3)})$, which incorporate such features as angular bonding between atoms. The three-body interactions also make the potential energy of each atom dependent on its environment. The finite element Hamiltonian can be written as the sum of the kinetic and potential energies in the elements, i.e.

$$H_{FE} = V_{FE} + K_{FE} \tag{7.68}$$

in which

$$V_{FE} = \frac{1}{2} \int_{\Omega} \epsilon(\mathbf{r}) \cdot \mathbf{C} \cdot \epsilon(\mathbf{r}) d\Omega \tag{7.69}$$

$$K_{FE} = \frac{1}{2} \int_{\Omega} \rho(\mathbf{r})(\dot{\mathbf{u}})^2 d\Omega \tag{7.70}$$

where ϵ is the strain tensor, $\mathbf{C}$ is the stiffness tensor, ρ is the material density and $\dot{\mathbf{u}}$ are the nodal velocities. Thus the potential energy contribution to the FE Hamiltonian, V_{FE}, is the integral of the strain energy, while the kinetic energy depends upon the nodal velocities. The TB total energy may be written as

$$V_{TB} = \sum_{n=1}^{N_{occ}} \epsilon_n + \sum_{i<j} V^{rep}(r_{ij}) \tag{7.71}$$

This energy can be interpreted as having contribution from an attractive part ϵ_n and a repulsive part V^{rep}. N_{occ} are the number of occupied states. While a detailed overview of tight binding methods is beyond the scope of this work, further details can be found in.[168]

The overlapping regions (FE/MD and MD/TB) are termed "handshake" regions, and each makes a contribution to the total energy of the system. The handshake potentials are combinations of the potentials given above, with weight factors chosen depending on whether the atomic bond crosses over the given interface. The three equations of motion (TB/FE/MD) are all integrated forward using the same timestep. This method was applied successfully to the simulation of brittle fracture by Abraham *et al.*[5]

An approach related to the TB/MD/FE approach of Abraham *et al.* was developed by Rudd and Broughton[392] called coarse-grained molecular dynamics (CGMD). This approach removes the TB method from the TB/MD/FE method and instead couples only FE and MD. Again, the FE mesh is graded down to the atomic scale. A key development was that in recognizing that degrees of freedom were missing from the system due to the coarse-graining approximation. A total energy was derived for CGMD using statistical mechanics principles, which is stated to be

$$E(\mathbf{u}_k, \dot{\mathbf{u}}_k) = U_{int} + \frac{1}{2} \sum_{j,k} (M_{jk} \dot{\mathbf{u}}_j \dot{\mathbf{u}}_k + \mathbf{u}_j K_{jk} \mathbf{u}_k) \tag{7.72}$$

where $U_{int} = 3(N - N_{node})kT$. The energy is comprised of the average kinetic and potential energies as well as a thermal term from the coarse grained (eliminated) degrees of freedom. It was demonstrated that elastic wave reflection measured by a reflection coefficient using CGMD was smaller than the previous TB/MD/FE method of Abraham.[392]

7.3.2 MD/FE Coupling - 1D Example

We now discuss the coupling between MD and FEM. We restrict our attention on the system that there is only one physical law in different scales, e.g. Newton's equations of motion

$$\mathbf{f} = \mathbf{Ma} . \tag{7.73}$$

Therefore *both* the MD and FE systems obey the equations of motion (7.73).

First we must define the force vector $\mathbf{f}$ and mass matrix $\mathbf{M}$ for each system. For a MD system, the force $\mathbf{f}_{MD}$ is computed by differentiating a potential energy function Φ, which is typically a function of the atomic positions, i.e.

$$\mathbf{f}_{MD} = -\nabla \Phi(\mathbf{r}_1, ..., \mathbf{r}_N) \tag{7.74}$$

where $\mathbf{r_i}$ is the distance between neighboring atoms. One of the most common interatomic potentials is the Lennard-Jones (LJ) 6-12 potential. The potential energy function for the LJ 6-12 is expressed as

$$\Phi(r_{ij}) = 4\epsilon \left((\frac{\sigma}{r_{ij}})^{12} - (\frac{\sigma}{r_{ij}})^6 \right) \tag{7.75}$$

where ϵ and σ are constants chosen to fit material properties and r_{ij} is the distance between two atoms i and j. The LJ 6-12 is termed a pair potential because the energy depends only upon the distance r_{ij} between two atoms. The $1/r_{ij}^{12}$ term is meant to model the repulsion between atoms as they approach each other, and is motivated by the Pauli principle in chemistry. The Pauli principle implies that as the electron clouds of the atoms begin to overlap, the system energy increases dramatically because two interacting electrons can not occupy the same quantum state. The $1/r_{ij}^{6}$ term adds cohesion to the system, and is meant to mimic van der Waals type forces. The van der Waals interactions are fairly weak in comparison to the repulsion term, hence the lower order exponential assigned to the term.

It is crucial to note that the LJ 6-12 is not a realistic potential, because of the pair interaction limitation. In accepting this limitation, the LJ 6-12 is most commonly used in simulations where a general class of effects is studied, instead of specific physical properties, and a physically reasonable yet simple potential energy function is desired. We may now derive the interatomic forces in 1D based on the LJ 6-12 potential by employing (7.74) to obtain

$$\frac{\partial \Phi}{\partial r_{ij}} = 4\epsilon \left(-12\frac{\sigma^{12}}{r_{ij}^{13}} + 6\frac{\sigma^{6}}{r_{ij}^{7}} \right) \tag{7.76}$$

The force is then the negative of the gradient of the potential energy. Assuming that $\sigma = 1$ and $\epsilon = 1$, the force f on atom i and is written in simplified form as

$$f_i = -\sum_{i \neq j} \frac{24}{r_{ij}^{7}} \left(1 - \frac{2}{r_{ij}^{6}} \right) \tag{7.77}$$

The force and potential energy for the LJ 6-12 are shown in Fig. (7.6). It should be mentioned that the axes of Fig. (7.6) are in terms of σ and ϵ, which are the LJ parameters. Furthermore, the equilibrium distance between two atoms interacting via a Lennard-Jones relation is $\sqrt[6]{2}\sigma$.

In our 1D coupling example, we assume that these atoms interact with their nearest neighbors via a harmonic potential. The harmonic potential energy can be written as

$$\Phi(r_{ij}) = \frac{1}{2}k(r_{ij} - r_0)^2 \tag{7.78}$$

where k is the spring constant, r_{ij} is the interatomic distance and r_0 is the equilibrium bond length. Taking the negative gradient of Φ with respect to r_{ij} gives the MD force displacement relationship

$$f_i = -k(r_{ij} - r_0) \tag{7.79}$$

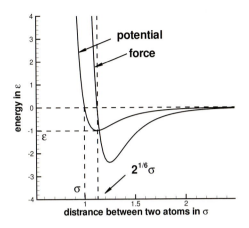

Fig. 7.6. Force and potential energy plot for Lennard-Jones 6-12 potential.

Eq. (7.79) can be rewritten in a different form by noting the following relationships. First, the equilibrium bond length is the difference in initial positions of two atoms, i.e. $r_0 = x_j - x_i$. The interatomic distance can then be written as a function of the initial positions and the displacements d of each atom as $r_{ij} = x_j + d_j - (x_i + d_i)$. Therefore, $r_{ij} - r_0 = d_j - d_i = \Delta x$, where Δx is the relative displacement between two neighboring atoms. We will use this notation for the remainder of this paper.

A useful analogy can be made by comparing the behavior of the harmonic potential to continuum linear elasticity. Note that unlike the Lennard-Jones potential, the harmonic potential cannot recognize bond breaking or separation, because the force is a continuous function of relative displacement. For the Lennard-Jones potential, the attractive force dies out quickly after about two interatomic distances, which allows bond breaking if the tensile force is strong enough. In this sense, the harmonic potential is akin to linear elasticity on the atomic level.

The mass matrix $\mathbf{M}_{MD}$ is a diagonal matrix with the individual atomic masses on the diagonal. For a two atom system, this would look as follows:

$$\mathbf{M}_{MD} = \begin{pmatrix} m_1 & 0 \\ 0 & m_2 \end{pmatrix} \tag{7.80}$$

where m_i are the masses of each individual atom. At this point, all the information needed to solve the equation of motion for the MD system has been defined, and we move onto defining the necessary finite element quantities. For a finite element system, the mass matrix $\mathbf{M}_{FE}$ can be defined as follows:

$$\mathbf{M}_{FE} = \int_{\Omega_0} \rho_0 \mathbf{N}^T \mathbf{N} d\Omega_0 \tag{7.81}$$

where ρ_0 is the initial material density, Ω_0 is the undeformed volume and $\mathbf{N}$ are the finite element shape functions, which are typically low order polynomials. If linear shape functions are used, the lumped mass matrix for a single element can be written as

$$\mathbf{M}_{FE} = \frac{\rho_0 A_0 l_0}{2} \begin{pmatrix} 1 & 0 \\ 0 & 1 \end{pmatrix} \tag{7.82}$$

where A_0 is the initial area of the finite element and l_0 is the initial length of the finite element. (7.82) states that half the mass in the finite element is assigned to each node. We use the lumped mass matrix in the following 1D MD/FE coupling example to preserve the diagonal quality of the global mass matrix.

For the FE force $\mathbf{f}_{FE}$, we shall assume that no external forces act upon the system, so that $\mathbf{f}_{FE} = \mathbf{f}_{FE}^{int}$, or the total force is equal to the internal force $\mathbf{f}_{FE}^{int}$. The internal force in a FE simulation is computed by multiplying the stiffness $\mathbf{K}$ by the nodal displacements $\mathbf{d}$

$$\mathbf{f_{FE}} = \mathbf{K_{FE}d} \tag{7.83}$$

For a linear elastic system, the stiffness matrix takes the familiar form

$$\mathbf{K}_{FE} = \frac{-kh_a}{l_0} \begin{pmatrix} 1 & -1 \\ -1 & 1 \end{pmatrix} \tag{7.84}$$

if it is assumed that the smallest element, i.e. that with nodal spacing equal to the atomic spacing h_a, acts as the parent element. The preceding finite element equations are derived and explained in detail in the text by Belytschko, Liu and Moran.[49]

Coupled FE/MD Equations of Motion. The key point here is how to couple the MD and FE systems. Suppose that there is one finite element in which the nodes exactly overlap the MD atoms. We refer to this as the "handshake" element. The question for this element is how to define the mass matrix and force vector. For the mass matrix, the procedure described above for (7.82) can be used. For the force vector, a different method must be utilized. The method used is to weight the contribution to the total force between the MD force and FE force. In this case, because the nodes and atoms overlap exactly, it is determined that the total force is equally weighted from the MD force and FE force.

In detail, the interaction force $\mathbf{f}_{MD}$ is calculated for the two "handshake" atoms, which are atoms 2 and 3 in Fig. 7.7. The force vector components f_2 and f_3 are then augmented as follows:

$$f_2 = f_2 + \frac{1}{2} f_{MD} \tag{7.85}$$

$$f_3 = f_3 + \frac{1}{2} f_{MD} \tag{7.86}$$

Because only half the MD force makes a contribution to the total force, the other half must come from the FE internal force. This contribution is made in the same manner as above. First, the FE force f_{FE} is computed for the boundary element

$$f_2 = f_2 + \frac{1}{2} f_{FE} \tag{7.87}$$

$$f_3 = f_3 + \frac{1}{2} f_{FE} \tag{7.88}$$

If the FE region does not exactly overlap the boundary MD region, different weighting combinations can be used.

For the 3-atom, 2 element (the handshake region counts as an element) example shown in Fig. 7.7, we now derive the coupled MD/FE equations of motion. The atoms are assumed to have mass m, as do both finite elements. The atomic spacing is h_a, as is the nodal spacing for both finite elements, which implies that the first finite element after the handshake element also has a nodal spacing which equals the interatomic spacing. The displacements of the atoms/nodes are assumed to be d_1, d_2, d_3 and d_4. Because the masses of the atoms and finite elements are equal, we can write the mass matrix $\mathbf{M}$ for this system as

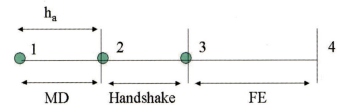

Fig. 7.7. Problem description for 1D MD/FE coupling.

$$\mathbf{M} = \begin{pmatrix} m & 0 & 0 & 0 \\ 0 & m & 0 & 0 \\ 0 & 0 & m & 0 \\ 0 & 0 & 0 & m \end{pmatrix} \tag{7.89}$$

We now turn to the details of constructing the force vector. For the first pair of atoms, the force can be written as

$$\mathbf{f} = -k\Delta\mathbf{x} = -k(d_2 - d_1) \tag{7.90}$$

where d_i are the displacements of each atom. The force is apportioned to each atom so the force vector is written as

$$\mathbf{f} = \begin{pmatrix} k(d_2 - d_1) \\ -k(d_2 - d_1) \\ 0 \\ 0 \end{pmatrix} \tag{7.91}$$

Note that the sum of the forces is zero. For the finite element with nodes 3 and 4, the force is calculated by multiplying the stiffness matrix by the displacements,

$$\begin{pmatrix} f_3 \\ f_4 \end{pmatrix} = -\frac{kh_a}{l_0} \begin{pmatrix} 1 & -1 \\ -1 & 1 \end{pmatrix} \begin{pmatrix} d_3 \\ d_4 \end{pmatrix} \tag{7.92}$$

Adding this to the global force vector $\mathbf{f}$ gives

$$\mathbf{f} = \begin{pmatrix} k(d_2 - d_1) \\ -k(d_2 - d_1) \\ k(d_4 - d_3) \\ -k(d_4 - d_3) \end{pmatrix} \tag{7.93}$$

Finally, the handshake region comprised of atoms/nodes 2 and 3 are considered. Because the FE nodes and MD atoms coincide for this case, it is assumed that each will contribute half of the total force. Mathematically, this says

$$\begin{pmatrix} f_2 \\ f_3 \end{pmatrix} = \frac{1}{2} \begin{pmatrix} f_{MD2} \\ f_{MD3} \end{pmatrix} + \frac{1}{2} \begin{pmatrix} f_{FE2} \\ f_{FE3} \end{pmatrix} \tag{7.94}$$

The MD forces are

$$\begin{pmatrix} f_{MD2} \\ f_{MD3} \end{pmatrix} = \frac{k}{2} \begin{pmatrix} d_3 - d_2 \\ -(d_3 - d_2) \end{pmatrix} \tag{7.95}$$

while the FE forces are

$$\begin{pmatrix} f_{FE2} \\ f_{FE3} \end{pmatrix} = \frac{k}{2} \begin{pmatrix} d_3 - d_2 \\ -(d_3 - d_2) \end{pmatrix} \tag{7.96}$$

Because the linear spring assumption is used for both the MD and FE systems, the MD forces are identical to the FE forces for this case. Thus, the complete system of equations for this 3 atom, 2 element system reads

$$\begin{pmatrix} m & 0 & 0 & 0 \\ 0 & m & 0 & 0 \\ 0 & 0 & m & 0 \\ 0 & 0 & 0 & m \end{pmatrix} \begin{pmatrix} \ddot{d}_1 \\ \ddot{d}_2 \\ \ddot{d}_3 \\ \ddot{d}_4 \end{pmatrix} = \begin{pmatrix} k(d_2 - d_1) \\ k(d_3 - 2d_2 + d_1) \\ k(d_4 - 2d_3 + d_2) \\ k(d_3 - d_4) \end{pmatrix} \tag{7.97}$$

MD/FE Coupling Numerical Examples. To further illustrate the direct coupling of FE and MD, we present a simplified 1D example of MAAD including coupling between finite elements and molecular dynamics, but excluding tight binding. The problem description is shown pictorially in Fig. 7.8. The problem is symmetric about x=0, as the MD region has 101 atoms from x=-2 to x=2. The finite element region has 100 elements. 50 elements are used between x=-10 and x=-1.96, and 50 elements are used between x=1.96 and x=10. Thus there are two handshake regions in which the atomic positions and the finite element nodes coincide. The first is the finite element with nodes at x=-2 and x=-1.96. The second is the finite element with nodes at x=1.96 and x=2.

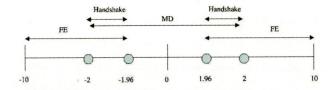

Fig. 7.8. Problem description for 1D MD/FE coupling.

As was mentioned above, only one system of equations is solved. Therefore, all the degrees of freedom are integrated in time using the same time integration algorithm. Typically for MD, a velocity Verlet or Gear sixth order time integrator is used e. g. [393] In the example, the equations of motion are integrated by using Ruth's symplectic leapfrog algorithm.[393] This is a two-stage algorithm to update the velocities and displacements and is implemented as follows:

$$\mathbf{v}_1 = \mathbf{v}_0 \tag{7.98}$$

$$\mathbf{d}_1 = \mathbf{d}_0 + \frac{1}{2}\mathbf{v}_1 \Delta t \tag{7.99}$$

$$\mathbf{v}_2 = \mathbf{v}_1 + \mathbf{f}(\mathbf{d}_1)\Delta t \tag{7.100}$$

$$\mathbf{d}_2 = \mathbf{d}_1 + \frac{1}{2}\mathbf{v}_2 \Delta t \tag{7.101}$$

The 0 subscripts indicate initial values, the 1 subscripts indicate values after the first stage, and the 2 subscripts are the values at the end of the timestep. The key point is that the force $\mathbf{f}$ is evaluated only once, which is crucial since the force calculation is the most expensive part of a MD simulation. The remaining parts of the two-stage update require little memory and computational expense.

A Gaussian-type wave which is symmetric about x=0 is applied to the MD system. The initial configuration of the problem is shown in Fig. 7.9. Two cases were tested. In the first case, all the finite elements had the same spacing as the atomic spacing. In the second case, the FE nodal spacing increased with the distance from the MD region. The atomic masses and spring constant were taken to be unity.

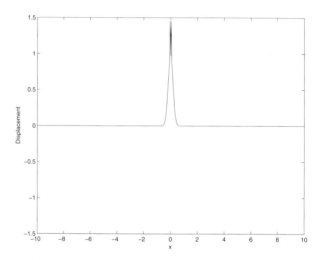

Fig. 7.9. Initial displacement for 1D FE/MD coupled problem.

In the first case, as depicted in Fig. 7.10(a), because the FE nodal spacing is the same as the atomic spacing everywhere, the coupled equations of motion are identical, and the wave sees the same system whether it is passing through the MD or FE regions. Thus, the transition from the MD to FE region is smooth as well, and no wave reflection occurs. In contrast, as shown in Fig. 7.10(b), if the FE mesh is graded as the distance from the MD region increases, wave reflection is immediately noticeable in the MD region. Due to the fact that the finite elements are unable to resolve the small wavelengths coming from the MD region, and because the formulation is energy conserving, the waves must go somewhere and are thus reflected back into the MD region.

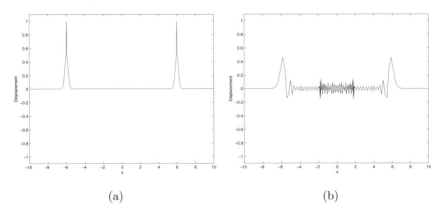

(a)   (b)

Fig. 7.10. Snapshot of displacement for 1D FE/MD coupled problem after wave has propagated into the FE region: (a) for the case where FE nodal spacing equals atomic spacing everywhere; (b) for the case where FE nodal spacing gradually increases with distance from MD region.

7.3.3 Quasicontinuum Method and Cauchy-Born Rule

A different approach to multi-scale modeling, the quasicontinuum method, was developed by Tadmor, Ortiz and Phillips.[421] The atomic to continuum link is achieved here by the use of the Cauchy-Born rule. The Born rule assumes that the continuum energy density W can be computed using an atomic potential, with the link to the continuum being the deformation gradient $\mathbf{F}$. To briefly review continuum mechanics, the deformation gradient $\mathbf{F}$ maps an undeformed line segment $d\mathbf{X}$ in the reference configuration onto a deformed line segment $d\mathbf{x}$ in the current configuration

$$dx = \mathbf{F}d\mathbf{X} \tag{7.102}$$

In general, $\mathbf{F}$ can be written as

$$\mathbf{F} = 1 + \frac{d\mathbf{u}}{d\mathbf{X}} \tag{7.103}$$

where $\mathbf{u}$ is the displacement vector. If there is no displacement in the continuum, the deformation gradient is equal to unity.

The major restriction and implication of the Cauchy-Born rule is that the continuum deformation must be homogeneous. This results from the fact that the underlying atomic system is forced to deform according to the continuum deformation gradient $\mathbf{F}$, as is illustrated in Fig. 7.11. By using the Born rule, the authors were able to derive a continuum stress tensor and tangent stiffness directly from the interatomic potential, which allowed the usage

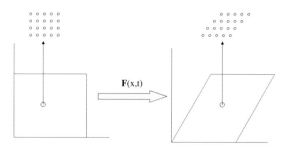

Fig. 7.11. Illustration of Cauchy-Born rule.

of the standard nonlinear finite element method. This can be done by the following relations

$$\boldsymbol{\mathcal{P}} = \frac{\partial W}{\partial \mathbf{F}^T} \tag{7.104}$$

$$\mathbf{C} = \frac{\partial^2 W}{\partial \mathbf{F^T} \partial \mathbf{F^T}} \tag{7.105}$$

where $\mathbf{C}$ is the Lagrangian tangent stiffness and $\boldsymbol{\mathcal{P}}$ is the first Piola-Kirchoff stress tensor.

Adaptivity criteria were used in regions of large deformation so that full atomic resolution could be achieved in these instances, i.e. near a dislocation. A non-local version of the Born rule was also developed so that non-homogeneous deformations such as dislocations could be modeled. The quasicontinuum method has been applied to quasi-static problems such as nano-indentation.[421]

Other related multi-scale approaches include that of Arroyo and Belytschko.[15] In this approach, a correction to the Born rule, the exponential map, was derived for application to the modeling of carbon nanotubes. It was shown that the exponential map made the Born rule valid for nanotube analysis. Extensions and analysis of the quasicontinuum method were performed by Diestler[132] and Shilkrot *et al.*[403]

Cauchy-Born Rule for Linear Springs. Two 1D examples are now done to illustrate how the Cauchy-Born rule is used. In the first case, we consider a quadratic potential energy function, which when minimized yields a linear spring force relation. The potential energy Φ is written as

$$\Phi = \frac{1}{2}k\Delta x^2 \qquad (7.106)$$

where k is the spring constant and $\Delta \mathbf{x}$ is the relative displacement between two atoms. Differentiating this energy function with respect to the relative displacement yields the MD force

$$f_{MD} = -\frac{\partial \Phi}{\partial \Delta x} = -k\Delta x \qquad (7.107)$$

In order to use the Cauchy-Born rule, we must make two modifications to (7.106). The first is to write Φ as a function of the deformation gradient F. The second is to modify the potential energy function Φ such that we can obtain the energy density W. In 1D, we can obtain the energy density directly by dividing the potential energy by the initial atomic bond length. This may be written as

$$W = \frac{\Phi}{r_0} \qquad (7.108)$$

where r_0 is the initial atomic bond length. One important property that must be conserved in using the Cauchy-Born rule is that the energy of the molecular system must equal the energy in the continuum system, which can be obtained by integrating the energy density and setting it equal to the summation of the bond energies

$$\sum_{i=1}^{nbond} \Phi_{bond} = \int_{\Omega_0} W\,dX \qquad (7.109)$$

To illustrate (7.109) via example, consider a three-atom molecular system with initial bond length r_0 and interacting by linear springs of stiffness k. The equivalent continuum system is then of length $2r_0$. This is shown in Fig. 7.12. The energy in the MD system is computed to be

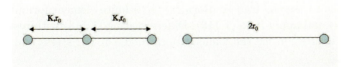

Fig. 7.12. Three atom system interacting via linear springs along with homogenized continuum.

$$E_{MD} = 2 \left(\frac{1}{2} k \Delta x^2 \right) = k \Delta x^2 \qquad (7.110)$$

In order to write the continuum energy density as a function of F, we note that using (7.102), the undeformed bond length r_0 can be related to the deformed bond length r through the deformation gradient by the relation

$$r = F r_0 \qquad (7.111)$$

The relative displacement Δx can then be written as a function of the deformation gradient F as

$$\Delta x = r_0 (F - 1) \qquad (7.112)$$

The expression for the continuum energy density becomes, by substituting (7.112) into (7.106) and normalizing by the initial bond length

$$W = \frac{1}{r_0} \left(\frac{1}{2} k \left(r_0 (F - 1) \right)^2 \right) \qquad (7.113)$$

Integrating over the continuum body of length $2r_0$ gives the energy of the continnum system

$$E_{CB} = k r_0^2 (F - 1)^2 \qquad (7.114)$$

If the continuum system is stretched to length $4r_0$, $F = 2$, and the continuum energy evaluates to $E_{CB} = k r_0^2$. The corresponding MD system energy is then obtained by evaluating (7.110), and yields the same as the continuum energy, $E_{MD} = k r_0^2$. It is *crucial* to note that in obtaining the MD energy, we assumed that the deformation of each bond was identical. This is a result of the homogeneous deformation assumption that underlies the Cauchy-Born rule, i.e. that the underlying atomic system deforms homogeneously like the continuum system.

In order to use the Cauchy-Born rule in a finite element formulation, we use (7.104) and apply it to (7.113). Differentiating (7.113) with respect to F gives the expression for the Cauchy-Born force (in 1D, stress and force are equivalent)

$$f_{CB} = k r_0 (F - 1) . \qquad (7.115)$$

Cauchy-Born Rule for Lennard-Jones 6-12 Potential. Our second example utilizes a standard nonlinear interatomic potential, the Lennard-Jones 6-12 potential. As was previously given in (7.77), the MD force for the LJ 6-12 potential can be written as

$$f_{MD} = 48\epsilon \frac{\sigma^{12}}{r^{13}} - 24\epsilon \frac{\sigma^6}{r^7} \tag{7.116}$$

To derive the Cauchy-Born force from the LJ 6-12 potential, we again write an expression for the energy density W. Doing so, we obtain

$$W = \frac{1}{r_0} 4\epsilon \left(\left(\frac{\sigma}{r}\right)^{12} - \left(\frac{\sigma}{r}\right)^6 \right) \tag{7.117}$$

Substituting (7.111) into (7.117) and minimizing W with respect to F, we obtain the Cauchy-Born force for the LJ 6-12 potential

$$f_{CB} = \frac{24\epsilon\sigma^6}{F^7 r_0^7} - \frac{48\epsilon\sigma^{12}}{F^{13} r_0^{13}} \tag{7.118}$$

It becomes clear in 1D that the *only* difference between the Cauchy-Born and MD force expressions for the same atomic spacing lies in the ability of the finite element simulation to accurately calculate the deformation gradient. If the deformation gradient is calculated exactly and the finite element spacing equals the atomic spacing, then $f_{CB} = -f_{MD}$. In practice, however, the deformation gradient is calculated numerically using finite element shape functions, i.e.

$$F = 1 + \frac{d\mathbf{u}}{d\mathbf{X}} \, . \tag{7.119}$$

In finite elements, the displacement field $\mathbf{u}$ is approximated by shape functions which interpolate nodal values, i.e.

$$\mathbf{u} = \sum_\mathbf{I} \mathbf{N_I}(\mathbf{X})\mathbf{d_I} \tag{7.120}$$

where $\mathbf{d_I}$ are the nodal displacements and $\mathbf{N_I}(\mathbf{X})$ are the shape functions, which are functions of space and are typically low order polynomials. Substituting (7.120) into (7.119), one obtains the numerical form of F

$$F = 1 + \sum_I \frac{d\mathbf{N_I}}{d\mathbf{X}} \mathbf{d_I} \tag{7.121}$$

Clearly, the quality of the shape function derivatives controls the accuracy to which the deformation gradient can be calculated numerically, and hence controls the accuracy to which the Cauchy-Born rule can mimic the actual MD forces. In 1D, F is a constant for each finite element. Therefore, F can be calculated *exactly* using linear shape functions, as the derivatives of the shape functions will be constants, and thus matches the order of F. Thus in 1D, if the finite element spacing is the same as the atomic spacing, the force computed using the Cauchy-Born rule will be exactly the negative of the force computed using an interatomic potential.

In the finite element formulation, the first Piola-Kirchoff stress $\mathcal{P}$ is used in the calculation of the internal force $\mathbf{f^{int}}$, which is defined as

$$\mathbf{M\ddot{d}} = -\mathbf{f^{int}} \tag{7.122}$$

where $\mathbf{M}$ is the finite element mass matrix and $\mathbf{\ddot{d}}$ are the nodal accelerations, or the second derivatives of the nodal displacements with respect to time. If the internal force is calculated numerically by summing over a discrete set of quadrature points at locations $\mathbf{X_q}$, the FE semi-discrete equations of motion can be expressed as

$$\mathbf{M\ddot{d}} = -\sum_q \frac{d\mathbf{N_I}}{d\mathbf{X}}(\mathbf{X_q})\mathcal{P}(\mathbf{X_q})\mathbf{w_q} \tag{7.123}$$

where w_q are the integration weights associated with point $\mathbf{X_q}$. In practice, then, because the FE internal force term involves the negative of the Cauchy-Born stress (or force in 1D), if the FE nodal spacing equals the atomic spacing, the MD and FE equations of motion see the same absolute value of the internal force. Therefore, the only difference between the two simulations is the mass matrix used; in MD, a diagonal mass matrix is used. In FE, a consistent mass matrix is used, though a lumped mass matrix can also be used as was described above. If the FE lumped mass matrix is used and the FE nodal spacing equals the atomic spacing, then the MD and FE displacements are essentially identical.

7.3.4 Cauchy-Born Numerical Examples

A simple 1D example problem has been run using the LJ 6-12 potential energy function to calculate the first Piola-Kirchoff stress in the finite element formulation. The LJ potential parameters $\sigma = \epsilon = 1$, such that the equilibrium atomic spacing was $r_0 = 2^{\frac{1}{6}}$. 201 finite element nodes and atoms were used, and were initially spaced at the equilibrium atomic spacing r_0. The problem is symmetric about $x = 0$, hence only the results for $x > 0$ are shown. The FE nodes and MD atoms were given the same initial displacement in the form of a Gaussian-type wave with a fine scale perturbation. The initial MD configuration is shown in Fig. 7.13(a), and the result after 100 timesteps is shown in Fig. 7.13. The corresponding FE configuration after 100 timesteps is shown in Fig. 7.14(a).

As can be seen by comparing Figs. 7.13 (b) and 7.14(a), the displacements calculated are nearly identical for the FE/Cauchy-Born and MD cases

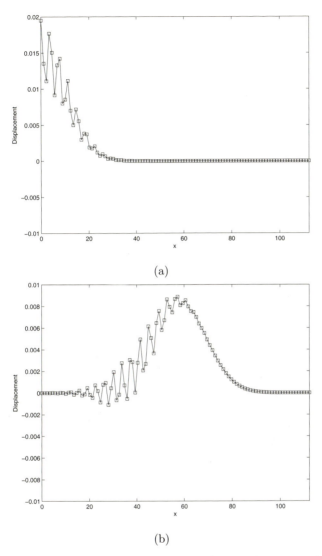

Fig. 7.13. (a) Initial MD configuration for Cauchy-Born example; (b) MD displacements after 100 timesteps.

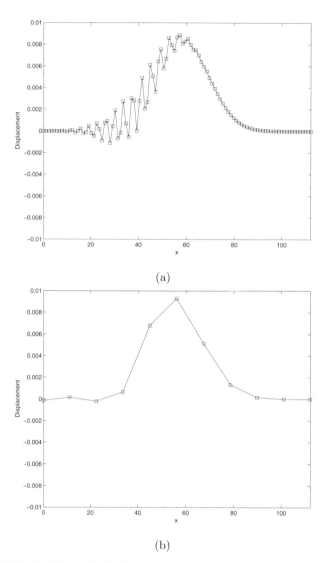

(a)

(b)

Fig. 7.14. (a) FE nodal displacements after 100 timesteps; (b) FE nodal displacements after 6 timesteps, reduced mesh size.

for the LJ 6-12 potential when the finite element nodal spacing is the same as the atomic spacing. This result agrees with the theoretical results discussed above. To further illustrate the Cauchy-Born rule, the same problem description was run, except using a FE nodal spacing of $10r_0$ was used, which means that 20 finite elements were used instead of 200. The configuration after 6 timesteps is shown in Fig. 7.14(b). As can be seen, even with the coarser finite element mesh, the major features of the Gaussian wave are captured. This example also illustrates one major shortcoming in using the Cauchy-Born rule, in that the coarse FE mesh does not allow resolution of the fine atomic details that are available from a full MD simulation. In using the coarser FE mesh here, because many fewer elements are used, the timestep needed is correspondingly larger, and is in fact five times larger than the MD timestep. This illustrates one major advantage of using the Cauchy-Born rule, in that the computational expense necessary to obtain a comparable solution to MD is much smaller, due to the smaller number of timesteps and finite elements necessary.

7.3.5 Multi-scale Algorithms

There is considerable challenge in developing an efficient yet accurate multi-scale method. One issue is the necessity of meshing the FE region down to the atomic scale. This presents two problems, one numerical and one physical. The numerical issue is that the timestep in an FE simulation is governed by the smallest element in the mesh. Thus, if the finite elements are meshed down to the atomic scale, many timesteps will be wasted simulating the dynamics in these regions. Furthermore, it seems unphysical that the variables of interest in the continuum region should evolve at the same time scales as the atomic variables. Thus, a multi-scale method that could incorporate larger timesteps for the continuum region would constitute a significant improvement in this area.

The physical issue in meshing the FE region down to the atomic scale lies in the FE constitutive relations. The constitutive relations typically used in FE calculations, e.g. for plasticity, are constructed based on the bulk behavior of many dislocations. Once the FE mesh size approaches the atomic spacing, the possibility of many dislocations becomes impossible, the bulk assumption disappears, and the constitutive relation is invalidated.

Another major problem in multi-scale simulations is that of pathological wave reflection, which occurs at the interface between the MD and FE regions. The issue is that the wavelength emitted by the MD region is considerably smaller than that which can be captured by the continuum FE region. Because of this and the fact that an energy conserving formulation is typically used, the wave must go somewhere and is thus reflected. This leads to spurious heat generation in the MD region, and a contamination of the simulation. One method used by Abraham and Rudd to eliminate this was to mesh the FE region down to the atomic scale so that the FE mesh is small enough that it can represent the short wavelengths emitted from the MD region. Despite this effort, other effects such as stiffness differences between the two regions still cause a small amount of wave reflection.

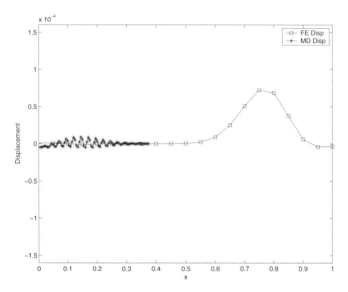

Fig. 7.15. Depiction of spurious wave reflection which results if MD and FEM regions are directly coupled with no special treatment.

The above example shows the necessity of accounting for and removing wave reflection. The example problem is that used by Wagner and Liu in their 1D bridging scale paper,[446] in which the harmonic potential is used to simulate the atomic interactions. In the bridging scale, as will be discussed in later sections, the MD region constitutes only a small portion of the domain, while the finite element representation is *everywhere* in the domain. The example problem was run with two cases, and the results are shown in Fig. 7.15 and 7.16. In the first case, the MD region is directly coupled to the FE region. The resulting wave reflection can clearly be seen in the MD region. In the second example, the boundary physics are correctly accounted for using a technique described below, the Generalized Langevin Equation (GLE). In comparing the MD displacement after the wave has propagated out of the MD region, it is clear there is almost no reflection in the MD region.

Fig. 7.17 shows a more quantifiable measure of wave reflection, by measuring the energy remaining in the MD system after the wave has passed through. In this example and for all examples to come in this work regarding energy transfer, the initial energy is the sum of the initial kinetic and potential energy in the MD region, while the final energy is the total kinetic

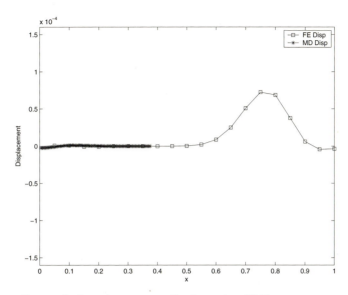

Fig. 7.16. Removal of spurious wave reflection using GLE.

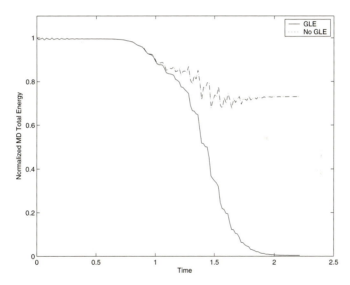

Fig. 7.17. Energy remaining in MD system using different boundary conditions between MD/FE regions.

and potential energy remaining in the MD region. If no boundary condition between the MD and FE regions is imposed, it is seen in Fig. 7.17 that only 30 percent of the total energy is transferred to the surrounding continuum. However, if the GLE is used, about 99.9 percent of the total energy is transferred.

7.3.6 Generalized Langevin Equation

Having demonstrated the effect of the GLE in multi-scale simulations, we now qualitatively describe the effect of using this method. Fig. 7.18 illustrates one possible decomposition scheme that develops from using the GLE. Originally, an entire molecular system exists. However, we would like to keep the effects of all the atomic degrees of freedom while not solving for them explicitly. The reduced MD system that results from using the GLE is then shown. It is shown that the full MD lattice can be reduced into a portion of that lattice along with external forces that act on the boundaries of the reduced lattice which represent the combined effects of all the atomic degrees of freedom that have been mathematically accounted for. Thus, the GLE can also be used as a boundary condition on an MD simulation; this is the basis for the work in[238].[447] Because the external forces on the reduced lattice are derived from

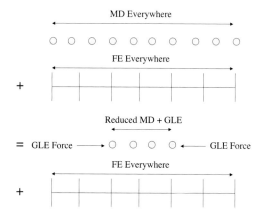

Fig. 7.18. Schematic of bridging scale method utilizing GLE.

the mathematically accounted for MD degrees of freedom, the effect of using the GLE in conjunction with the FEM is that small wavelengths which are on the order of the atomic spacing can be dissipated cleanly into the surrounding continuum, while the FEM can capture any longer wavelength information that is on the order of the FE mesh spacing or larger. The specific details showing the GLE and its relation to the bridging scale will be shown in a later section.

Derivation of Generalized Langevin Equation. Following Adelman and Doll[139],[7] we now derive the GLE. Assuming a harmonic lattice that vibrates with frequency $\omega = \sqrt{k/m}$, where m is the atomic mass and k is the spring constant, the equations of motion for the atoms can be written in matrix form as

$$\mathbf{M}\ddot{\mathbf{x}}(t) + \mathbf{K}\mathbf{x}(t) = 0 \qquad (7.124)$$

We define region 1 as the MD region where the degrees of freedom will be kept, and region 2 as the MD region to be mathematically accounted for as boundary forces acting on region 1. Decomposing (7.124) into those parts from regions 1 and 2 and defining $\mathbf{A} = \mathbf{M}^{-1}\mathbf{K}$, the equations of motion are written as

$$\begin{pmatrix} \ddot{\mathbf{x}}_1 \\ \ddot{\mathbf{x}}_2 \end{pmatrix} = -\begin{pmatrix} \mathbf{A}_{11} & \mathbf{A}_{12} \\ \mathbf{A}_{21} & \mathbf{A}_{22} \end{pmatrix} \begin{pmatrix} \mathbf{x}_1 \\ \mathbf{x}_2 \end{pmatrix} \qquad (7.125)$$

It should be noted that the terms $\mathbf{A}_{12}$ and $\mathbf{A}_{21}$ only play a role near the boundary between regions 1 and 2. For example, if nearest neighbor spring force interactions are considered, then each of those sub-matrices only has one term to account for the interaction of the boundary atom with its nearest neighbors.

The degrees of freedom $\mathbf{x}_2$ are eliminated by solving for them explicitly in (7.125) and substituting the result back into the equation for $\mathbf{x}_1$. This process is done by Laplace transforming the equation for $\mathbf{x}_2$. In doing so, we briefly review some basic equations related to Laplace transforms. The definition of the Laplace transform of a function $\mathbf{f}(t)$ is

$$\mathbf{F}(s) = \mathcal{L}(\mathbf{f}(t)) = \int_0^\infty \mathbf{f}(t)e^{-st}dt \qquad (7.126)$$

where the operator $\mathcal{L}$ transforms functions of time t into functions of Laplace space s. The inverse Laplace transform, which transforms functions of Laplace space s into functions of time t is defined to be

$$\mathbf{f}(t) = \mathcal{L}^{-1}(\mathbf{F}(s)) = \frac{1}{2\pi i}\int_{c-i\infty}^{c+i\infty} \mathbf{F}(s)e^{st}ds \qquad (7.127)$$

where c is a real constant greater than the real parts of all singularities of $\mathbf{F}(s)$. One useful property of Laplace transforms is the convolution property, which states that the convolution integral of two functions is equal to the product of the transforms of the individual functions. Mathematically, this says

$$\mathcal{L}(\int_0^t \mathbf{f}(t-t')\mathbf{g}(t')dt') = \mathbf{F}(s)\mathbf{G}(s) \tag{7.128}$$

A final relevant property of Laplace transforms concerns the transforms of time derivatives. Specifically relevant to this problem is the Laplace transform of a second derivative with respect to time, which can be written as

$$\mathcal{L}(\frac{d^2\mathbf{f}(t)}{dt^2}) = s^2\mathbf{F}(s) - s\mathbf{F}(t=0) - \frac{d\mathbf{f}}{dt}(t=0) \tag{7.129}$$

Using these properties to Laplace transform the equation for $\ddot{\mathbf{x}}_2$ in (7.125) gives

$$s^2\mathbf{x}_2(s) - s\mathbf{x}_2(0) - \dot{\mathbf{x}}_2(0) = -\mathbf{A}_{21}\mathbf{x}_1(s) - \mathbf{A}_{22}\mathbf{x}_2(s) \tag{7.130}$$

Solving this equation for $\mathbf{x}_2(s)$ gives

$$\mathbf{x}_2(s) = \boldsymbol{\theta}(s)(s\mathbf{x}_2(0) + \dot{\mathbf{x}}_2(0)) - \boldsymbol{\theta}(s)\mathbf{A}_{21}\mathbf{x}_1(s) \tag{7.131}$$

where the matrix $\boldsymbol{\theta}(s)$ can be written as

$$\boldsymbol{\theta}(s) = (s^2\mathbf{I} + \mathbf{A}_{22})^{-1} \tag{7.132}$$

Performing an inverse Laplace transform on (7.131) and using the convolution rule, an equation for $\mathbf{x}_2(t)$ can be found as

$$\mathbf{x}_2(t) = -\int_0^t \boldsymbol{\theta}(t-\tau)\mathbf{A}_{21}\mathbf{x}_1(\tau)d\tau + \boldsymbol{\theta}(t)\dot{\mathbf{x}}_2(0) + \dot{\boldsymbol{\theta}}(t)\mathbf{x}_2(0) \tag{7.133}$$

Substituting (7.133) into (7.125) yields an equation for $\ddot{\mathbf{x}}_1(t)$

$$\ddot{\mathbf{x}}_1(t) = -\mathbf{A}_{11}\mathbf{x}_1(t) + \mathbf{A}_{12}\int_0^t \boldsymbol{\theta}(t-\tau)\mathbf{A}_{21}\mathbf{x}_1(\tau)d\tau - \mathbf{A}_{12}\mathbf{x}_2^R(t) \tag{7.134}$$

where $\mathbf{x}_2^R(t)$ is

$$\mathbf{x}_2^R(t) = \boldsymbol{\theta}(t)\dot{\mathbf{x}}_2(0) + \dot{\boldsymbol{\theta}}(t)\mathbf{x}_2(0) \tag{7.135}$$

This can now be written in its final form

$$\ddot{\mathbf{x}}_1(t) = -\mathbf{A}_{11}\mathbf{x}_1(t) + \int_0^t \boldsymbol{\Theta}(t-\tau)\mathbf{x}_1(\tau)d\tau + \mathbf{R}(t) \qquad (7.136)$$

where

$$\mathbf{R}(t) = -\mathbf{A}_{12}\mathbf{x}_2^R(t) \qquad (7.137)$$

$$\boldsymbol{\Theta}(t) = \mathbf{A}_{12}\boldsymbol{\theta}(t)\mathbf{A}_{21} \qquad (7.138)$$

As can be seen in (7.136), the equations of motion for the region 1 atoms have been modified to include two additional terms. These two terms represent the mathematically accounted for degrees of freedom in region 2 in the form of external forces acting upon the boundary atoms of region 1. More specifically, the effects of the mathematically accounted for degrees of freedom in region 2 enter through the time history kernel $\boldsymbol{\Theta}(t)$ and the random force $\mathbf{R}(t)$. It is the time history kernel that mimics the collective behavior of the mathematically accounted for MD degrees of freedom in region 2, and thus is the key element to allowing small wavelengths to pass into the surrounding continuum.

The random force $\mathbf{R}(\mathbf{t})$ captures the exchange of energy between regions 1 and 2 due to temperature differences. The reason this external force is considered random is because this term depends on the initial conditions $\mathbf{x}_2(0)$ and $\dot{\mathbf{x}}_2(0)$, which in general are not known. The initial conditions in region 2 are not known because those degrees of freedom were mathematically eliminated in order to construct the Langevin equation. Furthermore, the temperature T of a solid is in general the only information known, hence the initial conditions can only be determined via a probability distribution. Therefore, a large number of initial conditions are possible, and the force is thus considered random.

Similar approaches have been undertaken by both Cai *et al.*[89] and E *et al.*[157] In both of these papers, the goal has been to numerically solve for the time history kernel $\boldsymbol{\Theta}$. In the method of E and Huang, the time history kernel is replaced by a truncated discrete summation. Weights are then chosen to minimize wave reflections. The method of Cai and coworkers involves using an MD simulation on a larger domain to characterize the time history kernel for the problem at hand. One issue with both methods is that they may not be transferable, which means that the method may not work for a general lattice structure, and instead work only for those on which they were originally computed. The GLE, in its original form, is also non-transferable. Techniques to resolve this issue are given in[238],[447]

7.3.7 Multiscale Boundary Conditions

The key issue of a concurrent simulation approach is the coupling between the coarse and fine scales. An approximation is necessary along the fine-coarse grain interface, due to fundamental incompatibility of the atomic and continuum descriptions, e.g.[131] This incompatibility is imposed by the mismatch of dispersion characteristics of the continuous and discrete media in dynamic simulations, and by nonlocal character of the atomic interaction in both dynamic and quasistatic simulations. Most of the concurrent approaches, excluding the bridging scale method, involve an artificial handshake or pad region,[128] where pseudoatoms are available on the continuum part of the interface and share physical space with finite elements. At the front end of the continuum interface, the finite elements have to be scaled down the chemical or ion bond lengths; that may call forth costly inversions of large and ill-conditioned stiffness ma-trices. The purpose of the handshake region is to assure a smoother coupling between the atomic and continuum regimes. The group of pseudoatoms serves for eliminating the non-physical surface in the atomic lattice structure, so that the real atoms along the interface have a full set of interactive neighbors in the continuum domain. In dynamic simulations, the handshake also serves as a damper/absorbent to reduce spurious reflection of high frequency phonons that can not pass into the coarse scale domain. In both dynamic and quasistatic simulations, an extremely fine finite element mesh is required in order to provide accurate positions of the pseudoatoms, as those are dictated by interpolation from the finite element nodal positions.[128] An alternative methodology has been proposed recently by Karpov et al[238, 239] and Wagner et al,[447] where positions of actual next-to-interface atoms from the coarse grain are computed at the intrinsic atomic level by means of a functional operator over the interface atomic displacements; that eliminates the need in a costly handshake domain. The sole purpose of a continuum model, when used in conjunction with multi-scale boundary conditions, is to represent effects of the peripheral (coarse grain) boundary conditions into the central atomic region of interest. Provided that this effect is ignorable, at least in the analytical sense, the multi-scale boundary conditions can also serve as a self-contained multiple scale method not to involve the Cauchy-Born rule and the consequent continuum model. Atomic resolution along the interface phase along with the intrinsic regularity of the internal structure of the crystalline solids allows calculating the structural response of the coarse scale on the atomic level, based on a group of lattice mechanics techniques.

Quasistatic Problems. The basic idea of the quasistatic multi-scale boundary conditions is explained in the 1D example problem depicted in Fig. 7.19. The boundary atom $n = 0$ of the MD domain is subjected to a load due to some atomic process on the left and the response of a coarse grain on the right. The solution for the interface atom can be computed, without solving the entire coarse scale, provided that a relationship between the displacements of atoms 1 and 0 is established,

$$\mathbf{u}_1 = A^{(a)}\{\mathbf{u}_0\} \tag{7.139}$$

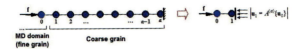

Fig. 7.19. An illustration to the concept of multi-scale boundary conditions: behavior of the MD boundary is governed by a deformable boundary equation, which accounts for the effect of a coarse scale domain.

Here, $A^{(a)}$ is a linear operator, whose form depends on the lattice properties and the coarse scale size parameter a. Only the first neighbor interaction is assumed in 7.139 for clarity. More general coarse grain boundary conditions $\mathbf{u}_a \neq 0$, rather than the shown case $\mathbf{u}_a = 0$, may also contribute to this solution, so that

$$\mathbf{u}_1 = A^{(a)}\{\mathbf{u}_0, \mathbf{u}_a\} \ . \tag{7.140}$$

Based on elementary arguments, one obtains, for the 1D problem depicted in Fig. 7.19,

$$\mathbf{u}_1 = \frac{a-1}{a}\mathbf{u}_0 + \frac{1}{a}\mathbf{u}_a \tag{7.141}$$

Relationships of the type (7.141) are referred to as the multi-scale boundary conditions. They are solved simultaneously with the MD equations for the fine grain to yield an atomic solution, which incorporates effects of the adjacent coarse scale domain. The corresponding position of atom 1 can be also viewed as a deformable boundary of the MD domain.

For more general multidimensional problems, the multi-scale boundary conditions can be obtained with the use of the Fourier analysis of periodic structures.[237] This approach was verified on a benchmark nano-indentation problem with the 3D FCC gold lattice,[239] as sche-matically shown in Fig. 7.20.

The bottom part of the substrate is considered as a bulk coarse scale that features almost homogeneous deformation patterns, and whose degrees of freedom can be eliminated from the explicit MD model. Periodic boundary conditions were applied along side-cut of the substrate. The atomic displacements along the deformable boundary layer in the reduced MD domain were corelated with the displacements of atoms in the adjacent layer through the discrete convolution operator

$$\mathbf{u}_{1,ml} = A^{(a)}\{\mathbf{u}_{0,m,l}\} = \sum_{m',l'} \Theta^{(a)}_{m-m',l-l'}\mathbf{u}_{0,m',l'} \ . \tag{7.142}$$

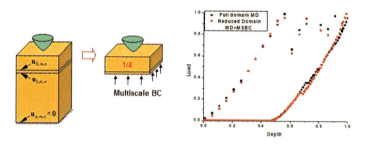

Fig. 7.20. Multi-scale boundary conditions for nano-indetation problem.

Here, the kernel matrix Θ depends solely on the choice of the interatomic potential and the size of the coarse scale, and indexes m, l show numbering of atoms on the given layer. Eq. (7.142) involves no coarse scale boundary conditions $\mathbf{u}_{a,m,l}$ (along the bottom layer of the substrate), as those are usually set trivial in nano-indentation simulations. The multi-scale boundary conditions according to (7.142) were performing well for a wide range of the indentation depths, where a good agreement with the benchmark full domain solution was observed. Most importantly, the approach adequately reproduces the plasticity phenomena in the substrate around the indenter tip. Those result in the discontinuous character of the load/indentation depth curve depicted in Fig. 7.20. Coarse scale lattice defects caused by the nano-indentation process are not restrained, because the method formulation assumes spatial regularity of the lattice structure only in the immediate vicinity of each given atom on the deformable boundary layer.

Dynamic Problems. The general dynamic formulation of the multi-scale boundary conditions is identical with the quasistatic case, i.e.

$$\mathbf{u}_1(t) = A^{(a)}\{\mathbf{u}_0(t), \mathbf{u}_a(t)\} \tag{7.143}$$

where $A\{\mathbf{u}_0(t)\}$ is some functional linear operator. In many dynamic problems, the coarse scale can often be viewed in the infinite sense, so that the effect of its boundary condition $\mathbf{u}_a$ is not present in the MD domain of interest. This situation is common for dynamic simulations due to the availability of a finite speed with which any mechanical excitation propagates through the molecular lattice as a progressive wave package.

Recall nano-indentation (Fig. 7.20), one observes the "one-way" wave flow from inside the domain of interest. Distant boundaries of a coarse grain domain then behave passively and usually remain stationary, unless the simulation time is large enough for the wave flow to reach the edges of the coarse grain. Due to physical arguments, and also for the sake of saving the computer efforts, it is then appropriate to assume that the progressive waves never reach the traction-free coarse scale boundaries, so that no inward flow of information occurs in the abovementioned problems, and

$$\mathbf{u}_1(t) = A\{\mathbf{u}_0(t)\} \ . \tag{7.144}$$

Here, the operator A no longer depends on the coarse scale size parameter a. This form of the dynamic multi-scale boundary conditions is also referred to as the impedance boundary conditions. From the knowledge of displacements $\mathbf{u}_0(t)$, $\mathbf{u}_1(t)$, and also the interatomic potential function, one can compute the force exerted by the coarse scale onto each given atom at the MD domain boundary. Note that this force will be analogous to the impedance force utilized in the dynamic bridging scale formulation.

As was shown by Karpov et al[237] and Wagner et al,[447] the form of operator A is particu-larly compact for the MD/coarse grain interface with a regular crystalline structure and harmonic character of the motion. For a plane-like interface, it acts as a time convolution integral and dis-crete spatial summation over the interface degrees of freedom. For the 2D lattice problem de-picted in Fig. 7.21,

$$\mathbf{u}_{l,m}(t) = \sum_{m'=m-m_c}^{m+m_c} \int_0^t \Theta_{m-m'}(t-\tau)\mathbf{u}_{0,m'}(\tau)d\tau \tag{7.145}$$

where the impedance kernel function Θ depends only on the form of the interatomic potential, and m_c is some critical difference $m - m'$ after which the summation is truncated. More complicated boundary shapes, such as a rectangle or parallelepiped, are assembled by combining sev-eral plane-like interfaces, where each face is treated according to (7.145). Calculation of the kernel matrix Θ involves a Laplace transform inversion, which can be accomplished numerically based on Weeks,[450] Papoulis,[368] and other algorithms. A crucial aspect is that the amplitude of this function decays in time and with the growth of the spatial parameter m, and it typi-cally behaves as shown in Fig. 7.22. The use of numeric Laplace inversion techniques[368, 450] normally implies a limited range for the arguments of the computed functions Θ of (7.145), from $t = 0$ to some critical value t_c for the difference $t - \tau$.

Therefore it is important to investigate the effect of temporal truncation for the convolution integral in (7.145) at various t_c. However, such a truncation considerably decreases the computational cost and computer memory

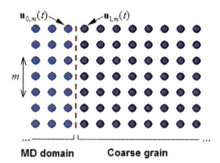

Fig. 7.21. Plane MD/coarse scale interface in a 2D cubic lattice. Index m shows atomic numbering along the interface.

requirements. Performance of the method, as depending on the choice the parameters m_c and t_c, was studied by Karpov et al[238] on an example 2D cubic lattice. A square boundary shape for the lattice was chosen. Fig. 7.22 presents the impedance kernel computed for this lattice. The authors introduced the reflectivity coefficient R as a measure of efficiency of the boundary conditions (7.145). The value R gives the ratio between the kinetic energy of wave flow reflected by the MD/coarse grain interface and the energy of incident waves,

$$R = \frac{E^{refl}}{E^{incid}} \tag{7.146}$$

In the ideal case of fully transparent boundary conditions, value R is trivial. The results of the calculations are presented in Fig. 7.23. Analysis of Figs 7.22 and 7.23 indicates that the use of 2 or 3 full oscillations of the kernel functions at $m_c = 4$ is sufficient to dissipate more than 99at the MD/coarse scale interface. As soon as the MD simulation is performed over a considerably small atomic domain, the effect of surrounding media can be taken into account with use the above techniques. Note that physical behavior and properties of simulated domains cannot be unambiguously attributed to a

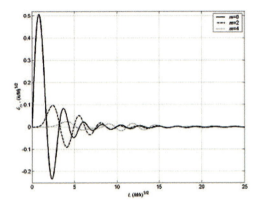

Fig. 7.22. Impedance kernel function for the 2D lattice.

corresponding macroscale system, unless the MD boundary conditions most rigorously describe this effect.

7.4 Introduction of Bridging Scale Method

The Bridging Scale Method[446] was recently developed by Wagner and Liu to couple atomic and continuum simulations. The fundamental idea is to decompose the total displacement field $\mathbf{u}(\mathbf{x})$ into coarse and fine scales

$$\mathbf{u}(\mathbf{x}) = \bar{\mathbf{u}}(\mathbf{x}) + \mathbf{u}'(\mathbf{x}) \tag{7.147}$$

This decomposition has been used before in solid mechanics by Hughes *et al.* in the variational multi-scale method.[212] The coarse scale $\bar{\mathbf{u}}$ is that part of the solution which can be represented by a set of basis functions, i.e. finite element shape functions. The fine scale $\mathbf{u}'$ is defined as the part of the solution whose projection onto the coarse scale is zero.

In order to describe the bridging scale, first imagine that a body in any dimension which is described by N_a atoms. The notation used here will mirror that used by Wagner and Liu.[446] The *total* displacement of an atom α is

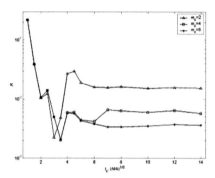

Fig. 7.23. Typical performance of the impedance boundary condition: dependence of the reflection coeffocoent on method parameters.

written as $\mathbf{u}_\alpha$. The coarse scale displacement is a function of the initial positions $\mathbf{X}_\alpha$ of the atoms. It should be noted that the coarse scale would at first glance be thought of as a continuous field, since it can be interpolated between atoms. However, because the fine scale is defined only at atomic positions, the total displacement and thus the coarse scale are discrete functions that are defined only at atomic positions. For consistency, Greek indices $(\alpha, \beta, ...)$ will define atoms for the remainder of this paper, and uppercase Roman indices $(I, J, ...)$ will define coarse scale nodes.

The coarse scale is defined to be

$$\bar{\mathbf{u}}(\mathbf{X}_\alpha) = \sum_I N_I^\alpha \mathbf{d}_I \tag{7.148}$$

Here, $N_I^\alpha = N_I(\mathbf{X}_\alpha)$ is the shape function of node I evaluated at point $\mathbf{X}_\alpha$, and $\mathbf{d}_I$ is the FE nodal displacement associated with node I.

As discussed above, the fine scale in the bridging scale decomposition is simply that part of the total displacement that the coarse scale cannot represent. Thus, the fine scale will be defined to be the projection of the coarse scale subtracted from the *total* solution $\mathbf{u}_\alpha$. We will select this projection operator to minimize the mass-weighted square of the fine scale, which can be written as

$$Error = \sum_{\alpha} m_{\alpha} \left(\mathbf{u}_{\alpha} - \sum_{I} N_I^{\alpha} \mathbf{w}_I \right)^2 \tag{7.149}$$

m_{α} is the atomic mass of an atom α and $\mathbf{w}_I$ are temporary nodal (coarse scale) degrees of freedom. It should be emphasized that (7.149) is only one of many possible ways to define an error metric. In order to solve for $\mathbf{w}$, the error is minimized with respect to $\mathbf{w}$, yielding the following result:

$$\mathbf{w} = \mathbf{M}^{-1} \mathbf{N}^T \mathbf{M}_A \mathbf{u} \tag{7.150}$$

where the coarse scale mass matrix $\mathbf{M}$ is defined as

$$\mathbf{M} = \mathbf{N}^T \mathbf{M}_A \mathbf{N} \tag{7.151}$$

$\mathbf{M}_A$ is a diagonal matrix with the atomic masses on the diagonal. The fine scale $\mathbf{u}'$ can thus be written as

$$\mathbf{u}' = \mathbf{u} - \mathbf{N}\mathbf{w} \tag{7.152}$$

or

$$\mathbf{u}' = \mathbf{u} - \mathbf{P}\mathbf{u} \tag{7.153}$$

where the projection matrix $\mathbf{P}$ can be defined to be

$$\mathbf{P} = \mathbf{N}\mathbf{M}^{-1}\mathbf{N}^T \mathbf{M}_A \tag{7.154}$$

The total displacement $\mathbf{u}_{\alpha}$ can thus be written as the sum of the coarse and fine scales as

$$\mathbf{u} = \mathbf{N}\mathbf{d} + \mathbf{u} - \mathbf{P}\mathbf{u} \tag{7.155}$$

The final term in the above equation is called the bridging scale. It is the part of the solution that must be removed from the total displacement so that a complete separation of scales is achieved, i.e. the coarse and fine scales are orthogonal to each other. This bridging scale approach was first used by Liu et al. to enrich the finite element method with meshfree shape functions.[306] Wagner and Liu[443] used this approach to consistently apply essential boundary conditions in meshfree simulations. Zhang et al.[465] applied the bridging scale in fluid dynamics simulations. Qian et al. recently used the bridging scale in quasi-static simulations of carbon nanotube buckling.[379] The bridging scale was also used in conjunction with a multi-scale constitutive law to simulate strain localization.[234]

Now that the details of the bridging scale have been laid out, some comments are in order. In equation (7.149), the fact that an error measure was defined implies that $\mathbf{u}_{\alpha}$ is the "exact" solution to the problem. This means that *any* atomic or molecular-level simulation tool could be used to generate the "exact" solution $\mathbf{u}_\alpha$, i.e. *ab initio*, Quantum Molecular Dynamics, etc. In our case, the atomic simulation method we choose to be our "exact" solution is molecular dynamics (MD). After determining that the MD displacements shall be referred to by the variable $\mathbf{q}$, equation (7.149) can be re-written as

$$Error = \sum_\alpha m_\alpha \left(\mathbf{q}_\alpha - \sum_I N_I^\alpha \mathbf{w}_I \right)^2 \tag{7.156}$$

where the MD displacements $\mathbf{q}$ now take the place of the total displacements $\mathbf{u}$. The equation for the fine scale $\mathbf{u}'$ can now be re-written as

$$\mathbf{u}' = \mathbf{q} - \mathbf{Pq} \tag{7.157}$$

The fine scale is now clearly defined to be the difference between the MD solution and its projection onto a pre-determined coarse scale basis. Finally, the equation for the total displacement $\mathbf{u}$ can be re-written as

$$\mathbf{u} = \mathbf{Nd} + \mathbf{q} - \mathbf{Pq} \tag{7.158}$$

It can be seen from (7.156) that the fine scale is simply the mass-weighted least-square error associated with projecting the MD solution onto a finite dimensional basis space. This is particularly useful in quasi-static zero temperature simulations, where atomic vibrations are absent. The fine scale can then be interpreted as a built in error estimator to the quality of the coarse scale approximation.

7.4.1 Multi-Scale Equations of Motion

The next step in the multi-scale process is to establish the coupled MD/ FE equations of motion. This is accomplished by first constructing a Lagrangian $\mathcal{L}$, which is defined to be the difference between the kinetic energy and the potential energy,

$$\mathcal{L}(\mathbf{u}, \dot{\mathbf{u}}) = \mathcal{K}(\dot{\mathbf{u}}) - V(\mathbf{u}) \tag{7.159}$$

Ignoring external forces, (7.159) can be written as

$$\mathcal{L}(\mathbf{u}, \dot{\mathbf{u}}) = \frac{1}{2} \dot{\mathbf{u}}^T \mathbf{M}_A \dot{\mathbf{u}} - U(\mathbf{u}) \tag{7.160}$$

where the $U(\mathbf{u})$ is the interatomic potential energy. Differentiating the total displacement $\mathbf{u}$ with respect to time gives

$$\dot{\mathbf{u}} = \mathbf{N}\dot{\mathbf{d}} + \mathbf{Q}\dot{\mathbf{q}} \tag{7.161}$$

where the complimentary projection operator $\mathbf{Q} \equiv \mathbf{I} - \mathbf{P}$. Substituting (7.161) into the Lagrangian (7.160) gives

$$\mathcal{L}(\mathbf{d}, \dot{\mathbf{d}}, \mathbf{q}, \dot{\mathbf{q}}) = \frac{1}{2}\dot{\mathbf{d}}^T \mathbf{M}\dot{\mathbf{d}} + \frac{1}{2}\dot{\mathbf{q}}^T \mathcal{M}\dot{\mathbf{q}} - U(\mathbf{d}, \mathbf{q}) \tag{7.162}$$

where the fine scale mass matrix $\mathcal{M}$ is defined to be $\mathcal{M} = \mathbf{Q}^T \mathbf{M}_A$. One elegant feature of (7.162) is that the total kinetic energy has been decomposed into the sum of the coarse scale kinetic energy plus the fine scale kinetic energy.

The multi-scale equations of motion are obtained from the Lagrangian by following the relations

$$\frac{d}{dt}\left(\frac{\partial \mathcal{L}}{\partial \dot{\mathbf{d}}}\right) - \frac{\partial \mathcal{L}}{\partial \mathbf{d}} = 0 \tag{7.163}$$

$$\frac{d}{dt}\left(\frac{\partial \mathcal{L}}{\partial \dot{\mathbf{q}}}\right) - \frac{\partial \mathcal{L}}{\partial \mathbf{q}} = 0 \tag{7.164}$$

Substituting the Lagrangian (7.162) into (7.163) and (7.164) gives

$$\mathbf{M}\ddot{\mathbf{d}} = -\frac{\partial U(\mathbf{d}, \mathbf{q})}{\partial \mathbf{d}} \tag{7.165}$$

$$\mathcal{M}\ddot{\mathbf{q}} = -\frac{\partial U(\mathbf{d}, \mathbf{q})}{\partial \mathbf{q}} \tag{7.166}$$

The two equations (7.165) and (7.166) are coupled through the derivative of the potential energy U, which can be expressed as functions of the interatomic force $\mathbf{f}$ as

$$\mathbf{f} = -\frac{\partial U(\mathbf{u})}{\partial \mathbf{u}} \tag{7.167}$$

By chain rule and utilizing (7.167) and (7.158) gives, one may recast (7.165) and (7.166) as

$$\mathbf{M}\ddot{\mathbf{d}} = -\frac{\partial U}{\partial \mathbf{u}}\frac{\partial \mathbf{u}}{\partial \mathbf{d}} = \mathbf{N}^T \mathbf{f} \tag{7.168}$$

$$\mathcal{M}\ddot{\mathbf{q}} = -\frac{\partial U}{\partial \mathbf{u}}\frac{\partial \mathbf{u}}{\partial \mathbf{q}} = \mathbf{Q}^T \mathbf{f} \tag{7.169}$$

Using the fact that $\mathcal{M} = \mathbf{Q}^T \mathbf{M}_A$, (7.169) can be rewritten as

$$\mathbf{Q}^T \mathbf{M}_A \ddot{\mathbf{q}} = \mathbf{Q}^T \mathbf{f} \tag{7.170}$$

Because $\mathbf{Q}$ can be proven to be a singular matrix, there are many uniquesolutions to (7.170). However, one solution which does satisfy (7.170) and is beneficial to us is

$$\mathbf{M}_A\ddot{\mathbf{q}} = \mathbf{f} \tag{7.171}$$
$$\mathbf{M}\ddot{\mathbf{d}} = \mathbf{N}^T\mathbf{f}(\mathbf{u}) \tag{7.172}$$

Eqs. (7.171) and (7.172) are the coupled multi-scale equations of motion. As can be seen, (7.171) is simply the MD equation of motion. Therefore, a standard MD solver can be used to obtain the MD displacements $\mathbf{q}$, while the MD forces $\mathbf{f}$ can be found by minimizing any relevant potential energy function. Furthermore, we can use standard finite element methods to find the solution to (7.172). One important point is that because the consistent mass matrix is used to decouple the kinetic energies of the coarse and fine scales, the finite element mass matrix $\mathbf{M}$ must be a consistent mass matrix. It is also crucial to note that while the MD equation of motion is only solved in the MD region, the FE equation of motion is solved *everywhere*.

The coupling between the two equations is through the coarse scale internal force $\mathbf{N}^T\mathbf{f}(\mathbf{u})$, which is a direct function of the MD internal force $\mathbf{f}$. In the region in which MD exists, the coarse scale force is calculated by interpolating the MD force. In the region in which MD has been mathematically accounted for, the coarse scale force can be calculated in multiple ways. Details are provided in a later section.

We would like to make a few comments here. The first is that the FE equation of motion is redundant for the case in which the MD and FE regions both exist everywhere in the domain, because the FE equation of motion is simply an approximation to the MD equation of motion. We shall remove this redundancy in the next section, when we create coupled MD/FE equations of motion for systems where the MD region is confined to a small portion of the domain.

The other relevant comment concerns the fact that the total solution $\mathbf{u}$ satisfies the same equation of motion as $\mathbf{q}$, i.e.

$$\mathbf{M}_A\ddot{\mathbf{u}} = \mathbf{f} \tag{7.173}$$

This is because that $\mathbf{q}$ and $\mathbf{u}$ satisfy the same initial conditions, and will be utilized in deriving the boundary conditions on the MD simulation in a later section.

7.4.2 Langevin Equation for Bridging Scale

We imagine the bridging scale method to be most applicable to problems in which the MD region is confined to a small portion of the domain, while the coarse scale representation exists everywhere. This coupled system is created by reducing the full system in which the MD region and the coarse scale exists everywhere in the domain; see Fig. 7.18 or 7.24 for illustrative examples. As was mentioned, one manner in which we can avoid the explicit solution of the many MD degrees of freedom is to utilize the Generalized Langevin Equation

(GLE). We now derive the connection between the GLE and the bridging scale.

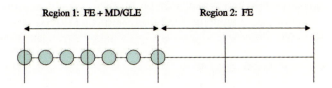

Region 1: FE + MD/GLE   Region 2: FE

Fig. 7.24. Separation of problem into two regions. Region 1 is FE + reduced MD, region 2 is FE.

The derivation is similar to the one presented above by Adelman and Doll, and mimics that given in.[446] Following the argument given in (7.173), we use the equality of the MD displacements $\mathbf{q}$ and the total displacement $\mathbf{u}$ to decompose the MD equation of motion as

$$\mathbf{M}_A\ddot{\mathbf{q}} = \mathbf{M}_A\ddot{\mathbf{u}} + \mathbf{M}_A\ddot{\mathbf{u}}' = \mathbf{f}(\mathbf{u}) \tag{7.174}$$

The force $\mathbf{f}(\mathbf{u})$ is then Taylor expanded about $\mathbf{u}' = 0$, giving

$$\mathbf{M}_A\ddot{\mathbf{u}} + \mathbf{M}_A\ddot{\mathbf{u}}' = \mathbf{f}(\bar{\mathbf{u}}) - \mathbf{K}\mathbf{u}' + \dots \tag{7.175}$$

where the stiffness $\mathbf{K}$ is defined as

$$\mathbf{K}_{\alpha\beta} = -\frac{\partial \mathbf{f}_\alpha}{\partial \mathbf{u}_\beta}\Big|_{\mathbf{u}'=0} \tag{7.176}$$

Three assumptions have been made in the preceding steps:

1. The Taylor expansion (linearization) of the force in (7.175) is truncated after linear terms in $\mathbf{u}'$;
2. Eq. (7.175) can be decomposed into two separate equations

$$\mathbf{M}_A\ddot{\mathbf{u}} = \mathbf{f}(\bar{\mathbf{u}}) \tag{7.177}$$
$$\mathbf{M}_A\ddot{\mathbf{u}}' = -\mathbf{K}\mathbf{u}'; \tag{7.178}$$

3. The stiffness matrix $\mathbf{K}$ is assumed not to vary on short time scales, i.e. the time scale of atomic vibrations.

The three assumptions given above are satisfied if the interatomic potential is harmonic, which means that $\mathbf{K}$ is a constant and the interatomic forces are linear in $\mathbf{u}'$. Another crucial point is that this assumption will be shown to only be necessary at the MD/FE boundary. Continuing with the derivation, the fine scale equation (7.178) is rewritten as

$$\ddot{\mathbf{u}}' = -\mathbf{A}\mathbf{u}' \tag{7.179}$$

where $\mathbf{A} = \mathbf{M}_A^{-1}\mathbf{K}$. The fine scale degrees of freedom are further partitioned into two vectors: $\mathbf{u}_1'$, the degrees of freedom to be simulated by MD, and $\mathbf{u}_2'$, the degrees of freedom which will be mathematically accounted for in the GLE (Generalized Langevin Equation). Eq. (7.179) is now written in partitioned matrix form as

$$\begin{pmatrix} \ddot{\mathbf{u}}_1' \\ \ddot{\mathbf{u}}_2' \end{pmatrix} = - \begin{pmatrix} \mathbf{A}_{11} & \mathbf{A}_{12} \\ \mathbf{A}_{21} & \mathbf{A}_{22} \end{pmatrix} \begin{pmatrix} \mathbf{u}_1' \\ \mathbf{u}_2' \end{pmatrix} \tag{7.180}$$

It should be noted that the terms $\mathbf{A}_{12}$ and $\mathbf{A}_{21}$ only play a role near the boundary. For example, if nearest neighbor spring force interactions are considered, then each of those sub-matrices only has one term to account for the interaction of the boundary atom with its nearest neighbors.

The degrees of freedom $\mathbf{u}_2'$ are eliminated by solving for them explicitly in (7.180) and substituting the result back into the equation for $\mathbf{u}_1'$. In the Laplace transformation domain, we have

$$s^2\mathbf{u}_2'(s) - s\mathbf{u}_2'(0) - \dot{\mathbf{u}}_2'(0) = -\mathbf{A}_{21}\mathbf{u}_1'(s) - \mathbf{A}_{22}\mathbf{u}_2'(s) \tag{7.181}$$

Rearranging this equation gives

$$\mathbf{u}_2'(s) = -\boldsymbol{\Theta}(s)\mathbf{A}_{21}\mathbf{u}_1'(s) + \boldsymbol{\Theta}(s)\left(s\mathbf{u}_2'(0) + \dot{\mathbf{u}}_2'(0)\right) \tag{7.182}$$

where

$$\boldsymbol{\Theta}(s) = (s^2\mathbf{I} + \mathbf{A}_{22})^{-1} . \tag{7.183}$$

Taking the inverse Laplace transform of (7.182) gives the desired expression for $\mathbf{u}_2'(t)$ as

$$\mathbf{u}_2'(t) = -\int_0^t \boldsymbol{\Theta}(t-\tau)\mathbf{A}_{21}\mathbf{u}_1'(\tau)d\tau + \dot{\boldsymbol{\Theta}}(t)\mathbf{u}_2'(0) + \boldsymbol{\Theta}(t)\dot{\mathbf{u}}_2'(0) \tag{7.184}$$

Substituting this equation into (7.180) yields

$$\ddot{\mathbf{u}}_1'(t) = -\mathbf{A}_{11}\mathbf{u}_1'(t) + \int_0^t \boldsymbol{\theta}(t-\tau)\mathbf{u}_1'(\tau)d\tau + \mathbf{R}(t) \tag{7.185}$$

where

$$\mathbf{R}(t) = -\mathbf{A}_{12}\left(\dot{\boldsymbol{\Theta}}(t)\mathbf{u}_2'(0) + \boldsymbol{\Theta}(t)\dot{\mathbf{u}}_2'(0)\right) \tag{7.186}$$

and

$$\boldsymbol{\theta}(t-\tau) = \mathbf{A}_{12}\boldsymbol{\Theta}(\mathbf{t}-\tau)\mathbf{A}_{21} \tag{7.187}$$

Recalling that the MD equation of motion can be scale decomposed into coarse and fine scale parts, i.e. (7.174), we add (7.185) and (7.177) to give

$$\ddot{\mathbf{q}}_1(t) = \mathbf{M}_{A1}^{-1}\mathbf{f}_1(\bar{\mathbf{u}}) - \mathbf{A}_{11}\mathbf{u}_1'(t) + \int_0^t \boldsymbol{\theta}(t - \tau)\mathbf{u}_1'(\tau)d\tau + \mathbf{R}(t) \qquad (7.188)$$

The final step to obtaining the coupled equations is to note that

$$\mathbf{M}_{A1}^{-1}\mathbf{f}_1(\bar{\mathbf{u}}) - \mathbf{A}_{11}\mathbf{u}_1'(t) = \mathbf{M}_{A1}^{-1}\mathbf{f}_1(\bar{\mathbf{u}}, \mathbf{u}_1', \mathbf{u}_2' = 0) \qquad (7.189)$$

Substituting (7.189) into (7.188) gives

$$\ddot{\mathbf{q}}_1(t) = \mathbf{M}_{A1}^{-1}\mathbf{f}_1(\bar{\mathbf{u}}, \mathbf{u}_1', \mathbf{u}_2' = 0) + \int_0^t \boldsymbol{\theta}(t - \tau)\mathbf{u}_1'(\tau)d\tau + \mathbf{R}(t) \qquad (7.190)$$

We write the final form of the MD equations of motion by noting that the fine scale component of the MD displacements can be written as

$$\mathbf{u}_1'(\tau) = \mathbf{q}_1(\tau) - \bar{\mathbf{u}}_1(\tau) \qquad (7.191)$$

The final form for the MD equations of motion then becomes

$$\ddot{\mathbf{q}}_1(t) = \mathbf{M}_{A1}^{-1}\mathbf{f}_1(\bar{\mathbf{u}}, \mathbf{u}_1', \mathbf{u}_2' = 0) + \int_0^t \boldsymbol{\theta}(t-\tau)(\mathbf{q}_1(\tau) - \bar{\mathbf{u}}_1(\tau))d\tau + \mathbf{R}(t) \quad (7.192)$$

This form for the MD equations of motions differs from that previously presented in[446] in the following manner. Firstly, only displacements and not velocities from the MD and FEM simulations are needed. This allows the usage of the simplest time integration algorithms for both simulations, i.e. velocity verlet and explicit central difference, as will be shown in a later section. The fact that only displacements are present in the time history kernel also lends itself nicely for numerical evaluation of the time history kernel, as explained in[238, 447].

The first term on the right hand side of (7.192) is simply the interatomic force calculated assuming that the fine scale in region 2 is zero. In simpler terms, this is just the standard interatomic force that is calculated in the MD simulation. Away from the MD boundary, this is the only term that remains from (7.192), and the standard MD equations of motion result.

The second term on the right hand side of (7.192) contains the time history kernel $\boldsymbol{\theta}(t - \tau)$, and acts to dissipate fine scale energy from the MD simulation into the surrounding continuum. The numerical result is a non-reflecting boundary between the MD and FE regions, as the time history kernel allows short wavelengths that cannot be represented by the surrounding continuum to leave the MD region. For a harmonic solid, the time history kernel can be evaluated analytically.

The final term on the right hand side is the random force $\mathbf{R}(\mathbf{t})$. As was described in a previous section, the random force arises due to temperature differences between the MD region and the surrounding coarse scale. In this work, we assume the random force to be zero, which implies that the temperature of the surrounding continuum is $0K$.

One issue in evaluating (7.192) is the time history integral involving $\theta(t - \tau)$. In the work by Wagner and Liu, this expression was evaluated in closed form. However, in multiple dimensions and a general lattice structure, a closed form solution may not be possible. Recently, Wagner, Karpov and Liu[447] have developed a means of numerically calculating the time history integral using numerical inverse Laplace transform techniques. The computational effort necessary to pursue such approaches in multiple dimensions, along with the relevant efficiency characteristics, have been explored by Karpov *et al.*[238] in the context of wave-transmitting boundary conditions for MD simulations.

The main ramification of using a GLE in the bridging scale method is that the atoms in region 2 are mathematically accounted for and act upon the atoms in region 1 through the external forces in (7.192). These external forces acting upon region 1 are the net resultant of collective behavior of the atoms in region 2. Computationally, this approach virtually eliminates spurious wave reflection between the MD and FE regions, which was one of the key factors mentioned previously in managing an accurate multiple-scale method.

Therefore, the coupled MD/FE equations of motion can be rewritten as

$$\ddot{\mathbf{q}}_1(t) = \mathbf{M}_{\mathbf{A1}}^1 \mathbf{f}_1(\bar{\mathbf{u}}, \mathbf{u}_1', \mathbf{u}_2' = 0)$$
$$+ \int_0^t \boldsymbol{\theta}(t - \tau)(\mathbf{q}_1(\tau) - \bar{\mathbf{u}}_1(\tau))d\tau + \mathbf{R}(\mathbf{t}) \tag{7.193}$$
$$\mathbf{M}\ddot{\mathbf{d}} = \mathbf{N}^T \mathbf{f}(\mathbf{u}) \tag{7.194}$$

As can be seen in (7.193), the MD equations of motion have been modified such that they depend on the FE displacements at the MD/FE interface. Now, the FE equation of motion is *not* redundant, as the time history integral depends on FE degrees of freedom. The MD degrees of freedom are coupled to the FE equations of motion everywhere, while the FE degrees of freedom affect the MD equations of motion by the terms which act on the boundary atom.

The major advantage of this is that finite-sized domains can be considered with this coupled approach. One example of this is in modeling dynamic crack propagation. If the cracks were modeled using a purely MD system, the major problem is that due to the number of atoms that would be required, the actual problem size cannot be simulated. If the coupled simulation is used, then the waves emitted from the crack tip can naturally propagate away into the far away continuum, such that a realistic problem size can be considered.

It appears that the bridging scale technique is well-suited for simulations of strain localization, as was discussed in the introduction. Because the waves emitting from the localized zone are typically of the elastic variety, the bridging scale boundary conditions derived above should allow the passage of those waves into the surrounding continuum. Furthermore, the evolution of the localized region can be on a smaller timescale than the surrounding continuum, which means that precious computational effort will not be wasted in updating the continuum variables at each timestep with the atomic variables. Kadowaki and Liu have applied the bridging multi-scale method to simulation strain localization .[234]

Comments on Time History Kernel. As was discussed above, one crucial element to the treatment of fine scale waves at the MD/FE boundary is the time history kernel $\boldsymbol{\theta}$. In,[446] $\boldsymbol{\theta}$ was derived in 1D assuming a harmonic lattice, and was shown to be

$$\boldsymbol{\theta}(t) = \frac{2k}{t} J_1(2\omega t) \tag{7.195}$$

where J_1 indicates a first order Bessel function, k is the spring stiffness, and the frequency $\omega = \sqrt{k/m_a}$. The spring stiffness can be determined in general by

$$k = \frac{\partial^2 \Phi(\mathbf{r})}{\partial \mathbf{r}^2} \tag{7.196}$$

For a harmonic solid, k is simply the spring stiffness. For the Lennard-Jones examples to be presented in a later section, the following definition will be used for k:

$$k = \frac{624\epsilon}{\sigma^2 y^{14}} - \frac{168\epsilon}{\sigma^2 y^8} \tag{7.197}$$

where $y = 2^{1/6}$.

7.4.3 Staggered Time Integration Algorithm

As was previously mentioned, one strength of the bridging scale lies in the ability to update the MD and FE equations of motion using appropriate time increments for each equation. In fact, both simulations are integrated through time using widely utilized integration algorithms; velocity verlet for MD, and explicit central difference for FE.

MD Update. The basic idea is that for each computational period, both simulations are advanced by a time step Δt. The MD simulation is advanced first by m steps of size $\Delta t/m$ while the FE simulation is advanced through a single time step of size Δt. A small modification in the standard MD velocity Verlet update is required because the MD simulation requires information from the FE simulation near the boundary (see (7.193)). The modification is that the FE boundary displacement and velocity will be interpolated at fractional timesteps, while the FE boundary acceleration will be assumed to be constant during the MD time subcycle. The FE boundary acceleration is assumed to be constant such that the actual FE equations of motion are not solved at each MD timestep. The stability of similar staggered time integration methods with subcycling was explored in[285][286]

$$\bar{\mathbf{u}}_{\Gamma}^{[j+1]} = \bar{\mathbf{u}}_{\Gamma}^{[j]} + \dot{\bar{\mathbf{u}}}_{\Gamma}^{[j]}\Delta t_m + \frac{1}{2}\ddot{\bar{\mathbf{u}}}_{\Gamma}^n\Delta t_m^2 \tag{7.198}$$

$$\dot{\bar{\mathbf{u}}}_{\Gamma}^{[j+1]} = \dot{\bar{\mathbf{u}}}_{\Gamma}^{[j]} + \ddot{\bar{\mathbf{u}}}_{\Gamma}^n\Delta t_m \tag{7.199}$$

$$\mathbf{q}^{[j+1]} = \mathbf{q}^{[j]} + \mathbf{p}^{[j]}\Delta t_m + \frac{1}{2}\mathbf{s}^{[j]}\Delta t_m^2 \tag{7.200}$$

$$\mathbf{p}^{[j+\frac{1}{2}]} = \mathbf{p}^{[j]} + \mathbf{s}^{[j]}\Delta t_m \tag{7.201}$$

$$\mathbf{s}^{[j+1]} = \mathbf{M}_{\mathbf{A}}^{-1}\mathbf{f}(\mathbf{q}^{[j+1]}, \bar{\mathbf{u}}_{\Gamma}^{[j+1]}, \mathbf{h}^{[j+1]}) \tag{7.202}$$

$$\mathbf{p}^{[j+1]} = \mathbf{p}^{[j+\frac{1}{2}]} + \frac{1}{2}(\mathbf{s}^{[j+1]} + \mathbf{s}^{[j]})\Delta t_m \tag{7.203}$$

where $\mathbf{p}$ is the MD velocity, $\bar{\mathbf{u}}_{\Gamma}$ is the FE boundary displacement, $\dot{\bar{\mathbf{u}}}_{\Gamma}$ is the FE boundary velocity, $\mathbf{q}$ is the MD displacement, $\mathbf{s}$ is the MD acceleration and $\mathbf{M_A}$ is the MD mass matrix.

FE Update. Once the MD quantities are obtained using the above algorithm at time $n + 1$, the FE displacements $\mathbf{d}$, velocities $\mathbf{v}$ and accelerations $\mathbf{a}$ are updated from time n to $n + 1$. A central difference scheme is adopted here.

$$\mathbf{d}^{n+1} = \mathbf{d}^n + \mathbf{v}^n\Delta t + \frac{1}{2}\mathbf{a}^n\Delta t^2 \tag{7.204}$$

$$\mathbf{a}^{n+1} = \mathbf{M}^{-1}\mathbf{N}^T\mathbf{f}(\mathbf{N}\mathbf{d}^{n+1} + \mathbf{Q}\mathbf{q}^{n+1}) \tag{7.205}$$

$$\mathbf{v}^{n+1} = \mathbf{v}^n + \frac{1}{2}(\mathbf{a}^n + \mathbf{a}^{n+1})\Delta t \tag{7.206}$$

where $\mathbf{M}$ is the consistent FE mass matrix. The internal force $\mathbf{f}$ is computed by combining the coarse scale part of the displacement $\mathbf{N}\mathbf{d}$ with the fine scale part of the MD simulation $\mathbf{Q}\mathbf{q}$.'

7.4.4 Bridging Scale Numerical Examples

Due to the fact that the verification of the bridging scale for linear problems was given in,[446] the numerical examples here utilize the Lennard-Jones 6-12

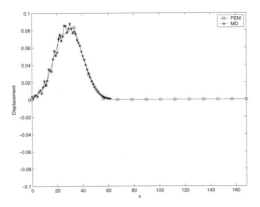

Fig. 7.25. Snapshot of MD/FE displacements showing fine scale MD displacement component.

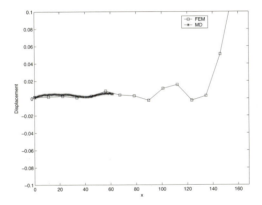

Fig. 7.26. Snapshot showing majority of fine scale MD information has passed through to coarse scale.

potential. For simplicity, σ and ϵ were assumed to be unity, and the atoms were given a prescribed displacement similar to the gaussian wave boundary condition used by Wagner and Liu.[446] 30 finite elements spanned the entire domain between $x = -150r_0$ and $x = 150r_0$, and 111 atoms were used between $x = -55r_0$ and $x = 55r_0$. The initial amplitude of the wave was 13 percent of the equilibrium spacing, and the FE nodal spacing was 10 times the MD atomic spacing. 55 FE timesteps were used, with 50 MD timesteps per FE timestep. The FE nodal forces in the coarse scale outside the coupled MD/FE region were calculated using the Cauchy-Born rule using the LJ 6-12 potential. Finally, the random force $\mathbf{R}(\mathbf{t})$ in (7.193) was taken to be zero, implying that the MD calculation was done at $0K$.

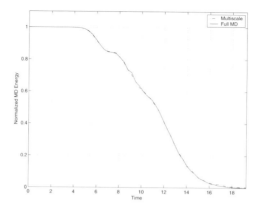

Fig. 7.27. Bridging scale vs. full MD energy comparison for high frequency displacement initial condition.

Figs. 7.25 and 7.26 show two snapshots of the bridging scale simulation. In comparing the two snapshots, it is clear that the majority of the high frequency (fine scale) component of the MD displacement has been dissipated cleanly into the surrounding continuum. The comparison between the bridging scale energy transfer and the actual MD energy transfer is shown in Fig. 7.27. It is shown that the bridging scale energy transfer is about 99 percent of the full MD energy transfer.

7.5 Applications

All MD calculations presented in this section utilize the Lennard-Jones (LJ) 6-12 potential (7.75). The examples were run with parameter values $\sigma = \epsilon = 1$ considering nearest-neighbor interactions only, while all atomic masses were chosen as $m = 1$. A hexagonal lattice structure was considered for the MD simulations, with the atoms initially in an equilibrium configuration. The interatomic distance r_{ij} which minimizes the potential energy Φ in (7.75) can be determined to be $2^{1/6}\sigma$.

We comment here on the usage of a general interatomic potential with parameters which do not match those of any real material. These choices were made in the interest of generality, such that large classes of realistic physical systems could be simulated without concentrating on a specific material. Future work could, of course, use specific values for σ and ϵ to match the behavior of a given material should a detailed study of a specific physical process be desired. Finally, the LJ potential was chosen to represent a model brittle material.

For the regions which satisfy a coarse scale-only description, the Cauchy-Born rule was utilized to calculate the coarse scale internal force following the description in an earlier section. The LJ 6-12 potential was used to describe the strain energy density W in coarse scale, i.e.

$$U(\mathbf{u}) = \sum_{\alpha} W_{\alpha}(\mathbf{u})\Delta V_{\alpha}, \tag{7.207}$$

such that the coarse scale internal force could be derived from the same potential that was used for the MD force calculations.

Fig. 7.28. Schematic illustrating the boundaries on which the MD boundary condition was applied.

The two-dimensional time history kernel was calculated using inverse Laplace transform, i.e. $\boldsymbol{\theta}_m(s) = \mathcal{L}^{-1}(\boldsymbol{\Theta}_m)$, for a hexagonal lattice structure interacting via an LJ 6-12 potential with nearest neighbor interactions and potential parameters $\sigma = 1$ and $\epsilon = 1$. For all numerical examples, the time history kernel was calculated only for the top and bottom boundary layers of the atomic lattice, while the right and left boundary layers were given no special treatment. This is demonstrated in a schematic plot in Fig. 7.28.

To efficiently caculate the impedance force at fine/coarse scale boundary, we use the following multi-scale MD formula,

$$\ddot{\mathbf{q}}_{0,m}(t) = \mathbf{M}_{\mathbf{a}}^{-1}\mathbf{f}_{0,m} + \sum_{\mathbf{m}'=-\mathbf{n}_{\mathrm{crit}}}^{\mathbf{n}_{\mathrm{crit}}} \int_{0}^{\mathbf{t}} \boldsymbol{\theta}_{\mathbf{m}-\mathbf{m}'}(\mathbf{t}-\tau)(\mathbf{q}_{0,\mathbf{m}'}(\tau) - \bar{\mathbf{u}}_{0,\mathbf{m}'}(\tau))\mathbf{d}\tau \tag{7.208}$$

where n_{crit} refers to a maximum number of neighbors which will be used to compute the impedance force. In the numerical examples presented later, n_{crit} is varied between zero and four, meaning that between one and nine neighbors are used to calculate the boundary force.

All units related to atomic simulations in this section, such as velocity, position and time, are given in reduced units. It should be noted that because of the choices of mass, σ and ϵ as unity, all normalization factors end up as unity. Finally, all numerical examples shown in this work were performed using the general purpose simulation code Tahoe, which was developed at Sandia National Laboratories (Tahoe, 2003).

7.5.1 Two-dimensional Wave Propagation

In this example, we demonstrate the effectiveness of the bridging scale in eliminating high frequency wave reflection between the FE and MD regions. To do so, a two-dimensional wave propagation example was run. A part of the MD region was given an initial displacement corresponding to a two-dimensional circular-type wave. The initial displacements given in polar coordinates were

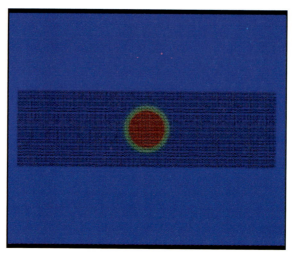

Fig. 7.29. Initial conditions for two-dimensional wave propagation example. Displacement magnitude shown.

$$\mathbf{u(r)} = \frac{\mathbf{A}}{\mathbf{A} - \mathbf{u_c}} \left(1 + \mathbf{b} \cos(\frac{\mathbf{2\pi r}}{\mathbf{H}}) \right) \left(\mathbf{A} e^{(-\frac{\mathbf{r}}{\sigma})^2} - \mathbf{u_c} \right) \tag{7.209}$$

$$\mathbf{u}(\theta) = \mathbf{0} \tag{7.210}$$

The corresponding parameters had values of $\sigma = 15$, $H = \sigma/4$, $A = 1.5e - 1$, $b = .1$, $r_c = 5\sigma$ and $u_c = Ae^{(-\frac{r_c}{\sigma})^2}$. A controls the wave amplitude, b controls the degree of high frequency content in the wave ($b = 0$ implies zero high frequency content) and r_c controls the cut-off distance of the initial displacements. The initial configuration for the problem is shown in Fig. 7.29.

In order to have a comparison for the bridging scale simulations, a larger MD simulation was performed, and taken to be the benchmark solution. In this simulation, the same initial displacements prescribed by (7.209) for the bridging scale simulation were prescribed for the MD lattice. For the full MD region to match the entire bridging scale region, 91,657 atoms were used. The

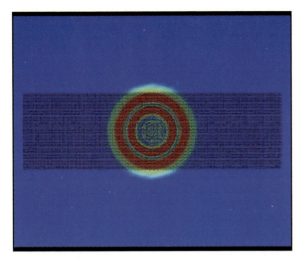

Fig. 7.30. A snapshot of wave propagation from the MD region into the continuum region. The figure shows contours of the displacement magnitude.

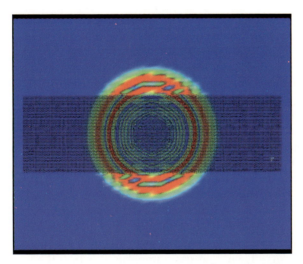

Fig. 7.31. A later snapshot of wave propagation from the MD region into the continuum region. The figure shows contours of the displacement magnitude.

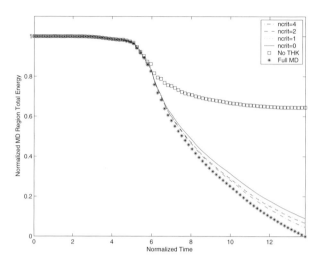

Fig. 7.32. Comparison of energy transfer for two-dimensional wave propagation example.

wave was allowed to propagate away from the center of the lattice until just before the domain boundaries were reached.

The corresponding bridging scale simulation contained 31,157 atoms and 1920 finite elements, of which 600 were in the coupled MD/FE region. 20 MD time steps were used for each FE time step. The bridging scale MD domain contained as many atoms in the x-direction as the full MD simulation, but only one-third the number of atoms in the y-direction. Because the MD boundary conditions were only enforced on the top and bottom of the bridging scale MD lattice, this would ensure that the waves reached and passed through the top and bottom boundaries before the left and right boundaries were reached by the wave.

In order to test the accuracy of the MD impedance force, five cases were run. The first case involved not applying the impedance force, which would expectedly lead to large amounts of high frequency wave reflection at the MD/FE boundary. Then, four cases were run in which the number of neighbors used in calculating the impedance force (n_{crit} in (7.208)) was increased. Snapshots showing the natural propagation of the wave which originated in the MD region into the surrounding continuum are shown in Figs. 7.30 and 7.31. It is important to note that while the continuum representation exists everywhere, that part of the FE mesh overlaying the MD region is not shown to better illustrate how the coarse scale captures the information originating in the MD region.

The resulting total energy (kinetic energy + potential energy) transferred from the MD region using the bridging scale is shown in Fig. 7.32. The full MD energy was measured only in the same region in which the bridging scale

MD existed, such that a valid comparison could be made. The full MD energy was then normalized to be the reference solution, such that it tends to zero in Fig. 7.32. Of course, because the wave has not fully exited the system, the actual energy in all systems does not go to zero, but this normalization is performed such that a percentage measurement and comparison between the full MD and bridging scale MD systems can be obtained.

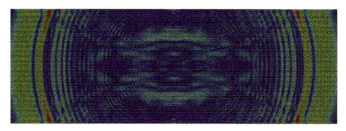

Fig. 7.33. Final displacements in MD region if MD impedance force is applied.

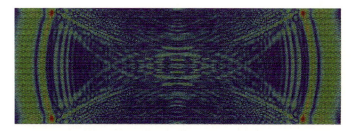

Fig. 7.34. Final displacements in MD region if MD impedance force is not applied.

As figure 7.32 shows, if the MD impedance force is not applied, only about 35 percent of the MD energy is transferred in comparison to the full MD. However, if the impedance force that is presented in this work is utilized, even if only one neighbor is used in calculating the boundary force (i.e. $n_{crit}=0$), about 91 percent of the MD energy is transferred in comparison to the full MD simulation. The percentage steadily increases until more than 95 percent of the energy is transferred if nine neighbors ($n_{crit} = 4$) are used to calculate the impedance force. These results show the necessity in correctly accounting for the mathematically eliminated fine scale degrees of freedom in the form of the time history convolution integral, i.e. Eq. 7.208.

We note that while the energy plots show the total energy (kinetic + potential), similar results can be shown which demonstrate that the system kinetic energy behaves similarly. Therefore, because kinetic energy and temperature are closely linked for an atomic ensemble (Leach 2001), the MD simulation can be easily extended to run at finite temperatures; the thermal energy will be dissipated away by the impedance force. However, no means of tracking the energy once it leaves the MD region has been established.

A parametric study to determine the influence of MD time step on the bridging scale solution is shown in Fig. 7.35. There, the MD time step was reduced by a factor of one-half three times to determine if a dramatic effect is seen by reducing the MD time step, while the FE time step was kept constant. As can be seen in the Figure, once a certain MD time step threshold is reached (in this case, $\Delta t = .014$), the energy transferred from the MD region essentially stops increasing. A corresponding study has been performed by

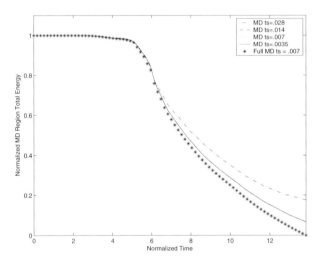

Fig. 7.35. Comparison of energy transfer for different MD time steps.

keeping the MD time step fixed while changing the FE time step. As also can be seen in Fig. 7.36, the results obtained by varying the FE time step while keeping the MD time step fixed are quite similar to those obtained by varying the MD time step and fixing the FE time step. As the Figure indicates, if the timestep is too large, a fairly inaccurate solution is obtained. However, once a threshold FE time step is reached, further reductions in time step gives essentially the same solution while continuing to increase the computational expense.

A final note of importance is made in further analyzing the bridging scale result shown in Fig. 7.33. One element of the bridging scale simulation that results is a long wavelength reflection back into the MD region. The reason for this long wavelength reflection is due to the fact that the FE internal force is calculated by two different means for the boundary nodes. The result of this is a system with slightly different stiffnesses. For the problem shown in this

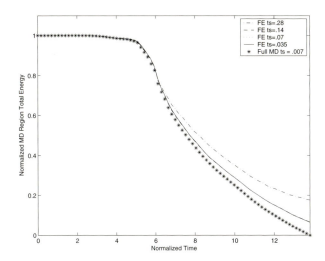

Fig. 7.36. Comparison of energy transfer for different FE time steps.

section, because most of the initial MD energy is concentrated in the high frequency waves, the majority of the energy is transferred into the continuum. In general, this may not be the case, and more of the system energy may be concentrated in the longer wavelengths.

There are multiple ways of reducing this effect. One approach is to split the boundary element such that it receives a contribution from both the MD forces and the Cauchy-Born forces. By doing so, a transition element is created whose properties are an average of the two systems. By creating this transition element, the long wavelength reflection is eliminated, as has been done in one-dimensional problems (Wagner and Liu 2003).

Another option is to use meshfree shape functions everywhere in the domain. Because of the non-local nature of the meshfree shape functions, the transition element described above will be naturally created without the need for special integration techniques for the boundary element. This issue will be addressed more carefully in a later work.

7.5.2 Dynamic Crack Propagation in Two Dimensions

The previous example dealt with a specific case in which all of the initial energy of the problem was in the MD domain, then is dissipated away into the surrounding continuum. This type of example, while useful for verifying the effectiveness of the derived MD impedance force, does not demonstrate all relevant facets of a generalized multiple scale simulation. In fact, it could be reasonably argued that for problems such as the wave propagation example in which the only goal is to allow passage of the fine scale waves out of the MD region without causing internal reflection, using techniques such as those introduced in Wagner, Karpov and Liu (2003) would be sufficient, thereby rendering the coarse scale redundant. However, one case in which

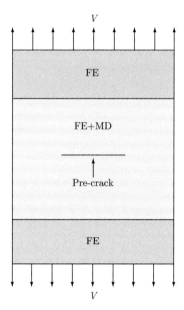

Fig. 7.37. Configuration for two-dimensional dynamic crack propagation example.

methods such as Wagner, Karpov and Liu (2003) would fail are those in which the flow of information is reversed, e.g. the initial/boundary conditions are applied on the coarse scale, and the information eventually reaches the MD domain and is passed as a boundary condition to the MD region via the coarse/fine coupling established in (7.208). The reason for failure is the fact that the assumption is implicitly made that the excitation of interest originates within the MD region; the MD boundary condition is only designed as a one-way filter of information, i.e. out of the MD region, and thus is unable to represent an outside excitation propagating into the MD region. To demonstrate the ability of the bridging scale to overcome this potential limitation, we solve a two-dimensional dynamic crack propagation in which the boundary conditions are applied to the coarse scale. The problem schematic is shown in Fig. 7.37.

The pre-crack is generated by preventing the interaction of two rows of atoms. In this way, the atoms on the faces of the pre-crack effectively behave as if on a free surface, and the crack opens naturally in tension. A ramp velocity is applied to the top and bottom nodes of the continuum region such that the atomic fracture occurs in a mode-I fashion. The application of the ramp velocity is shown in Fig. 7.38.

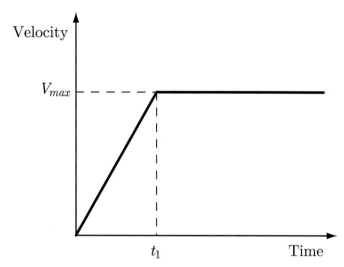

Fig. 7.38. Ramp velocity boundary condition that is applied on FE region for dynamic fracture examples.

A full MD simulation was also run in which the entire domain was comprised of atoms. The bridging scale simulation consisted of 91 051 atoms and 1800 finite elements, of which 900 were in the coupled MD/FE region. Correspondingly, the full MD simulation consisted of 181 201 atoms. The identical velocity boundary condition as shown in Fig. 7.38 was applied on the full MD simulation as was the bridging scale simulation, with the peak velocity $V_{max} = .04$. For all bridging scale simulations shown in this section, only one neighbor was utilized in evaluating the time history force, i.e. $n_{crit} = 0$.

A comparison between the full MD simulation and bridging scale simulations is shown in Figs. 7.39 and 7.40. In these Figures, the potential energy of the MD domain is shown. As can be seen, both simulations show the same dominant characteristics, notably the size and intensity of the process zone immediately ahead of the crack tip, and also in the high-frequency radiation emitted from the crack tip. This high-frequency radiation, which appears as concentric circles radiating away from the crack tip, is emitted each time a single atomic bond is broken by the propagating crack. The opening of the crack is shown clearly by magnifying the y-component of the displacement by a factor of three. It should be noted that while the interatomic interactions have been restricted to nearest neighbors, the potential is not truncated at any point such that the potential energy and force are fully continuous functions of interatomic distance.

If a larger peak velocity V_{max} is chosen for the velocity boundary condition or the simulation is run for a sufficiently lengthy period of time, then complete fracture of the atomic lattice into two sections will occur. This is

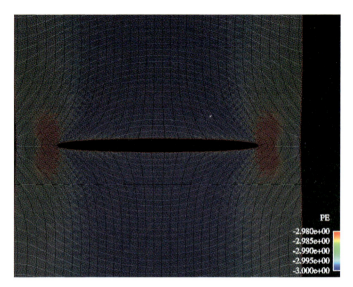

Fig. 7.39. Potential energy contours of full MD fracture simulation.

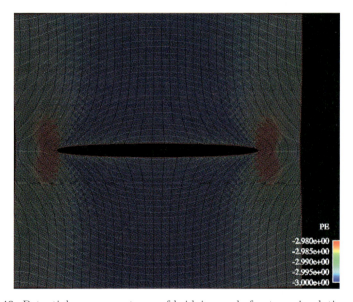

Fig. 7.40. Potential energy contours of bridging scale fracture simulation.

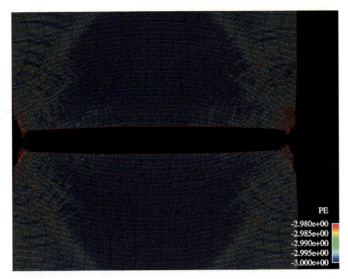

Fig. 7.41. Potential energy contours of full MD simulation after complete fracture of lattice has occured.

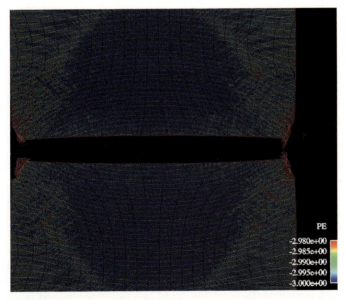

Fig. 7.42. Potential energy contours of bridging scale simulation after complete fracture of lattice has occurred.

demonstrated in Figs. 7.41 and 7.42, where the peak velocity was chosen as $V_{max} = .06$.

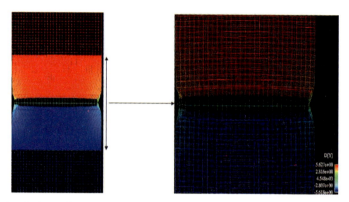

Fig. 7.43. Left: y-displacements of entire structure. Note that MD exists only in small part of the domain, while FE exists everywhere. Right: zoom in of y-displacements in coupled MD/FE region.

As can be seen in Figs. 7.41 and 7.42, the bridging scale simulation agrees very well with the full MD simulation. It is also noteworthy that complete fracture of the underlying MD lattice is allowed in the coupled simulation - this is because the finite element simulation in that region is simply carried along by the MD simulation. Furthermore, because the finite element forces in that region are calculated directly from the underlying MD forces via (7.172), no calculation of deformation gradients are necessary. Because of this, the finite elements which overlay the MD region can deform in a manner that finite elements governed by traditional constitutive laws cannot. This is exemplified by the pinched nature of the deformation of the finite elements at the edge of the cracked specimen in Fig. 7.43, which shows the deformed FE mesh and separated MD lattice plotted together. Again, the crack opening is shown in the Figures by magnifying the y-component of the displacement by a factor of three.

Another useful measure of comparison between the bridging scale and a full MD simulation is in tracking the initiation times and subsequent position of the crack tips. In our simulations, because the location of the pre-crack is known, the location of the crack tip could be easily ascertained by comparing the y-displacements of the atoms ahead of the pre-crack, and checking if they had exceeded a critical value. The comparison between the full MD and two different bridging scale simulations using different MD domain sizes is shown in Fig. 7.44, for the case $V_{max} = .04$. As can be seen, the bridging scale simulations predict the identical crack initiation time as the full MD simulation as well as the position of the crack tip as it evolves through time.

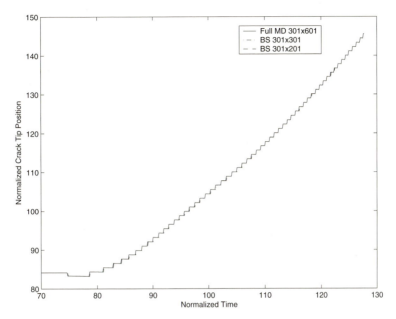

Fig. 7.44. Comparison of crack position with respect to time for full MD and two different bridging scale simulations.

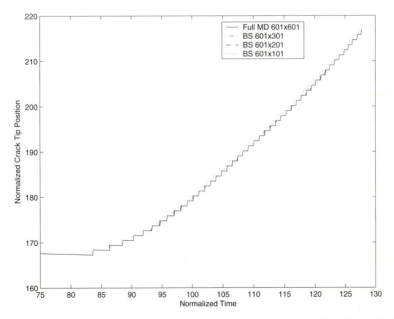

Fig. 7.45. Comparison of crack position with respect to time for full MD and three different bridging scale simulations.

A slightly different system was run for further comparisons. In this case, the full MD system contained 362 101 atoms (601 atoms in the x-direction by 601 atoms in the y-direction). Three different bridging scale simulations with varying MD system sizes (601 × 301, 601 × 201, 601 × 101) along with 3600 finite elements were run. The crack tip history is shown in Fig. 7.45. As can be seen, the first two bridging scale simulations match the crack initiation time and time history exactly. However, for the smallest MD region within a bridging scale simulation (the 601 × 101 atom case), the crack initiates at a slightly earlier time than in the full MD case. After initiation, the velocity of the crack appears to match the velocity of the crack in the full MD case. We note that other simulations were run in which the number of atoms in the y-direction were reduced to less than 101. For these simulations, the crack either initiated much earlier or later than expected, or did not initiate at all. It appears as though the incorrect physics demonstrated in these cases reflects the assumption of linearity at the MD boundary being violated if the MD region is too small.

Simulation	Normalized Time
Full MD : 362101 atoms = 724202 $DOFs$	1.0
Bridging Scale: 121,901 atoms + 3,721 nodes = 251,244 DOF's	0.51
Bridging Scale: 61,851 atoms + 3,721 nodes = 131,144 DOF's	0.33

Table 7.1. Comparison of simulation times using bridging scale vs. full MD. $n_{crit} = 0$ for bridging scale simulations.

Finally, a comparison of the computational expense incurred utilizing the bridging scale versus a pure atomic simulation is shown in Table 7.5.1. The full MD simulation of 362 101 atoms was set as the benchmark simulation in terms of computational time. This benchmark full MD simulation was then compared to two bridging scale simulations. The first used approximately one-third the number of atoms, 121,901 along with 3721 FE nodes. The second used approximately one-sixth the number of atoms, 61 851 along with 3721 FE nodes. As can be seen, computational speedups of two and three times were observed, respectively, using the bridging scale for the case in which all computations were performed in serial. The bridging scale simulation times do not scale exactly as the fraction of MD degrees of freedom due to the additional expense of the terms introduced by the bridging scale coupling, and also because certain optimization tools, such as the truncation of the number of displacement histories stored per boundary atom, have not yet been implemented.

7.5.3 Simulations of Nanocarbon Tubes

In this section, we introduce the multi-scale simulations of nanocarbon tubes, which was conducted By Qian et al.[379] Three benchmark problems were solved to verify the proposed multi-scale projection method. Localized deformation is involved in all of the problems. We partitioned the problem domains into two regions: the coarse scale region and the enrichment region. The problem is treated as quasi-static, and the energy minimum is obtained using a limited-memory quasi-Newton algorithm.[354] A (10,10) carbon nanotube with length of 115.6 $\circ A$ and 1900 carbon atoms is considered for the first two problems. Note that these are very small-scale simulations because of the number of atoms involved. The scalability and efficiency of the multi-scale scale method are illustrated in the third case.

Twisting of a (10,10) Nanotube. The nanotube surface is discretized with 380 particles. Twisting angles are imposed incrementally at the two ends of the nanotube while holding the cross-sectional circular shape unchanged. The loading step is 0.25^o per-step and is imposed until a total twisting angle of 50° is reached. Plotted in Fig. 7.46 are the initial and final configurations of the twist. At the point of buckling due to twist, it is expected that there is a transition from the uncollapsed section to the fully collapsed section. In the initial configuration, it can be seen that a nanotube structure of 25 hexagons along the axial direction is embedded to account for the fine scale during the collapse. Correspondingly, we performed molecular dynamics computation and coarse scale computations on the same carbon nanotube with the same twisting angle. Plotted in Fig. 7.47 are the deformation patterns from the molecular dynamics, meshfree method and multi-scale method, respectively. The deformed molecular structure in the meshfree method and coarse scale region of the multi-scale method are interpolated with the use of shape function. One can see that there is no significant difference in the results obtained from the three different approaches.

In contrast, significant differences are found in the energy comparison of these three different approaches, which is plotted in Fig. 7.48. The average energy is defined as the change in the potential energy for each atom at each loading step as compared to its relaxed state. As indicated from the molecular dynamics calculation, the collapse due to twist takes place at a twisting angle of 30°. Before collapse occurs, the results from the three approaches match well. This is expected since the deformation at this stage is almost homogeneous. After the collapse, the coarse scale method becomes inaccurate, and large discrepancies in the energy result. On the other hand, the multi-scale method still can accurately capture the energy in the collapsed stage.

Bending of a (10,10) Nanotube. The same meshfree discretization as in the last section is used for this case. Incremental bending angles are imposed at the two ends of the nanotube at 0.25^o per step. The simulation is carried out for 50 steps, which corresponds to a total bending angle of 25°. The

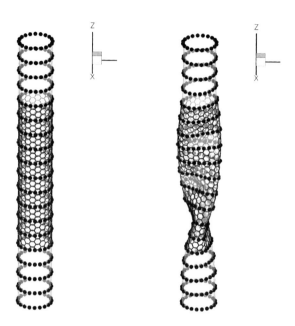

Fig. 7.46. (Left) Initial particle distribution and embedded molecular structure of the carbon nanotube in the case of twist. (Right) Final deformation of the particle and molecular structure at a twist angle of 50°.

buckling due to bending takes place in the middle of the tube at a bending angle of 9° . Plotted in Fig. 7.49 are the initial and final configurations of the bending. In the regions of buckling, a nanotube structure of 17 hexagons along the axial direction is added. Plotted in Fig. 7.50 are the comparisons in terms of the post-buckling molecular structure from three different methods. The molecular structure is obtained in the same way as in the twisting case. Significant differences between the coarse scale and multi-scale methods can be found in the zoom-in plot in Fig. 7.50. The coarse scale method produces a non-symmetric pattern, due to the fact that it is not capable of approximating the buckling pattern, while the multi-scale method capture this because of the coupling with MD. As in the twisting case, we also compared the average bending energy with the results obtained from the other two approaches: molecular dynamics and purely coarse scale method. The average bending energy is defined as the change in the potential energy due to bending for each atom. The energy comparison is plotted in Fig 7.51 . As in the twisting case, it can be seen that in the pre-buckling region, the results from the multi-scale method fits well with the results from molecular dynamics, as do the results from the purely coarse scale method. However, at the onset of buckling, the coarse scale method is no longer able to track the local

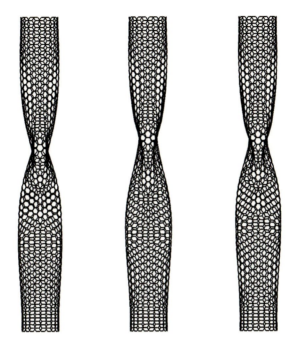

Fig. 7.47. Comparison of the deformation in the y-z plane from three different simulation approach. From left to right are (a) molecular dynamics simulation; (b) meshfree method; (c) proposed multi-scale method.

deformation precisely. In contrast, the multi-scale method yields excellent agreement with MD in the post-buckling region. This indicates that the local deformation can be well-captured using the multi-scale method due to the embedded molecular structure. The estimation on the energy from the coarse scale method is typically higher, and therefore results in a stiffer behavior.

Bending of 15-walled Carbon Nanotube. A 15-walled MWCNT is considered with the outer most shell being a (140, 140) nanotube, and all inner shells of the (n, n) type; from the outer most shell, n reduces by 5 every layer. The length of the tube is 90 nm and the original MD system contains approximately 3 million atoms. This is discretized with a system of 27,450 particles. In addition, the meshfree discretization is enriched with molecular structures in selected regions as shown in Fig. 7.52

The total number of atoms used in the MD part of the multi-scale simulation is 340,200. The same loading condition as in the previous section is applied. Pure meshfree simulation on a similar problem has been presented in.[380] In Fig. 7.53, it is shown that buckling develops at two local regions. The detailed atomic deformations for the atoms in these regions are traced by the added molecular structure. In Fig. 7.54, a comparison on the final buckling

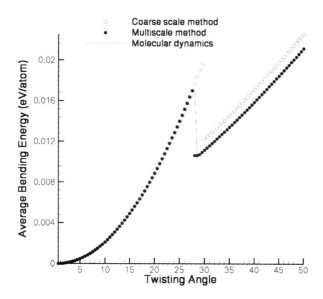

Fig. 7.48. Comparison of average twisting energy between molecular dynamics and multi-scale method.

pattern with the experimental image is made. As can be seen, the computation yields results that fit well with the observation. The advantage of the simulation is that a 3D buckling pattern is shown with full details, as compared with a 2D image obtained from experimental observation. Finally, the relation between the average bending energy and the bending angle is plotted in Fig. 7.55. A snap-through in the curve is observed at the angle of 25°, which signifies the onset of instability. Detailed buckling and post-buckling analysis for such system will be presented in the future.

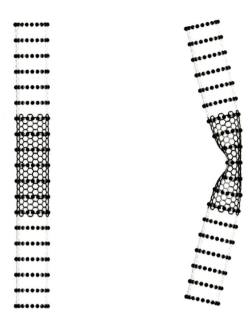

Fig. 7.49. (left) Initial particle distribution and embedded molecular structure of the carbon nanotube in the case of bending. (right) Final deformation of the particle and molecular structure at the final stage of bending.

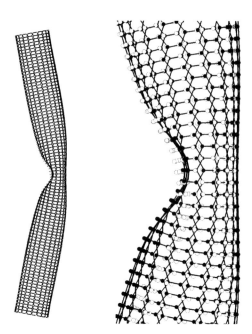

Fig. 7.50. Comparison of the post-buckling pattern from three different method. Left (a) The computed molecular structure for the (10,10) CNT. The bonds computed from MD are plotted as the black lines. The atomic positions from the coarse scale method are plotted as the empty dots, and from the multi-scale method are plotted as the filled dots. Right (b). A zoom-in plot of the post-buckling zoon. Note the differences in the results between the coarse scale and multi-scale method as compared with MD.

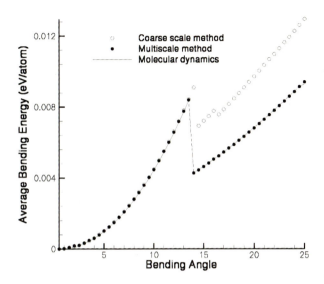

Fig. 7.51. Comparison of average bending energy between molecular dynamics and multi-scale method.

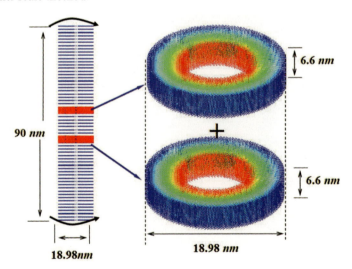

Fig. 7.52. Multiscale discretization scheme for bending of a 15-walled carbon nanotube.

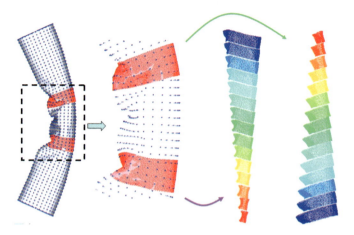

Fig. 7.53. Multiscale description of the buckling in MWCNT. The buckling region is zoomed in and the detailed deformation for the enriched 15 layers of molecular structure is shown.

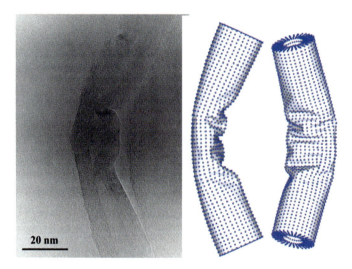

Fig. 7.54. (From left) (a) Experimental observation of buckling of multi-walled carbon nanotubes (from[380]). (b) Simulated buckling pattern of MWCNT in 2D. (c) Buckling pattern of the nanostructure in 3D.

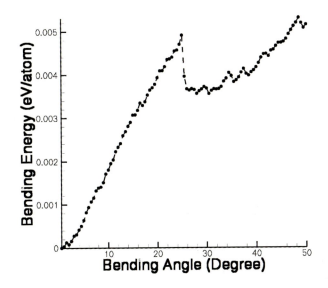

Fig. 7.55. Average bending energy as a function of the angle for the case of bending of 15-walled CNT.

8. Immersed Meshfree/Finite Element Method and Applications

8.1 Introduction

For the past few decades, tremendous research efforts have been directed to the development of modeling and simulation approaches for fluid-structure interaction problems. Methods developed by Tezduyar and his coworkers are widely used in the simulations of fluid-particle (rigid) and fluid-structure interactions.[428,430] To accommodate the complicated motions of fluid-structure boundaries, we often use adaptive meshing or the arbitrary Lagrangian Eulerian (ALE) techniques.[207,216,290-292] Recently, such approaches have also been adopted by Hu, *et al.* in the modeling of fluid-particle (rigid) systems.[206] Nevertheless, mesh update or remeshing algorithms can be time consuming and expensive and a detailed discussion on this issue is presented in Ref.[222]

In the 1970's, Peskin developed the Immersed Boundary (IB) method[369] to study flow patterns around heart valves. The mathematical formulation of the IB method employs a mixture of Eulerian and Lagrangian descriptions for fluid and solid domains. In particular, the entire fluid domain is represented by a uniform background grid, which can be solved by finite difference methods with periodic boundary conditions; whereas the submerged structure is represented by a fiber network. The interaction between fluid and structure is accomplished by distributing the nodal forces and interpolating the velocities between Eulerian and Lagrangian domains through a smoothed approximation of the Dirac delta function. The advantage of the IB method is that the fluid-structure interface is tracked automatically, which removes the costly computations due to various mesh update algorithms.

Nevertheless, one major disadvantage of the IB method is the assumption of the fiber-like one-dimensional immersed structure, which may carry mass, but occupies no volume in the fluid domain. This assumption also limits accurate representations for immersed flexible solids which may occupy finite volumes within the fluid domain. Furthermore, uniform fluid grids also set limitations in resolving fluid domains with complex shapes and boundary conditions.

Recently, an alternative approach, the Immersed Finite Element Method (IFEM) was developed by Zhang et al.[466] This method is able to eliminate the aforementioned drawbacks of the IB method and adopt parts of the work on the Extended Immersed Boundary Method (EIBM) developed by Wang and Liu.[448] With finite element formulations for both fluid and solid domains, the submerged structure is solved more realistically and accurately in comparison with the corresponding fiber network representation. The fluid solver is based on a stabilized equal-order finite element formulation applicable to problems involving moving boundaries,[210, 428, 430] one of the main techniques in using finite element methods to simulate fluid flow. This stabilized formulation prevents numerical oscillations without introducing excessive numerical dissipations. Moreover, in the proposed IFEM, the background fluid mesh does not have to follow the motion of the flexible fluid-structure interfaces and thus it is possible to assign sufficiently refined fluid mesh within the region around which the immersed deformable structures in general occupy and move.

Unlike the Dirac delta functions in the IB method which yield C^1 continuity,[371, 440] the discretized delta function in IFEM is the C^n shape function often employed in the meshfree Reproducing Kernel Particle Method (RKPM). Because of the higher order smoothness in the RKPM delta function, the accuracy is increased in the coupling procedures between fluid and solid domains.[448] Furthermore, the RKPM shape function is also capable of handling nonuniform fluid grids.

The outline of this paper is as follows. We first give a review of the IFEM formulations in Section 8.2. Section 8.3 summarizes the discretized governing equations and provides an outline of the IFEM algorithm. Application of IFEM to the modeling of biological systems is given in Section 8.4, where IFEM is coupled with protein molecular dynamics to study the aggregation of RBCs.

8.2 Formulations of Immersed Finite Element Method

The IFEM was recently developed by Zhang, Gerstenberger, Wang, and Liu[466] to solve complex fluid and deformable structure interaction problems. Because much of the following has already been derived in detail in Zhang, et al.,[466] we refer the interested reader to those works for further details.

Let us consider an incompressible three-dimensional deformable structure in Ω^s completely immersed in an incompressible fluid domain Ω^f. Together, the fluid and the solid occupy a domain Ω, but they do not intersect:

$$\Omega^f \cup \Omega^s = \Omega, \tag{8.1a}$$

$$\Omega^f \cap \Omega^s = \emptyset. \tag{8.1b}$$

In contrast to the IB formulation, the solid domain can occupy a finite volume in the fluid domain. Since we assume both fluid and solid to be incompressible, the union of two domains can be treated as one incompressible continuum with a continuous velocity field. In the computation, the fluid spans over the entire domain Ω, thus an Eulerian fluid mesh is adopted; whereas a Lagrangian solid mesh is constructed on top of the Eulerian fluid mesh. The coexistence of fluid and solid in Ω^s requires some considerations when developing the momentum and continuity equations.

In the computational fluid domain Ω, the fluid grid is represented by the time-invariant position vector $\mathbf{x}$; while the material points of the structure in the initial solid domain Ω_0^s and the current solid domain Ω^s are represented by $\mathbf{X}^s$ and $\mathbf{x}^s(\mathbf{X}^s, t)$, respectively. The superscript s is used in the solid variables to distinguish the fluid and solid domains.

In the fluid calculations, the velocity $\mathbf{v}$ and the pressure p are the unknown fluid field variables; whereas the solid domain involves the calculation of the nodal displacement $\mathbf{u}^s$, which is defined as the difference of the current and the initial coordinates: $\mathbf{u}^s = \mathbf{x}^s - \mathbf{X}^s$. The velocity $\mathbf{v}^s$ is the material derivative of the displacement $d\mathbf{u}^s/d\mathbf{t}$.

As in Refs.,[448, 466] we define the fluid-structure interaction force within the domain Ω^s as $f_i^{\text{FSI,s}}$, where FSI stands for Fluid-Structure Interaction:

$$f_i^{\text{FSI,s}} = -(\rho^s - \rho^f)\frac{dv_i}{dt} + \sigma_{ij,j}^s - \sigma_{ij,j}^f + (\rho^s - \rho^f)g_i, \quad \mathbf{x} \in \Omega^s. \qquad (8.2)$$

Naturally, the interaction force $f_i^{\text{FSI,s}}$ in Eq. (8.2) is calculated with the Lagrangian description. Moreover, a Dirac delta function δ is used to distribute the interaction force from the solid domain onto the computational fluid domain:

$$f_i^{\text{FSI}}(\mathbf{x}, t) = \int_{\Omega^s} f_i^{\text{FSI,s}}(\mathbf{X}^s, t)\, \delta(\mathbf{x} - \mathbf{x}^s(\mathbf{X}^s, t))\mathrm{d}\Omega. \qquad (8.3)$$

Hence, the governing equation for the fluid can be derived by combining the fluid terms and the interaction force as:

$$\rho^f \frac{dv_i}{dt} = \sigma_{ij,j}^f + f_i^{\text{FSI}}, \quad \mathbf{x} \in \Omega. \qquad (8.4)$$

Since we consider the entire domain Ω to be incompressible, we only need to apply the incompressibility constraint once in the entire domain Ω:

$$v_{i,i} = 0. \qquad (8.5)$$

To delineate the Lagrangian description for the solid and the Eulerian description for the fluid, we introduce different velocity field variables v_i^s and v_i to represent the motions of the solid in the domain Ω^s and the fluid within the entire domain Ω. The coupling of both velocity fields is accomplished with the Dirac delta function:

$$v_i^s(\mathbf{X}^s, t) = \int_\Omega v_i(\mathbf{x}, t)\, \delta(\mathbf{x} - \mathbf{x}^s(\mathbf{X}^s, t))\mathrm{d}\Omega. \tag{8.6}$$

Let us assume that there is no traction applied on the fluid boundary, *i.e.* $\int_{\Gamma_{h_i}} \delta v_i h_i \mathrm{d}\Gamma = 0$, applying integration by parts and the divergence theorem, we can get the final weak form (with stabilization terms):

$$\int_\Omega (\delta v_i + \tau^m v_k \delta v_{i,k} + \tau^c \delta p_{,i}) \left[\rho^f (v_{i,t} + v_j v_{i,j}) - f_i^{\mathrm{FSI}} \right] \mathrm{d}\Omega + \int_\Omega \delta v_{i,j}\sigma_{ij}^f \mathrm{d}\Omega$$

$$- \sum_e \int_{\Omega_e} (\tau^m v_k \delta v_{i,k} + \tau^c \delta p_{,i})\sigma_{ij,j}^f \mathrm{d}\Omega + \int_\Omega (\delta p + \tau^c \delta v_{i,i})v_{j,j}\mathrm{d}\Omega = 0. \tag{8.7}$$

The nonlinear systems are solved with the Newton-Raphson method. Moreover, to improve the computation efficiency, we also employ the GMRES iterative algorithm and compute the residuals based on matrix-free techniques.[397, 465]

Note that for brevity we ignore the fluid stress within the solid domain. The transformation of the weak form from the updated Lagrangian to the total Lagrangian description is to change the integration domain from Ω^s to Ω_0^s. Since we consider incompressible fluid and solid, and the Jacobian determinant is 1 in the solid domain, the transformation of the weak form to total Lagrangian description yields

$$\int_{\Omega_0^s} \delta u_i \left[(\rho^s - \rho^f)\ddot{u}_i^s - \frac{\partial P_{ij}}{\partial X_j} - (\rho^s - \rho^f)g_i + f_i^{\mathrm{FSI,s}} \right] \mathrm{d}\Omega_0^s = 0, \tag{8.8}$$

where the first Piola-Kirchhoff stress P_{ij} is defined as $P_{ij} = JF_{ik}^{-1}\sigma_{kj}^s$ and the deformation gradient F_{ij} as $F_{ij} = \partial x_i / \partial X_j$.

Using integration by parts and the divergence theorem, we can rewrite Eq. (8.8) as

$$\int_{\Omega_0^s} \delta u_i(\rho^s - \rho^f)\ddot{u}_i^s \mathrm{d}\Omega_0^s + \int_{\Omega_0^s} \delta u_{i,j}P_{ij}\mathrm{d}\Omega_0^s - \int_{\Omega_0^s} \delta u_i(\rho^s - \rho^f)g_i\mathrm{d}\Omega_0^s$$

$$+ \int_{\Omega_0^s} \delta u_i f_i^{\mathrm{FSI,s}}\mathrm{d}\Omega_0^s = 0. \tag{8.9}$$

Note again that the boundary integral terms on the fluid-structure interface for both fluid and solid domains will cancel each other and for brevity are not included in the corresponding weak forms.

For structures with large displacements and deformations, the second Piola-Kirchhoff stress S_{ij} and the Green-Lagrangian strain E_{ij} are used in the total Lagrangian formulation:

$$S_{ij} = \frac{\partial W}{\partial E_{ij}} \quad \text{and} \quad E_{ij} = \frac{1}{2}(C_{ij} - \delta_{ij}), \tag{8.10}$$

where the first Piola-Kirchhoff stress P_{ij} can be obtained from the second Piola-Kirchhoff stress as $P_{ij} = S_{ik}F_{jk}$.

Finally, in the interpolation process from the fluid onto the solid grid, the discretized form of Eq. (8.6) can be written as

$$v_{iI}^s = \sum_J v_{iJ}(t)\phi_J(\mathbf{X}_J - \mathbf{x}_I^s), \quad \mathbf{x}_J \in \Omega_{\phi I}. \tag{8.11}$$

Here, the solid velocity $\mathbf{v}_I^s$ at node I can be calculated by gathering the velocities at fluid nodes within the influence domain $\Omega_{\phi I}$. A dual procedure takes place in the distribution process from the solid onto the fluid grid. The discretized form of Eq. (8.3) is expressed as

$$f_{iJ}^{\text{FSI}} = \sum_I f_{iI}^{\text{FSI,s}}(\mathbf{X}^s, t)\phi_I(\mathbf{x}_J - \mathbf{x}_I^s), \quad \mathbf{x}_I^s \in \Omega_{\phi J}. \tag{8.12}$$

By interpolating the fluid velocities onto the solid particles in Eq. (8.11), the fluid within the solid domain is bounded to solid material points. This ensures not only the no-slip boundary condition on the surface of the solid, but also stops automatically the fluid from penetrating the solid, provided the solid mesh is at least two times denser than the surrounding fluid mesh. This heuristic criterion is based on the numerical evidence and needs further investigation.

8.3 Computational Algorithm

The governing equations of IFEM in discretized form (except the Navier-Stokes equations, for convenience) are summarized as follows:

$$f_{iI}^{\text{FSI,s}} = -f_{iI}^{\text{inert}} - f_{iI}^{\text{int}} + f_{iI}^{\text{ext}}, \qquad \text{in} \quad \Omega^s, \tag{8.13a}$$

$$f_{iJ}^{\text{FSI}} = \sum_I f_{iI}^{\text{FSI,s}}(\mathbf{X}^s, t)\phi_I(\mathbf{x}_J - \mathbf{x}_I^s), \qquad \mathbf{x}_I^s \in \Omega_{\phi J}, \tag{8.13b}$$

$$\rho^f(v_{i,t} + v_j v_{i,j}) = \sigma_{ij,j} + \rho g_i + f_i^{\text{FSI}}, \qquad \text{in} \quad \Omega, \tag{8.13c}$$

$$v_{j,j} = 0, \qquad \text{in} \quad \Omega, \tag{8.13d}$$

$$v_{iI}^s = \sum_J v_{iJ}(t)\phi_J(\mathbf{x}_J - \mathbf{x}_I^s), \qquad \mathbf{x}_J \in \Omega_{\phi I}. \tag{8.13e}$$

An outline of the IFEM algorithm with a semi-explicit time integration is illustrated as follows:

1. Given the structure configuration $\mathbf{x}^{s,n}$ and the fluid velocity $\mathbf{v}^n$ at time step n,
2. Evaluate the nodal interaction forces $\mathbf{f}^{FSI,s,n}$ for solid material points, using Eq. (8.13a),
3. Distribute the material nodal force onto the fluid mesh, from $\mathbf{f}^{FSI,s,n}$ to $\mathbf{f}^{FSI,n}$, using the delta function as in Eq. (8.13b),
4. Solve for the fluid velocities $\mathbf{v}^{n+1}$ and the pressure p^{n+1} implicitly using Eqs. (8.13c) and (8.13d),
5. Interpolate the velocities in the fluid domain onto the material points, i.e., from $\mathbf{v}^{n+1}$ to $\mathbf{v}^{s,n+1}$ as in Eq. (8.13e), and
6. Update the positions of the structure using $\mathbf{u}^{s,n+1} = \mathbf{v}^{s,n+1}\Delta t$ and go back to step 1.

Note that even though the fluid is solved fully implicitly, the coupling between fluid and solid is explicit. If we rewrite the fluid momentum equation (for clarity only discretized in time), we have

$$\rho^f \left[\frac{v_i^{m+1} - v_i^m}{\Delta t} + v_j^{m+1} v_{i,j}^{m+1} \right] = \sigma_{ij,j}^{f,m+1} + f_i^{FSI,m}. \tag{8.14}$$

It is clear that the interaction force is not updated during the iteration, i.e., the solid equations are calculated with values from the previous time step. For a fully implicit coupling, this force must be a function of the current fluid velocity and the term $f_i^{FSI,m+1}$ should be included into the linearization of the fluid equations.

8.4 Application to Biological Systems

In this section, we present the simulations of biological fluid flow problems with deformable cells using a newly developed modeling technique by Liu, Zhang, Wang, and Liu[313] with a combination of IFEM[466] and protein molecular dynamics. The effects of cell-cell interaction (adhesive/repulsive) and hydrodynamic forces on RBC aggregates are studied by introducing equivalent protein molecular potentials into the IFEM. For a detailed description of the IFEM coupled with cell interactions and its applications to hemodynamics, we refer the readers to Liu et al.[313]

The human blood circulatory system has evolved to supply nutrients and oxygen to and carry the waste away from the cells of multicellular organisms through the transport of blood, a complex fluid composed of deformable cells, proteins, platelets, and plasma. Overviews of recent numerical procedures for the modeling of macro-scale cardiovascular flows are available in Refs.[423]

While theories of suspension rheology generally focus on homogeneous flows in infinite domains, the important phenomena of blood flows in microcirculation depend on the combined effects of vessel geometry, cell deformability, wall compliance, flow shear rates, and many micro-scale chemical, physiological, and biological factors.[138] There have been past studies on shear flow effects on one or two cells, leukocyte adhesion to vascular endothelium, and particulate flow based on continuum enrichment methods.[444] However, no mature theory is yet available for the prediction of blood rheology and blood perfusion through micro-vessels and capillary networks. The different time and length scales as well as large motions and deformations of immersed solids pose tremendous challenges to the mathematical modeling of blood flow at that level.

In the following part of this section, we concentrate on the rheological aspects of flow systems of arterioles, capillaries, and venules which involve deformable cells, cell-cell interactions, and various vessels. The demonstrating problems are: rigid/soft spheres falling in a tube, deformation of balloon/airfoil under flow, shear of a cluster of deformable RBCs, normal and sickle RBCs passing through capillary vessels, a single cell squeezing through a micro-vessel constriction, RBCs deposition, and finally a flexible valve-viscous fluid interaction problem with experimental comparison.

8.4.1 Three Rigid Spheres Falling in a Tube

In this section, we simulate three spheres placed with non-equal distance between each other falling in a tube as shown in Fig. 8.1. The spheres are 1.35 and 0.65 cm apart from each other. In this example, we again consider the structure to be rigid. It is ideal to set the material modulus to be high. However, doing that would require a relatively small time step, which might not be computationally efficient in this case. Therefore, special treatment is considered to impose the rigidity constraint. Here, we calculate the average velocity of the entire solid domain with the conservation of the total linear momentum (the angular momentum is ignored for simplicity) and then reassign the average velocity to all the solid nodes, i.e. $v_{avg} = \sum_i m_i v_i / \sum_i m_i$.

This rigidity treatment yields the zero internal force. The properties used in this problem can be found in Table 8.1.

Table 8.1. Properties used for three rigid spheres falling in a tube.

fluid	9508 nodes	$\rho^f = 1 \text{ g/cm}^3$
	51448 elements	$\mu = 0.02 \text{ g} \cdot \text{cm/s}$
solid	3×997 nodes	$\rho^s = 3 \text{ g/cm}^3$
	3×864 elements	$D = 1.0 \text{ cm}$
		$g = 980 \text{ cm/s}^2$

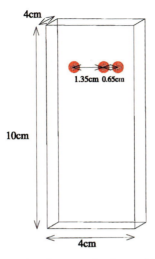

Fig. 8.1. Problem statement for the 3 rigid spheres falling in a tube.

As shown in Fig. 8.2, when the three spheres just start to fall, vortices at the middle plane (y=2 cm) form around both sides of the spheres, except in between spheres 2 and 3 (numbering from left to right), where the spheres are placed closer to each other. As time progresses, the pressure generated between spheres 2 and 3 increases to the extent that sphere 2 begins to repel from sphere 3, and pushes 3 to fall faster in the fluid domain. At the same time, sphere 2 is moving towards sphere 1, they then start to interact with each other.

8.4.2 20 Soft Spheres Falling in a Channel

This example solves 20 soft spheres falling in a channel. These same-sized spheres are placed randomly near the top of the channel. The spheres experience the gravitational and internal forces as well as the interaction forces with the surrounding fluid. The properties used in this problem can be found in Table 8.2.

The movements of the spheres at different time steps are shown in Fig. 8.3. To clearly illustrate the deformations of the spheres, we present in Fig. 8.4 the enlarged images of the spheres and the pressure distributions at different time steps.

8.4.3 Fluid-flexible Structure Interaction

The first example shows the deformation of an airfoil with a very flexible tail under flow field. The flow angle is 14° (2D setting, 3D tetrahedral elements).

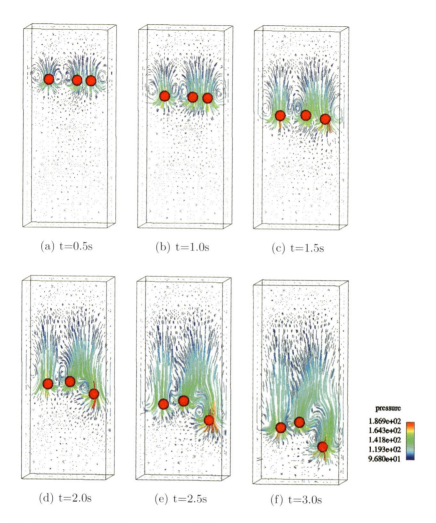

Fig. 8.2. Fluid velocity vectors at different time steps for 3 rigid spheres falling in a tube.

Inflow velocity is around 10.3 cm/s with a viscosity of 0.2 - 0.1 gcm/s. The Reynolds number is 200 - 400 for this example.

The usual vortex originating at the tail deforms the tail and the tail starts to oscillate while the vortices appear. This example illustrates the advantage of having no fluid mesh deformation as in ALE methods. The severe mesh deformation would require a frequent remeshing of parts of the fluid domain.

Another example shows a 3D calculation of a balloon expansion on a coarse tetrahedral mesh. The inflow from the sidewall (not visible) inflates the membrane structure. The membrane is modeled with several layers of 3D tetrahedral solid elements.

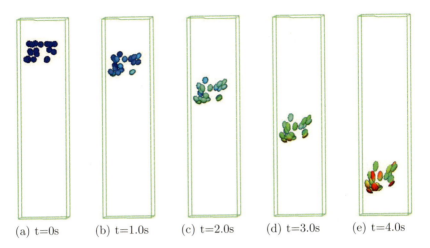

Fig. 8.3. The movement of 20 spheres falling in a channel at different time steps.

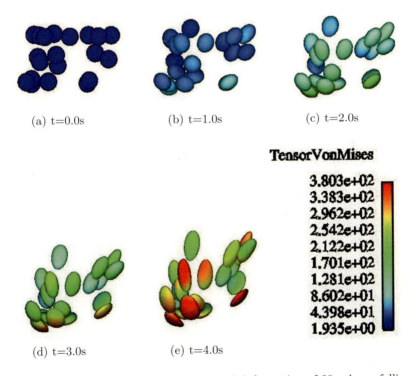

Fig. 8.4. The Von Mises stress distribution and deformation of 20 spheres falling in a channel at different time steps.

Table 8.2. Properties for 20 soft spheres falling in a tube.

fluid	4861 nodes	$\rho^f = 1$ g/cm^3	
	22557 elements	$\mu = 10.0$ g · cm/s	
solid	20×997 nodes	$\rho^s = 2$ g/cm^3	$C_1 = 2.93 \times 10^1$ g/(cm · s^2)
	20×864 elements	$D = 0.5$ cm	$C_2 = 1.77 \times 10^1$ g/(cm · s^2)
		$g = 980$ cm/s^2	

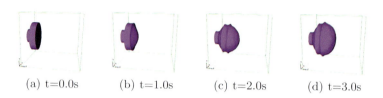

(a) t=0.0s   (b) t=1.0s   (c) t=2.0s   (d) t=3.0s

Fig. 8.5. The deformation of the balloon at different time steps.

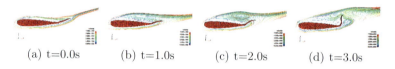

(a) t=0.0s   (b) t=1.0s   (c) t=2.0s   (d) t=3.0s

Fig. 8.6. The deformation of the airfoil under flow at different time steps. Development of vorticity near the airfoil is also shown in the figures

8.4.4 IFEM Coupled with Protein Molecular Dynamics

Discrete RBC Model. The RBC is modeled as a flexible membrane enclosing an imcompressible fluid. As shown in Fig. 8.7, to account for both bending and membrane rigidities, RBC membrane is modeled with a three-dimensional finite element formulation using the Lagrangian description. In this work, a typical membrane is discretized with 1043 nodes and 4567 elements. The static shape of a normal RBC is a biconcave discoid. The material behavior of the RBC membrane is depicted by the Neo-Hookean strain energy function.

RBC Aggregation. Cell-cell adhesion plays an important role in various physiological phenomena including the recognition of foreign cells. Although the exact physiological mechanisms of RBC coagulation and aggregation are still ambiguous, it has been found that both the RBC surface structure and membrane proteins are key factors in producing adhesive/repulsive forces. The basic behavior of the interaction forces between two RBCs is simply illustrated as the weak attractive force at far distances and strong repulsive force at short distances. Due to the complexity of the aggregation process, we accumulate the intermolecular force, electrostatic force, and protein dynam-

(a) 3D RBC Model   (b) RBC Cross-Section   (c) RBC Mesh

Fig. 8.7. A three-dimensional finite element mesh of a single RBC model.

ics into a potential function, similar to an intermolecular potential. In this work, we adopt the Morse potential, which is found to be capable of quantitatively predicting the aggregation behaviors consistent with experimental observations:

$$\phi(r) = D_e \left[e^{2\beta(r_o - r)} - 2e^{\beta(r_o - r)} \right], \tag{8.15}$$

$$f(r) = -\frac{\partial \phi(r)}{\partial r} = 2D_e \beta \left[e^{2\beta(r_o - r)} - e^{\beta(r_o - r)} \right], \tag{8.16}$$

where r_o and D_e are the zero force length and surface energy, respectively, and β is a scaling factor.

The potential function is chosen such that the RBCs will de-aggregate at the shear rate above 0.5 s^{-1}. After the finite element discretization of the solid domain, a sphere with the diameter of the cut-off length is used to identify the cell surface $\mathbf{Y^c}$ within the influencing domain around the cell surface $\mathbf{X^c}$. Hence, a typical cell-cell interaction force can be denoted as $\mathbf{f^c}(\mathbf{X^c}) = - \int_{\Gamma(\mathbf{Y^c})} \frac{\partial \phi(r)}{\partial r} \frac{\mathbf{r}}{r} d\Gamma$, where $\mathbf{r} = \mathbf{X^c} - \mathbf{Y^c}$, $r = \|\mathbf{X^c} - \mathbf{Y^c}\|$, and $\Gamma(\mathbf{Y^c})$ represents the cell surface area within the influencing domain surrounding surface $\mathbf{X^c}$.

To incorporate IFEM with protein molecular dynamics, the cell-cell interaction force is applied on the surfaces of cells:

$$\sigma_{ij}^s n_j = f_i^c, \tag{8.17}$$

Blood Viscoelasticity. Blood plasma can be accurately modeled with a Newtonian fluid model, yet blood flows do exhibit non-Newtonian or viscoelastic behaviors, in particular under low Reynolds numbers. The shear rate-dependence of blood viscoelasticity may be characterized as follows: initially the shear rate increases the blood viscosity decreases, and eventually

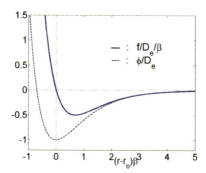

Fig. 8.8. Non-dimensionalized Morse potential and force.

blood viscosity reaches a plateau marking the plasma viscosity, and the blood elasticity continues to decrease.

On the microscopic level, RBCs play an important role in the viscoelastic behavior of blood.[172] In the quiescent state, normal RBCs tend to aggregate. Under low shear rate, aggregates are mainly influenced by cell-cell interaction forces; in the mid-shear rate region, RBC aggregates start to disintegrate and the influence of the deformability gradually increases; and under high shear rate, RBCs tend to stretch, align with the flow, and form layers. The illustration of these three different stages is shown in Fig. 8.9.

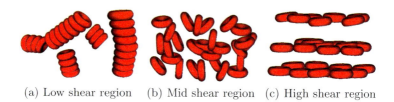

(a) Low shear region   (b) Mid shear region   (c) High shear region

Fig. 8.9. Blood microscopic changes under different shear rates.

8.4.5 Cell-cell Interaction and Shear Rate Effects

RBC aggregation is one of the main causes of the non-Newtonian behaviors of blood flows. Due to the presence of the cross-linking proteins fibrinogen on cell membranes and globulin in the plasma, RBCs tend to form aggregates called rouleaus, in which RBCs adhere loosely like a stack of coins. The presence of massive rouleaus can impair blood flow through micro- and capillary vessels and cause fatigue and shortness of breath. The variation in the level of RBC aggregation may be an indication of thrombotic disease. In general, cell-cell interaction forces are not sufficient to deform cell membranes. However, the ensuing aggregate could alter the surrounding fluid significantly.

Fig. 8.10. The shear of a four-RBC cluster at the shear rate of 0.25, 0.5, and 3.0 s^{-1} respectively. The vectors represent the fluid interaction force.

In a set of numerical experiments, we subject the RBC aggregate and viscous fluid mixer to low shear rates and observe that RBC aggregate rotates as a bulk, as shown in Fig. 8.10. With an intermediate shear rate, our numerical simulation demonstrates that after the initial rotations the RBC aggregate aligns with the shear direction and then de-aggregates. At higher shear rate, the RBC aggregate completely disintegrates and the cells begin to orient themselves into parallel layers. The disintegration of RBC aggregates with the increase of the shear rate is an indication of the decrease of the macroscopic viscosity, which is consistent with the experimental observation.

8.4.6 Micro- and Capillary Vessels

Red blood cells are important for blood flows in microcirculation. The typical diameter of a micro-vessel is $1.5 \sim 3$ times larger than that of a cell. On the other hand, a capillary vessel diameter is about $2 \sim 4 \ \mu m$, which is significantly smaller. The pressure gradient which drives the flow is usually around $3.2 \sim 3.5$ KPa. For the chosen diameter and pressure, the Reynolds number in a typical capillary is around 0.01. In fact, in the process of squeezing through capillaries, large deformations of red blood cells not only slow down the blood flow, but also enable the exchange of oxygen through capillary vessel walls.

Sickle cell anemia occurs from genetic abnormalities in hemoglobin. When sickle hemoglobin loses oxygen, the deoxygenated molecules form rigid rods which distort the cell membrane into a sickle or crescent shape. The sickle-shaped cells are both rigid and sticky and tend to block capillary vessels and cause blood flow blockage to the surrounding tissues and organs. To relate blood rheology to sickle cell anemia, we consider the normal and sickle RBCs passing through a micro-vessel contraction. The strong viscous shear introduced by such a flow contraction leads to some interesting phenomena of the RBC aggregation with respect to cell-cell interaction forces and cell deformability. Furthermore, the modeling of this complex fluid-solid system also demonstrates the capability of the coupling of the Navier-Stokes equations with protein molecular dynamics.

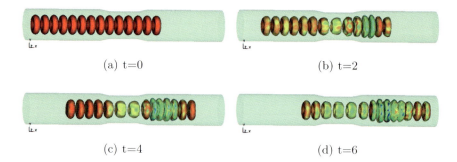

Fig. 8.11. The normal red blood cell flow with the inlet velocity of 10 μm/s at different time steps.

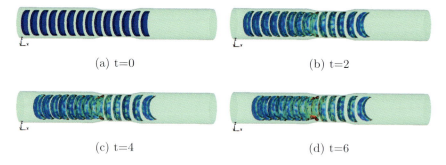

Fig. 8.12. The sickle cell flow with the inlet velocity of 10 μm/s at different time steps.

It is shown in Fig. 8.11, as RBCs pass the diffuser stage of the contraction, the deceleration of the RBCs forms blockage for the incoming RBCs. Therefore, dilation of RBCs is coupled with the pile-up of RBCs at the outlet of the vessel constriction. As also confirmed in Fig. 8.12, under the similar flow conditions, rigid and sticky sickle cells eventually block the micro-vessel entrance which will certainly result in de-oxygenation of surrounding tissues.

To demonstrate the effect of vessel constriction more clearly, we present a three-dimensional simulation of a single red blood cell squeezing through a capillary vessel. The RBC diameter is 1.2 times larger than that of the capillary vessel, which leads to the divergence of the cytoplasm (internal liquid) to the two ends of the capsule by deforming into a slug during the squeezing process. During the exiting process, there is a radial expansion of the slug due to the convergence of the cytoplasm, which deforms the capsule into an acaleph (or jellyfish) shape. In Fig. 8.13, four snapshots illustrate various stages of the red blood cell with respect to the capillary vessel.

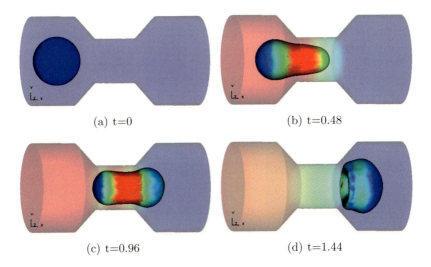

(a) t=0   (b) t=0.48

(c) t=0.96   (d) t=1.44

Fig. 8.13. Three-dimensional simulation of a single red blood cell (essentially a hollow sphere for simplicity) squeezing through a capillary vessel.

8.4.7 Adhesion of Monocytes to Endothelial Cells

Fig. 8.14 illustrates some preliminary results of a study of the adhesion of monocytes to endothelial cells near the expansion of a large blood vessel, a condition which may be the result of a poorly matched vascular graft. The vessel section upstream of the graft may differ in size, creating a geometry similar to a diverging duct. This geometry results in a classical flow recirculation region into which particles suspended in the fluid may become entrained, allowing them to approach and deposit on the vessel wall. We use tools previously developed in Refs.[313, 466] to model the interactions of RBCs suspended in a fluid, extended to include cell-vessel wall attraction/repulsion via a similar potential approach. The eventual goal of the study is the development of a predictive tool which would be of use during the design and evaluation of engineered grafts, stents, etc. and to provide a model beyond the typical continuum formulation with growth prediction model.

In order to demonstrate the ability of the IFEM/Protein dynamics modeling scheme to represent physically realistic phenomena, the results of a simulation with conditions meant to mimic the experimental, have been compared to the published result by Chiu et al. The qualitative behavior of the two systems are compared. Notice that as the simulation progresses, there comes a point after which no new particles are introduced into the recirculating region, though particles do leave the area. Further, notice that the bulk of the cells which have adhered to the lower wall are near the reattachment region, the expected result, as adhesion should (and does, physically) occur in

regions of low wall shear stress, which corresponds to small tangential velocities. Further, this is illustrated by the comparison of the cell concentration
plots, presented as 3d contours. It can be seen that there is a large increase in
number of particles present in the region near the reattachment point, illustrated by the peak in the concentration plot, again, here, as the simulation
results are preliminary, the comparison is qualitative. However, we noticed
that the distribution of particles along the wall are similar to those seen in
the experiments.

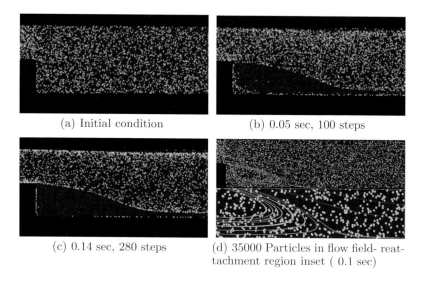

(a) Initial condition   (b) 0.05 sec, 100 steps

(c) 0.14 sec, 280 steps   (d) 35000 Particles in flow field- reat-
                                         tachment region inset (0.1 sec)

Fig. 8.14. Preliminary result of Monocytes in separated flow.

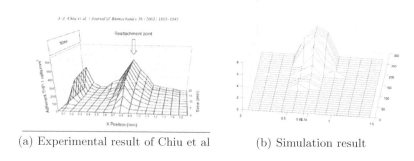

(a) Experimental result of Chiu et al   (b) Simulation result

Fig. 8.15. Comparison of the cell concentration of monocytes in separated flow.

8.4.8 Flexible Valve-viscous Fluid Interaction

In order to validate the IFEM, results from a simple flexible valve-viscous fluid interaction analysis are compared with experiments performed at ABIOMED recently. As shown in Fig. 8.16, water was pulsed through a column with a square cross section (5cm x 5cm) at a frequency of 1 Hz. A rubber sheet was located inside this column. Analysis of Fig. 8.16 reveals excellent correlation between experimental observations of structural displacement and the simulation. A detailed description of this comparison will be available in Refs.[449]

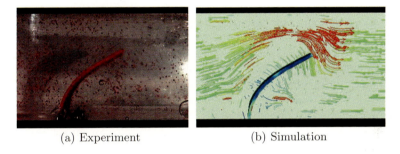

(a) Experiment   (b) Simulation

Fig. 8.16. Comparison of experimental observation and simulation result of a rubber sheet deflecting in a column of water. Pulsatile flow through the column (square cross section) is from left to right at a frequency of 1 Hz. Velocity vectors and beam stress concentration can be seen in the simulation.

9. Other Meshfree Methods

9.1 Natural Element Method

9.1.1 Construction of Natural Neighbor

To construct natural neighbor interpolation, the first step is to establish the natural neighbor coordinate. The concepts of the nearest neighbors and neighboring nodes are contained in the first-order Voronoi diagram. However, in the first-order Voronoi diagram, the nearest neighbors and neighboring nodes are only identified for nodal points, $x_I, I \in \Lambda$, not for arbitrary point $x \in \mathbb{R}^d$. To define the nearest neigbbor and neighboring nodes for an arbitrary point in a domain of interests, we need the second-order Voronoi diagram [1]

The second order Voronoi cell is defined as a set for two points x and x_I,

$$T(\bar{x}, x_I) = \left\{ x \in \mathbb{R}^2 \ \middle|\ d(x, \bar{x}) < d(x, x_I) \leq d(x, x_J), \quad \forall\, x_J \in \Lambda \right\} \quad (9.1)$$

where $\bar{x}$ can be nodal point or any point in the domain of interests.

If $\bar{x} = x_J (J \neq I)$ is a nodal point, then its natural neighbors have been identified by the first-order Voronoi diagram. For any node, $I \in \Lambda$, its neighbors are those nodes whose Voronoi cell share a commom edge with the I-th Voronoi cell, or, those nodes share a common triangle in corresponding Delaunay triangulation. When $\bar{x}$ is not a node, one has to contruct a Voronoi cell for point $\bar{x}$, i.e., $T(\bar{x}, x_I)$ followed the similar rule in construction of the first-order Voronoi cells. In Fig. 9.1, the perpendicular bisectors from x to its natural neighbors are constructed, which form a polygon $abcd$. It consists of four non-zero second order Voronoi cells: $T(\bar{x}, x_1)$, $T(\bar{x}, x_2)$, $T(\bar{x}, x_3)$, $T(\bar{x}, x_4)$. And the first order Voronoi cell for point $\bar{x}$ is

$$T_{\bar{x}} = \sum_{I \in \Lambda} T(\bar{x}, x_I) \qquad (9.2)$$

from which one may find that point $\bar{x}$ has four natural neighbors, they are nodes, 1, 2, 3, and 4.

---

[1] By extending the 1st-order Voronoi diagram, one can construct higher order $(k - order, k > 1)$ Voronoi diagrams in a plane.

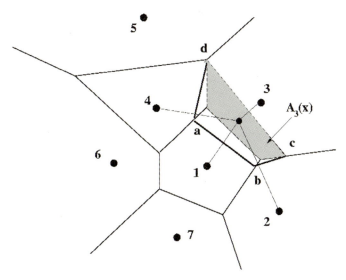

Fig. 9.1. Construction of natural neighbors for a non-nodal point

9.1.2 Natural Neighbor Interpolation

Sibson Interpolation. To this end, we can define natural neighbor coordinate. In a two-dimensional case, let $A_I(x)$ be the area of $T(x, x_I)$ and $A(x)$ be the area of $T(x)$. The natural neighbor coordinates of x with respect to a natural neighbor I is defined as

$$\Phi_I(x) = \frac{A_I(x)}{A(x)} , \quad I \in \Lambda \tag{9.3}$$

where $A(x) = \sum_{I \in \Lambda} A_I(x)$.

The four regions shown in Fig. 9.1 are the second-order Voronoi cell, whereas their union (polygon abcd) is the first order Voronoi cell. The natural neighbor coordinates of x with respect to a natural neighbor I is defined as the ratio of the area of the second order Voronoi cell $T(x, x_I)$ and the first order Voronoi cell $T(x)$. If $x = x_I$, $T(x_I, x_I) = T_I$ and $T(x_I) = T_I$, therefore, $\Phi_I(x_I) = 1$. Furthermore, $T(x_J, x_I) = \emptyset$, $I \neq J$, therefore, $\Phi_I(x_J) = 0$, $I \neq J$.

Since area is a positive quantity, the Sibson interpolant is a partition of unity,

1. $0 \leq \Phi_I(x) \leq 1$;
2. $\Phi_I(x_J) = \delta_{IJ}$;
3. $\sum_{I \in \Lambda} \Phi_I(x) = 1$.

The Sibson interpolant is also isoparametric, i.e.

$$x = \sum_{I \in \Lambda} \Phi_I(x) x_I \tag{9.4}$$

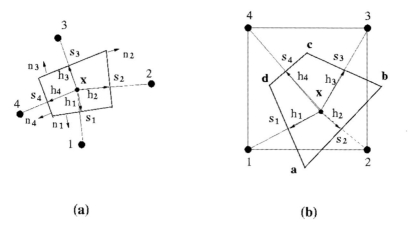

(a)   **(b)**

Fig. 9.2. Non-Sibsonian interpolation: (a) on irregular particle distribution, (b) on regular grid.

Non-Sibsonian Interpolation. A non-Sibsonian interpolation over scattering particles is proposed by Belikov and his co-workers.[50] The idea is to use the ratio between the edges of the first order Voronoi cell at an arbitrary point and its distances to nearest neighbors to form an interpolant.

A graphic construction is shown in Fig. 9.2. In Fig. 9.2, the given point x has four neighbors, $x_{I_1}, bx_{I_2}, x_{I_3}$, and x_{I_4}. The first order Voronoi cell thus has four edges, and obviously they are functions of the coordinate x, $s_{I_1}(x), s_{I_2}(x), s_{I_3}(x)$, and $s_{I_4}(x)$. It is always assumed that the edges of the first order Voronoi cell T_x to the non-neighboring particles are zero.

Denote the distances between the point x to its four neighbors as $2h_{I_1}(x), 2h_{I_2}(x), 2h_{I_3}(x)$, and $2h_{I_4}(x)$. The non-Sibsonian interpolant proposed by Belikov et al.[50] is defined as follows,

$$\Phi_I := \frac{\alpha_I(x)}{\displaystyle\sum_{J \in \Lambda} \alpha_J(x)}, \quad \alpha_J(x) := \frac{s_J(x)}{h_J(x)} \tag{9.5}$$

where I is the global index.

We know show that the non-Sibsonian interpolant is a partition of unity and it preserves linear completeness, i.e.

$$\sum_{I \in \Lambda} \Phi_I(x) = 1 \tag{9.6}$$

$$\sum_{I \in \Lambda} (x - x_I)\Phi_I(x) = 0 \tag{9.7}$$

For the example shown in Fig. 9.2, we assume that every particle in the domain of interests has only four neighbors (which is not essential to the proof). It then can be easily verified that

$$\sum_{I \in \Lambda} \Phi_I(\boldsymbol{x}) = \sum_{I \in \Lambda} \frac{\alpha_I(\boldsymbol{x})}{\sum_{J \in \Lambda} \alpha_J(\boldsymbol{x})} = \sum_{i=1}^{4} \frac{s_{I_i}(\boldsymbol{x})/h_{I_i}(\boldsymbol{x})}{\sum_{i=1}^{4} s_{I_i}(\boldsymbol{x})/h_{I_i}(\boldsymbol{x})} = 1, \quad \forall \, \boldsymbol{x} \in \Omega$$

(9.8)

To show (9.7) in $\mathbb{R}^d$, we consider the vector identity (Green's theorem):

$$\int_{\Omega} \nabla f d\Omega = \oint_{S} f d\mathbf{S}$$

(9.9)

where $d\mathbf{S} = \mathbf{n} dS$ and $\mathbf{n}$ is the outward normal of surface S. Let $f = 1$ in above equation and consider the example in Fig. 9.2. We obtain

$$\int_{\Omega} \nabla f d\Omega = \mathbf{0} = \oint dS = \oint \mathbf{n} dS$$

$$= \sum_{i=1}^{4} \frac{(\boldsymbol{x} - \boldsymbol{x}_{I_i})}{|\boldsymbol{x} - \boldsymbol{x}_{I_i}|} s_{I_i}(\boldsymbol{x}) = \sum_{I \in \Lambda} \frac{(\boldsymbol{x} - \boldsymbol{x}_I)}{|\boldsymbol{x} - \boldsymbol{x}_I|} s_I(\boldsymbol{x})$$

(9.10)

where $(\boldsymbol{x} - \boldsymbol{x}_{I_i})/|\boldsymbol{x} - \boldsymbol{x}_{I_i}| = \mathbf{n}_{I_i}$ is the outward normal for the edge I_i (i=1,2,3,4). Again the number of edges in the first order Voronoi cell is not essential.

Since $h_I(\boldsymbol{x}) = |\boldsymbol{x} - \boldsymbol{x}_I|/2$, we have

$$\frac{1}{2} \sum_{I \in \Lambda} (\boldsymbol{x} - \boldsymbol{x}_I) \frac{s_I(\boldsymbol{x})}{h_I(\boldsymbol{x})} = 0$$

$$\Rightarrow \sum_{I \in \Lambda} (\boldsymbol{x} - \boldsymbol{x}_I) \frac{s_I(\boldsymbol{x})/h_I(\boldsymbol{x})}{\sum_{J \in \Lambda} s_J(\boldsymbol{x})/h_J(\boldsymbol{x})} = 0$$

$$\Rightarrow \sum_{I \in \Lambda} (\boldsymbol{x} - \boldsymbol{x}_I) \Phi_I(\boldsymbol{x}) = 0$$

(9.11)

9.1.3 Examples of Natural Neighbor Interpolant

9.2 Free Mesh Method

9.3 Meshfree Finite Difference Methods

One of the earliest meshfree methods is meshfree finite difference method, which was proposed by Liszka and Orkisz.[280, 281] The conventional finite difference method is a mesh based method. It requires a stencil, or a grid, to establish approximation of the spatial derivatives for unknown functions.

The objective of meshfree finite difference method is to establish a rule to approximate the spatial derivatives of an unknown function among randomly distributed nodal points.

To illustrate the construction process of meshfree finite difference scheme, we consider a two-dimensional example.

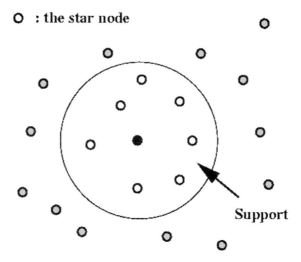

Fig. 9.3. Selection stars based on local support

Like other meshfree methods, every particle is associated with a support in meshfree finite difference method. Consider particle I as a central node having coordinate (x_0, y_0) with a local index 0. Inside the particle I's support, there are other particles, which are called as "*stars*" whose coordinates are arranged in the order of the local index, $(x_1, y_1), (x_2, y_2), \cdots, (x_m, y_m)$ as shown in Fig. 9.3.

Assume that an unknown scalar function is a sufficiently differentiable at the point (x_0, y_0). Using Taylor series expansion, for each star, i, $1 \le i \le m$, one may write

$$f_i = f_0 + h_i \frac{\partial f_0}{\partial x} + k_i \frac{\partial f_0}{\partial y} + \frac{h_i^2}{2} \frac{\partial^2 f_0}{\partial x^2}$$

$$+ \frac{k_i^2}{2} \frac{\partial^2 f_0}{\partial y^2} + h_i k_i \frac{\partial^2 f_0}{\partial x \partial y} + \mathcal{O}(\Delta^3), \quad 1 \le i \le m \tag{9.12}$$

where

$$h_i = x_i - x_0, \quad k_i = y_i - y_0, \quad \Delta = \max_{1 \le i \le m} \left\{ \sqrt{h_i^2 + k_i^2} \right\} \tag{9.13}$$

Let $\{\delta f\} = \{f_1 - f_0, f_2 - f_0, \cdots, f_m - f_0\}$. We shall have a set of linear algebraic equations $(m \ge 5)$

$$[A]\{Df\} = \{\delta f\} \tag{9.14}$$

with

$$[A] = \begin{bmatrix} h_1 & k_1 & h_1^2/2 & k_1^2/2 & h_1k_1 \\ h_2 & \cdots & \cdots & \cdots & \cdots \\ \vdots & \ddots & \cdots & \cdots & \cdots \\ \vdots & \cdots & \ddots & \cdots & \cdots \\ \vdots & \cdots & \cdots & \ddots & \cdots \\ h_m & \cdots & \cdots & \cdots & h_mk_m \end{bmatrix} \tag{9.15}$$

where the five unknown derivatives at the point (x_0, y_0) are

$$\{Df\}^T = \left\{ \frac{\partial f_0}{\partial x}, \frac{\partial f_0}{\partial y}, \frac{\partial^2 f_0}{\partial x^2}, \frac{\partial^2 f_0}{\partial y^2}, \frac{\partial^2 f_0}{\partial x \partial y} \right\} \tag{9.16}$$

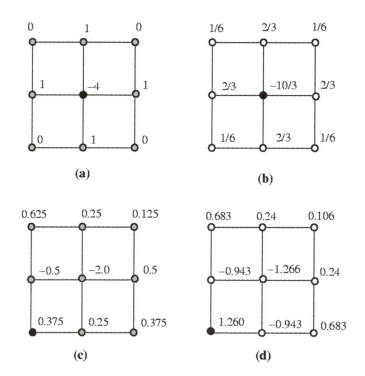

Fig. 9.4. Comparison between finite difference (a,c) and meshfree finite difference (b,d)

The matrix $[A] = [A]^{m \times 5}$ is not a square matrix. The problem ⁀ᵃʳ
if $m \geq 5$. Even if $m \geq 5$, the matrix A can still be singular or ill-c `ᵉᵃʳ
the particle distribution is degenerated. If the particle distr⁀`
and $m > 5$ in this case, the system (9.14) is an overdeterᵣ`

algebraic equations. Its solution may be obtained from the weighted least square minimization of the following objective

$$\mathbf{E} = \sum_{i=1}^{m}\left[\left(f_0 - f_i + \frac{\partial f_0}{\partial x}h_i + \cdots\right)w_i\right]^2 \to \min \tag{9.17}$$

where weight $w_i = 1/\Delta_i^3$ and $\Delta_i = \sqrt{h_i^2 + k_i^2}$.

By the stationary condition,

$$\frac{\partial \mathbf{E}}{\partial \{Df\}} = \{0\} \tag{9.18}$$

one may derive five independent linear algebraic equations with five unknowns.

In principle, the method should work when particle distribution is uniform as well. For uniform particle distribution, meshfree difference method is compared with regular finite difference method (see Fig. 9.4). It is found that meshfree finite difference method provides a more accurate approximation than regular finite difference method in unform particle distribution.

9.4 Vortex-in-cell Methods

In computational fluid mechanics, most of the numerical algorithms for the Navier-Stokes equations are based on the velocity-pressure formulation. An alternative to velocity-pressure formulation is the vorticity-velocity formulation:

$$\frac{\partial \boldsymbol{\omega}}{\partial t} + (\mathbf{u} \cdot \nabla)\boldsymbol{\omega} = (\boldsymbol{\omega} \cdot \nabla)\mathbf{u} + \nu \Delta \boldsymbol{\omega} \tag{9.19}$$

$$\Delta \mathbf{u} = -\nabla \times \boldsymbol{\omega} \tag{9.20}$$

where vorticity $\boldsymbol{\omega} = \nabla \times \mathbf{u}$.

The Lagrangian form of the above equations are

$$\frac{d\mathbf{x}_I}{dt} = \mathbf{u}(\boldsymbol{x}_I, t) \tag{9.21}$$

$$\frac{d\boldsymbol{\omega}}{dt} = [\nabla \mathbf{u}(\boldsymbol{x}_I, t)]\boldsymbol{\omega}_I + \nu \Delta \boldsymbol{\omega}(\boldsymbol{x}_I, t) \tag{9.22}$$

the velocity field can be obtained from the Poisson's equation (9.20). pressed by the Biot-Savart integral,

where th

$$\cdots - \boldsymbol{y}) \times \boldsymbol{\omega} d\boldsymbol{y} \tag{9.23}$$

$$G(\mathbf{z}) = \begin{cases} -\dfrac{1}{2\pi} \dfrac{z}{|z|^2} & 2D \\ \dfrac{1}{4\pi} \dfrac{z}{|z|^3} & 3D \end{cases} \tag{9.24}$$

The essence of the vortex method is to discretize above the Lagrangian description by the finite number of moving material particles. Following the movement of these particles, one may construct or evaluate the velocity field as well as the vorticity field.

In early approaches, a point (singular) vortex method was employed to represent the vorticity field,

$$\boldsymbol{\omega}(\boldsymbol{x}) = \sum_I \Gamma_I \delta(\boldsymbol{x} - \boldsymbol{x}_I) \tag{9.25}$$

For example, the 2D discrete velocity field is

$$\frac{d\mathbf{x}_I}{dt} = \frac{1}{2\pi} \sum_J \frac{(\boldsymbol{x}_I - \boldsymbol{x}_J) \times \mathbf{e}_z \Gamma_J}{|\mathbf{x}_I - \mathbf{x}_J|^2} \tag{9.26}$$

Today, most researchers use vortex blob, or smooth vortex methods. It implies that a smoothing kernel function is used to eliminate singularities so that the algorithm may be more stable. The resulting equation becomes,

$$\boldsymbol{\omega}_\rho(\boldsymbol{x}) = \sum_I \Gamma_I \boldsymbol{\omega}_I \gamma_\rho(\boldsymbol{x} - \boldsymbol{x}_I) \tag{9.27}$$

where $\gamma_\rho(\boldsymbol{x}) = \rho^{-d}\gamma(\boldsymbol{x}/\rho)$ is the smoothing kernel. It may be noted that the idea of the vortex blob method is very similar to that of SPH, or RKPM. When using the vortex blob method, the velocity field in 2D may be written as

$$\frac{d\mathbf{x}_I}{dt} = -\frac{1}{2\pi} \sum_J \frac{(\boldsymbol{x}_I - \boldsymbol{x}_J) \times \mathbf{e}_z \Gamma_J g(|\boldsymbol{x}_I - \boldsymbol{x}_J|/\rho_J)}{|\boldsymbol{x}_I - \boldsymbol{x}_J|^2} \tag{9.28}$$

where $G(y) = 2\pi \int_0^y \gamma(z)z dz$.

The vortex method was first used in computations of incompressible and inviscid flow, e.g.[11,34] Later, it was applied to solve viscous flow problems,[111,126,165] and show that the method has the ability to provide accurate simulation of complex high Reynolds number flows.[62,277,453] Two versions of vortex methods were used in early implementation: Chorin's random walk[111,112] and Leonard's core spreading technique.[260,261] Today, most people use the following re-sampling scheme:

$$\frac{d\boldsymbol{x}_I}{dt} = \sum_J V_J K_\varrho(\boldsymbol{x}_I - \boldsymbol{x}_J) \times \boldsymbol{\omega}_J \tag{9.29}$$

$$\frac{d\boldsymbol{\omega}_I}{dt} = \left[\sum_J V_J \nabla K_\rho(\boldsymbol{x}_I - \boldsymbol{x}_J) \times \boldsymbol{\omega}_J \right]$$

$$+ \nu\rho^{-2} \sum_J V_J [\boldsymbol{\omega}_J - \boldsymbol{\omega}_I] \gamma_\rho(|\boldsymbol{x}_I - \boldsymbol{x}_J|) \tag{9.30}$$

9.5 Material Point Method (Particle-in-cell Method)

Like the vortex-in-cell approach, the particle-in-cell method is a dual description (Lagrangian and Eulerian) method. The main idea is to trace the
motions of a set of material points, which carry the information of all the
state variables, in a Lagrangian manner; whereas the spatial discretization,
hence the displacement interpolation, is made with respect to spatial coordinate detached from the material body as an Eulerian description. At the
beginning of each time step, one may first find the velocities and accelerations
at each spatial nodal point based on the information of surrounding material
points. In the same manner, internal and external forces on a specific spatial
nodal point at each time step are calculated by summing up the contribution
from the surrounding material points. The method was first used in computational fluid dynamics by Brackbill.[72, 76–78, 88] It was reformulated by Sulsky
and co-workers for solid mechanics applications. Some very good illustrations
such as the Taylor bar impact problem and ring collision problem have been
shown by Sulsky et al.[28, 418, 419]

In the particle-in-cell method, the total mass or total volume of the continuum is divided among N particles

$$\rho(\boldsymbol{x}, t) = \sum_I M_I \delta(\boldsymbol{x} - \mathbf{X}_I(t)) \tag{9.31}$$

Consider a weak formulation of the momentum equation

$$\int_\Omega \rho \mathbf{w} \cdot \mathbf{a} d\Omega = - \int_\Omega \rho \boldsymbol{\sigma} : \nabla \mathbf{w} d\Omega + \int_{\partial \Gamma_t} \mathbf{w} \cdot \mathbf{t} dS + \int_\Omega \rho \mathbf{w} \cdot \mathbf{b} d\Omega \tag{9.32}$$

Substituting (9.31) into (9.32), a Lagrangian type of discretization can be
achieved

$$\sum_{I=1}^{N_p} M_I \mathbf{w}(\mathbf{X}_I(t), t) \cdot \mathbf{a}(\mathbf{X}_I(t), t) = - \sum_{I=1}^{N_p} M_I \boldsymbol{\sigma}(\mathbf{X}_I(t), t) : \nabla \mathbf{w}(\boldsymbol{x}, t) \Big|_{\boldsymbol{x} = \mathbf{X}_I(t))}$$

$$+ \int_{\Gamma_t} \mathbf{w} \cdot \mathbf{t} dS + \sum_{I=1}^{N_p} M_I \mathbf{w}(\mathbf{X}_I(t), t) \cdot \mathbf{b}(\mathbf{X}_I(t), t) \tag{9.33}$$

Since the kinematic variables are discretized in an Eulerian grid, the accelerations are governed by the discrete equation of motion at spatial nodal
points,

$$\sum_{j=1}^{N_n} m_{ij} \mathbf{a}_j = \mathbf{f}_i^{int} + \mathbf{f}_i^{ext} \tag{9.34}$$

The exchange of information between the particles and spatial nodal points is
described in.[418] The main advantage of the particle-in-cell method is to avoid

using a Lagrangian mesh and to automatically track material boundaries. Recent applications of the particle-in-cell method are plasma physics (such as magneto-hydrodynamics, Maxwell-Lorentz equations), astrophysics, and shallow-water/free-surface flow simulations.[79, 127, 201, 345, 346]

9.6 Lattice Boltzmann Method

There have been several excellent reviews on Lattice Boltzmann method (LBM).[109, 383, 415] The discussion presented here is intended to put the method in comparison with its "peers", and look at it from a different perspective. The ancestor of LBM is the Lattice Gas Cellular Automaton (LGCA) method, which is also regarded as a special case of molecular dynamics.[390] LBM is designed to improve its statistical "resolution".

Currently, LBM is a very active research front in computational fluid dynamics because of its easy implementation and parallelization. The LBM technology has been used in simulations of low Mach number combustion,[163] multiphase flow and Rayleigh-Taylor instability,[198] flow past a cylinder,[319] flow through porous media,[409] turbulent flow, and thermal flow. One may also find some related references in[57, 317, 320, 473] and a convergence study of LBM in[391]

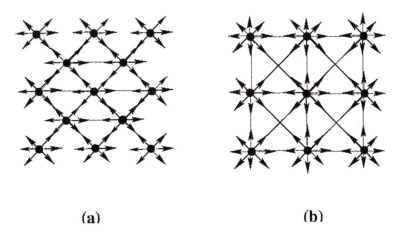

(a)   **(b)**

Fig. 9.5. Lattice and velocity directions: (a) triangular lattice; (b) square lattice.

The basic equation, or the kinetic equation, of the lattice Boltzmann method is

$$f_i(\boldsymbol{x} + \mathbf{e}_i \Delta \boldsymbol{x}, t + \Delta t) - f_i(\boldsymbol{x}, t) = \Omega_i\Big(f(\boldsymbol{x}, t)\Big), \qquad i = 0, 1, 2, \cdots, M \quad (9.35)$$

where f_i is the particle velocity distribution function along the i-th direction, and Ω_i is the collision operator that represents the rate change of f_i during the collision.

Note that in the lattice Boltzmann method, for a particle at a given node, there are only a finite number of velocity directions ($\mathbf{e}_i$ $i = 0, 1, \cdots, M$) that the particle can have. Fig. 9.5 illustrates examples of plane lattice, and the discrete velocity paths. Fig. 9.6 shows a 3D lattice with the associated discrete velocity set. Viewing Eq. (9.35) as a discrete meso-scale model, one can average (sum) the particle distribution over the discrete velocity space to obtain the macro-scale particle density at nodal position i,

$$\rho = \sum_{i=1}^{M} f_i . \tag{9.36}$$

The particle velocity momentum at macro-scale can also be obtained by averaging the meso-scale variables

$$\rho \mathbf{u} = \sum_{i=1}^{M} f_i \mathbf{e}_i . \tag{9.37}$$

Unlike most of the other particle methods, the lattice Boltzmann method is a mesh based method. In the LBM, the spatial space is discretized in a way that it is consistent with the kinetic equation, i.e. the coordinates of the nearest points around $\boldsymbol{x}$ are $\boldsymbol{x} + \mathbf{e}_i$. Therefore, it requires not only grid, but also the grid has to be uniform. This actually causes problems at generally curved boundaries. Recently, efforts have been made to extend LBM to irregular grids,[236, 410] and specific techniques are developed to enforce boundary conditions.[317, 321] During a simulation, a particle moves from one lattice node to another. Most likely, there is a probability that the next node is also occupied by other particles. The non-zero density of particle distribution at that point indicates the possibility of collision.

There are two approaches to choosing collision operator Ω_i. Using Chapman-Enskog expansion, or multi-scale singular perturbation,[170] one may find the following continuum form of the kinetic equation,

$$\frac{\partial f_i}{\partial t} + \mathbf{e}_i \cdot \nabla \cdot f_i + \epsilon \left(\frac{1}{2} \mathbf{e}_i \mathbf{e}_i : \nabla \nabla f_i \mathbf{e}_i \nabla \frac{\partial f}{\partial t} + \frac{1}{2} \frac{\partial^2 f_i}{\partial t^2} \right) = \frac{\Omega_i}{\epsilon} \tag{9.38}$$

is consistent with the discrete kinetic equation (9.35) up to the second order of ϵ — a small number proportional to the Knudsen number. By choosing a proper collision operator, for instance using the lattice BGK theory (after Bhatnagar, Gross and Krook in continuum kinetic theory[70]),

$$\frac{\Omega_i}{\epsilon} = -\frac{\delta_{ij}}{\epsilon \tau}(f_j - f_j^{eq}) \tag{9.39}$$

Eq. (9.38) may recover Navier-Stokes hydrodynamics equations, provided the equilibrium state of particle density is well defined, e.g. that of Qian et al.,[381]

$$f_i^{eq} = \rho w_i \left(1 + 3\mathbf{e}_i \cdot \mathbf{u} + \frac{9}{2}(\mathbf{e}_i \cdot \mathbf{u})^2 - \frac{3}{2}u^2\right) \tag{9.40}$$

The alternative is to consider Eq. (9.35) as the discrete version of the continuum Boltzmann equation, and one may derive the discrete collision operator by discretizing the Maxwell-Boltzmann equilibrium distribution.[109,415] The resulting difference equations may reproduce Navier-Stokes hydrodynamic equations in the limit of small Knudsen number, i.e. particle mean-free path much smaller than typical macroscopic variation scales.[57]

In principle, Lattice Boltzmann method is a bona fide computational meso-mechanics paradigm. It has both "micro-mechanics" part, the statistical movement of the molecules—Boltzmann equation, and the "homogenization" part, the assemble or averaging in the phase (velocity) space. In fact, to extend the Boltzmann Lattice method to irregular lattice, or quasi-lattice structure is the current research topic. In 1997, Succi[415] wrote:

"Most of the excitement behind LGCA was driven by the 'Grand-dream':

LGCA : Turbulence = Ising Model : Phase Transitions.

Ten years later, all reasonable indications are that the "Grand-dream' has turned into a 'Grand-illusion' (but, who knows the future ?).

LBE was born on a much less ambitious footing: just provide a useful tool to investigate fluid dynamics and, maybe mesoscopic phenomena, on parallel machines. And in that respect, it appears hard to deny that, even though much remains to be done, the method has indeed lived up to the initial expectations. ... "

This assessment has been both accurate and modest, considering the recent development of LBM.

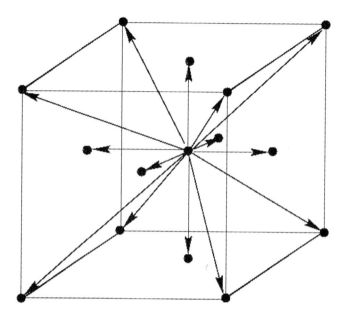

Fig. 9.6. Cubic Lattice with 15 molecular speeds (D3Q15).

References

1. Abraham, F.F. (1996) "Dynamics of brittle fracture with variable elasticity," *Physical Review Letters*, Vol. 77, pp. 869-872.
2. Abraham, F.F., Brodbeck, D., Rudge, W.E. and Xu, X. (1997) "A molecular dynamics investigation of rapid fracture mechanics," *Journal of the Mechanics and Physics of Solids*, Vol. 45, pp. 1595-1619.
3. Abraham, F.F., Brodbeck, D., Rudge, W.E., Broughton, J. Q., Schneider, D., Land, B., Lifka, D., Gerber, J., Rosenkrantz, M., Skovira, J. and Gao, H. (1998) "*Ab initio* Dynamics of Rapid Fracture," *Modeling and Simulation in Materials Science and Engineering*, Vol. 6, pp. 639-670.
4. Abraham, F. F., Broughton, J., Bernstein, N. and Kaxiras, E. (1998) "Spanning the continuum to quantum length scales in a dynamic simulation of brittle fracture," *Europhysics Letters*, Vol. 44, pp. 783-787.
5. Abraham, F. F., Broughton, J. Q., Bernstein, N. and E Kaxiras, E. (1998) "Spanning the length scales in dynamic simulation," *Computers in Physics*, Vol. 12, pp. 538-546.
6. Acioli, P.H. (1997) "Review of quantum Monte Carlo methods and their applications," *Journal of Molecular Structure (Theochem)*, Vol. 394, pp. 75-85.
7. Adelman, S.A. and Doll, J.D. (1976) "Generalized Langevin equation approach for atom/solid-surface scattering: General formulation for classical scattering off harmonic solids," *Journal of Chemical Physics*, Vol. 64, pp. 2375-2388.
8. Allen, M. P., Tildesley, D.J. (1987) *Computer Simulation of Liquids*, Oxford University Press, Oxford, UK.
9. Allen, M. P., Tildesley, D.J. (1993) *Computer Simulation of Chemical Physics*, Kluwer Academic Publishers, Dordrecht.
10. Aluru, N.R. (2000) "A point collocation method based on reproducing kernel approximations," *International Journal for Numerical Methods in Engineering*, Vol. 47, pp. 1083-1121.
11. Anderson, C., and Greengard, C. (1985) "On Vortex Methods," *SIAM Numer. Anal.*, Vol. 22, pp. 413-440.
12. Anderson, P.W. (1968) "Self-consistent pseudo-potentials and ultralocalized functions for energy bands," *Physics Review Letter*, Vol. 21, pp. 13-.
13. Anderson, P.W. (1969) "Localized orbitals for molecular quantum theory. I The Hückel theory," *Physics Review*, Vol. 181, pp. 25-.
14. Argyris, J. H., Fried, I. and Scharpf, D. W. [1968], "The TUBA family of plate elements for the matrix displacement method," *The Aeronautical of the Royal Aeronautical Society*, **72**, pp. 701-709.
15. Arroyo, N. and Belytschko, T. (2002) "An atomistic-based finite deformation membrane for single layer crystalline films," *Journal of Mechanics and Physics of Solids*, Vol. 50, pp. 1941-1977.

16. Atluri, S. N., and Zhu, T. L. (1998) "A new meshless local Petrov-Galerkin (MLPG) approach to nonlinear problems in computer modeling and simulation," *Computer Modeling and Simulation in Engineering*, Vol. 3, pp. 187-196.

17. Atluri, S. N., Kim, H.G., and Cho, J.Y. (1998) "A critical assessment of the truly meshless local Petrov-Galerkin (MLPG) and local boundary integral equation (LBIE) methods," *Computational Mechanics*, Vol. 24, pp. 348-372.

18. Atluri, S. N., Cho, J.Y. and Kim, H.G. (1998) "Analysis of thin beams, using the meshless local Petrov-Galerkin method, with generalized moving least square interpolations," *Computational mechanics*, Vol. 24, pp. 334-347.

19. Atluri, S. N., and Zhu, T. L. (2000) "The meshless local Petrov-Galerkin (MLPG) approach for solving problems in elasto-statics," *Computational Mechanics*, Vol. 25 , pp. 169-179.

20. Atluri, S.N., Sladek, J., Sladek, V. S. N., Zhu, T. L. (2000) "The local boundary integral equation (LBIE) and it's meshless implementation for linear elasticity," *Computational Mechanics*, Vol. 25, pp. 180-198.

21. Atluri, S. N., and Zhu, T. L. (2000) "New concepts in meshless methods," *International Journal for Numerical methods in Engineering*, Vol. 47, pp. 537-556.

22. Babuška, I., Oden, J. T., and Lee, J.K. (1977) "Mixed-hybrid finite element approximations of second-order elliptic boundary-value problems," *Computer Methods in Applied Mechanics and Engineering*, Vol. 11, pp. 175-206.

23. Babuška, I., and Melenk, J.M. (1997) "The partition of unity method," *International Journal for Numerical Methods in Engineering*, Vol. 40, pp. 727-758.

24. Babuška, I., and Zhang, Z. (1998) "The partition of unity method for the elastically supported beam," *Computer Methods in Applied Mechanics and Engineering*, Vol. 152, pp. 1-18.

25. Babuška, I., and Banerjee, U., and Osborn, J. E. (2003) "Survey of meshless and generalized finite element methods: A unified approach," *Acta Numerica*, Vol. 12, pp. 1-125.

26. Baer, R. (2000) "Ab-initio molecular deformation barriers using auxiliary-field quantum Monte Carlo with application to the inversion barrier of water," *Chemical Physics Letters*, Vol. 324, pp. 101-107.

27. Bar-Lev, M. and Yang H. T. (1975), "Initial flow field over an impilsively started circular cylinder," *Journal of Fluid Mechanics*, Vol. 72, pp. 625-

28. Bardenhagen, S. G., Brackbill, J.U., and Sulsky, D. (2000) "The Material-point Method for Granular Materials," *Computer Methods in Applied Mechanics and Engineering*, Vol. 187, pp. 529-541.

29. Barnhill, R. E., Birkhoff, G. and Gordon, W. J. (1973), "Smooth interpolation in triangles," *Journal of Approximation Theory*, Vol. 8, pp. 114-128.

30. Baskes, M., Daw, M., Dodson, B., Foiles, S. (1988) "Atomic-scale simulation in materials science," *MRS BULLETIN*, Vol. 13, pp. 28-35.

31. Batra, R. C. and N. V. Nechitailo, N. V. (1997) "Analysis of failure modes in impulsively loaded pre-notched steel plates," *International Journal of Plasticity*, Vol. 13, pp. 291-308.

32. Batra, R. C. and Stevens, J. B. [1998] "Adiabatic shear bands in axisymmetric impact and penetration problems," *Computer Methods in Applied Mechanics and Engineering*, Vol. 151, pp. 325-342.

33. Bazeley, G. P., Cheung, Y. K., Irons, B. M., and Zienkiewicz, O. C. (1965), "Triangle elements in bending: conforming and nonconforming solutions," *Proc. 1st Conference on matrix methods in structural mechanics*, Wright-Patterson, AFB, Ohio .

34. Beale, J. T. (1986) "A Convergent 3-D Vortex Method with Grid-Free Stretching," *Mathematics of Computation*, Vol. 46, pp. 401-424.

35. Bell, K. (1969), "A refined triangular plate bending finite element," *International Journal for Numerical Methods in Engineering*, **1**, pp. 101-122.

36.] Beissel, S., Belytschko, T. (1996) "Nodal integration of the element-free Galerkin method", *Computer Methods in Applied Mechanics and Engineering* Vol. 139, pp. 49-74.

37. Belytschko, T., Fish, J. and A. Bayliss, A. (1990) "The spectral overlay on finite elements for problems with high gradients," *Computer Methods in Applied Mechanics and Engineering*, Vol. 81, pp. 71-89.

38. Belytschko, T. and Tabbara, M. (1993) "H-adaptive finite element methods for dynamic problems with emphasis on localization," *International Journal for Numerical Methods in Engineering*, Vol. 36, pp. 4245-4265.

39. Belytschko, T., Chiang, H.-Y. and Plaskacz, E. (1994) "High resolution two-dimensional shear band computations: imperfections and mesh dependence," *Computer Methods in Applied Mechanics and Engineering*, Vol. 119, pp. 1-15.

40. Belytschko, T., Lu, Y.Y. and Gu, L. (1994) "Element free Galerkin methods," *International Journal for Numerical Methods in Engineering*, Vol. 37, pp. 229-256.

41. Belytschko, T., Lu, Y.Y. and Gu, L. (1994) "Fracture and crack growth by Element-free Galerkin methods, *Model. Simul. Sci. Comput. Engrg.*, Vol. 2, pp. 519-534.

42. Belytschko, T., Lu, Y.Y. and Gu, L. (1995) "Element-free Galerkin methods for static and dynamic fracture," *International Journal of Solids and Structures*, Vol. 32, pp. 2547-2570.

43. Belytschko, T., Lu, Y.Y. and Gu, L. (1995) "Crack propagation by Element-free Galerkin Methods," *Engineering Fracture Mechanics*, Vol. 51, pp. 295-315.

44. Belytschko, T., Krongauz, Y., Organ, D., Flaming, M. and Krysl, P. (1996) "Meshless methods: An overview and recent developments", *Computer Methods in Applied Mechanics and Engineering* Vol. 139, pp. 3-48.

45. Belytschko, T., Krongauz, Y., Fleming, M., Organ, D., and Liu, W.K. (1996) "Smoothing and accelerated computations in the element free Galerkin method," *Journal of Computational and Applied Mathematics*, Vol. 74, pp. 111-126.

46. Belytschko, T. and Tabbara, M. (1997) "Dynamic fracture using element-free Galerkin methods," *Journal of Computational and Applied Mathematics*, Vol. 39, pp. 923-938.

47. Belytschko, T., Krongauz, Dolbow, J., and Gerlach, C (1998) "On the completeness of meshfree particle methods," *International Journal for Numerical Methods in Engineering*, Vol. 43, pp. 785-819

48. Belytschko, T., Guo, Y., Liu, W.K. and Xiao, P (2000) "A unified stability analysis of meshless particle methods," *International Journal for Numerical Methods in Engineering*, Vol. 48, pp. 1359-1400.

49. Belytschko, T., Liu, W. K., and Moran, B. (2000) *Nonlinear Finite Elements for Continua and Structures*, Wiley, England.

50. Belikov, V. V., Ivanov, V. D., Kontorivich, V. K. Korytnik, S. A., Semenov, A. Y. (1997) "The non-Sibsonian interpolation: a new method of interpolation of the values of a function on an arbitrary set of points," *Computational Mathematics and Mathematical Physics*, Vol. 37, pp. 9-15.

51. Belosludov, R.V. and M, Sluiter, M., Li, Z.Q., and Kawazoe, Y. (1999) "*Ab initio* and lattice dynamics studies of the vibrational and geometrical properties of the molecular complex of hydroquinone and C_{60}," *Chemical Physics Letters*, Vol. 312, pp. 299-305.

52. Belytschko, T., Liu, W.K., Moran, B. (2000) "Nonlinear Finite Elements for Continua and Structures," John Wiley and Sons, New York.

53. Benson, B. J. and Hallquist, J. O. (1990), "A single surface contact algorithm for the post-buckling analysis of structures," *Computer Methods in Applied Mechanics and Engineering*, Vol. 78, pp. 141-163.

54. Benson, B. J. (1997) "A mixture theory for contact in multi-material Eulerian formulations," *Computer Methods in Applied Mechanics and Engineering*, Vol. 140, pp. 59-86.

55. Benz, W (1990) "Smooth particle hydrodynamics: a review," In *Numerical Modeling of Non-linear Stellar Pulsation: Problems and Prospects*, Kluwer Academic, Boston.

56. Benz, W. and Asphaug, E. (1995) "Simulations of brittle solids using smooth particle hydrodynamics," *Computer Physics Communications*, Vol. 87, pp. 253-265

57. Benzi, R. and Succi, S. and Vergassola, M. (1992) "The lattice Boltzmann equation: Theory and applications," *Phys. Rep.*, Vol. 222, pp. 145-

58. Berczik, P., and Kolesnik, I.G. (1993) "Smoothed particle hydrodynamics and its application to astrophysical problems," *Kinematics and Physics of Celestial Bodies*, Vol. 9, pp. 1-11.

59. Berczik, P., and Kolesnik, I.G. (1998) "Gasdynamical model of the triaxial protogalaxy collapse," *Astronomical and Astrophysical Transactions*, Vol. 16, pp. 163-185.

60. Berczik, P. (2000) "Modeling the Star Formation in Galazies Using the Chemodynamical SPH code," *Astronomy and Astrophysics*, Vol. 360, pp. 76-84.

61. Berendsen, H. J. C., Postma, J. P. M., van Gunsteren, W. F., DiNola, A., Haak, J. R. (1984) "Molecular dynamics with coupling to an external bath," *Journal of Chemical Physics*, Vol. 81, 3684-3690.

62. Bernard, P.S. (1995) "A Deterministic vortex sheet method for boundary layer flow," *Journal of Computational Physics*, Vol. 117, pp. 132-145.

63. Binder, K. (1988) "The Monte Carlo Method in Condensed Matter Physics," Springer, Berlin,Heidelberg.

64. Binder, K. (1992) "The Monte Carlo Simulation in Statistical Physics," Springer, Berlin,Heidelberg.

65. Bird, G. A. (1994) "Molecular Gas Dynamics and the Direct Simulation of Gas Flow," Oxford University Press, Oxford, U.K.

66. Birkhoff, G. and de Boor, C. (1965), "Piecewise polynomial interpolation and approximation," in *Approximation of functions*, edited by H. L. Garabedian, Elsevier, New York.

67. Birkhoff, G., Schultz, G. M. H. and Varga, R. S. (1968), "Piecewise Hermite interpolation in one and two variables with applications to partial differential equations," *Numer. Math*, **11**, pp. 232-256.

68. Biswasa, R. and Hamann, D. R. (1985) "Interatomic potential for silicon structural energies," *Phys. Rev. B*, Vol. 55, pp. 2001-2004.

69. Beylkin, G., Coifman, R., and V. Rokhlin, V. (1991) "Fast Wavelet Transforms and Numerical Algorithm I," *Communications on Pure and Applied Mathematics*, Vol. XLIV, pp. 141-183.

70. Bhatnagar, P. and Gross, EP and Krook, MK (1954) "A model for collision processes in gases. I. Small amplitude processes in charged and neutral one-component systems," *Phys. Rev.*, Vol. 94, pp. 511-

71. Bloch, F. (1928) "Über die quantenmechanik der elektronen in kristallgittern," *Zeitschrift für Physik*, Vol. 52, pp. 555-600.

72. Brackbill, J.U., and Ruppel, H.M. (1986) "FLIP: A method for adaptively zoned, particle-in-cell calculations in two dimensions," *Journal of Computational Physics*, Vol. 65, pp. 314-343.

73. Bonet, J. and Lok, T.S. (1999) "Variational and momentum preservation aspects of smooth particle hydrodynamic formulation," *Computer methods in Applied Mechanics and Engineering*, Vol. 180, pp. 97-115.

74. Bonet, J. and Kulasegaram, S. (2000) "Correction and stabilization of smooth particle hydrodynamic methods with applications in metal forming simulations," *International Journal for Numerical Methods in Engineering*, Vol. 47, pp. 1189-1214.

75. Born, M. and Oppenheimer, J.R. (1927) "Zur Quantentheorie," *Ann. Physik*, Vol. 84, pp. 457-

76. Brackbill, J. U. (1987) "On modeling angular momentum and velocity in compressible fluid flow," *Comput. Phys. Commun.*, Vol. 47, pp. 1-

77. Brackbill, J. U. and Kothe, D. B. and Ruppel, H.M. (1988) "FLIP: A low-dissipation, particle-in-cell method for fluid flow," *Comput. Phys. Commun.*, Vol. 48, pp. 25-38.

78. Brackbill, J. U. (1988) "The ringing instability in particle-in-cell calculation of low speed flow," *Journal of Computational Physics*, Vol. 75, pages 469

79. Brackbill, J.U. (1991) "FLIP-MHD: A particle-in-cell method of magnetohydrodynamics," *Journal of Computational Physics*, Vol. 96, pages 163-192.

80. Breitkopf, P., Touzot, G., and Villon, P. (1998) "Consistency approach and diffuse derivation in element free methods based on moving least squares approximation," *Comp. Assist. Mech. Eng. Sci.*, Vol. 5, pp. 479-501

81. Breitkopf, P., Touzot, G., Villon, P.(2000) "Double grid diffuse collocation method," *Computational Mechanics*, Vol. 25, No. 2/3, pages 180-198

82. Brenner, D.W. (1990) "Empirical potential for hydrocarbons for use in simulating chemical vapor deposition of diamond films," *Physical Review B*, Vol. 42, pp. 9458-9471.

83. Brenner, D.W., Shenderova, O. A., Harrison, J. A., Stuart, S. J., Ni, B., Sinnott, S. B. (2002) "A second-generation reactive empirical bond order (REBO) potential energy expression for hydrocarbons," *Journal of Physics: Condensed Matter*, Vol. 14, pp. 783-802.

84. Brenner, S. C. and Scott, L. R. (1994) *The Mathematical Theory of Finite Element Methods*, Springer, New York.

85. Breitkopf, P., Rassineux, Touzot, G., and Villon, P.(2000) "Explicit Form and Efficient Computation of MLS Shape Function and Their Derivatives," *International Journal for Numerical Methods in Engineering*, Vol. 48, pp. 451-466.

86. Broughton, J. Q., Abraham, F. F., Bernstein, N. and Kaxiras, E. (1999) "Concurrent coupling of length scales: Methodology and applications," *Physics Review* B, Vol. 60, pp, 2391–2403.

87. Bueche, D., Sukumar, N., Moran, B. (2000) "Dispersive properties of the natural element method," *Computational Mechanics*, Vol. 25, pp. 207-219

88. Burgess, D. and Sulsky, D. and Brackbill, J. U. (1992) "Mass matrix formulation of the FLIP particle-in-cell method," *Journal of Computational Physics*, Vol. 103, pp. 1-15.

89. Cai, W. and DeKoning, M. and Bulatov, V. V. and Yip, S. (2000), "Minimizing boundary reflections in coupled-domain simulations," *Physical Review Letter*, Vol. 85, pp. 3213–3216.

90. Car, R. and Parrinello, M. (1985) "Unified approach for molecular dynamics and density-functional theory," *Physical Review Letters*, Vol. 55, pp. 2471-2474.

91. Catlow, C.R.A., Parker, S.C., Allen, M.P. (1990) *Computer Modelling of Fluids Polymers and Solids*, Kluwer Academic Publishers, Dordrecht.

92. Chen, J. K., Beraun, J. E., Jih, C.J. (1999) "Completeness of corrective smoothed particle method for linear elastodynamics," *Computational Mechanics*, Vol. 24, pp. 273-285.

93. Chen, J. K., Beraun, J. E., Jih, C.J. (1999) "An improvement for tensile instability in smoothed particle hydrodynamics," *Computational Mechanics*, Vol. 23, pp. 279-287

94. Chen, J. K., Beraun, J. E., Carney, T. C. (1999) "A corrective smoothed particle method for boundary value problems in heat condution," *Int. J. Numer. Methods Engng.*, Vol. 46, pp. 231-252

95. Chen, J. K., Beraun, J. E. (2000) "A generalized smmothed particle hydrodynamics method for nonlinear dynamic problems," *Computer Methods in Applied Mechanics and Engineering*, Vol. 190, pp. 225-239

96. Chen, J. S. Pan, C., Wu, C.-T., and Liu, W. K. (1996) "Reproducing kernel particle methods for large deformation analysis of non-linear structures," *Computer Methods in Applied Mechanics and Engineering* Vol. 139, pp.195-228

97. Chen, J.S., Pan, C., and Wu, C.T. (1997) "Large deformation analysis of rubber based on a reproducing kernel particle method," *Computational Mechanics*, Vol. 19, pp. 153-168

98. Chen, J.S., Pan, C., Wu, C.T., and Roque, C. (1997) "A Lagrangian reproducing kernel particle method for metal forming analysis," *Computational Mechanics*, Vol. 21, pp. 289-307.

99. Chen, J.S., Pan, C., and Wu, C.T. (1998) "Application of reproducing kernel particle methods to large deformation and contact analysis of elastomers," *Rubber Chemistry and Technology*, Vol. 7, pp. 191-213.

100. Chen, J.S., Roque, C., Pan, C., and Button, S.T. (1998) "Analysis of metal forming process based on meshless method," *Journal of Materials Processing Technology*, Vol. 80-81, pp. 642-646.

101. Chen, J. S. Wang, H. P., Yoon, S., You, Y. (2000) "Some recent improvements in meshfree methods for incompressible finite elasticity boundary value problems with contact," *Computational Mechanics*, Vol. 25, pp. 137-156.

102. Chen, J.S., Wu, C.T. and Belytschko, T. (2000) "Regularization of material instabilities by meshfree approximations with intrinsic length scales," *International Journal for Numerical Methods in Engineering*, Vol. 47, pp. 1303-1322

103. Chen, J.S., Wu, C.T., Yoon, S. and You, Y. (2001) "A stabilized conforming nodal integration for Galerkin meshfree methods," *International Journal for Numerical Methods in Engineering*, Vol. 50, pp. 435-466.

104. Chen, J.S., and Wang, H.P. (2000) "New boundary condition treatments in meshfree computation of contact problems," *Computer Methods in Applied Mechanics and Engineering*, Vol. 187, pp. 441-468.

105. Chen, J.S., Yoon, S., and Wu, C.T. (2002) "Nonlinear version of stabilized conforming nodal integration for Galerkin meshfree methods," *International Journal for Numerical Methods in Engineering*, Vol. 12, pp. 2587-2615.

106. Chen, J.S., Han, W., You, Y., and Meng, X. (2003) "A reproducing kernel method with nodal interpolation property," *International Journal for Numerical Methods in Engineering*, Vol. 56, pp. 935-960.

107. Chen, S., Wang, Z., Shan, X.W., Doolen, G.D. (1992) "Lattice Boltzmann computational fluid dynamics in three dimensions," *J. Stat. Phys.*, vol. 68, pp. 379-400.

108. Chen, S., Martinez, D., Mei, R. (1996), "On boundary conditions in lattice Boltzmann methods," *Physics Fluids*, Vol.8, pp. 2527-2536.

109. Chen, S., Doolen, G.D. (1998) "Lattice Boltzmann method for fluid flows," *Annual Review of Fluid Mechanics*, Vol. 30, pp. 329-364.

110. Ciarlet, P. G. (1978) *The Finite Element Method for Elliptic Problems*, North Holland, Amsterdam.

111. Chorin, A.J. (1973), "Numerical study of slightly viscous flow," *Journal of Fluid Mechanics*, Vol. 57, pp. 785-796.

112. Chorin, A. J. (1973), "Discretization of a Vortex Sheet, with an example of roll-up," *Journal of Computational Physics*, Vol. 13, pp. 423-429.
113. Chorin, A. J. (1978) "Vortex sheet approximation of boundary layers," *Journal of Computational Physics*, Vol. 27, pp. 428-442.
114. Chui, C.K. (1992) *An Introduction to Wavelets*, Academic Press, Inc., Boston.
115. Christon, M. A., Roach, D. W. (2000) "The numerical performance of wavelets for PDEs: the multi-scale finite element," *Computational Mechanics*, Vol. 25, pp. 230-244.
116. Ciccotti, G., Hoover, W.G. (1986) *Molecular Dynamics Simulation of Statistical Mechanical Systems*, North-Holland, Amsterdam.
117. Ciccotti, G., Frenkel, D., McDonald, I.R. (1987) *Simulation of Liquids and Solids. Molecular Dynamics and Monte Carlo Methods in Statistical Mechanics*, North-Holland, Amsterdam.
118. Clementi, E. (1988) "Global scientific and engineering simulations on scalar, vector and parallel LCAP-type supercomputer," *Philosophical Transactions of the Royal Society of London, A*, Vol. 326, pp. 445-470.
119. Clementi, E., Chin, S., Corongiu, G., Detrich, J.H., Dupuis, M., Folsom, D., Lie, G.C., Logan, D., and Sonnad, V. (1989) "Supercomputing and supercomputers for science and engineering in general and for chemistry and biosciences in particular," *International Journal of Quantum Chemistry*, Vol. 35, pp. 3-89.
120. Clementi, E. (2000) "*Ab initio* computations in atoms and molecules," *IBM Journal of Research and Development*, Vol. 44, pp. 228-245.
121. Clough, R. W. and Tocher, J. (1965), "Finite element stiffness matrices for the analysis of plate bending," *Proc. 1st Conference on matrix methods in structural mechanics*, Wright-Patterson, AFB, Ohio.
122. Coifman, R. and Meyer, Y. (1990) "Othogonormal Wavelet Packet Bases," preprint, Yale University, New Haven, CT.
123. Colling, W. M. and Dennis, S. C. R. (1973), "The initial flow past an impulsively started circular cylinder," *Q. J. Mech. Appl. Math.*, Vol. 26, 53
124. Colling, W. M. and Dennis, S. C. R. (1973), "Flow past an impulsively started circular cylinder," *Journal of Fluid Mechanics*, Vol. 60, 105
125. Cordes, L. W. and Moran, B. (1996) "Treatment of material discontinuity in the Element-free Galerkin method," *Computer Methods in Applied Mechanics and Engineering*, Vol. 139, pp. 76–90
126. Cotte, G.H. and Koumoutsakos, P. and Salihi, M.L.O. (2000) "Vortex Methods with Spatially Varying Cores," *Journal of Computatioal Physics*, Vol. 162, pages 164-185.
127. Cushman-Roisin, B., Esenkov, O.E., Mathias, B.J. (2000) "A particle-in-cell method for the solution of two-layer shallow-water equations," *International Journal for Numerical Methods in Fluids*, Vol. 32, pp. 515-543.
128. Curtin, W. A. and Miller, R. E. (2003) "Atomistic/continuum coupling in computational material science," *Model. Simulat. Mater. Sci. Engrg.* Vol. 11, R33-R68.
129. Daubechies, I. (1992) *Ten Lectures on Wavelets*, Society for Industrial and Applied Mathematics, Philadelphia, Pennsylvania
130. Daubechies, I. (1993) "Orthonormal Bases of Compactly Supported Wavelets II Variations on a Theme," *SIAM Journal on Mathematical Analysis*, Vol. 24, pp. 499-519.
131. Deymier, P. A., Vasseur, J. O. (2002) "Concurrent multiscale model of an atomic crystal coupled with elastic continua," *Phys. Rev.* B, Vol. 66, Art. No. 134106
132. Diestler, D. J. (2002) "Coarse-grained descriptions of multiple scale processes in solid systems," *Physics Review* B, Vol. 66, pages 184104-1–184104-7.

133. Duarte, C.A., Hamzeh, O.N., Liszka, T.J., Tworzydlo, W.W. (2001) "A generalized finite element method for the simulation of three dimensional dynamic crack propagation," *Computer Methods in Applied Mechanics and Engineering*, Vol. 190, pp. 2227-2262.

134. Daux, C., Moes, N., Dolboaw, J., Sukumar, N., and Belytschko, T. (2000) "Arbitrary branched and intersecting cracks with the extended finite element method," *International Journal for Numerical Methods*, Vol. 48, pp. 1741-1760

135. Danielson, K. T., Adely, M. D. (2000) "A meshless treatment of three-dimensional penetrator targets for parallel computation," *Computational Mechanics*, Vol. 25, No. 2/3, pages 257-266

136. De, S., and Bathe, K.J. (2000) "The method of finite spheres," *Computational Mechanics*, Vol. 25, pp. 329-345.

137. DeRose Jr., G. C., Diaz, A. R. (2000) "Solving three-dimensional layout optimization problems using fixed scale wavelets," *Computational Mechanics*, Vol. 25, No. 2/3, pages 274-285

138. Diamond, S. L. (2001) "Reaction Complexity of Flowing Human Blood," *Biophysical Journal*, Vol. 80, pages 1031-1032.

139. Doll, J. D. and Dion, D. R. (1976) "Generalized Langevin equation approach for atom/solid-surface scattering: Numerical techniques for Gaussian generalized Langevin dynamics," *Journal of Chemical Physics*, Vol. 65, pp. 3762–3766.

140. de Veubeke, Fraeijs, B. (1965), "Bending and stretching of plates – special models for upper and lower bounds," *Proc. 1st Conference on matrix methods in structural mechanics*, Wright-Patterson, AFB, Ohio.

141. de Veubeke, Fraeijs, B. (1968), "A Conforming finite element for plate bending," *International Journal of Solids and Structures*, **4**, 95-108.

142. Danielson, K., Hao, S., Liu, W.K., Uras, A. and Li, S. (2000) "Parallel computation of meshless methods for explicit dynamic analysis," *International Journal for Numerical Methods in Engineering*, Vol. 47, pp. 1323-1341.

143. Daw, M.S. and Baskes, M.I. (1984) "Embedded-atom method: Derivation and application to to impurities, surfaces, and other defects in solids," *Physical Review B*, Vol. 29, pages 6443-6453.

144. Daw, M. S., Foiles, S. M., Baskes, M. I. (1993) "The embedded-atom method: a review of theory and applications," *Mater. Sci. Rep.*, Vol. 9, pp. 251-310.

145. Dilts, G. (1999) "Moving-least-square-partilce hydrodynamics—I. consistency and stability," *Int. J. Numer. Meth. in Engng.* Vol. 44, pages 1115-1155

146. Dilts, G. (2000) "Moving least-square particle hydrodynamics II: conservation and boundaries," *International Journal for Numerical Methods in Engi neering*, Vol. 48, pp. 1503-1524.

147. Dirac, P.A.M. (1958) *The Principles of Quantum Mechanics*, Oxford University Press, London.

148. Delves, L. and Mohamed, J. (1985) *Computational Methods for Integral Equations*, Cambridge University Press, New York.

149. Dolbow, J. and Belytschko, T. (1999) "Volumetric locking in the element-free Galerkin Method," *International Journal for Numerical Methods in Engineering*, Vol. 46, pp. 925-942.

150. Dolbow, J. and Belytschko, T. (1999) "Numerical integration of the Galerkin weak form in meshfree methods," Computational Mechanics, Vol. 23, pp. 219-230.

151. Dolbow, J., Moës, N., Belytschko, T. (2000) "Discontinuous enrichment in finite elements with a partition of unity method," *Finite Elements in Analysis and Design*, Vol. 36, pp. 235-260

152. Drovetsky, B. Y., Chu, J.C., Mak, C.H. (1998) "Computer simulations of self-avoiding polymerized membranes," *Journal of Chemical Physics*, Vol. 108, pp. 6554-6557.

153. Duarte, C. A., Oden, J. T. (1996) "An h-p adaptive method using clouds," *Computer Methods in Applied Mechanics and Engineering* Vol. 139, pages 237-262

154. Duarte, C. A., Oden, J. T. (1996) "An hp adaptive method using clouds," *Computer Methods in Applied Mechanics and Engineering*, Vol. 139, pp. 237-262

155. Dyka, C. T., Ingel, R. P., (1995) "An approach for tension instability in smoothed particle hydrodynamics", *Computer and Structures* Vol. 57, pages 573-580

156. Dyka, C. T., Randles, P. W., Ingel, R. P. (1997) "Stress points for tensor instability in SPH," *Inter. J. Numer. Methods Engng.* Vol. 40, pages 2325-2341

157. E, W. and Huang, Z. Y. (2002) "A dynamic atomistic-continuum method for the simulation of crystalline materials," *Journal of Computational Physics*, Vol. 182, pp. 234–261.

158. Erdogan, F. and Sih, G. C. (1963) "On the crack extension in plates under plane loading and transverse shear," *Journal of Basic Engineering*, Vol. 85D, pages 519-527.

159. Falk, M.L. and Langer, J.S. (1998) "Dynamics of viscoplastic deformation in amorphous solids," *Physical Review E*, Vol. 57, pages 7192-7205.

160. Falk, M.L. (1999) "Molecular-dynamics study of ductile and brittle fracture in model noncrystalline solids," *Physical Review B (Condensed Matter)*, Vol. 60, pages 7062-7070.

161. Felippa, C. and Clough, R. W. [1970], "The finite element in solid mechanics," in *SIAM-AMS Proceedings*, **2**, 210-252, Amer. Math. Soc. Providence, RI

162. Feynman, R.D. and Hibbs, A.R. (1965) *Quantum mechanics and path integrals*, McGraw-Hill, New York.

163. Filippova, O. and Hänel,, D. (2000) "A Novel Lattice BGK Approach for Low Mach Number Combustion," *Journal of Computational Physics*, Vol. 158, pages 139-160.

164. Finnis, M. W. and Sinclair, J. E. (1984) "A simple empirical N-body potential for transition metals," *Philosophical Magazine*, Vol. 50, pp. 45-55.

165. Fishelov, D. (1990) "A New Vortex Scheme for Viscous Flows," *Journal of Computational Physics*, Voil. 86, pages 211-224.

166. Fleming, M., Chu, Y. A., Moran, B., Belytschko, T. (1997) "Enriched element-free Galerkin methods for crack tip fields," *International Journal for Numerical Methods in Engineering*, Vol. 40, pages 1483-1504.

167. Fock, V. (1930) "Näherungsmethode zur Lösung des quantenmechanischen Mehrkörperproblems," *Z. Physik*, Vol. 61, pages 126

168. Foulkes, W. M. C. and Haydock, R. (1989) *Physics Review* B, Vol. 39, pages 12520-.

169. Frisch, U., Hasslacher, B., Pomeau, Y. (1986) "Lattice gas cellular automata for the Navier-Stokes equations," *Physics Review Letter*, Vol. 56, pp. 1505

170. Frisch, U., d'Humiéres, D., Lallemand, P., Pomeau, Y., Rivet, J.P. (1987) *Lattice gas hydrodynamics in two and three dimensions*, *Complex Systems*, Vol. 1, pp. 649-707.

171. Fulk, D. A. (1994) "A numerical analysis of smooth particle hydrodynamics," Ph.D. Thesis, Air Force Institute of Technology.

172. Y.C. Fung, Y. C. (Ed.) (1996) *Biomechanics: Circulation*, Springer Verlag

173. Furukawa, T., Yang, C. Yagawa, G. and Wu, CC (2000) "Quadrilateral approaches for accurate free mesh method," *International Journal for Numerical Methods in Engineering*, Vol. 47, pp. 1445-1462.

174. Gallagher, R. H., Cellatly, R. A., Padlog, J. and Mallett, R. H. (1967), "A discrete element procedure for thin-shell instability analysis," *AIAA Journal*, **5**, 138-145.

175. Garcia-Senz, D., Bravo, E., and Woosley, S.E. (1999) "Single and multiple detonations in white dwarfs," *Astronomy and Astrophysics*, Vol. 349, pp. 177-188.

176. Gingold R. A., Monaghan J. J. (1977) "Smoothed particle hydrodynamics: theory and application to non-spherical stars", *Mon. Not. Roy. Astr. Soc.* Vol. 181, pp. 375-389.

177. Gingold, R. A. and Monaghan, J. J. (1982) "Kernel estimates as a basis for general particle methods in hydrodynamics," *Journal of Computational Physics*, Vol. 46, pp. 429-453.

178. Gioia, G. and Ortiz, M. (1996) "The two-dimensional structure of dynamic boundary layers and shear bands in thermoviscoplastic solids," *Journal of Mechanics and Physics of Solids*, Vol. 44, pages 251-291.

179. Giovanola, J. (1988) "Adiabatic shear banding under pure shear loading. 1. Direct observation of strain localization and energy measurements," *Mechanics of Materials*, Vol. 7, pp. 59-71.

180. Giovanola, J. (1988) "Adiabatic shear banding under pure shear loading. 2. Fractographic and metallographic observations," *Mechanics of Materials*, Vol. 7, pp. 72-87.

181. Given, J.A. and Clementi, E. (1989) "Molecular dynamics and Rayleigh-Benard convection," *Journal of Chemical Physics*, Vol. 90, pp. 7376-7383.

182. Gosz, J., and Liu, W.K. (1996) "Admissible approximations for essential boundary conditions in the reproducing kernel particle method," *Computational Mechanics*, Vol. 19, pp. 120-135.

183. Gross, W.J., Vasileska, D., Ferry, D.K. (1999) "A novel approach for introducing the electron-electron and electron-impurity interactions in particle-based simulations," *IEEE Electron Device Letters*, Vol. 20, pp. 463-465.

184. Grossmann, A. and Morlet, J. (1984) "Decomposition of Hardy function into square integrable wavelets of constant shape," *SIAM Journal of Mathematical Analysis*, Vol. 15, pp. 723-736.

185. Günther, F. and Liu, W.K. (1998) "Implementation of boundary conditions for meshless methods," *Computer Methods in Applied Mechanics and Engineering*, Vol. 163, pp. 205-230

186. Günther, F., Liu, W. K., Diachin, D. and Christon, M. (2000) "Multi-scale meshfree parallel computations for viscous compressible flows," *Computer Methods in Applied Mechanics and Engineering*, Vol. 190, pp. 279-303.

187. Guduru, P. and Rosakis, A. J. and Ravichandran, G. (2001) "Dynamic shear bands: An investigation using high speed optical and infrared diagnostics," *Mechanics of Materials*, Vol. 23, pp. 371-402.

188. Hadamard, J. (1903) *Leçons sur la Propagation des Onders et les Équations de l'Hydrodynamique*, Hermann, Pairs.

189. Haile, J. M. (1992) *Molecular Dynamics Simulation*, Wiley & Sons, New York.

190. Han, W. and Meng, X. (2001) "Error analysis of the reproducing kernel particle method," *Computer Methods in Applied Mechanics and Enhineering*, Vol. 190, pp. 6157-6181.

191. Han, W., Meng, X., (2002) "Some studies of the reproducing kernel particle method," In *Lecture Notes in Computational Science and Engineering*, Vol. 26, Springer, pp. 193-210.

192. Han, W., Wagner, G. J., and Liu, W. K. (2002) "Convergence analysis of a hierarchical enrichment of Dirichlet boundary condition in a meshfree method," *International Journal for Numerical Methods in Engineering*, Vol. 53, pp. 1323-1336.

193. Hao, S. and Park, H.S. and Liu, W.K. (2002) "Moving particle finite element method," *International Journal for Numerical Methods in Engineering*, Vol. 53, pp. 1937-1958.

194. Hartree, D.R. (1928) "The wave mechanics of an atom with a non-Coulomb central field. Part I, Theory and Methods," *Proc. Cambridge Phil. Soc.*, Vol. 24, pp. 89-

195. Hartree, D.R. (1932) "A practical method for the numerical solution of differential equations," *Mem. Manchester Phil. Soc.*, Vol. 77, pp. 91-106.

196. Harris, J. (1985) "Simplified method for calculating the energy of weakly interacting fragments," *Physical Review B*, Vol. 31, pages 1770-1779.

197. Hedman, F., Laaksonen, A. (2000) "Parallel aspects of quantum molecular dynamics simulations of liquids," *Computer Physics Communications*, Vol. 128, pp. 284-294.

198. He, X. and Chen, S. and Zhang, R. (2000) "A lattice Boltzmann scheme for incompressible multiphase flow and its application in simulation of Rayleigh-Taylor instability," *Journal of Computational Physics*, Vol. 152, pp. 642-663.

199. Henon, M. (1987) "Viscosity of a lattice gas," *Complex Systems*, Vol. 1, page 763

200. Hinton, E., Rock, T. and Zienkiewicz, O. C. (1976) "A note on mass lumping and related process in finite element method," *Earthquake Engineering and Structural Dynamics*, Vol. 4, pages 245-249.

201. Hockney, R. and Eastwood, J. (1988) *Computer Simulation Using Particles*, Adam Hilger, Bristol.

202. Hohenberg, P. and Kohn, W. (1964) "Inhomogeneous electron gas," *Physical Review*, Vol. 136, pages B864.

203. Hong, J., Zhao, X.S. (2000) "New propagators for quantum-classical molecular dynamics simulations," *Journal of Chemical Physics*, Vol. 113, pp. 930-935.

204. Hughes, T.J.R. and Winget, J. (1980) "Finite rotation effects in numerical integration of rate constitutive equations arising in large-deformation analysis," *International Journal of Numerical Methods in Engineerings*, Vol. 15, pp. 1862-1867.

205. Brooks, A. N. and Hughes, T. J. R. (1982) "Streamline upwind/Petrov-Galerkin formulations for convection dominated flows with particular emphasis on the incompressible Navier-Stokes equations," *Computer Methods in Applied Mechanics and Engineering*, Vol. 32, pages 199-259.

206. Hu, H. H. and Patankar, N. A. and Zhu, M. Y. (2001) "Direct numerical simulations of fluid-solid systems using the arbitrary Lagrangian-Eulerian technique," *Journal of Computational Physics*, Vol. 169, pp. 427-462.

207. Hughes, T. J. R., W. K. Liu, and T. K. Zimmerman, (1981) "Lagrangian-Eulerian finite element formulations for incompressible viscous flows," *Computer Methods in Applied Mechanics and Engineering*, Vol. 29, pp. 329-349.

208. Hughes, T. J. R. (1984) "Numerical implementation of constitutive models: rate-independent deviatoric plasticity," *Theoretical Foundations for Large Scale Computations of Nonlinear Material Behavior*, pages 29-57, Edited by Nemat-Nasser, S. Martinus Nijhoff Publishers

209. Hughes, T. J. R., (1987) *The Finite Element Method: Linear Static and Dynamic Finite Element Analysis*, Prentice Hall Inc., Englewood Cliffs, NJ

210. Hughes, T.J.R. and Franca, L. P. and Balestra, M. (1986) "A new finite element formulation for computational fluid dynamics: V. Circumventing the

Babuška-Brezzi condition: A stable Petrov-Galerkin formulation of the Stokes problem accommodating equal-order interpolations," *Comupter Methods in Applied Mechanics and Engineering*, Vol. 59, pages 85-99.

211. Hughes, T.J.R. and Franca, L. P. (1987) "A new finite element formulation for computational fluid dynamics: V II. The Stoke problem with various well-posed boundary conditions: Symmetric formulations that converge for all velocity/pressure space," *Comupter Methods in Applied Mechanics and Engineering*, Vol. 65, pages 85.

212. Hughes, T.J.R. and Feijoo, G. and Mazzei, L. and Quincy, J. (1998) "The variational multiscale method - a paradigm for computational mechanics," *Comupter Methods in Applied Mechanics and Engineering*, Vol. 166, pages 3–24.

213. Hughes, T. J. R. and Engel, G. and Mazzei, L. and Larson, M. G. (2000) "The continuous Galerkin method is locally conservative," *Journal of Computational Physics*, **163**, pp. 467-488.

214. Hulbert, G. M. (1996) "Application of Reproducing kernel Particle Methods in Electromagnetics," *Computer Methods in Applied Mechanics and Engineering*, Vol. 139, pp. 229-238

215. Hultman, J., and Pharayn, A. (1999) "Hierarchical, dissipative formation of elliptical galaxies: is thermal instability the key mechanism ? Hydrodynamical simulations including supernova feed back multi-phase gas and metal enrichment in CDM: structure and dynamics of elliptical galaxies," *Astronomy and Astrophysics*, Vol. 347, pp. 769-798.

216. Huerta, A., Liu, W.K. (1988) "Viscous flow with large free surface motion," *Computer Methods in Applied Mechanics and Engineering*, Vol. 69, pages 277-324.

217. , Huerta, A., and Fernandez-Mendez, S. (2000), "Enrichment and coupling of finite element and meshless methods," *International Journal for Numerical Methods in Engineering*, Vol. 48, pp. 1615-1536.

218. Irons, B. M. (1969), "A Conforming quartic triangular element for plate bending," *International Journal for Numerical Methods in Engineering*, **1**, 29–45.

219. Irons, B. M. and Razzaque, A. (1972), "Experience with the patch test for convergence of finite element methods," In *Mathematical Foundations of the Finite Element Method with Applications to Partial Differential Equations*, Edited by A. K. Aziz, Academic Press. pp. 557-587.

220. Ishikawa, Y., Binning Jr., R.C., and Shramek, N.S. (1999) "Direct ab initio molecular dynamics study of $NO_2^+ + (H_2O)_4 to HNO_3(H_7O_3)^+$," *Chemical Physics Letters*, Vol. 313, pages 341-350.

221. Jin, X., Li, G., and Aluru, R. N. (2001) "On the equivalence between least-square and kernel approximations in meshless methods," *Computer Modeling in Engineering & Science*, Vol. 2, pp. 447-462.

222. Johnson, A. and Tezduyar, T. E. (1999) "Advanced Mesh Generation and Update Methods for 3D Flow Simulations," *Computational Mechanics*, Vol. 23, pp. 130-143.

223. Johnson, G. R., Petersen, E. H., and Stryk, R. A. (1993) "Incorporation of an SPH Option into EPIC code for a Wide Range of High Velocity Impact Computations," *Int. J. Impact Engng.*, Vol. 14, pages 385-394

224. Johnson, G. R., Stryk, R. A., and Beissel, S. R. (1996) "SPH for high velocity impact computations," *Computer Methods in Applied Mechanics and Engineering*, Vol. 139, pp. 347-374

225. Johnson, G. R., Beissel, S. R. (1996) "Normalized Smoothing Functions for SPH Impact Computations," *Int. J. Numer. Meth. Engng.* Vol. 39, pp. 2725-2741

226. Johnson, G. R., Beissel, S. R., and Stryk, R. A. (2000) "A generalized particle algorithm for high velocity impact computations," *Computational Mechanics*, Vol. 25 , No. 2/3, pp. 245-256

227. Jun, S., Liu, W.K., and Belytschko, T. (1998) "Explicit Reproducing Kernel Particle Methods for Large Deformation Problems," *International Journal for Numerical Methods in Engineering*, Vol. 41, pp. 137-166.

228. Jun, S., Im, S. (2000) "Multiple-scale meshfree adaptivity for the simulation of adiabatic shear band formation," *Computational Mechanics*, Vol. 25 , No. 2/3, pp. 257-266

229. Jones, J.E. (1924) "On the determination of molecular fields-I. From the variation of the viscosity of a gas with temperature," *Proceedings of the Royal Society (London)*, Vol. 106A, pages 441-462.

230. Jones, J. E. (1924) "On the determination of molecular fields-II. From the equation of state of a gas," *Proceedings of the Royal Society (London)*, Vol. 106A, pages 463-.

231. Kadanoff, L. (1986) "On two levels," *Physics today*, Vol. 39, pp. 7-9.

232. Kadanoff, L., McNamara, G.R., Zanetti, G. (1987) "A Poiseuille viscometer for lattice gas automata," *Complex Systems*, Vol. 1, pp. 791-

233. Kadanoff, L., McNamara, G.R., Zanetti, G. (1989) "From automata to fluid flow: comparisons of simulation and theory," *Physics Review A*, Vol. 40, pp. 4527-

234. Kadowaki, H. and Liu, W. K. (2003) "Bridging multi-scale method for localization problems," *Computer Methods in Applied Mechanics and Engineering*, to appear.

235. Kaljevic, I. and Saigal, S. (1997) " An improved element free Galerkin formulation," *International Journal of Numerical Methods in Engineering*, Vol. 40, pp. 2953-2974.

236. Karlin, IV and Succi, S., and Orszag, S. (1999) *Lattice Boltzmann method for irregular grids*, *Physical Review Letters*, Vol. 26, pp. 5245-5248.

237. Karpov, E. G., Stephen, N. G., and Liu, W. K., (2003) "Initial tension in randomly disordered periodic lattice," *International Journal of Solids and Structures*, Vol. 40, pp. 5371-5388.

238. Karpov, E. G., Wagner, G. J. and Liu, W. K. (2003) "A Green's function approach to deriving wave-transmitting boundary conditions in molecular dynamics simulations," *Computational Material Science*, to appear.

239. Karpov, E. G., Yu, H. Park, H. S., Liu, W. K., and Wang, J. (2003) "Multiscale boundary conditions in crystalline solids: theory and application to nanoindentation," Submitted for publication.

240. G. Kaiser, G. (1994) *A Friendly Guide to Wavelets*, Birkhäuser, Boston.

241. Kohn, W. and Sham, L.J. (1965) "Self-consistent equations including exchange and correlation effects," *Physics Review*, Vol. 140, pages 1133-.

242. Krause, C. D. and Raftenberg, M. N. (1993) "Metallographic observations of rolled-homogeneous-armor specimens from plates perforated by shaped charged jets," Army Research Labortary, ARL-TR-68.

243. Kihe, C., Yildirim, T., Mehrez, H., Ciraci, S. (2000) "A first-principles study of the structure and dynamics of C_8H_8, Si_8H_8, and Ge_8H_8 moleculars," *Journal of Physical Chemistry A*, Vol. 104, pp. 2724-2728.

244. Kosloff, D. and Frazier, G. A. "Treatment of hourglass patterns in low order finite element codes," *Numerical and Analytical Methods in Geomechanics*, Vol. 2, pp. 57-72.

245. Kim, N. H., Choi, K. K., Chen, J. S., Park, Y. H. (2000) "Meshless shape design sensitivity analysis and optimization for contact problem with friction," *Computational Mechanics*, Vol. 25 , pp. 157-168.

246. Klaas, O., and Shephard, M. S. (2000) "Automatic generation of octree-based three-dimensional discretizations for Partition of Unity methods," *Computational Mechanics*, Vol. 25 , pp. 296-304.

247. Kobrak, M.N., Bittner, E.R. (2000) "Quantum molecular dynamics study of polaron recombination in conjugated polymers," *Physics Review B*, Vol. 62, pp. 11473-11486.

248. Krongauz, Y. and Belytschko, T. (1996) "Enforcement of essential boundary conditions in meshless approximations using finite elements," Computer Methods in Applied Mechanics and Engineering Vol. 131, pp. 133-145

249. Krumrine, J. R., Jang, S., Alexander, M.H., Voth, G.A. (2000) "Quantum molecular dynamics and spectral simulation of a boron impurity in solid para-hydrogen," *Journal of Chemical Physics*, Vol. 113, pp. 9079-9089.

250. Krysl, P. and Belytschko, T. (1996) "Analysis of Thin Shells by the Element-free Galerkin Method," *International Journal of Solids and Structures*, Vol. 33, pp. 3057-3080.

251. Krysl, P. and Belytschko, T. (1996) "Element-free Galerkin method: convergence of the continuous and discontinuous shape function," *Computer Methods in Applied Mechanics and Engineering*, Vol. 148, pp. 257-277.

252. Krysl, P. and Belytschko, T. (1999) "The element free Galerkin method for dynamic propagation of arbitrary 3-D cracks," *International Journal of Solids and Structures*, Vol. 44, pp. 767-800.

253. Kum, O., Hoover, W.G. and Posch, H.A. (1995) "Viscous conducting flows with smooth-particle applied mechanics," *Physics Review E*, Vol. 109, pp. 67-75.

254. La Paglia, F. L. (1971), *Introduction Quantum Chemistry*, Harper & Row Publishers, New York.

255. Lancaster, P. , Salkauskas, K. (1981) "Surface generated by moving least square methods," *Math. Comput.* Vol. 37, pp. 141-158.

256. Landau, L. D. and Lifshitz, E.M. (1965) *Quantum Mechanics: non-relativistic theory*, Pergmon, Oxford, London.

257. Landau, L. D. and Lifshitz, E.M. (1978) *Mechanics*, Pergamon Press, Oxford, New York.

258. Lee, W.H. (1998) "Newtonian hydrodynamics of the coalescence of black holes with neutron stars II. Tidally locked binaries with a soft equation of state," *Monthly Notices of the Royal Astronomical Society*, Vol. 308, pp. 780-794

259. Lee, W.H. (2000) "Newtonian hydrodynamics of the coalescence of black holes with neutron stars III. Irrotational binaries with a stiff equation of state," *Monthly Notices of the Royal Astronomical Society*, Vol. 318, pp. 606-624.

260. Leonard, A. (1980) "Vortex methods for flow simulation," *Journal of Computational Physics*, Vol. 37, pp. 289-335.

261. Leonard, A. (1985) "Computing three-dimensional incompressible flows with vortex elements," *Ann. Rev. Fluid Mech.*, Vol. 17, pp. 523-529.

262. LeMonds, J. and Needleman, A. (1986) "Finite element analysis of shear localization in rate and temperature dependent solids," *Mechanics of Materials*, Vol. 5, pages 339-361.

263. LeMonds, J. and Needleman, A. (1986) "An analysis of shear band development incorporating heat conduction," *Mechanics of Materials*, Vol. 5, pp. 363-

264. Li, S., and Liu, W. K. (1996) "Moving least-square reproducing kernel particle method. Part II: Fourier analysis," *Computer Methods in Applied Mechanics and Engineering*, Vol. 139, pp. 159-194

265. Li, S. and Liu, W. K. (1998) "Synchronized reproducing kernel interpolant via multiple wavelet expansion," *Computational Mechanics*, Vol. 28, pages 28-47.

266. Li, S. and Liu, W. K. [1999] "Reproducing kernel hierarchical partition of unity Part I: Formulations" *International Journal for Numerical Methods in Engineering* Vol. 45, pp. 251-288

267. Li, S. and Liu, W. K. [1999] "Reproducing kernel hierarchical partition of unity Part II: Applications" *International Journal for Numerical Methods in Engineering* Vol. 45, pp. 289-317

268. Li, S., W. Hao, and Liu, W. K. (2000) "Numerical simulations of large deformation of thin shell structures using meshfree methods," *Computational Mechanics*, Vol. 25 No. 2/3, pages 102-116

269. Li, S. and Liu, W.K. "Numerical simulations of strain localization in inelastic solids using mesh-free methods," *International Journal for Numerical Methods in Engineering*, Vol. 48, pp. 285-1309.

270. Li, S., W. Hao, and Liu, W. K. (2000) "Meshfree simulations shear banding under large deformation", *International Journal of Solids and Structures* Vol. 37, pp. 7185-7206.

271. Li, S., Qian, Q., Liu, W.K., and Belytschko, T. (2000) "A meshfree contact-detection algorithm," *Computer Methods in Applied Mechanics and Engineering*, Vol. 190, pp. 7185-7206.

272. Li, S., Liu, W. K., Qian, D., Guduru, P. and Rosakis, A. J. (2001) "Dynamic shear band propagation and micro-structure of Adiabatic Shear Band," *Computer Methods in Applied Mechanics and Engineering*, Vol. 191, pp 73-92.

273. Li, S., Liu, W. K., Rosakis, A. J., Belytschko, T., and Hao, W. (2002), "Mesh-free Galerkin simulations of dynamic shear band propagation and failure mode transition," *International Journal of Solids and Structures*, Vol. 39, pp. 1213-1240.

274. Li, S. and Liu, W. K. (2002) "Meshfree and particle methods and their applications," *Applied Mechanics Review*, Vol. 55, pp. 1-34.

275. Li, S., Lu, H., Han, W., Liu, W. K., and Simkins, D. C. Jr. (2004) "Reproducing kernel element, Part II. Globally conforming I^m/C^n hierarchies," *Computer Methods in Applied Mechanics and Engineering*, Vol. 193, pp. 953-987.

276. Li, X., Millam, J.M. and Sohlegel, H.B. (2000) "Ab initio molecular dynamics studies of the phoyodissociation of formaldehyde, $H_2CO - H_2 + CO$: Direct classical trajectory calculations by MP2 and density function theory," *Journal of Chemical Physics*, Vol. 113, pages 10062-10067.

277. Lin, H. and Vezza, M. (1996) "A pure vortex sheet method for simulating unsteady, incompressible, separated flows around static and pitching aerofoils," *Proceedings of the 20th Congress of the International Council of the Aeronautical Sciences*, Sorento, Italy, pp. 2184-2193.

278. Libersky, L. D., Petschek, A. G. (1991) "Smoothed particle hydrodynamics with strength of materials", in Proceedings *The Next Lagrange Conference* Springer-Verlag, New York, pp. 248-257.

279. Libersky, L. D., Petschek, A. G., Carney, T.C., and Hipp. J. R. (1993) "High strain Lagrangian hydrodynamics", *J. Comput. Phys.* Vol. 109, pp. 67-75.

280. Liszka, T. and Orkisz, J. (1980) "The finite difference method at arbitrary irregular grids and its application in applied mechanics," *Comp. Struct.*, Vol. 11, pp. 83-95.

281. Liszka, T. (1984) "An Interpolation method for an irregular net of nodes," *International Journal for Numerical Methods in Engineering*, Vol. 20, pp. 1599-1612.

282. Liszka, T. J., Duarte, C. A. M., and Tworzydlo (1996) "hp-Meshless cloud method" *Computer Methods in Applied Mechanics and Engineering*, Vol. 139, pp. 263-288.

283. Liu, G. R., and Gu, Y. T. (2000) "Meshless local Petrov-Galerkin (MLPG) method in combination with finite element and boundary element approaches," *Computational Mechanics*, Vol. 26, pp. 536-546.

284. Lucy, L. B. (1977) "A numerical approach to the testing of the fission hypothesis", *The Astronomical Journal* Vol. 82, pp. 1013-1024.

285. Liu, W. K. (1983) "Development of mixed time partition procedures for thermal analysis of structures," *International Journal of Numerical Methods in Engineering*, Vol. 19, pp. 125–140.

286. Liu, W. K. and Belytschko, T. (1982) "Mixed-time implicit-explicit finite elements for transient analysis," *Computers and Structures*, Vol. 15, pp. 445–450.

287. Liu, W. K., Belytschko, T., Chang, H. (1986) "An arbitrary Lagrangian-Eulerian finite element method for path-dependent materials," *Computer Methods in Applied Mechanics and Engineering*, Vol. 58, pages 227-246.

288. Liu, W. K., Chang, H., Chen, J. S., Belytschko, T. (1988) "Arbitrary Lagrangian and Eulerian Petrov-Galerkin finite elements for nonlinear problems," *Computer Methods in Applied Mechanics and Engineering*, Vol. 68, pages 259-310.

289. Liu, W. K., Chen, J. S., Belytschko, T., Zhang, Y.F. (1991) "Adaptive ALE finite elements with particular reference to external work rate on frictional interface," *Computer Methods in Applied Mechanics and Engineering*, Vol. 93, pp. 189-216.

290. Liu, W. K. and Ma, D. C. (1982) "Computer implementation aspects for fluid-structure interaction problems," *Computer Methods in Applied Mechanics and Engineering*, Vol. 31, pp. 129-148.

291. Liu, W. K. and Gvildys, J. (1986) "Fluid-structure interaction of tanks with an eccentric core barrel," *Computer Methods in Applied Mechanics and Engineering*, Vol. 58, pp. 51-77 .

292. Liu, W. K. (1981) "Finite element procedures for fluid-structure interactions and application to liquid storage tanks," *Nuclear Engineering Design*, Vol. 65, pp. 221-238.

293. Liu, W. K., Chang, H., Chen, J. S., and Belytschko, T., (1988) "Arbitrary Lagrangian-Eulerian Petrov-Galerkin finite elements for nonlinear continua," *Computer Methods in Applied Mechanics and Engineering*, Vol. 68, pp. 259-310.

294. Liu, W.K., Zhang, Y, Ramirez, M. R. (1991) "Multiple scale finite element methods," *Int. J. Num. Methods Engng.*, Vol. 32, pp. 969-990

295. Liu, W.K., Adee, J., Jun, S. (1993) "Reproducing kernel particle methods for elastic and plastic problems," In: Benson D. J., Asaro, R. A. (eds.) *Advanced Computational Methods for Material Modeling*, AMD 180 and PVP 33, ASME, pp. 175-190.

296. Liu, W.K. and Oberste-Brandenburg, C (1993), "Reproducing kernel and wavelets particle methods," In *Aerospace Structures: Nonlinear Dynamics and System Response*, edited by Cusumano, J. P., Pierre, C., and Wu, S. T. AD 33 ASME, pp. 39-56.

297. Liu, W.K., Jun, S., and Zhang, Y.F. (1995) "Reproducing kernel particle methods," *International Journal of Numerical Methods in Fluids*, Vol. 20, pp. 1081-1106.

298. Liu, W.K., Jun, S., Li, S., Adee, J. and Belytschko, T. (1995), "Reproducing kernel particle methods for structural dynamics," *International Journal for Numerical Methods in Engineering*, Vol. 38, pp. 1655-1679.

299. Liu, W.K. and Chen, Y. (1995) "Wavelet and multiple scale reproducing kernel method," *International Journal of Numerical Methods in Fluids*, Vol. 21, pp. 901-933.

300. Liu, W. K., Chen, Chang, C. T., and Belytschko (1996) "Advances in multiple scale kernel particle methods," *Computational Mechanics*, Vol. 18, pp. 73-111.

301. Liu, W. K., Chen, Y., Uras, R. A. and Chang, C. T. (1996), "Generalized multiple scale reproducing kernel particle methods," *Computer Methods in Applied Mechanics and Engineering*, Vol. 139, pp. 91-158.

302. Liu, W. K., Chen, Y., Jun, S., Chen, JS, Belytschko, T., Uras, R.A., Chang, C.T. (1996) "Overview and applications of the reproducing kernel particle methods," *Archives of Computational Mechanics in Engineering: State of the reviews*, Vol. 3, pp. 3-80.

303. Liu, W.K., Li, S., and Belytschko, T. (1997) "Moving least square reproducing kernel method Part I: methodology and convergence," *Computer Methods in Applied Mechanics and Engineering*, Vol. 143, pp. 422-453.

304. Liu, W. K., Jun, S., Sihling, D.T., Chen, Y., Hao, W. (1997) "Multiresolution reproducing kernel particle method for computational fluid dynamics," *International Journal for Numerical Methods in Fluids*, Vol. 24, pp. 1391-1415.

305. Liu, W.K., Hao, W., Chen, Y., Jun, S., and Gosz, J. (1997) "Multiresolution reproducing kernel particle methods," *Computational Mechanics*, Vol. 20, pp. 295-309.

306. Liu, W.K., Uras, R.A. and Chen, Y. (1997) "Enrichment of the finite element method with reproducing kernel particle method," *Journal of Applied Mechanics*, Vol. 64, pp. 861-870.

307. Liu, W.K., and Jun, S. (1998) "Multiple scale reproducing kernel particle methods for large deformation problems," *International Journal for Numerical Methods in Engineering*, Vol. 141, pp. 1339-1362.

308. Liu, W. K., Li, S., and Hao, W. (1999) "Simulations of fluids and solids by multi-scale meshfree methods," *International Conference on Enhancement and Promotion of Computational Methods in Engineering and Science*, edited by J. Bento and E. A. Oliviera and E. Pereira, Vol. 1, pp. 43-52, Macao.

309. Liu, W.K., Hao, S., Belytschko, T. Li, S. and Chang, C.T. (1999) "Multiple scale meshfree methods for damage fracture and localization," *Computational Materials Science*, Vol. 16, pp. 197-205.

310. Liu, W.K., Hao, S., Belytschko, T. Li, S., and Chang, T. (2000) "Multi-scale methods," *International Journal for Numerical Methods in Engineering*, Vol. 47, pp. 1343-1361.

311. Liu, W.K., Han, W., Lu, H., Li, S., and Cao, J. (2004) "Reproducing kernel element method: Part I. Theoretical formulation," *Computer Methods in Applied Mechanics and Engineering*, Vol. 193, pp. 933-951.

312. Liu, W. K., Karpov, E. G., Zhang, S. and Park, H. S. (2004) "An introduction to computational nanomechanics and materials," *Computer Methods in Applied Mechanics and Engineering*, Vol. 193, pp. 1529-1578.

313. Liu, Y. and Zhang, L. and Wang, X. and Liu, W. K. (2003) "Coupling of Navier-Stokes equations with protein molecular dynamics and its application to Hemodynamics," Submitted to *International Journal for NumericalMethods in Fluids*.

314. Lu, H., Li, S., Simkins, Jr., D. C. Liu, W.K. and Cao, J. (2004), "Reproducing kernel element method Part III. Generalized enrichment and applications," *Computer Methods in Applied Mechanics and Engineering*, Vol. 193, pp. 989-1011.

315. Lu, Y.Y., T, Belytschko, T., and Tabbara, M. (1995), "Element-free Galerkin method for wave propagation and dynamic fracture," *Computer Methods in Applied Mechanics and Engineering*, Vol. 126, pp. 131-153.

316. MacNeal, R. H. and R.L. Harder, R. L. (1985) "A proposed standard set of problems to test finite element accuracy," *Finite Element Analysis and Design*, Vol. 11, pp. 3-20.

317. Maier, RS (1986) "Boundary conditions for the lattice Boltzmann method," *Physics Fluids*, Vol. 8, pp. 1788-1801.

318. Marchand, A. and Duffy, J. (1988) "An experimental study of the formation process of adiabatic shear bands in a structural steel," *Journal of Mechanics and Physics of Solids*, Vol. 38, pp. 251-283.

319. Mazzocco, F. and Arrighetti, C. (2000) "Multiscale lattice Boltzmann schemes: A preliminary application to axial turbomachine flow simulations," *International Journal of Modern Physics*, Vol. 11, pp. 233-245.

320. McNamara, G. and Zanetti, G. (1988) "Use of the Boltzmann equation to simulate lattice-gas automata," *Physical Review Letter*, Vol. 61, pp. 2332-

321. Mei, R. and Shyy, W. and Yu, D. and Luo, LS (2000) "Lattice Boltzmann Method for 3-D flows with curved boundary," *Journal of Computational Physics*, Vol. 161, pp. 680-699.

322. Mendonaca, P. de T. R., de Barcellos, C. S., Duarte, A (2000) "Investigations on the hp-Cloud method by solving Timoshenko beam problems," *Computational Mechanics*, Vol. 25, pp. 286-295.

323. Melenk, J. M. and Babuska, I. (1996) "The partition of unity finite element method: Basic theory and applications," *Computer Methods in Applied Mechanics and Engineering*, Vol. 139, pp. 289-314.

324. Meyer, Y. (1992) *Wavelets and Operators*, Cambridge University Press, Note that the French edition was published in 1990 under the name *Ondelettes et Operateurs*.

325. Meyer, Y. and Ryan, R. D. (1993) *Wavelets: Algorithms & Applications*, SIAM, Philadelphia.

326. Milstein, F. (1982) *Crystal Elasticity*, In Mechanics of Solids, pp. 417-452, Pergamon, Oxford.

327. Mirsky, L. (1990), *An Introduction to Linear Algebra*, Dover Publications, Inc.

328. Molinari, A. and Cliffton, R. J. (1987) "Analyical characterization of shear localization in thermoviscoplastic materials," ASME *Journal of Applied Mechanics*, Vol. 54, pp. 806-812.

329. Monaghan, J. J. (1982) "Why particle methods work (Hydrodynamics)," *SIAM Journal on Scientific and Statistical Computing*, Vol. 3, pp. 422-433.

330. Monaghan J. J., Gingold, R. A. (1983) "Shock simulation by the particle method SPH," *J. Comput. Phys.* Vol. 52, pp. 374-389.

331. Monaghan, J. J. (1985) "Particle methods for hydrodynamics," *Computer physics Report*, Vol. 3, pp. 71-124.

332. Monaghan, J.J. and Lattanzio, J. C. (1985) "A refined particle method for astrophysical problems," *Astronomy and Astrophysics*, Vol. 149, pp. 135-143.

333. Monaghan, J. J. and Pongracic, H. (1985) "Artificial viscosity for particle methods," *Applied Numerical Mathematics*, Vol. 1, pp. 187-194.

334. "An introduction to SPH," *Computer Physics Communications*, Vol. 48, pp. 89-96.

335. Monaghan J. J. (1989) "On the problem of penetration in particle methods," *J. Comput. Phys.* Vol. 82, pp. 1-

336. Monaghan J. J. (1990) "Modeling the universe," *Proceedings of the Astronomical Society of Australia*, Vol. 18, pp. 233-237.

337. Monaghan, J.J., and Lattanzio, J. C. (1991) "A simulation of the collapse and fragmentation of cooling molecular clouds," *Astrophysical Journal*, Vol. 375, pp. 177-189.

338. Monaghan J. J. (1992) "Smoothed particle hydrodynamics," *Annu. Rev. Astron. Astrophys.* Vol. 30, pp. 543-574

339. Monaghan J. J. (1994) "Simulating free surface flows with SPH" *J. Comput. Phys.* Vol. 110, pp. 399-

340. Monaghan, J. J. (1997) "SPH and Riemann solvers," *Journal of Computational Physics*, Vol. 136, pp. 298-307.

341. Monaghan, J.J. (1999) "Implicit SPH drag and dust gas dynamics," *Journal of Computational Physics*, Vol. 138, pp. 801-820.

342. Monaghan, J. J. (2000) "SPH without a tensile instability," *Journal of Computational Physics*, Vol. 159, pp. 290-311.

343. Morris, J. P. (1996) "Stability properties of SPH," *Publ. Astron. Soc. Aust.* Vol. 13, pp. 97-

344. Morris, J. P., Fox, P. J., Zhu, Y. (1997) "Modeling low Reynolds number incompressible flows using SPH," *J. Comput. Phys.* Vol. 136, pp. 214-226.

345. Munz, C.D., Schneider, R., Sonnendrücker, E., Stein, E., Voss, U., and Westermann, T. (1999) "A finite-volume Particle-in-cell method for the numerical treatment of Maxwell-Lorentz equations on boundary-fitted meshes," *International Journal for Numerical Methods in Engineering*, Vol. 44, pp. 461-487.

346. Munz, C.D., Schneider, R., and Voss, U. (1999) "A finite-volume particle-in-cell method for the numerical simulation of devices in pulsed-power technology," *Survey on Mathematics for Industry*, Vol. 8, pp. 243-257

347. Nagtegaal, J. C., Parks, D. M. and Rice, J. R. (1974) "On numerical accurate finite element solutions in the fully plastic range," *Computer Methods in Applied Mechanics and Engineering*, Vol. 4, pp. 153-177.

348. Nayroles, B., Touzot, G. and Villon, P. (1992) "Generalizing the finite element method: Diffuse approximation and diffuse elements," *Computational Mechanics*, Vol. 10, pp. 307-318.

349. Needleman, A (1988) "Material rate dependent and mesh sensitivity in localization problems," *Computer Methods in Applied Mechanics and Engineering*, Vol. 67, pp. 68-85.

350. Needleman, A (1989) "Dynamic shear band development in plane strain," *Journal of Applied Mechanics*, Vol. 56, pp. 1-9.

351. Nemat-Nasser, S., Chung, D.-T. and Taylor, L.M. (1989) "Phenomenological modelling of rate-dependent plasticity for high strain rate problems," *Mechanics of Materials*, Vol. 7, pp. 319-344.

352. Nemat-Nasser, S. and Chung, D. T. (1992) "An explicit constitutive algorithm for large-strain, large-strain-rate elastic-viscoplasticity," *Computer Methods in Applied Mechanics and Engineering*, Vol. 95, pp. 205-219.

353. Niederreiter, H. (1978) "Quasi-Monte Carlo methods and pseudo-random numbers," *Bull. Amer. Math. Soc.* (N.S.), Vol. 84, pages 957-1041.

354. Nocedal, J. (1980) "Updating quasi-newton matrices with limited storage," *Math. Comput.*, Vol. 35, pp. 773-782.

355. Noguchi, H (1997) "Application of element free Galerkin method to analysis of Mindlin type plate/shell problems," *Proceedings of ICE97*, pp. 918-923.

356. Noguchi, H., and Kawashima, T. and Miyamura, T. (2000) "Element free analysis of shell and spatial structures," *International Journal for Numerical methods in Engineering*, Vol. 47, pp. 1215-1240.

357. Oden, J.T., Duarte, CAM, and Zienkiewicz, O.C. (1998) "A new Cloud-based hp finite element method," *Computer Methods in Applied Mechanics and Engineering*, Vol. 153, pp. 117-126

358. Ohno, K., Esfarjani, K. and Kawazoe, Y. (1999) "Computational materials science: from Ab initio to Monte Carlo methods," Springer, Berlin.

359. Oñate, E. and Idelsohn, S. Zienkiewicz, O.C. and Taylor, R.L. (1996) "A stabilized finite point method for analysis of fluid mechanics problems," *Computer Methods in Applied Mechanics and Engineering*, Vol. 139, pp. 315-347

360. Oñate, E. and Idelsohn, S. Zienkiewicz, O.C. and Taylor, R.L. (1996) "A finite point method in computational mechanics. application to convective transport and fluid flow," *International Journal for Numerical Methods in Engineering*, Vol. 39, pp. 3839-3866

361. Oñate, E. and Idelsohn, S. (1998) "A mesh-free point method for advective-diffusive transport and fluid flow problems," *Computational Mechanics*, Vol. 21, pp. 283-292.

362. Oran, E.S., Oh, C.K., and Cybyk, B.Z. (1998) "Direct simulation Monte Carlo: recent advances and applications," *Annual Review of Fluid Mechanics*, Vol. 30, pp. 403-441.

363. Ortiz, M., Leroy, Y. and A. Needleman, A. (1987) "A finite element method for localized failure analysis," *Computer Methods in Applied Mechanics and Engineering*, Vol. 61, pages 189-214.

364. Ortiz, M. and Quigley, J. J. IV (1991) "Adaptive mesh refinement in strain localization problems," *Computer Methods in Applied Mechanics and Engineering*, Vol. 90, pages 781-987.

365. Pan, J. (1983) "Perturbation analysis of shear strain localization in rate sensitive materials," *International Journal of Solids and Structures*, Vol. 19, pages 153-164.

366. Park, H. S. and Liu, W.K. (2004) "An introduction and tutorial on multi-scale analysis in solids," *Communications in Numerical Methods in Engineering*, Vol. 193, pp. 1713-1732.

367. Pang, Z (2000), "Treatment of point loads in element free Galerkin method (EFGM)," *Communications in Numerical Methods in Engineering*, Vol. 16, pp. 335-341.

368. Papoulis, A. (1956) "A new method of inversion of the Laplace transform," *Quart. Appl. Math*, Vol. 14, pp. 405-414.

369. Peskin, C. S. (1977) "Numerical analysis of blood flow in the heart," *Journal of Computational Physics*, Vol. 25, pages 220-252.

370. Peskin, C. S. and McQueen, D. M. (1996) "Fluid dynamics of the heart and its valves," In *Case Studies in Mathematical Modeling-Ecology, Physiology, and Cell Biology*, Edited by Othmer, H. G. and Adler, F. R. and Lewis, M.A. and Dallon, J. C.

371. Peskin, C. S. (2002) "The immersed boundary method," *Acta Numerica* Vol. 11, pages 479-517.

372. Petschek, A. G. and Libersky, L. D., (1993) "Cylinderical smoothed particle hydrodynamics", *J. Comput. Phy.* Vol. 109, pages 76-83

373. Peirce, D., Asaro, J. and Needleman, A. (1982) "An analysis of nonuniform and localized deformation in ductile single crystals," *Acta Metallurgica*, Vol. 30, pages 1087-1119.

374. Peirce, D., Shih, C. F., and Needleman, A., (1983) "A tangent modulus method for rate dependent solids," *Computer & Structure*, Vol. 5, pages 875-887.

375. Pilar, F. L. (1990) *Elementrary Quantum Chemistry*, McGraw-Hill, New York.

376. Posch, H.A., Hoover, W.G., and Kum, O. (1995) "Steady-state shear flows via nonequilibrium molecular dynamics and smooth-particle applied mechanics," *Physics Review E*, Vol. 52, pp. 1711-1719.

377. Przemieniecki, J. S. (1968), *Theory of matrix structural analysis*, McGraw-Hill

378. Qian. D., Wagner, G. J., Liu, W.K., Yu, M.-F., Ruoff, R. S. (2002) "Mechanics of carbon nanotubes," *Applied Mechanics Review*, Vol. 55, pp. 495-533.

379. Qian, D., Wagner, G. J. and Liu, W. K. (2004) "A multi-scale projection method for the analysis of carbon nanotubes," *Computer Methods in Applied Mechanics and Engineering*, Vol. 193, pp. 1603-1632.

380. Qian, D., Liu, W. K., Subranmoney, S. and Ruoff, R.S. (2003) "Effect of interlayer interaction on the mechanical deformation of multiwalled carbon nanotube," *Journal of Nanoscience and Nanotechnology*, Vol. 3, pp. 185-191.

381. Qian, Y.H.,d'Humiéres D., Lallemand, P. (1992), "Lattice BGK models for the Navier-Stokes equation," *Europhys. Lett.*, Vol. 17, pp. 479-484.

382. Qian, Y.H., Orszag, S.A. (1993) "Lattice BGK models for the Navier-Stokes equation: nonlinear deviation in compressible regimes," *Europhys. Lett.*, Vol. 21, pp. 255-259.

383. Qian, Y.H., Succi, S., Orszag, S.A. (2000), "Recent advances in lattice Boltzmann computing," *Annual Reviews of Computational Physics*, Vol. III, edited by D. Stauffer, pp. 195-242, World Scientific, Singapore.

384. Rachford Jr., H.H. and Wheeler, M. F. (1974) "An H^{-1}-Galerkin procedure for the two-point boundary value problem," in *Mathematical Aspects of Finite Elements in Partial Differential Equations*, Carl de Boor, ed., Academic Press, Inc. pages 353–382.

385. Raftenberg, M. N. (2000) "A shear banding Model for penetration calculations," *Technical Report of Army Research Labortary*, ARL-TR-2221

386. Randles, P. W., Carney, T. C., Libersky, L. D., and Petschek, A. G. (1995) "Calculation of oblique impact and fracture of tungsten cubes using smoothed particle hydrodynamics," *Int. J. Impact Engng.* Vol. 17

387. Randles, P. W., Libersky, L. D. (1996) "Smoothed particle hydrodynamics: Some recent improvements and applications," *Computer Methods in Applied mechanics and Engineering*, Vol. 139, pp. 375-408

388. Randles, P. W., Libersky, L. D. (2000) "Normalized SPH with stress points," *International Journal for Numerical Methods in Engineering*, Vol. 48, pp. 1445-1462.

389. Rao, B. N. and Rahman, S. (2000) "An efficient meshless method for fracture analysis of crack," *Computational Mechanics*, Vol. 26, pp. 398-408.

390. Rapaport, D.C. (1995) *The art of molecular dynamics simulation*, Cambridge University Press, Cambridge, UK.

391. Reider, MB and Sterling, JD (1995) "Accuracy of discrete-velocity BGK models for the simulation of the impressible Navier-Stokes equations," *Computer Fluids*, Vol. 24, pages 459-467.

392. Rudd, R. E. and Broughton, J. Q. (1998) "Coarse-grained molecular dynamics and the atomic limit of finite elements," *Physics Review* B, Vol. 58, pages 5893–5896.

393. Ruth, R. D. (1983) *IEEE Trans. Nucl. Sci.*, Vol. NS-30, pages 2669-.

394. Rvachev, V. L., Sheiko, T. I., Shapiro, V., Tsukanov, I. (2000) "On completeness of RFM solution structures," *Computational Mechanics*, Vol. 25, No. 2/3, pages 305-317

395. Ryckaert, J.P., Ciccotti, G. and Berendsen, H.J.C. (1977) "Numerical integration of the Cartesian equations of motion of a system with constraints: molecular dynamics of n-alkanes," *Journal of Computational Physics*, Vol. 23, pages 327-341.

396. Shawki, T. G. and Clifton, R. J. (1989) "Shear band formation in thermal viscoplastic materials," *Mechanics of Materials*, Vol. 8, pages 13-43.

397. Saad, Y. and Schultz, M. H. (1986), "GMRES: A generalized minimal residual algorithm for solving nonsymmetric linear systems," SIAM *Journal on Scientific and Statistical Computing*, Vol. 7, No. 3, pages 856-869.

398. Saigal, S, Barry, W. (2000) "A slices based element free Galerkin formulation," *Computational Mechanics*, Vol. 25, No. 2/3, pages 220-229

399. Schuller, I.K. (1988) "Molecular dynamics simulation of epitaxial growth," *MRS BULLETIN*, Vol. 13, pages 23-27.

400. Szabó, B. and Babuška, I. (1991) *Finite Element Analysis*, John Wiley & Sons, Inc., New York.

401. Schweitzer, M. A. (2003) *A Parallel Multilevel Partition of Unity Method*, *Lecture Notes in Computational Science and Engineering*, Vol. 29, Springer

402. Shirazaki, M., and Yagawa, G. (1999) "Large-scale parallel flow analysis based on free mesh method: a virtually meshless method," *Computer Methods in Applied Mechanics and Engineering*, Vol. 174, pp. 419-431

403. Shilkrot, L. E., Curtin, W. A. and Miller, R. E. (2002) "A coupled atomistic/continuum model of defects in solids," *Journal of Mechanics and Physics of Solids*, Vol. 50, pages 2085–2106.

404. Simkins, Jr. D.C., Li, S., Lu, H. and Liu, W. K. (2004), "Reproducing kernel element method. Part IV. Globally compatible $C^n(n \geq 1)$ triangle hierarchy," *Computer Methods in Applied Mechanics and Engineering*, Vol. 193, pp. 1013-1034.

405. Simo, J. C., Oliver, J. and Armero, F. (1993) "An analysis of strong discontinuities induced by strain-softening in rate-independent inelastic solids," *Computational Mechanics*, Vol. 12, pages 277-296.

406. Simo, J. C. and Rifai, M. S. (1990) "A class of mixed assumed strain methods and the methods of incompatible modes," *international journal of Numerical Methods in Engineering*, Vol. 29, pp. 1595-1638

407. Simo, J. C. and Hughes, T. J. R. (1998) *Computational Inelasticity*, Springer, New York.

408. Slater, J.C. and Koster, G.F. (1954) "Simplified LCAO method for the periodic potential problem," *Physics Review*, Vol. 94, pages 1498-.

409. van der Sman, RGM (1997) "Lattice Boltzmann scheme for natural convection in porous media," *International Journal of Modern Physics*, Vol. 8, pages 879-888.

410. van der Sman, RGM and Ernst, MH (2000) "Convection-diffusion lattice Boltzmann scheme for irregular lattices," *Journal of Computational Physics*, Vol. 160, pp. 766-782.

411. Steinmann, P. and Willam, K. (1991) "Performance of enhanced finite element formulations in localized failure computations," *Computer Methods in Applied Mechanics and Engineering*, Vol. 90, pp. 845-867.

412. Stillinger, F. H. and Weber, T. A. (1985) "Computer simulation of local order in condensed phases of silicon," *Phys. Rev. B*, Vol. 31, 5262-5271.

413. Sladek, J., Sladek, S., Atluri, N. (2000) "Local boundary integral equation (LBIE) method for solving problems of elasticity with nonhomogeneous material properties," *Computational Mechanics*, Vol. 24 6, pages 456-462

414. Starikov, E.B. (2000) "Nucleic acids as objects of material science: importance of quantum chemical and quantum mechanical studies," *International Journal of Quantum Chemistry*, Vol. 77, pages 859-870.

415. Succi, S. (1997) "Lattice Boltzmann equation: failure or success ?," *Physica A*, Vol. 240, pages 221-228.

416. Sukumar, N. and Moran, B. and Belytschko, T. (1998) "The natural element method in solid mechanics," *International Journal for Numerical Methods in Engineering*, Vol. 43, pp. 839-887.

417. Sukumar, N. and Moran, B. (1999) "C^1 Natural neighbor interpolant for partial differential equations," *Numerical Methods for Partial Differential Equations*, Vol. 15, pp. 417-447.

418. Sulsky, D., Zhou, S. H., Schreyer, H. L. (1995) "Application of a particle-in-cell method to solid mechanics," *Computer Physics Communications* Vol. 87, pages 236-252

419. Sulsky, D., Schreyer (1996), "Axisymmetric form of the material point method with applications to upsetting and Taylor impact problem," *Computer Methods in Applied mechanics and Engineering*, Vol. 139, pp. 409-430

420. Swegle, J. W., Hicks, D. L., Attaway, S. W., (1995) "Smoothed particle hydrodynamics stability analysis", *J. Comput. Phys.* Vol. 116, pages 123-134

421. Tadmor, E., Ortiz, M. and Phillips, R. (1996) "Quasicontinuum analysis of defects in solids," *Philosophical Magazine A*, Vol. 73, pp. 1529-1563.

422. Takashima, H., Kitamura, K., Tanabe, K., and Nagashima, U. (2000) "Is large-scale Ab initio Hartree-Fock calculation chemically accurate ? Towards improved calculation of biological molecule properties," *Journal of Computational Chemistry*, Vol. 20, pp. 443-454.

423. Taylor, C. A. and Hughes, T.J.R. and C.K. Zarins, C. K. (1998) "Finite element modeling of blood flow in arteries," *Computer Methods in Applied Mechanics and Engineering*, Vol. 158, pp. 155-196.

424. Taylor, G. I. and Quinney, H. (1934) "The latent energy remaining in a metal; after cold working," *Proc. R. Soc.* **A**, Vol. 143, pp. 307-326.

425. Takeda, H., Miyama, S. M., Sekiya, M. (1994) "Numerical simulation of viscous flow by smoothed particle hydrodynamics", *Prog. Theor. Phys.*, Vol. 92, pp. 939-

426. Tersoff, J. (1988) "New empirical approach for the structure and energies," *Phys. Rev.* B, Vol. 37, pp. 6991-7000.

427. Tersoff, J. (1988) "Empirical interatomic potential for carbon, with application to amorphous carbon," *Phys. Rev. Lett.*, Vol. 61, pp. 2879-2882.

428. Tezduyar, T. E. (1992) "Stabilized finite element formulations for incompressible-flow computations," *Advances in Applied Mechanics*, Vol. 28, pp. 1-44.

429. Tezduyar, T. E. and Osawa, Y. (2000), "Finite element stabilization parameters computed from element matrices and vectors," *Computer Methods in Applied Mechanics and Engineering*, Vol. 190. pp. 411-

430. Tezduyar, T. E. (2001) "Finite Element Methods for Flow Problems with Moving Boundaries and Interfaces," *Archives of Computational Methods in Engineering*, Vol. 8, pp. 83-130.

431. Tvergaard, V., Needleman, A. and Lo, K. K. (1981) "Flow localization in the plane strain tensile test," *Journal of Mechanics and Physics of Solids*, Vol. 29, pp. 115-142.

432. Tvergaard, V. (1988) "3D-analysis of localization failure in a ductile material containing two size-scales of spherical particles," *Engineering Fracture Mechanics*, Vol. 31, pp. 421-436.

433. Turner, M. J., Clough, R. W., Martin, R. W., and Topp, L. J. (1956) "Stiffness and deflection analysis of complex structures," *Journal of the Aeronautical Sciences*, Vol. 23, pp. 805-823.

434. Tunon, I., Martins-Costa, M.T.C., Millot, C., Ruiz-Lopez, M.F., Rivail, J.L. (1996) "A coupled density functional-molecular mechanics Monte Carlo simulation: the water molecule in liquid water," *Journal of Computational Chemistry*, Vol. 17, pp. 19-29.

435. Timoshenko, S. P. and Woinowsky-Krieger, S. (1959), *Theory of Plates and Shells*, McGRAW-HILL, New York, Second Edition.

436. Tu, Y. and Laaksonen, A. (2000) "Combined Hartree-Fock quantum mechanical and molecular mechanical dynamics simulations of water at ambient and

supercritical conditions," *Journal of Chemical Physics*, Vol. 133, pp. 11264-11269.

437. Ugural, A. C. (1999) *Stresses in Plates and Shells*, Second Edition, McGraw-Hill, Boston

438. Verlet, L. (1967) "Computer experiments on classical fluids. I. Thermodynamical properties of Lennard-Jones molecules," *Phys. Rev.*, Vol. 159, pp. 1029-

439. Wang, Y., Tomanek, D., Bertsch, G. F. (1991) "Stiffness of a Solid Composed of C60 Clusters," *Phys. Rev.*, B, Vol. 44, pp. 6562-6565.

440. Wang, X. (2003) "On the discretized delta function and force calculation in the immersed boundary method," In *Computational Fluid and Solid Mechanics*, Edited by K.J. Bathe, K. J., pages 2164-2169.

441. Wagner, G.J. and W. K. Liu (2000) "Turbulence simulation and multiple scale subgrid models," *Computational Mechanics*, Vol. 25 No. 2/3, pages 117-136

442. Wagner, G. J. and Liu, W. K. (2000) "Application of Essential Boundary Conditions in Mesh-free Methods: a corrected collocation method," *International Journal for Numerical Methods in Engineering*, Vol. 47, pp. 1367-1379.

443. Wagner, G. J. and Liu, W. K. (2000) "Hierarchical enrichment for bridging scales and meshfree boundary conditions," *International Journal for Numerical Methods in Engineering*, Vol. 50, pp. 507-524.

444. Wagner, G.J., Moes, N., Liu, W.K., Belytschko, T. (2001) "The extended finite element method for rigid particles in Stokes flow," *International Journal for Numerical Methods in Engineering*, Vol. 51, pp. 293-

445. Wagner, G. J., Ghosal, S., Liu, W.K. (2002) "Particular flow simulations using lubrication theory solution enrichment," Accepted for publication in *International Journal for Numerical Methods in Engineering*

446. Wagner, G. J. and Liu, W. K. (2003) "Coupling of atomistic and continuum simulations using a bridging scale decomposition," *Journal of Computational Physics*, Vol. 190, pp. 249-274.

447. Wagner, G. J., Karpov, E. G. and Liu, W. K. (2004) "Molecular dynamics boundary conditions for regular crystal lattices," *Computer Methods in Applied Mechanics and Engineering*, Vol. 193, pp. 1579-1601.

448. Wang, X. and Liu, W. K. (2003) Extended Immersed Boundary Method Using FEM and RKPM," To appear in *Computer Methods in Applied Mechanics and Engineering*.

449. Wang, X. and Liu, Y. and Corbett, S. and Zarinetchi, F. and Liu, W. K. (2003) "An Experimental Validation of Immersed Finite Element Method," Preprint

450. Weeks, W. T. (1966), "Numerical inversion of Laplace transforms using Laguerre functions," *J. Assoc. Comput. Math*, Vol. 13, pp. 419-426.

451. Watanabe, O., Zbib, H. M., Takenouchi, E. (1998) "Crystal plasticity: microshear banding in polycrystals using voronoi tessellation," *International Journal of Plasticity*, Vol. 14, pp. 771-788.

452. White, F. (1986) *Fluid Mechanics*, McGraw-Hill, New York.

453. Winckelmans, GS and Leonard, A. (1993) "Contributions to vortex particle methods for the computation of three-dimensional incompressible unsteady flows," *Journal of Computational Physics*, Vol. 109, pp. 247-273.

454. Wozniakowski, H. (1991) "Average Case Complexity of Multivariate Integration", *Bull. Amer. Math. Soc.*, Vol. 24, pp. 185-194

455. Wright, T. W. and Batra, R. C., (1985) "The initiation and growth of adiabatic shear band," *International Journal of Plasticity*, Vol. 1, pp. 205-212.

456. Wright, T. W. and Walter, J. W. (1987) "On stress collapes in adiabatic shear band," *Journal of Mechanics and Physics of Solids*, Vol. 35, pp. 701-720.

457. Wright, T. W. (1995) "Scaling laws for adiabatic shear bands," *International Journal of Solids and Structures*, Vol. 32, pp. 2745-2750.

458. Wright, T. W. and Walter, J. W. (1996) "The asymptotic structure of an adibatic shear band in antiplane motion," *Journal of Mechanics and Physics of Solids*, Vol. 44, pp. 77-97.

459. Wu, F. H. and Freund, L. B. (1984) "Deformation trapping due to thermoplastic instability in one-dimensional wave propagation," *Journal of the Mechanics and Physics of Solids*, Vol. 32, pp. 119-130.

460. Yagawa, G. and Yamada, T. (1996) "Free mesh method: a new meshless finite element method," *Computational Mechanics*, Vol. 18, pp. 383-386.

461. Yagawa, G., and Yamada, T. (1998) "Meshless method on massively parallel processors with application to fracture mechanics," *Key Engineering Materials*, Vol. 145-149, pp. 201-210.

462. Yagawa, G. and Furukawa, T. (2000) "Recent development of free mesh method," *International Journal for Numerical Methods in Engineering*, Vol. 47, pp. 1419-1417.

463. Zbib, H. M. and Jubran, J. S. (1992) "Dynamic shear banding: a three-dimensional analysis," *International Journal of Plasticity*, Vol. 8, pages 619-641.

464. Zeng, Q. and Combescure, A. (1998) "A new one-point quadrature, general nonlinear quadrilateral shell element with physical stabilization," *International Journal for Numerical Methods in Engineering*, Vol. 42, pages 1307-1338

465. Zhang, L. T., Wagner, G. J. and Liu, W. K. (2002) "A parallel meshfreee method with boundary enrichment for large-scale CFD," *Journal of Computational Physics*, Vol. 176, pages 483–506.

466. Zhang, L. and Gerstenberger, A. and Wang, X. and W.K. Liu, W.K. (2003) "Immersed Finite Element Method," To appear in *Computer Methods in Applied Mechanics and Engineering*.

467. Zhou, M., Rosakis, A. J., and Ravichandran, G. (1996) "Dynamically propagating shear bands in impact-loaded prenotched plates—I. experimental investigations of temperature signatures and propagation speed," *Journal of Mechanics of Physics and Solids*, Vol. 44, pages981-1006.

468. M. Zhou and G. Ravichandran and A. J. Rosakis, (1996) "Dynamically Propagating Shear Bands in Impact-loaded Prenotched Plates—II. Numerical Simulations," *Journal of Mechanics of Physics and Solids*, Vol. 44, pages 1007-1032.

469. Zhou, M. and Rosakis, A. J. and Ravichandran, G. (1998) "On the growth of shear bands and failure-mode transition in prenotched plates: A comparision of singly and doubly notched specimens," *International Journal of Plasticity*, Vol. 14, pages 435-451.

470. Zhu, T. and Atluri, S.N. (1998) "A Modified Collocation method and a penalty formulation for enforcing the essential boundary conditions in the element free Galerkin method," *Computational Mechanics*, Vol. 21, pp. 211-222.

471. Zhu, T., Zhang, J., and Atluri, S.N. (1999) "A Meshless Numerical Method Based on the Local Boundary Integral Equation (LBIE) to solve linear and non-linear boundary value problems," *Engineering Analysis with Boundary Elements*, Vol. 23, pp. 375-389.

472. Zhu, T. (1999) "A New Meshless Regular Local Boundary Integral Equation (MRLBIE) Approach," *International Journal for Numerical Methods in Engineering*, Vol. 46, pp. 1237-1252.

473. Ziegler, DP (1993) "Boundary conditions for lattice Boltzmann simulations," *Journal of Statistical Physics*, Vol. 71, pages 1171

474. Zienkiewicz, O. C., Rojek, J., Taylor, R. L. and Pastor, M. (1998) "Triangles and tetrahedra in explicit dynamic codes for solids," *International Journal for Numerical Methods in Engineering*, Vol. 43, pages 565-583.

475. Zienkiewicz, O. C. and Hung, M. (1995) "Localization problems in plasticity using finite elements with adaptive remeshing," *International Journal for Numerical and Analytical Methods in Geomechanics*, Vol. 19, pages 127-148.

10. Program Listings

```
      program main
c*********************************************************
c
c  This is a program  to solve the 1D Helmholtz equation
c  of a Dirichlet-Fixed bar problem by using the RKPM.
c
c  In order to enforce the essential boundary condition,
c  a special technique is used, which is described in
c  chapter 3.
c
c  Subroutines, dgeco.f and dgesl.f are standard LINPACK
c  programs that can be downloaded from public domain.
c
c
c  Model Equation:
c
c       d^2 F/dx^2  + k^2 * F = 0
c
c  with
c       F(0) = gZ;   F(1) = gL;
c
c  where,
c
c   h  : dx; input.
c
c  dhp ---- array of dilation parameter
c
c  dxp ---- array of integration weight (Quadrature weight)
c
c  xp  ---- array of particle's coordinates
c
c  xgk ---- array of Gauss points
c
c  xwk ---- array of weights for each Gauss point
c
c  mnode:=  maximum nodal points
c
c  max  :=  maximum number of Gauss quadrature points
c
c  Lg   ---- local gauss point
```

```
c
c  Wg    ---- weight for local Gauss point
c
c  Lmap ---- connectivity array for each Gauss point
c
c  cc    ---- output array for the correct function
c
c  fn    ---- output function
c
c  fndx ---- output 1st derivative
c
c  shpi --- array for shape function group i
c
c  shpj --- array for shape function group j
c
c
c************************************************
      implicit double precision (a-h,o-z)
c
      include 'parameter.h'
c
      dimension xpj(mnode),dxp(mnode),dhp(mnode)
      dimension stiff(mnode,mnode),force(mnode)
      dimension xgk(max),xwk(max)
      dimension shpbZ(mnode),shpbL(mnode)
      dimension b(2),bdx(2),shpi(0:1),shpj(0:1)
      dimension shpZ(0:1),shpL(0:1)
      dimension bZ(2),bZdx(2),bL(2),bLdx(2)
      dimension Cg(mnint),Wg(mnint)
      dimension Lmap(mnode),Lmapb(2,mbp)
      dimension ipvt(mnode),z(mnode),det(2)
c
      character*72 title
c
      open(10,file='input')
      open(11,file='output')
c
c.....input data
c
c.....1. input number of particles and number of
c        integration point
c
      read(10,*) title
      read(10,*) np,ngp
c
c.....2. input non-dimensional dilation parameter,tolerance
c
      read(10,*) title
      read(10,*) af,eps
c
c.....3. input definition of domain
c
      read(10,*) title
```

```
      read(10,*) xZ,xL
c
c--------end of input----------------
c
      mgk =  np * ngp
      mp  =  mnode
      dx  = (xL- xZ)/(np-1)
c
      gZ  = 1.0d0
      gL  = 0.0d0
c
c.....initialization
c
      do i = 1, np
  shpbZ(i) = 0.00d0
  shpbL(i) = 0.00d0
      enddo
c
      do i = 1,2
  b(i)       = 0.00d0
  bZ(i)      = 0.00d0
  bL(i)      = 0.00d0
c
  bdx(i)     = 0.00d0
  bZdx(i)    = 0.00d0
  bLdx(i)    = 0.00d0
c
  shpi(i-1) = 0.00d0
  shpj(i-1) = 0.00d0
c
  shpZ(i-1) = 0.00d0
  shpL(i-1) = 0.00d0
      enddo
c
c.....calculate the coordinates xpj(i) and
c     the radius of compact support
c
c.....For uniform spacing
c
      do i= 1, np
  xpj(i)  = xZ + (i-1)*dx
  dhp(i)  = af * dx
      enddo
c
c.... and
c
      xjacob = 0.50d0 * dx
      radius = 3.00d0 * af * dx
c
c.....Calculation of the integration weight (Trapezodail rule)
c
      dxp(1)  = 0.50d0*(xpj(2)  - xpj(1))
      dxp(np) = 0.50d0*(xpj(np) - xpj(np-1))
```

```
c
      do i = 2, np-1
 dxp(i) = 0.50d0*(xpj(i+1) - xpj(i-1))
      enddo
c
c.... Calculate Gauss Point Array
c
      call gauss(Cg, Wg, ngp)
c
      mGauss = 0
      do i = 1, np - 1
         do  j = 1, ngp
    mGauss = mGauss + 1
    xgk(mGauss) = xpj(i) +  Cg(j)*xjacob
    xwk(mGauss) = Wg(j)*xjacob
         enddo
      enddo
c----------------------------------------
c
c  Find boundary connectivity, and assign
c  the value for array shpb1, and shpbn
c
c----------------------------------------
      ibpZ = 0
      ibpL = 0
      rbZ  = 3.0d0*dhp(1)
      rbL  = 3.0d0*dhp(np)
c
      call crgo1a(bZ,bZdx,dhp,dxp,mp,np,xpj,xZ)
      call crgo1a(bL,bLdx,dhp,dxp,mp,np,xpj,xL)
c
      do i = 1, np
c
 ha  = dhp(i)
 dxj = dxp(i)
 xj  = xpj(i)
c
 if (dabs(xj - xZ) .lt. rbZ) then
    ibpZ = ibpZ + 1
    Lmapb(1,ibpZ) = i
c
    call shrgo1a(shpi,bZ,bZdx,dxj,ha,xj,xZ)
c
              shpbZ(i) = shpi(0)
c
 elseif(dabs(xj - xL) .lt. rbL) then
    ibpL = ibpL + 1
    Lmapb(2,ibpL) = i
c
    call shrgo1a(shpj,bL,bLdx,dxj,ha,xj,xL)
    shpbL(i) = shpj(0)
 endif
c
```

```
        enddo
c
c
c.....Initialization
c
        tgZ = gZ/shpbZ(1)
        tgL = gL/shpbL(np)
c
        haZ = dhp(1)
        haL = dhp(np)
c
        dxZ = dxp(1)
        dxL = dxp(np)
c
c--------------------------------
c
c    1st Main Loop: No: 200
c
c--------------------------------
c
        do 200 wk = 1,30
c
 dwk = wk * wk
c
 do i = 1, np
    do j = 1, np
        stiff(i,j) = 0.0d0
    enddo
    force(i) = 0.0d0
 enddo
c
c===========================
c
c    2nd  Main Loop: No. 300
c
c===========================
c
c
        do 300 k = 1, mGauss
c
c.....compute connectivity array
c
        ip = 0
c
 do i = 2, np-1
c
    if (dabs(xpj(i) - xgk(k)) .le. radius) then
        ip = ip + 1
        Lmap(ip) = i - 1
    endif
c
        enddo
c
```

```
 xpt  = xgk(k)
 call crgo1a(b,bdx,dhp,dxp,mp,np,xpj,xpt)
c
c---------------------
c.....secondary loop
c---------------------
c
c===================================
c.....Assemble Stiffness Matrix
c===================================
c
 call shrgo1a(shpZ,b,bdx,dxZ,haZ,xZ,xpt)
 call shrgo1a(shpL,b,bdx,dxL,haL,xL,xpt)
c
c...............................
c
        do i = 1, ip
   ii  = Lmap(i)
   iii = ii + 1
c
           call shrgd1a(shpi,shpbZ,shpbL,b,bdx,
   &                    dxp,dhp,iii,mp,np,xpj,xpt)
c
           force(ii) = force(ii)
   &       - (tgZ*shpZ(0) + tgL*shpL(0))*dwk*shpi(0)*xwk(k)
   &       + (tgZ*shpZ(1) + tgL*shpL(1))*shpi(1)*xwk(k)
c
c
   do j = 1, ip
      jj  = Lmap(j)
      jjj = jj + 1
c
              call shrgd1a(shpj,shpbZ,shpbL,b,bdx,
   &                       dxp,dhp,jjj,mp,np,xpj,xpt)
c
      stiff(ii,jj) = stiff(ii,jj)
   &                     + ( dwk*shpi(0)*shpj(0)
   &                     -   shpi(1)*shpj(1) ) * xwk(k)
   enddo
 enddo
c
c...............................
c
c....Remark: for non-uniform spacing, be carefule about using
c....xjacob. A proper xjacob should be chosen.
c
  300 continue
c
c=========================================
c    end of the loop 300
c=========================================
c.....Solve the Algebraic Equation
c===================================
```

```
c
c
        print *, '*** solve the system equation ***'
c
        nn = np - 2
c
        do i = 1, nn
  ipvt(i) = 0
  z(i)    = 0.0d0
        enddo
c
        call dgeco(stiff,mnode,nn,ipvt,rcond,z)
        print *, wk, rcond
        call dgesl(stiff,mnode,nn,ipvt,force,0)
c
c==========================================
c
c.....Output
c
c==========================================
c
c.... shift the coefficients back to normal
c
c==========================================
c
        do i = 1,np-2
  force(np-i) = force(np-i-1)
        enddo
c
        force(1)  = tgZ
        force(np) = tgL
c
        do i = 2, np-1
  force(1) = force(1)  - force(i)*shpbZ(i)/shpbZ(1)
  force(np)= force(np) - force(i)*shpbL(i)/shpbL(np)
        enddo
c
c......Reproducing and Sampling the midpoint value
c
           nout = (np-1)/2 + 1
  xi = xpj(nout)
           call crgo1a(b,bdx,dhp,dxp,mp,np,xpj,xi)
c
c............................
c
  fx  = 0.0d0
  fxd = 0.0d0
c
  do j = 1, np
c
      dxj = dxp(j)
      hj  = dhp(j)
      xj  = xpj(j)
```

```
c
      call shrgo1a(shpj,b,bdx,dxj,hj,xj,xi)
c
              fx  = fx   + force(j)*shpj(0)
      fxd = fxd  + force(j)*shpj(1)
c
   enddo
c
          dfx = log(abs(fx))
c
   gx   = dsin(wk*(1.0d0-xi))/dsin(wk)
   dgx  = log(abs(gx))
c
c.....Output the middle point displacement vs. frequency
c
          hk   =  wk * dhp(nout)
c
          write(11,9220)  hk, dfx, dgx
c
   200  continue
c====================================
c   End of The Main Loop
c====================================
c      ----------------------
c....Standard Output Format
c      ----------------------
c
 9220 format(2x, 3(x,e14.7))
c
      close(10)
      close(11)

      stop
      end


=======================================================================
      subroutine crgo1a(b,bdx,dhp,dxp,mp,np,xpj,xpt)
c
c****************************************************************
c
c $1.  This subroutine is to calculate the b vector and its
c      derivatives.
c      This code is only offering b vector and its' 1st
c      derivatives for 1-D case.
c
c $2.  The mathematical formualtion for b vector is:
c
c      B := (1/D1) [ M0, -M1]^{t}
c
c      The first order derivative is:
c
c      d/dx(B) := - M^{-1}* d/dx(M) *B
c
```

```
c
c  $3.    Arguments:
c
c  i       np: the numbers of particles
c
c  i       xpj(np):  the array that stores all the particle's
c                       global coordinations
c
c  i       dhp(np):  the array that stores all the dilation parameters
c                       of each every particle
c
c  i       dxp(np):  the array that stores dxj,
c                       ( The calculation are done by using Traperzoidal
c                       rule )
c
c  i       xpt: the point at where the shape function is
c                   evaluated
c
c
c    o    b  : the b vector at point xpt
c
c
c    o    bdx: the d/dx(b) at point xpt
c
c
c  l    ha := dhp(j)  : dilation parameter; a scale
c
c    c   gm(2,2)   : the origianl M matrix; never used in calculation
c
c  l    gminv(2,2): the inverse of the gm, i.e. gm^{-1}
c
c  l    gmdx(2,2):  the 1st derivatives of gm: d/dx (gm)
c
c
c  $4.  Remark:
c    ------------
c    The shape function is generated by the linear polynomial vector
c
c         P[(y-x)/ha] := [ 1, (y-x)/ha ]
c
c
c    The subroutine calls a window function subroutine:
c                        window1d.f
c
c*****************************************************************
      implicit double precision (a-h,o-z)
      dimension xpj(mp),dhp(mp),dxp(mp)
      dimension b(2),bdx(2)
      dimension gminv(2,2),gmdx(2,2),ad(2,2)
c
c.....set the initial value for moment and its derivatives:
c
      am0       = 0.00d0
```

```fortran
      am1       = 0.00d0
      am2       = 0.00d0
c
      am0dx    = 0.00d0
      am1dx    = 0.00d0
      am2dx    = 0.00d0
c
c.....set the initial value for all array:
c
      do i = 1, 2
 do j = 1, 2
    gminv(i,j)   = 0.00d0
    gmdx(i,j)    = 0.00d0
    ad(i,j)      = 0.00d0
        enddo
 b(i)    = 0.00d0
 bdx(i) = 0.00d0
      enddo
c
c
      xx    = xpt
c
c....main loop: Calculate moments by Traperzodial rule
c
      do 30 j = 1, np
c
c........define intermediate variable
c
 ha    =  dhp(j)
       xj    =  xpj(j)
 xnorm =  dabs(xj -xx)
 rr    =  3.0d0 * ha
 if(xnorm .gt. rr) go to 30
c
 dx1  = -1.0d0/ha
 dx2  =  dx1 * dx1
 dxj  =  dxp(j)   !! this corresponds to the definition
c                            of window function:
c                               Phi_r := (1/rho) Phi(x/rho)
c
 x1   = (xj - xx)/ha
 x2   =  x1 * x1
c
       call window1d(aw,awdx,awddx,ha,xj,xx)
c
       aw    = aw  * dxj
 awdx  = awdx* dxj
c
       am0   = am0    + aw
       am1   = am1    + x1 * aw
       am2   = am2    + x2 * aw
c
c
```

```
         am0dx = am0dx  + awdx
  am1dx = am1dx  + dx1 * aw + x1 * awdx
  am2dx = am2dx  + 2.0d0*dx1*x1*aw + x2*awdx
c
c
  30  continue
c
c
c.....end of the main loop
c
c.....Assemble the cofactor matrices ( a(i,j) )
c
      a11 =   am2
      a12 =   am1
      a21 =   a12
      a22 =   am0
c
c
c.....calculate the determinat det
c
      det = am0*am2 - am1*am1
c
      zero = 0.00d0
      if (det .le. zero ) then
  print *, 'det = ', det
  print *, 'STOP! the determinat det < 0 '
  print *, 'xpt=', xpt
      else
      endif
c
c.....assemble the gminv(i,j)
c
      cdet        =   1.0d0/det
      gminv(1,1) =   a11*cdet
      gminv(1,2) =  -a12*cdet
      gminv(2,1) =   gminv(1,2)
      gminv(2,2) =   a22*cdet
c
c.....construct the b vector
c
c
      b(1) = gminv(1,1)
      b(2) = gminv(1,2)
c
c.....calculate the 1st derivative of gm
c
      gmdx(1,1) = am0dx
      gmdx(1,2) = am1dx
      gmdx(2,1) = am1dx
      gmdx(2,2) = am2dx
c
c
c.....Calculating d/dx(B)
```

```
c
c
      do i = 1, 2
 do j = 1, 2
    do k = 1, 2
       ad(i,j) = ad(i,j) + gminv(i,k)*gmdx(k,j)
    enddo
 enddo
      enddo
c
      do i = 1,2
 do j = 1,2
    bdx(i) = bdx(i) - ad(i,j)*b(j)
 enddo
      enddo
c
c
      return
      end

===============================================================================
      subroutine shrgo1a(shp,b,bdx,dxj,ha,xj,xx)
c
c*****************************************************************
c
c $ 1.
c         This subroutine is to calculate the 1D shape functions
c         and their derivatives at particle ip
c
c         The mathematical formualtion for shape function is:
c
c         N_{ip}(ha,xpt,xpj) := C(ha,xpt,xpj) * \Phi((xpt - xpj)/ha)
c         * dxp(ip);
c
c         (d/dx)N_{ip} := (d/dx C)*\Phi*dxp(ip) + C*(d/dx \Phi)*dxp(ip);
c
c
c
c $ 2.    Arguments:
c
c h       np: total numbers of particles
c
c
c h       ip: the integer index for shape function, i.e.
c             we are computing N_{ip}
c
c       With the hidden variable ip, on the above level, the following
c       input data are specified:
c
c
c i       b(2)   : the b vector calculated in crgo1a.f
c
c i       bdx(2) : the 1st derivative of b vector
```

```
c
c  i       dxj:      the integration weight
c
c  i       xj := xpj(ip):  the array stores all the particle's
c                             global coordinates
c
c  i       dxj:= dxp(ip):  the array stores all the increment's
c                             of particle's coordinates
c
c  i       ha:= dhp(ip) :   dilation parameter
c
c  i       xx:   the point at where the shape function is
c                   evaluated
c
c
c     o   shape function N_{ip} and its 1st derivative
c                       at point xx
c
c     o   shp(0):   shape function N_{ip} at point xx
c
c     o   shp(1):   the first order derivative of shape function
c                       N_{ip},x  at point xx
c
c
c     o   css(1) = coref :   the correct function C(x,ip)
c
c         shp(1) = dc/dx * \phi + c * d\phi/dx
c
c
c     Remark:
c     -------
c     The shape function is generated by the polynomial vector
c
c         P[(y-x)/ha] := [ 1, (y-x)/ha ];
c
c     l   p0    := (1,0);
c
c     l   pv    := (1,x);
c
c     l   pvdx := [0, - 1/ha ]^{t}
c
c     The subroutine calls a window function subroutine:
c                       window_1d.f
c
c**************************************************************
      implicit double precision (a-h,o-z)
      dimension b(2),bdx(2)
      dimension pv(2),pvdx(2),shp(0:1)
c
c.....define polynomial vector pv:
c
      pv(1) = 1.0d0
      pv(2) = (xj - xx)/ha
```

```
c
c
      pvdx(1) =    0.0d0
      pvdx(2) = - 1.0d0/ha
c
c.....Computing the correction function
c
      coref = 0.0d0
      do i = 1, 2
         coref = coref + pv(i)*b(i)
      enddo
c
c
c.....the first order derivative
c
      cdx  = 0.0d0
      do j = 1, 2
 cdx  = cdx + pvdx(j)*b(j) + pv(j)*bdx(j)
      enddo
c
c
c......Check whether or not xx is within the compact support
c
      xnorm  = dabs((xj - xx))
c
      rr = 3.0d0 * ha
      if (xnorm .gt. rr ) then
  shp(0) = 0.0d0
  shp(1) = 0.0d0
      else
c
      call window1d(aw,aw1d,aw2d,ha,xj,xx)
c
  aw      = aw    * dxj
  aw1d    = aw1d * dxj
c
  shp(0) = coref * aw
  shp(1) = cdx * aw + coref * aw1d
c
c
      endif
c
c
      return
      end

==========================================================================
      subroutine shrgd1a(shp,shpbZ,shpbL,b,bdx,
     &                    dxp,dhp,jp,mp,np,xpj,xx)
c
c*************************************************************
c
c  $1.
```

```
c          This subroutine is to calculate the 1D shape functions
c          and their derivatives at particle jp.
c          The program is particularly designed to modify the original
c          shape function according to boundary condition adjustment
c          Dirichlet problem.
c
c
c  $ 2.    Arguments:
c
c  h        np: total numbers of particles
c
c  i        mp: the physical range of the array xpj, dxp
c
c  h        jp: the integer index for shape function, i.e.
c              we are computing N_{jp}
c
c     With the hidden variable ip, on the above level, the following
c     input data are specified:
c
c
c  i        b(2)   : the b vector
c
c  i        bdx(2) : the 1st derivative of b vector
c
c  i        dxj:     the integration weight at point, jp
c
c  i        xj := xpj(jp):  the array stores all the particle's
c                                   global coordinations
c
c  i        dxj:= dxp(jp):  the array stores all the increment's
c                                   of particle's coordinates
c
c  i        ha:=dhp(ip) :   dilation parameter
c
c  i        xx:    the point at where the shape function is
c                     evaluated
c
c
c     o     shape function N_{ip} and its 1st derivative
c                          at point cpt;
c
c     o     shp(0):   shape function N_{ip} at point xx
c
c     o     shp(1):   the first order derivative of shape function
c                          N_{ip},x   at point xx
c
c     o     coref :   the correct function C(x,ip)
c
c
c     c     shpbZ : the array store the values of all the shape
c                      function at the end point x1
c
c     c     shpbL : the array store the values of all the shape
```

```
c                      function at the end point xn
c
c     Remark:
c     -------
c     The shape function is generated by the polynomial vector
c
c          P[(y-x)/ha] := [ 1, (y-x)/ha ];
c
c    l    p0   := (1,0);
c
c    l    pv   := (1,x);
c
c    l    pvdx := [0, - 1/ha ]^{t}
c
c****************************************************************
      implicit double precision (a-h,o-z)
      dimension dxp(mp),xpj(mp),dhp(mp)
      dimension shpbZ(mp),shpbL(mp)
      dimension b(2),bdx(2),pv(2),pvdx(2)
      dimension pvZ(2),pvL(2),shp(0:1)
c
      xZ = xpj(1)
      xL = xpj(np)
      xj = xpj(jp)
c
      dxZ = dxp(1)
      dxL = dxp(np)
      dxj = dxp(jp)
c
      haZ = dhp(1)
      haL = dhp(np)
      haj = dhp(jp)
c
c.....define polynomial vector pv:
c
      pv(1)  = 1.0d0
      pv(2)  = (xj - xx)/haj

      pvZ(1) = 1.0d0
      pvZ(2) = (xZ - xx )/haZ
c
      pvL(1) = 1.0d0
      pvL(2) = ( xL - xx )/haL
c
c
      pvdx(1) =   0.0d0
      pvdx(2) = - 1.0d0/haj
c
c.....Computing the correction function
c
      coref  = 0.0d0
      corefZ = 0.0d0
      corefL = 0.0d0
```

```
      cdx    = 0.0d0
      cZdx   = 0.0d0
      cLdx   = 0.0d0
c
      do i = 1, 2
         coref  = coref  + pv(i)*b(i)
         corefZ = corefZ + pvZ(i)*b(i)
         corefL = corefL + pvL(i)*b(i)
  cdx    = cdx    + pvdx(i)*b(i) + pv(i)*bdx(i)
  cZdx   = cZdx   + pvdx(i)*b(i) + pvZ(i)*bdx(i)
  cLdx   = cLdx   + pvdx(i)*b(i) + pvL(i)*bdx(i)
      enddo
c
c......Check whether or not xx is within the compact support
c
      xnorm  = dabs(xj - xx)
      xnormZ = dabs(xZ - xx)
      xnormL = dabs(xL - xx)
c
      rrZ = 3.0d0 * haZ
      rrL = 3.0d0 * haL
      rrj = 3.0d0 * haj
c
      if (xnorm .gt. rrj ) then
  shp(0)  = 0.0d0
  shp(1)  = 0.0d0
      elseif ((xnormZ .le. rrZ) .and. (xnorm .le. rrj)) then
  call window1d(aw,awdx,awddx,haj,xj,xx)
  call window1d(awZ,awZdx,awZddx,haZ,xZ,xx)
c
  shp(0) = coref*aw*dxj
  shp(1) = (cdx*aw + coref*awdx)*dxj
c
  shpZ   = corefZ*awZ*dxZ
  shpZd  = (cZdx*awZ+ corefZ*awZdx)*dxZ

  shp(0) = shp(0) - shpbZ(jp)*shpZ/shpbZ(1)
  shp(1) = shp(1) - shpbZ(jp)*shpZd/shpbZ(1)
      elseif ((xnormL .le. rrL ) .and. (xnorm .le. rrj)) then
  call window1d(aw,awdx,awddx,haj,xj,xx)
  call window1d(awL,awLdx,awLddx,haL,xL,xx)
c
  shp(0) = coref*aw*dxj
  shp(1) = (cdx*aw + coref*awdx)*dxj
c
  shpL   = corefL*awL*dxL
  shpLd  = (cLdx*awL + corefL*awLdx)*dxL
c
  shp(0) = shp(0) -  shpbL(jp)*shpL/shpbL(np)
  shp(1) = shp(1) -  shpbL(jp)*shpLd/shpbL(np)
      else
  call window1d(aw,awdx,awddx,haj,xj,xx)
c
```

```
      shp(0) = coref*aw*dxj
      shp(1) = (cdx*aw + coref*awdx)*dxj
          endif
c
          return
          end

========================================================================
          subroutine gauss(s,w,ngp)
C************************************************************
c
c     This subroutine offers the coordinates of Gauss quadrature
c     points and their associated weights according to number of integration
c     points.
c
c     The standard domain is taken as [0, 2] instead of
c     [-1, 1].
c
c
c     Arguments:
c
c     o  s(ngp): the array contains the coordinates of
c                gauss quadrature;
c     o  w(ngp): the array contains the weights of
c                gaus quadrature points;
c
c     i ngp    : the number of integration points

C************************************************************
          implicit double precision (a-h,o-z)
c
          dimension s(*),w(*)
c
c.....check the array storage limitation
c
          nmax = 10
          if (ngp .gt. nmax) then
  go to 999
          else
              go to (10,20,30,40,50,60,70,80,90,100) ngp
          endif
c
  10    continue
          s(1) = 1.0d0
          w(1) = 2.0d0
c
          return
  20    continue
          s(1) = 1.0d0 - 0.577350269189626d0
          s(2) = 2.0d0 - s(1)
          w(1) = 1.0d0
          w(2) = 1.0d0
c
```

```
      return
30    continue
      s(1) = 1.0d0 - 0.774596669241483d0
      s(2) = 1.0d0
      s(3) = 2.0d0 - s(1)
      w(1) = 0.5555555555555556d0
      w(2) = 0.8888888888888889d0
      w(3) = w(1)
c
      return
40    continue
      s(1) = 1.0d0 - 0.861136311594053d0
      s(2) = 1.0d0 - 0.339981043584856d0
      s(3) = 2.0d0 - s(2)
      s(4) = 2.0d0 - s(1)
      w(1) = 0.347854845137454d0
      w(2) = 0.652145154862546d0
      w(3) = w(2)
      w(4) = w(1)
c
      return
50    continue
      s(1) = 1.0d0 - 0.906179845938664d0
      s(2) = 1.0d0 - 0.538469310105683d0
      s(3) = 1.0d0
      s(4) = 2.0d0 - s(2)
      s(5) = 2.0d0 - s(1)
      w(1) = 0.236926885056189d0
      w(2) = 0.478628670499366d0
      w(3) = 0.568888888888889d0
      w(4) = w(2)
      w(5) = w(1)
c
      return
60    continue
      s(1) = 1.0d0 - 0.932469514203152d0
      s(2) = 1.0d0 - 0.661209386466265d0
      s(3) = 1.0d0 - 0.238619186083197d0
      s(4) = 2.0d0 - s(3)
      s(5) = 2.0d0 - s(2)
      s(6) = 2.0d0 - s(1)
      w(1) = 0.171324492379170d0
      w(2) = 0.360761573048139d0
      w(3) = 0.467913934572691d0
      w(4) = w(3)
      w(5) = w(2)
      w(6) = w(1)
c
      return
70    continue
      s(1) = 1.0d0 - 0.949107912342759d0
      s(2) = 1.0d0 - 0.741531185599394d0
      s(3) = 1.0d0 - 0.405845151377397d0
```

```
         s(4) = 1.0d0
         s(5) = 2.0d0 - s(3)
         s(6) = 2.0d0 - s(2)
         s(7) = 2.0d0 - s(1)
         w(1) = 0.129484966168870d0
         w(2) = 0.279705391489277d0
         w(3) = 0.381830050505119d0
         w(4) = 0.417959183673469d0
         w(5) = w(3)
         w(6) = w(2)
         w(7) = w(1)
c
         return
  80     continue
         s(1) = 1.0d0 - 0.960289856497536d0
         s(2) = 1.0d0 - 0.796666477413627d0
         s(3) = 1.0d0 - 0.525532409916329d0
         s(4) = 1.0d0 - 0.183434642495650d0
         s(5) = 2.0d0 - s(4)
         s(6) = 2.0d0 - s(3)
         s(7) = 2.0d0 - s(2)
         s(8) = 2.0d0 - s(1)
         w(1) = 0.101228536290374d0
         w(2) = 0.222381034453374d0
         w(3) = 0.313706645877887d0
         w(4) = 0.362683783378362d0
         w(5) = w(4)
         w(6) = w(3)
         w(7) = w(2)
         w(8) = w(1)
c
         return
  90     continue
         s(1) = 1.0d0 - 0.968160239507626d0
         s(2) = 1.0d0 - 0.836031107326636d0
         s(3) = 1.0d0 - 0.613371432700590d0
         s(4) = 1.0d0 - 0.324253423403809d0
         s(5) = 1.0d0
         s(6) = 2.0d0 - s(4)
         s(7) = 2.0d0 - s(3)
         s(8) = 2.0d0 - s(2)
         s(9) = 2.0d0 - s(1)
         w(1) = 0.081274388361574d0
         w(2) = 0.180648160694857d0
         w(3) = 0.260610696402935d0
         w(4) = 0.312347077040003d0
         w(5) = 0.330239355001260d0
         w(6) = w(4)
         w(7) = w(3)
         w(8) = w(2)
         w(9) = w(1)
c
         return
```

```
  100 continue
      s(1) = 1.0d0 -0.973906528517172d0
      s(2) = 1.0d0 -0.865063366688985d0
      s(3) = 1.0d0 -0.679409568299024d0
      s(4) = 1.0d0 -0.433395394129247d0
      s(5) = 1.0d0 -0.148874338981631d0
      s(6) = 2.0d0 -s(5)
      s(7) = 2.0d0 -s(4)
      s(8) = 2.0d0 -s(3)
      s(9) = 2.0d0 -s(2)
      s(10)= 2.0d0 -s(1)
      w(1) = 0.066671344308688d0
      w(2) = 0.149451349150581d0
      w(3) = 0.219086362515982d0
      w(4) = 0.269266719309996d0
      w(5) = 0.295524224714753d0
      w(6) = w(5)
      w(7) = w(4)
      w(8) = w(3)
      w(9) = w(2)
      w(10)= w(1)
c
      return
  999 write(*, 2000) ngp
 2000 format(2x,'**** ERROR ****',5x,/'No',i3,2x,
     &    'Point Integration excess maximum array storage'/)
c
      end


================================================================================
      subroutine window1d(aw,awdx,awddx,ha,xj,xx)
C*************************************************************
c
c   This is a subroutine to compute 1D cubic spline window
c   function and its first and second derivatives.
c
c
c   The window function is based  on the formula provided in
c   ''Ten lectures on Wavelets'' by Ingrid Daubechies [1991];
c    Page 79.
c
c   arguments:
c
c   i          ha:  dilation parameter in x direction
c
c   i          xx:  variable's x coordinate, or center
c
c   i          xj:  x coordinate for center, or the j-th
c                   window function
c
c       o      aw:  1D cubic spline window function
c                   Phi(( xj - xx)/ha)
c
```

```
c       o    awdx:  d/dx(aw);
c                   d/dx ( Phi(( xj - xx)/ha))
c
c       o    awddx: d^2/dx^2(aw);
c                   d^2/dx^2 ( Phi(( xj - xx)/ha))
c
c
c    Remark:
c    -------
c
c    (1) The window function output is already normalized as
c        ( \int \Phi dy = 1 ).
c
c****************************************************************
      implicit double precision (a-h,o-z)
c
c.....normalize the argument
c
      x1  = (xj-xx)/ha
      x2  = x1*x1
      x3  = x1*x2
      hv  = 1.0d0/ha
c
      two1 = -2.00d0
      one1 = -1.00d0
      zero =  0.00d0
      one2 =  1.00d0
      two2 =  2.00d0
c
c.....dx := d(xr)/dx;
c
      dx1 = -1.00d0/ha
      dx2 = dx1*dx1
c
      if((x1.ge.two1) .and. (x1.lt.one1)) then
        aw    = (1.0d0/6.0d0)*(2.0d0 + x1)**3.
        awdx  = 0.50d0*dx1*(2.0d0 + x1)**2.
        awddx = dx2*( 2.0d0 + x1)
      elseif ((x1 .ge. one1) .and. (x1 .lt. zero)) then
        aw    = 2.0d0/3.0d0 - x2 - 0.50d0*x3
        awdx  = - dx1*(2.0d0*x1 + 1.50d0*x2)
        awddx = - dx2*(2.0d0 + 3.0d0*x1 )
      elseif((x1.ge.zero).and.(x1.lt.one2)) then
        aw    = 2.0d0/3.0d0 - x2 + 0.50d0*x3
        awdx  = -dx1*(2.0d0*x1 - 1.50d0*x2)
        awddx = -dx2*(2.0d0 - 3.00d0*x1 )
      elseif((x1.ge.one2).and.(x1.le.two2)) then
        aw    = (1.0d0/6.0d0)*(2.0d0 - x1)**3.
        awdx  = - 0.50d0*dx1*(2.0d0 -  x1)**2.
        awddx =   dx2*(2.0d0 -  x1)
      else
        aw    = 0.00d0
        awdx  = 0.00d0
```

```
          awddx = 0.00d0
        endif
c
c

        aw    = aw*hv
        awdx  = awdx*hv
        awddx = awddx*hv
c

        return
        end
```

```
==========================================================================
C*******************************************
c
c     parameter.h
c     -----------
c
c     Definition of upper limits of all working arrays
c
c
c     max : maximun number of all points;
c
c     mnp : maximun number of all particle number;
c
c     mnint: maximun number of integration point
c            (by adopt Gauss qudarture )
c
c
c
C*******************************************
        integer max
        integer mnode
        integer mnint
        integer mbp
c
        parameter ( mnode = 1001 )
        parameter ( mnint = 7 )
        parameter ( max = mnode*mnint )
        parameter ( mbp = 20 )
```

```
===================================================================
                    ***  makefile  ***

opts = -c -a -w -C
objs= main.o\
      crgo1a.o\
      shrgo1a.o\
      shrgd1a.o\
      dgedi.o\
      dgefa.o\
      dgesl.o\
      dgeco.o\
      ddot.o\
```

```
         daxpy.o\
         dswap.o\
         dscal.o\
         dasum.o\
         idamax.o\
         gauss.o\
         solver.o\
         window1d.o

wave1d:     $(objs); f77 -o wave1d  $(objs)
```

```
====================================================================
                    *** input   file   ***

'input np and ngp'
11  5

'input: dilation coefficeint, epsilon '
1.1    0.000001

'input domain definition'
0.0  1.0

'input boundary conditions '
1.0  0.0
```